1st International Conference on Advances in Environmental Engineering (AEE 2017)

IOP Conference Series: Earth and Environmental Science
Volume 92

Ostrava, Czech Republic
28 - 30 November 2017

Editors:

Nadezda Stevulova
Adriana Estokova
Vojtech Vaclavik

ISBN: 978-1-5108-5112-2
ISSN: 1755-1307

Printed from e-media with permission by:

Curran Associates, Inc.
57 Morehouse Lane
Red Hook, NY 12571

Some format issues inherent in the e-media version may also appear in this print version.

Printed by Curran Associates, Inc. (2017)

For permission requests, please contact the Institute of Physics
at the address below.

Institute of Physics
Dirac House, Temple Back
Bristol BS1 6BE UK

Phone: 44 1 17 929 7481
Fax: 44 1 17 920 0979

techtracking@iop.org

Additional copies of this publication are available from:

Curran Associates, Inc.
57 Morehouse Lane
Red Hook, NY 12571 USA
Phone: 845-758-0400
Fax: 845-758-2633
Email: curran@proceedings.com
Web: www.proceedings.com

Table of contents

Volume 92

1st International Conference on Advances in Environmental Engineering (AEE 2017)

28–30 November 2017, Ostrava, Czech Republic

Accepted papers received: 24 October 2017
Published online: 2 November 2017

Preface

Papers

PREFACE

The 1st International Scientific Conference on „Advances in Environmental Engineering" (AEE 2017) was held in Ostrava, Czech Republic from November 28 – 30, 2017. The Conference was organised by the Faculty of Mining and Geology of the VŠB-Technical University of Ostrava, Czech Republic and the Faculty of Civil Engineering of the Technical University of Košice, Slovakia. The idea of founding new series of environmentally oriented conferences was based on a need of scientists and expert from praxis in V4 region to have a common platform for exchange of knowledge and expertise on various environmental topics.

The AEE 2017 offered an international platform for the dissemination of original research results, new ideas and practical development and discovered advances in the field of environmental engineering and related interdisciplinary topics.

The Conference assembled more than 80 delegates, from 5 countries, including PhD students' presentations. The papers presented involved a wide range of research fields: water and wastewater management, sediment pollution control and remediation, soil degradation, conservation and remediation, waste management, landscape management, air quality, environmental impacts of buildings and materials, durability of building materials, and use of alternative raw materials in building materials production. This issue provides an integrated view of the trends in the development of environmentally friendly materials and solving the problems associated with the achievement of sustainability.

79 articles have been selected to be published in the volume of Journal IOP Conference Series: Earth and Environmental Science. The organizing committee appreciates the enormous work of both the editorial board and reviewers in the preparation of this volume.

Adriana Eštoková
Nadežda Števulová
Vojtěch Václavík

IOP Conf. Series: Earth and Environmental Science 92 (2017) 011001 doi:10.1088/1755-1315/92/1/011001

COMMITTEES

Editors

Nadežda ŠTEVULOVÁ	Technical University of Košice
Adriana EŠTOKOVÁ	Technical University of Košice
Vojtěch VÁCLAVÍK	VSB – Technical university of Ostrava

Scientific Committee

Chairperson

Nadežda ŠTEVULOVÁ	Technical University of Košice

Co-chairpersons

Jan VALÍČEK	VSB – Technical university of Ostrava
Vojtěch VÁCLAVÍK	VSB – Technical university of Ostrava
Magdaléna BÁLINTOVÁ	Technical University of Košice
Adriana EŠTOKOVÁ	Technical University of Košice

Members

Kestutys BALTAKYS, Kaunas University of Technology, Lithuania
Edita BALTRENAITE, Vilnius Gediminas Technical University, Lithuania
Endre DOMOKOS, University of Pannonia, Hungary
Konstiantyn DYADYURA, Sumy State University, Ukraine
Radmila KUČEROVÁ, VSB – Technical university of Ostrava, Czech Republic
Miroslav KYNCL, VSB – Technical university of Ostrava, Czech Republic
Antonio MESSINEO, University of Enna „Kore", Italy
NGUYEN THI MAI LINH - Ton Duc Thang University, Vietnam
Eva PERTILE, VSB – Technical university of Ostrava, Czech Republic
PHAM ANH DUC - Ton Duc Thang University, Vietnam
Janusz RAK, Rzeszow University of Technology, Poland
Mindaugas RIMEIKA, Vilnius Gediminas Technical University, Lithuania
Khrystyna SOBOĽ, Lviv Polytechnic National University, Ukraine
Serkan SAHINKAYA, Nevsehir Haci Bektas Veli University, Turkey
Alena SIČÁKOVÁ, Technical University of Košice, Slovakia
Tomáš SVĚRÁK, Brno University of Technology, Czech Republic
Miloslav ŠLEZINGR, Mendel University, Czech Republic
Silvia VILČEKOVÁ, Technical University of Košice, Slovakia
Martins VILNITIS, Riga Technical Univerity, Latvia
Tadeusz ZABOROWSKI, Poznan Technical University, Poland
Martina ZELEŇÁKOVÁ, Technical University of Košice, Slovakia

Organising committee

Tomáš DVORSKÝ - VSB – Technical university of Ostrava
Marta HARNIČÁROVÁ - VSB – Technical university of Ostrava
Dalibor SUROVKA - VSB – Technical university of Ostrava
Eva SINGOVSZKÁ - Technical University of Košice
Viola HOSPODÁROVÁ - Technical University of Košice
Michaela SMOLÁKOVÁ - Technical University of Košice

IOP Conf. Series: Earth and Environmental Science **92** (2017) 011001 doi:10.1088/1755-1315/92/1/011001

LIST OF REVIEWERS:

prof. Ing. Milan Oravec, PhD
prof. RNDr. Nadežda Števulová, PhD.
prof. RNDr. Adriana Eštoková, PhD.
prof. RNDr. Peter Andráš, PhD.
prof. RNDr. Pavol Eliáš, CSc.
prof. RNDr. Magdaléna Bálintová, PhD.
prof. RNDr. Jaroslav Briančin, CSc.
prof. RNDr. Janka Sedláková Kaduková, PhD.
prof. Mgr. Juraj Ladomerský, CSc.
prof. Ing. Dagmar Juchelkova, Ph.D.
prof. Ing. Zuzana Vranayová, PhD.
prof. Ing. Vojtěch Dirner, PhD.
prof. Ing. Martin Palou, PhD.
prof. Ing. Jaroslava Vrábliková, CSc.
prof. Ing. Eva Michaeli, PhD.
prof. Ing. Danica Fazekašová, PhD.
prof. Ing. Dagmar Samešová, PhD.
prof. Dr. Ing. Miroslav Kyncl
prof. Dr. Ing. Miloslav Šlezingr
assoc. prof. Ing. Emília Hroncová, PhD.
assoc. prof. RNDr. Zuzana Melichová, PhD.
assoc. prof. RNDr. Stanislav Jeleň, PhD.
assoc. prof. RNDr. Pavol Purcz, PhD.
assoc. prof. RNDr. Martin Pipíška, PhD.
assoc. prof. RNDr. Kušnerová Milena, Ph.D.
assoc. prof. RNDr. Jana Dadová, PhD.
assoc. prof. RNDr. Eva Bulánková, CSc.
assoc. prof. Mgr. Mária Kaňuchová, PhD.
assoc. prof. Ing. Zdenka Lhotáková,CSc.
assoc. prof. Ing. Vojtěch Václavík, PhD.
assoc. prof. Ing. Viliam Bárek, CSc.
assoc. prof. Ing. Tomáš Svěrák, CSc.
assoc. prof. Ing. Šárka Kročová, Ph.D.
assoc. prof. Ing. Silvia Vilčeková, PhD.
assoc. prof. Ing. Ružena Králiková, PhD.
assoc. prof. Ing. Roman Vodička, PhD.
assoc. prof. Ing. Petr Vráblík, PhD.
assoc. prof. Ing. Peter Kapalo, PhD.
assoc. prof. Ing. Orest Voznyak
assoc. prof. Ing. Milan Macák, CSc.
assoc. prof. Ing. Martina Zeleňáková, PhD.
assoc. prof. Ing. Marián Schwarz, PhD.
assoc. prof. Ing. Marek Sokáč, PhD.
assoc. prof. Ing. Ľubica Kozáková, PhD.
assoc. prof. Ing. František Hnilička, PhD.
assoc. prof. Ing. Eva Kormaníková, PhD.
assoc. prof. Ing. Daniela Kaposztásová, PhD.
assoc. prof. Ing. Andrej Kapjor, PhD.
assoc. prof. Ing. Alena Sičáková, PhD.
assoc. prof. Ing. Alena Pauliková, PhD.
assoc. prof. Dr. Ing. Radmila Kučerová
RNDr. Yvonne Němcová, PhD.
RNDr. Tatiana Durmeková, PhD.
RNDr. Lenka Demková, PhD.
RNDr. Ján Buša, CSc.
RNDr. Eva Terpáková, PhD.
Mgr. Patrik Kabátník, Ph.D.
Ing. Marcela Ondová, PhD.
Ing. Vlasta Ondrejka Harbuľáková, PhD.
Ing. Víťazoslav Krúpa, DrSc.
Ing. Veronika Veľková, PhD.
Ing. Tomáš Mandičák, PhD.
Ing. Štefan Demčák
Ing. Slávka Harabínová, PhD.
Ing. Silvia Dolinská, PhD.
Ing. Róbert Figmig
Ing. Radim Ptáček, Ph.D.
Ing. Oľga Šestinová, PhD.
Ing. Natália Junáková, PhD.
Ing. Milan Gomboš, PhD.
Ing. Matej Špak, PhD.
Ing. Martin Sisol, PhD.
Ing. Markéta Bouchalová, Ph.D.
Ing. Mariana Eliášová, PhD.
Ing. Marián Holub, PhD.
Ing. Lenka Nevřivová, PhD.
Ing. Kristýna Červená, Ph.D.
Ing. Július Strigáč, PhD.
Ing. Jozef Mitterpach, PhD.
Ing. Jozef Junák, PhD.
Ing. Natália Junáková, PhD.
Ing. Jozef Hančuľák, PhD.
Ing. Jitka Fialová, PhD.
Ing. Jiří Fiedor, Ph.D.
Ing. Helena Hybská, PhD.
Ing. Marta Harničárová, PhD.
Ing. Hana Uhmannová, CSc.
Ing. Eva Singovszká, PhD.
Ing. Eva Mačingová, PhD.
Ing. Eva Krídlová Burdová, PhD.
Ing. Alena Šoltísová, PhD.
Ing. Alena Struhárová, PhD.
Ing. Alena Luptaková, PhD.
Dr. inz. Katarzyna Pietrucha-Urbanik
Dr. Grzegorz Wisz, Ph.D.
Dr. Csaba Szabó
Dr. Agnieszka Stec, PhD., Eng.
Dr. Agnieszka Barcik
Dr. Anna Jasińska - Biliczak

Peer review statement

All papers published in this volume of *IOP Conference Series: Earth and Environmental Science* have been peer reviewed through processes administered by the proceedings Editors. Reviews were conducted by expert referees to the professional and scientific standards expected of a proceedings journal published by IOP Publishing.

IOP Conf. Series: Earth and Environmental Science **92** (2017) 012001 doi:10.1088/1755-1315/92/1/012001

Verification of vermural stabilization of ash from biomass and sewage sludge

L Adamkova[1], D Kucerova[1], B Lyckova[1], R Kucerova[1] and D Takac[1]

[1]VSB – Technical University of Ostrava, Faculty of Mining and Geology Czech Republic

E-mail: lucie.adamkova.st@vsb.cz

Abstract. The aim of this study was to find dependence on biofuels and sludge from sewage treatment plants in the vermicomposting process. In the framework of the research carried out at our workplace, a project aimed at finding an appropriate method for the reprocessing of problematic biodegradable waste and asphalt from combustion biomass was used as a raw material for the production of rectification substrate and sludge from sewage treatment plants that could be used as Secondary raw material.

1. Introduction

Soil degradation increases lately. The reason is primarily the development of civilized human society, agricultural development, and the increasing need to feed a large number of people. High productivity leads to imbalances in natural processes. Soil is the main medium of energy exchange and its resources are constantly being devalued by human activity, on this issue.

One of the best processes for processing biodegradable waste for to obtain good organic material is vermicomposting. This process, which processes the organic material by earthworms, has thanks to the resulting product a very positive effect on the quality of the soil and microbiological activity, when it changes a mineral substance the influence of the chemical, but also of biochemical transformations.

All member countries of the European Union are according to the European directive no 1999/31/EC on the landfill of waste required by 2020, gradually reduce the amount of deposited biodegradable municipal waste to 35 % of the original weight of this type of waste generated in 1995. Therefore, it is very important to find for these wastes, appropriate methods of processing and subsequent use.

However, due to the strategic EU documents (e.g. the waste Framework directive (98/2008/EC) prefers material to be used before the so-called removal of waste), is undoubtedly more appropriate and desirable solution to prioritize the evaluation and utilization of evaluated materials from landfills. For the purposes of putting into circulation, storage and use of fertilizers, auxiliary soil substances, auxiliary herbal preparations and substrates used in the present time in the Czech Republic act no. 156/1998 Sb.

2. Vermicompost

Vermicomposting is a natural process that can‘t be rushed. However, if the natural raw materials back to nature, it can have a very positive effect. One of the reasons why fertilize vermicompostem is that in comparison with chemical fertilizers leads to a better regeneration of the nutrient exhausted the soil system, reducing the cost of production and transportation of chemical fertilizers.

The process of vermicomposting is based on the decomposition of organic material using worms, the process is also one of the oldest and also the more advanced ways. The vermicomposting is utilized by earthworms („Eisenia foetida"), or also as in this case specially bred californian hybrids („Eisenia Andrei").

Using earthworms is a material intensively processed, aerated and there is no need for outside intervention. The resulting product, vermicompost is a high-quality fertilizer, which contains large amounts of enzymes, organic substances, growth substances and nutrients. Enzymes which comes from the secretions of the earthworms, are an important component of vermicompost and thanks to the fact is in comparison with classic compost of better quality. The high amount of enzymes increases the bioavailability of mineral elements in the soil by almost 90%.

Other positive features of vermicompost as the ability to suppress the in plants the formation of certain types of diseases and fungi. Increases the pH values which are particularly suitable for acidic soils. Last but not least, the digestive tract of earthworms disposed of toxic, mutagenic or pathogenic substances.[1,2].

It is important to maintain these basic conditions during vermicompost [6]:

- Temperature 19-22 °C (below 7 °C and above 30 °C the earthworms are apathetic and die at temperatures below 0 ° C and above 40 °C).
- Optimal substrate humidity ranging from 78 to 82% (humidity below 60% and more than 90% killing earthworms).
- Neutral pH (pH above 8 and below 6 has lethal effects on earthworms).

Other conditions that need to be adhered to are oxygen, weather protection (sun, wind rain) and protection from predators (birds, frogs, rodents).

3. Biodegradable waste

Biowaste are any aerobically or anaerobically degradable wastes in which the respiratory activity after four days (AT4) is greater than 10 mg O_2 / g of dry matter, and the value of the dynamic respiratory index exceeds 1000 mg of oxygen per kilogram of combustible matter per hour [7]. In the total amount of municipal waste, the volume and weight are very important groups which, due to natural degradation, burden the environment with greenhouse gases as well as leachate containing hazardous residues [8]. The analysis of the Ministry of the Environment (MoE) shows that in municipal waste almost half of the volume is made up of biodegradable materials. Also included are municipal greenery and public catering facilities, which make up a significant part of the total volume of municipal waste [9,10].

Figure 1. Vermicompost beds / vermicompost of biowaste.

4. Testing the efficiency of fly ash from combustion of biomass and sludge from WWTP

The aim of the experiment was vermicomposition of ash from biomass and sludge combustion from WTPs. For this purpose, vermicompost beds (figure 1) were established for the processing of ash and sludge in different ratios as seen on table 1. These ratios were also observed when feeding the Eisenia andrei earthworms. The properties of the output materials of these processing methods (As, Cd, Cr, Cu, Hg, Mb, Ni, Pb, Zn, dry matter, pH, Ca, Mg, K, F, combustion, total nitrogen). As a base layer (figure 2) for the establishment of the vermicomposting process, fertilized horse manure was used up to about 10 cm.

Table 1. Different ratios of vermicompost beds.

	1	2	3
Horse manure	50%	50%	50%
Ashes	-	20%	10%
Sludge from WWTP	50%	30%	40%

On the base layer of the compost were laid substrates with red calcareous earthworms (figure 1) and then ashes and sludge added in the given ratios. The final layers had a thickness of about 25 cm and were covered with a thick foil to prevent drying of the material. This film also provided sufficient protection for earthworms from direct sunlight and a temperature of about 20 °C for proper earthworm activity. The substrate with Californian earthworms contained about 25,000 individuals.

4.1 Entrance values of organic matter

Table 2. Input materials values.

Index	Manure Otice	Asbestos from biomass combustion	Sludge from sewage treatment plants	Unit
As	0.73	12.4	1.42	mg / kg in dry matter
Cd	0.49	8.09	1.46	mg / kg in dry matter
Cr	<2.50	42.9	38.2	mg / kg in dry matter
Cu	12.4	150	307	mg / kg in dry matter
Hg	0.023	0.244	2.73	mg / kg in dry matter
Mo	<0.50	x	x	mg / kg in dry matter
Ni	4.05	42.4	29.7	mg / kg in dry matter
Pb	4.35	54.9	30.6	mg / kg in dry matter
Zn	53.3	664	1160	mg / kg in dry matter
Ashes	21.92	84.25	26.77	%
pH	8	12.7	7.8	
Ca	7.93	87.1	40.7	g / kg in dry matter
Mg	3.61	10.7	4.65	g / kg in dry matter
K	20.5	33.4	1.7	g / kg in dry matter
F	3.6	3.88	30.5	g / kg in dry matter
Flammable substances	83.8	16.1	57.8	% in dry matter
Nitrogen total	1.5	0.22	4.33	% in dry matter
Ratio C: N	28	37	7	

The input materials were analyzed by an accredited laboratory MORAVA s.r.o. in Studenka. The base layer, which was composed of horse manure, was processed and imported from the riding section, which is located in Otice near Opava. The fly ash from biomass burning, which was necessary for the experiment, was imported from the biomass incinerator. Stabilized sewage sludge was brought from the wastewater treatment plant in Olomouc.

4.2 Resultant vermicompost values

The resulting values of the vermicomposts are shown in Table III. Vermicomposts were divided into three categories (50/40/10, 50/30/20 and 50/50/0). The experiment ran from 17 May 2016 to 20 September 2016 near Ostrava on private land. As mentioned above, the hybrid Californian earthworms, ("Eisenia andrei") which are best suited to the vermicompost process, have been used for this experiment. At the time of the experiment, these earthworms could be found throughout the vermicompost area, even in the other two vermicompost beds, although they were only fed with ash and slurry in different ratios, see table 1. In order to compare these three vermicomponent beds, tests were performed and the resulting values are listed in table 3 below.

Table 3. Resulting values with different input material ratios.

Index	50/40/10	50/30/20	50/50/0	Unit
As	2,13	2,91	1,78	mg / kg in dry matter
Cd	2,53	3,91	1,26	mg / kg in dry matter
Cr	29,10	28,40	22,20	mg / kg in dry matter
Cu	168,00	160,00	156,00	mg / kg in dry matter
Hg	1,04	0,87	1,24	mg / kg in dry matter
Mo	3,09	2,22	3,30	mg / kg in dry matter
Ni	16,10	31,40	18,00	mg / kg in dry matter
Pb	30,00	32,70	30,20	mg / kg in dry matter
Zn	695,00	668,00	652,00	mg / kg in dry matter
dry matter	21,60	21,48	20,63	%
pH	7,60	7,70	7,30	
Ca	52,80	60,40	43,30	g / kg in dry matter
Mg	4,71	5,23	3,42	g / kg in dry matter

5. Discussion

Veristabilization of sewage sludge and fly ash from biomass took place without any difficulty. Due to vermicomposting, the sludge and fly ash values were significantly reduced, which is a favorable result for further sludge and fly ash handling. The pH value after completion of vermicomposting did not differ significantly in all three vermicomposts and ranged from 7.3-7.7. It was expected that the content of hazardous substances and elements in sludge and fly ash after vermicompact will decrease compared to the quantity of hazardous substances and elements in untreated sludge and fly ash. This theory was confirmed in all three vermicomposts. Vermicompled sewage sludge and fly ash can be used on agricultural land in terms of the content of hazardous substances and elements in sludge as it complies with Decree No. 437/2016 Coll., Requirements and Criteria for Evaluation and Control of Outputs from Biowaste Facilities, See table 4.

The Decree of the Ministry of Agriculture No. 474/2000 Coll., States: "Limit values of risk elements in fertilizers, auxiliary soil substances, auxiliary plant preparations and substrates" [10] This decree compared the resulting values of vermicompost beds according to ČSN 46 5735 for industrial composts (table 5).

Comparison of the limit values according to the table for organic and farmyard fertilizers with dry matter is above 13%. If we compare these values, Table V, it is clear that the concentration of some elements is slightly exceeded (Cu, Zn, Hg, Cd). The reason for this exceedance can be found in the nature of stabilized sludge, where the concentration of these elements is higher.

Table 4. Comparison of the limit concentration of selected risk elements with Decree No. 437/2016 Coll., With output values of vermicomposts.

Index	Limit concentrations - 2 groups, III. class	Limit concentrations - 3 group, stabilized biowaste [mg/kg]	50/40/10 [mg/kg]	50/30/20 [mg/kg]	50/50/0 [mg/kg]
As	30	40	2.13	2.91	1.78
Cd	4	5	2.53	3.91	1.26
Cr	300	600	29.10	28.40	22.20
Cu	300	600	168.00	160.00	156.00
Hg	2	5	1.04	0.87	1.24
Ni	120	150	16.10	31.40	18.00
Pb	400	500	30.00	32.70	30.20
Zn	1500	1800	695	668	652

Table 5. Comparison of values with Decree 474/2000 Coll., With values of vermicompostes.

Index	value [mg/kg]	50/40/10 [mg/kg]	50/30/20 [mg/kg]	50/50/0 [mg/kg]
As	20	2.13	2.91	1.78
Cd	2	2.53	3.91	1.26
Cr	100	29.10	28.40	22.20
Cu	150	168.00	160.00	156.00
Hg	1.0	1.04	0.87	1.24
Ni	50	16.10	31.40	18.00
Pb	100	30.00	32.70	30.20
Zn	600	695	668	652
Mo	20	3.09	2.22	3.30

Acknowledgments
This article was created with the support of the student grant competition of the Mining University - Technical University of Ostrava (Evaluation of Selected Biowaste Using Biological Methods, SP2016 / 33).

References
[1] Hanc A and Pliva P 2013 Water Biomassing Prague Czech University of Agriculture
[2] Waste Management Plan of the Czech Republic as amended. In: 197/2003 2003
[3] Slejska A 1999 Vermicomposting
[4] Hanc A, Plíva P 2013 Vermicomposting - a Perspective Method for Biowaste Processing Proceedings of IX. Conference - Biodegradable waste 1-11
[5] Hlavatá M 2004 Waste management Ostrava VŠB – Technická univerzita 174
[6] Kotoulova Z, Vana J 2001 Manual for Handling Municipal Biowaste Prague Ministry of the Environment in cooperation with the Czech Environmental Institute 70
[7] Decree on the details of the treatment of biodegradable waste and amending Decree No. 294/2005 Coll. On the conditions of landfilling and landfill waste and amending Decree No. 383/2001 Coll. On waste management details (Decree on details of the treatment of biodegradable waste) 2008 Czech Republic
[8] Ministry of the environment 2015 Analysis of existing strategies for the prevention of waste Environmental Operational Program Ernst & Young 21
[9] Decree No. 341/2008 Coll. Decree on the details of the treatment of biodegradable waste (Decree on details of the treatment of biodegradable waste) 2017 Praha Ministry of the Environment 38

Importance of calcium and magnesium in water - water hardening

D Barloková[1], J Ilavský[1], O Kapusta[2] and V Šimko[3]

[1]Department of Sanitary & Environmental Engineering, Faculty of Civil Engineering STU, Radlinského 11, 810 05 Bratislava, Slovakia
[2]The Central Slovak Water Company, Partizánska cesta 5, 974 01 Banská Bystrica, Slovakia
[3]Water Progress, Račianske mýto 10990/1C, 831 02 Bratislava, Slovakia

Email: danka.barlokova@stuba.sk

Abstract. Basic information about importance of calcium and magnesium in water, about their properties, effect to human health, problems what can cause under the lower (< 1 mmol/L) and higher (> 5 mmol/L) concentrations in water supply distribution systems, the most commonly used methods of water hardening are presented. The article contains the water hardening results carried out during the pilot plant experiments in WTP Hriňová and WTP Turček. For water hardening, treated water at the end of the process line, i.e., after coagulation, sedimentation and filtration, saturated with CO_2 and filtrated through half-burnt dolomite material (PVD) was used. The results show that the filtration rate is 17.1 m/h in the case of WTP Hriňová and 15.2 m/h in the case of WTP Turček to achieve the recommended concentration of Ca and Mg in the treated water after the addition of CO_2 and filtration through PVD. The longer the water contact time with PVD (depending on the CO_2 content), the more water is enriched with magnesium, but the calcium concentration has not so much increased.

1. Introduction

Drinking water is an important source of minerals and trace elements that are necessary for a person's life, because the body can´t create them themselves and must be received by food and water. Drinking water contains essential macro elements (e.g. calcium, magnesium), as well as trace elements (e.g, chromium, cobalt, copper, zinc, selenium, manganese, iron, sodium, potassium, iodine) necessary for the proper function of the individual organs of the human body. Drinking water comes naturally from the geological base through which water passes or is added to water to increase its concentration in water wherever it is needed. In drinking water they are usually in ionic form, perfectly dissolved and therefore easily absorbable and usable (compared to foods).

The drinking water supplied to the water supply network has to comply with the legal regulations, currently it is Government Regulation No. 354/2006. There are 81 parameters (excluding radiological indicators) with precisely defined limit values. These are based on EU Directive No. 98/83/EC, as well as data from the World Health Organization (WHO) [1].

In GR No. 354/2006 are determined only recommended values (table 1) for calcium more than 30 mg/L, for magnesium more than 10 mg/L. Some areas of Slovakia comply with legislation, Ca and Mg values in water are exceeded and water softening (especially Western Slovakia) is required, but in other areas the Ca and Mg concentrations are low and water needs to be hardened. Such areas include, for example, the whole area of the High Tatras, Central Slovakia between Zvolen and Tisov, the

Slovak Ore Mountains and the north-eastern part of Eastern Slovakia. Large surface water reservoirs such as Turček, Hriňová, Málinec, Klenovec, Bukovec, and Starina have been built in these regions. This is due to the hydrogeological composition of the substrate, which is made up of rocks of magmatic origin, deficient in calcium and magnesium content. Such water causes health and technological problems.

Table 1. Values of Water hardness and concentrations of Ca and Mg in drinking water by the Government Regulation No. 354/2006.

Parameter	Symbol	Unit	Drinking water	
			Limit value	Recommended value
Calcium + Magnesium	Ca + Mg	mmol/L	-	1.1 - 5.0
Calcium	Ca	mg/L	-	> 30
Magnesium	Mg	mg/L	125	10 - 30

The term of water hardness is used in relation to calcium and magnesium contents. In general, the water hardness is understood as a sum of material concentrations of Ca, Mg, Sr, and Ba or just the sum of Ca and Mg.

Table 2 contains water "hardness" scales used both in the domestic and foreign literature. Expressing hardness in degrees (e.g. German, French) has already been abandoned; in the Anglo-Saxon literature we can find hardness expressed as the equivalent of $CaCO_3$ (mg/L) or, more rarely, of Ca (mg/L).

Table 2. Water hardness scales.

Scale of hardness	Ca + Mg [mmol/L]	Hardness [^{o}dH]	Hardness Ca [mg/L]	Hardness $CaCO_3$ [mg/L]
Very soft	0 - 0.71	0 - 4	0 - 20	0 - 50
Soft	0.71 - 1.42	4 - 8	20 - 40	50 - 100
Medium hardness	1.42 - 2.14	8 - 12	40 - 60	100 - 150
Higher hardness	2.14 - 3.20	12 - 18	60 - 80	150 - 200
Hard	3.20 - 5.40	18 - 30	80 - 120	200 - 300
Very hard	> 5.40	> 30	> 120	> 300

Both extremes of hardness are undesired from both the health and technological points of view; it is complicated to determine optimal concentrations of Ca and Mg in drinking water and the health requirements need not necessarily comply with the technological ones.

1.1. Calcium and magnesium - effects to human health

Calcium and magnesium are necessary for healthy growth and protection of bones from decalcification, they reduce nerve-muscle excitability, stress, and fatigue, support immunity, and affect blood coagulation [2]. Water from the water main is thus the simplest every-day source of calcium and magnesium for our organisms (we can absorb almost 60% of magnesium contained in water compared to only 30-40% of magnesium contained in the food).

If the calcium supply is insufficient for a longer period of time, a continuous osteoporosis and pain of joints can occur with a consequent more frequent fragilitas ossium. Lack of calcium is demonstrated by headaches, migraine, and brittle nails; you can experience eczemas. If the calcium supply is excessive for a longer period of time (especially in the form of artificial preparations with excessive content of calcium), formation of uratomas or kidney stones and faster vessel calcification could occur, especially when vitamin D is not co-administered [3].

Magnesium together with calcium Ca^{2+} represents one of the main controllers of muscle contraction including tension of smooth muscles and lack of it could result in cramps, heartbeat, hypertension, nervousness, caprice, sleep disorders, and improper functioning of the nervous system or endocrine glands (digestive system, kidneys, thyroid). Excessive content of magnesium in our organism is very rare, since magnesium can be easily excreted in urine. That is why a sufficient, not excessive supply of both minerals for a healthy human provides not only optimal mineralization of skeleton and prevention from osteoporosis, but also the ability of a human to cope with active physical performance, regular exercises, and increased daily stress, and provides proper functioning of almost every metabolic or enzymatic process running in the organism as well as proper functioning of the nervous system.

The health significance of water hardness (calcium and magnesium contents) should be taken into account not later than when selecting a water source for drinking water supply (the health significance should be preferred to the technical point of view, so the water hardness treatment should only be accepted at a significant excess of the recommended values). From the health point of view, we prefer harder waters. Statistical results show that in the regions with drinking-water containing higher concentrations of calcium and magnesium, the number of inhabitants suffering from cardiovascular diseases and the risk of sudden cardiac death is lower [3, 4].

1.2. Calcium and magnesium in the distribution network

From the technological point of view, neither too soft water being aggressive and causing pipelines corrosion nor hard water reducing service life of pipelines and basins due to formation of incrustations are appropriate. It is reported that, in dependence on the interaction with other factors such as pH and alkalinity, incrustations are caused by water with hardness (as equivalent of $CaCO_3$) higher than 200 mg/L (WHO, 1993). When water contains hydrogen carbonates, water warming will cause elimination of CO_2 and transformation of hydrogen carbonate onto (calcium) carbonate to be precipitated on the walls of boiling utensils, pipelines, and boilers in the form of firm water scale. Due to these reasons, calcium and magnesium are removed from water.

1.3. Water hardening (recarbonization)

Hardness is increased at water deacidification and stabilization. Remineralization is performed by filtration through appropriate materials (marble, dolomite, half-burnt dolomite, magnesite, and others) or by direct dosing of calcium compounds (lime). From the hygienic point of view, the trouble-less situation will be when wholesome raw materials approved for getting in contact with drinking water are used, when the recommended ratio Mg:Ca is adhered to, and when water is acceptable from the sensory point of view. Since solubility of the raw materials used is low, it should be increased by adding CO_2 in water.

Poorly mineralized water (water with the value $ANC_{4.5} < 0.8$ mmol/L) is not suitable either from the health point of view for its potential of washing up mineral substances from the organism in the volume greater than its own supply [6] or from the technological point of view. When treating such waters, concentrations of Ca^{2+} and Mg^{2+} in the technological procedure should be increased for minimizing its corrosive or aggressive impacts.

The most frequently used procedures of water recarbonization include recarbonization using lime and carbon dioxide, carbonates and carbon dioxide, carbonates and strong acids and PVD and carbon dioxide [5]. Each of the aforementioned possibilities of recarbonization has negative and positive aspects (such as clogging of pipes at the lime dosing, cleanliness of used raw materials, sufficient contact time with the filtration materials, temperature and pH effects, presence of Fe and Mn, organic substances, filters washing). The chosen recarbonization technology is affected by the amount of water to be treated and its mineralization, operating costs, treatment plant equipment, and so on.

2. Experimental part

The pilot-plant experiments were aimed at checking the possibility of enriching water of magnesium in the water treatment plant (WTP) Hriňová. Currently, the contents of calcium and magnesium in water at the outlet from the WTP Hriňová are 27 mg/L of Ca and 4 mg/l of Mg; the average value of Ca + Mg is 0.84 mmol/L.

The experiments were performed on the model equipment for filtration speeds 17 m/h corresponding with the real needs in the case of implementation of this technological level during the planned reconstruction of the WTP Hriňová. For the experiments, we used water treated in accordance with the current water treatment technology (UV radiation, dosing of CO_2, dosing of coagulant PIX113, fast mixing using a hydro mixer, slow mixing,sedimentation, dosing of lime, filters with sand. The aim of the other pilot-plant experiments in WTP Turček was to verify the influence of different filtration rates on increasing the calcium and magnesium content in treated water in WTP Turček. For water hardening, treated water was used at the end of the process line, i.e., after coagulation, lamellar settling tanks and filtration, enriched with CO_2 and filtered through PVD. At present, the calcium and magnesium content in the water at the outflow from the WTP Turček district is about 21 mg/L Ca and 3.5 mg/L Mg, the average Ca + Mg value is 0.55 mmol/L.

The water from the filters was led through a control valve to the CO_2 injection reaction column and then led to the two columns with half-burnt dolomite (PVD). The columns contained filtration material Everzit dol from Germany (particle size 2-4 mm, height 110 cm, column area with PVD 95 cm^2, medium volume 10,454 dm^3). Each column had a separate water outlet and a different flow rate was set in each column. Samples were collected before the reaction tank, after the addition of CO_2 through the reaction tank and the exit of the PVD columns. In table 3 shows the filtration conditions of the experiment and the contact time of the water with the PVD material.

Table 3. Filtration conditions of the experiment in WTP Turček.

	Experiment 1	Experiment 2	Experiment 3	Experiment 4	Experiment 5
Filtration rate [m/h]	6.1	7.2	8.0	12.9	15.2
Water flow [l/min]	0.96	1.14	1.267	2.042	2.407
Contact time [min]	10.88	9.17	8.25	5.12	4.34

3. Results and discussion

In figure 1 is presented the progress of water hardening and change of individual indicators during model tests at WTP Hriňová (at a filtration rate of 17.1 m/h).

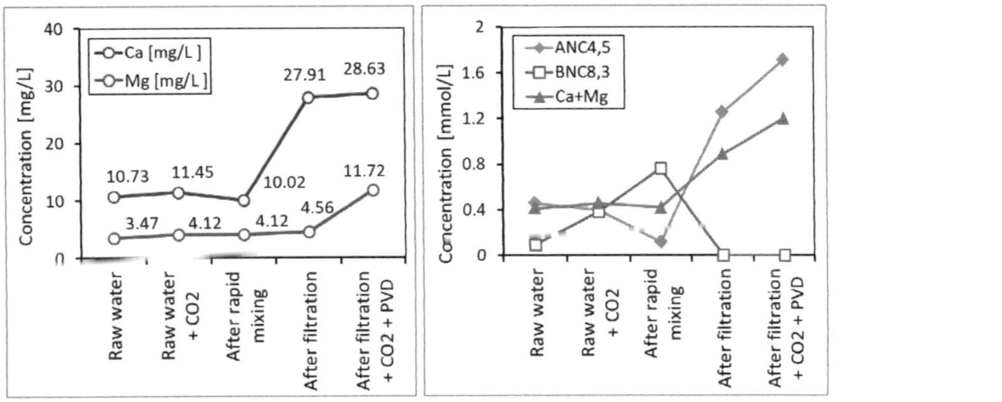

Figure 1. The course of changes of concentrations of Ca, Mg (left), Ca+Mg, $ANC_{4.5}$ and $BNC_{8.3}$ in particular water treatment processes during model tests in the WTP Hriňová.

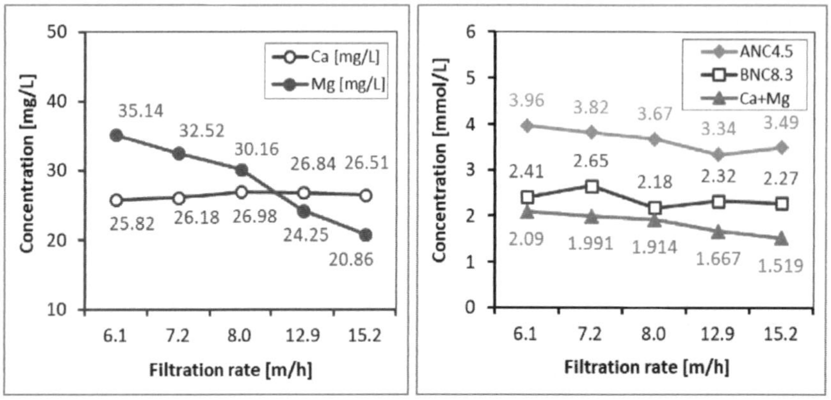

Figure 2. The course of change of concentration of Ca and Mg (left), Ca+Mg, $ANC_{4,5}$ and $BNC_{8,3}$ in particulat water treatment processes during model tests in the WTP Turček.

Figure 2 shows the results of the experiments for the different filter rates and change of concentration of Ca and Mg during the experiments in WTP Turček.

The results show that the filtration rate 17.1 m/h in the case of WTP Hriňová and 15.2 m/h in the case of Turček to achieve the recommended concentration of Ca and Mg in the treated water after the addition of CO_2 and filtration through PVD. At this rate, the total hardness value of Ca + Mg 1.2 and respectively, 1.52 mmol/L which is compared to GR No. 354/2006 which recommends 1.1 to 5 mmol/L. The longer the water contact time with PVD (depending on the CO_2 content), the more water is enriched with magnesium, but the calcium concentration has not so much increased.

4. Conclusion

The article shows the health significance of the presence of calcium and magnesium in drinking-water and the significance of water recarbonization from the health and technological points of view. Drinking water should only be softened when contents of calcium and magnesium significantly exceed the recommended upper hardness limit (5 mmol/l). In the case of soft waters, recarbonization should mainly be used for increasing the content of magnesium. Both in the distribution network and at the consumer, water should be in the balanced state without corrosive and incrustation properties. As shown by the pilot-plant tests, using half-burnt dolomite is suitable even in large water treatment plants. WTP Hriňová, Turček, Klenovec, and Málinec have the content of magnesium in distributed water 3.5-5.5 mg/l and it is a very low value as compared to the recommended value of 10-30 mg/l according to GR No. 354/2006 Coll. as amended.

Acknowledgements

The experimental measurements were performed with the financial support of the APVV-15-0379 and VEGA 01/0400/15 projects.

References

[1] WHO 2011 Guidelines for Drinking-water Quality (GDWQ) Geneva
[2] WHO 2009 Calcium and magnesium in drinking water – Public health significance Geneva p 194
[3] http://www.meda.sk/sk/Pacient/Terapeuticke-oblasti/Vyznam-horcika-pre-ludsky-organizmus.html
[4] Kožíšek F 2003 The health significance of the "hardness" of drinking water National Institute of Public Health in Prague
[5] Olejko Š 2000 Technology of drinking water hardening The 4th Int. Con. Proc. Zlín pp 119-124

IOP Conf. Series: Earth and Environmental Science **92** (2017) 012003 doi:10.1088/1755-1315/92/1/012003

Analysis of post-tensioned girders structural behaviour using continuous temperature and strain monitoring

Ł Bednarski[1], R Sieńko[2] and T Howiacki[2]

[1] AGH University of Science and Technology in Cracow, Mickiewicza Al. 30, 30-059, Poland

[2] Cracow University of Technology, Warszawska St. 24, 31-155 Kraków, Poland

E-mail: rsienko@pk.edu.pl

Abstract. This article presents the possibility of using structural health monitoring system data for the analysis of structure's operation during its life cycle. Within the specific case study it was proved, that continuous, automatic and long term monitoring of selected physical quantities such as strains and temperatures, can significantly improve the assessment of technical condition by identifying hazardous phenomena. In this work the analysis of structural behaviour of post-tensioned girders within the roofing of sport halls in Cracow, Poland, was performed based on measurement results and verified by numerical model carried out in SOFiSTiK software. Thanks to the possibility of performing calculations in real time and informing the manager of the object about abnormalities it is possible to manage the structure in effective way by, inter alia, planning the renovations or supporting decisions about snow removal.

1. Introduction

Application of restressed concrete technology in the opinion of professor Olszak “undoubtedly means the higher level of dominating the nature by technique" [1]. The use of stressed tendons in concrete allow for increasing the length of spans, reducing the cross sections of structural members and taking more efficient advantage of material strengths.

Already in the middle of last century in Poland, lightweight, prefabricated KBO and KBOS post tensioned concrete girders were developed. They were introduced into catalogues [2] and recommended for widespread use. In a short time about 850 objects were built in the country, which means that more than 10,000 girders have been put into operation as structural members of roofs [3]. Unfortunately, mainly due to the low quality of performance on sites as well as the aggressive chloride environment [4] and high humidity in the production halls, the technical parameters of the girders was significantly reduced. Unfavorable operating conditions and the lack of adequate tendon protection have caused pitting corrosion of steel, which in turn caused the breaking of the tendons, sometimes even one year after the girder was built into the structure! Several structural failures in the 1960s caused a decrease in trust top restressed concrete structures.

Nowadays, due to decades-long operation, variable environmental conditions, influence of humidity, progressive corrosion of restressing steel and other factors, one can expect a decrease in the overall bearing capacity of this girders. Furthermore, it should be noticed that the current standard requirements for the structural safety are stricter than before. Thus, it is necessary to properly control the condition of these structures to provide their safe operation. Periodic inspections are carried out, but thanks to advanced structural health monitoring (SHM) we are able to observe the girders behavior continuously. In addition, by estimating the bearing capacity of specific girder [5] and preparing the appropriate calculation procedures, it is possible to assess the stress-strain state in the material in real time.

This article presents analysis performed for real case: girders within roofing of "Study Centre for Physical Education and Sport" of Jagiellonian University, Cracow, Poland (see figures 1, 2).

Figure 1. The monitored hall - view from outside.

Figure 2. The view of monitored post-tensioned girders.

Thanks to the appropriate selection of beam profiles, it is possible to reduce the consumption of concrete and steel. The KBO and KBOS girders have been designed in such a way that the longitudinal configuration has been adapted to the course of bending moments. Thanks to this, materials are used effectively in every cross section, which allow us to include them into the group of economical girders [6].

Considered post-tensioned girders are consisted of eccentrically-compressed upper flange, eccentrically-tensioned bottom flange and vertical braces, which cause bending moments in flanges. They work should be analysed in two stages: before and after bonding with roof panels. In the second case the girders cooperate with panels, providing additional spatial stiffness, which cannot be considered in the 2D model. Obviously, three-dimensional analysis will more accurately reflect the actual state of the structure under operation, and the calculated bearing capacity will be greater than in the case of two-dimensional analysis [6,7]. In view of the threats presented above, the resulting supply is particularly important. In some cases, percentage difference between load capacities obtain from theoretical 2D and 3D model is even more than 50%.

2. Structural health monitoring

Structural health monitoring system of considered girders is based on continuous control of their work by automated and remote measurements of strain and temperature changes in selected measuring points within the structure, in reference to zero readings.

The main purpose of SHM system is to control changes in girders stress-strain state, mainly caused by the snow load. This allows for more effective management of the objects, including supporting those responsible for making decisions about snow removal. Other possible causes of changes in structural effort of girders are corrosion of restressing tendons, the breakdown of roof panels, the loss of bonding between panels and upper flanges of girders or the deformation caused by the temperature. The temperature load, in particular the temperature difference between the girders and the roof panels, has significant impact on girders operation [3]. During traditional geodetic measurements of displacement, special attention has to be paid to the thermal conditions, which should be as close as possible to each other during consecutive measurement cycles. Assessment of girders operation in this case becomes an extremely difficult task. The superiority of the structural health monitoring system lies mainly in performing measurements continuously over long period of time, what allow for analysis of monthly and annual cycles i.e. when changes in strains are clearly visible against the background of temperature changes. In considered sport halls measurements are made with a frequency of 15 minutes and calculated automatically according to designed procedures. The location of measuring points within monitored objects is presented at figure 3.

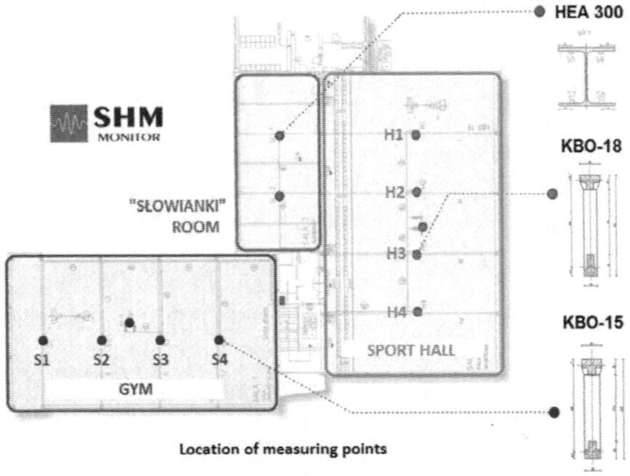

Figure 3. Configuration of monitored structures with location of measuring points [courtesy of SHM System].

Within considered sport halls following sensors were installed:

- 28 vibrating wire strain sensors [14], integrated with internal thermistors;
- 20 additional thermistors.

Measuring devices were connected with structural membersin a way providing durability of the connection during long-term operation. The location of the sensors within specific KBO-18 girder is shown at figure 4.

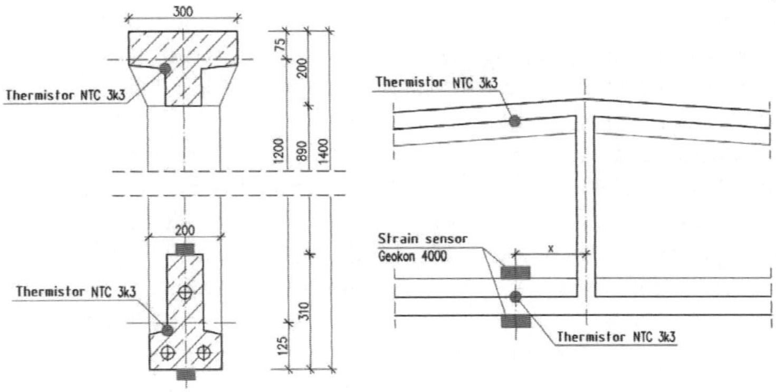

Figure 4. Sensors location within KBO-18 girders [courtesy of SHM System].

3. Exemplary results and discussion

At figures 5 and 6 the changes in strain and temperature values were presented between 01.03.2015 and 16.09.2017 for measuring points S2 and S3 within Gym (see figure 3). The blue line corresponds to strains measured at upper edge of bottom flange, and green line to the lower edge of bottom flange. Sign "+" corresponds to the increase of tensile strains in reference to zero reading. The values of strains are presented in microdeformations $\mu\varepsilon = 10^{-6}$. The temperature changes were marked by dotted lines. Values of temperature are similar within upper and lower edge of bottom flange, as well as the upper flange. The differences did not exceed 1°C in considered period of time, which allows in numerical modelling for assumption of uniform temperature load.It should be noticed, that strain values calculated and presented in figures 5 and 6 corresponds only to strains, which cause the stress of structural member (they do not include free, thermal strains).

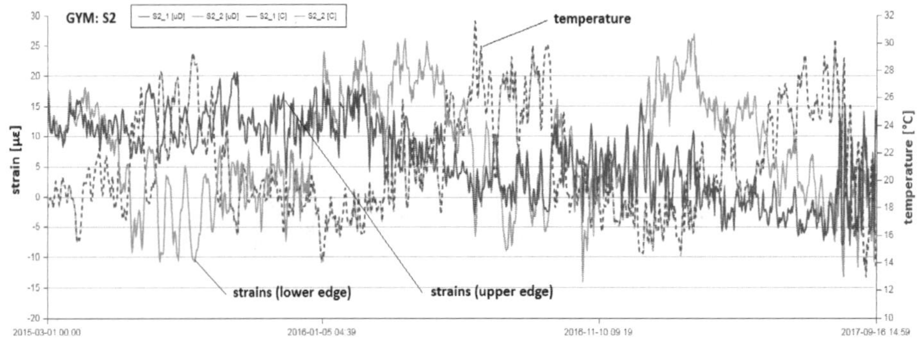

Figure 5. Strain and temperature changes between 01.03.2015 and 16.09.2017 – Gym, S2.

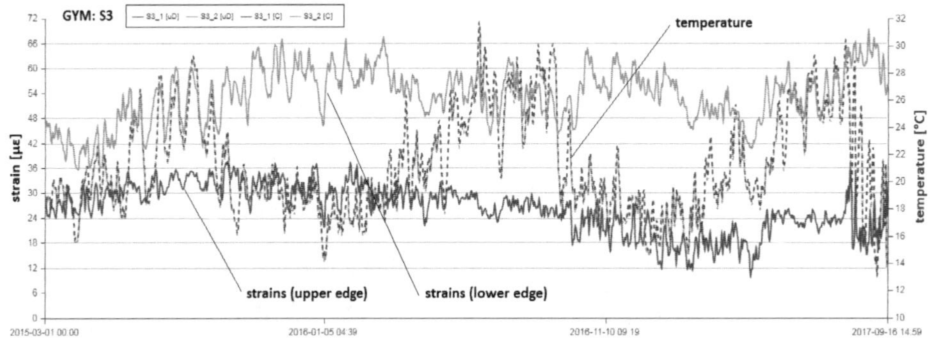

Figure 6. Strain and temperature changes between 01.03.2015 and 16.09.2017 – Gym, S3.

In point S2 we can clearly observe the influence of temperature on strains measured in the bottom flange of girder in annual cycle. This is common situation for almost all analysed girders, expect of girder S3 (see figure 6), where strains are insensitive to thermal effects. The reason of this phenomena could be explained by numerical simulations.

The model was created in SOFiSTiK software to analyze girders work in two configurations: as statically determined beam (2D model) and including spatial cooperation with roof panels and other girders (3D model). In the 3D model ribbed roof panels were assumed as areas with equal, averaged thickness, and the upper flanges of girders as monolithically connected with concrete overlay. The visualization of structure's deformation under the uniform snow and temperature load was respectively presented at figures 7 and 8 (at figure 8 roof panels were deleted to provide appropriate clarity).
Figure 7 shows the advantages of bonding girders with roof panels, which provide spatial stiffness and significantly reduces deflections under the snow load in comparison to the free, single beam. On the other hand, while loading the structure with uniform temperature, we can observe that it does not affect the stress in statically determined beam (2D model). This corresponds to the situation presented for girder S3 and suggests the possibility of loss the bonding between the upper flange of girder and roof panels. This could be dangerous for load bearing capacity, because the supplies resulting from spatial cooperation of structural members are reduced.

Getting information about structural behavior in real time based on SHM data analysis is the basis for effective decision making. Abnormalities are visualized for the User in serviceable way, so he is able to start prevent actions immediately, regarding for example with: snow removal, hiring an Expert for making engineering calculations and estimate the actual condition of the building,organizing*in-situ*inspections or setting a schedule of renovations in long term.

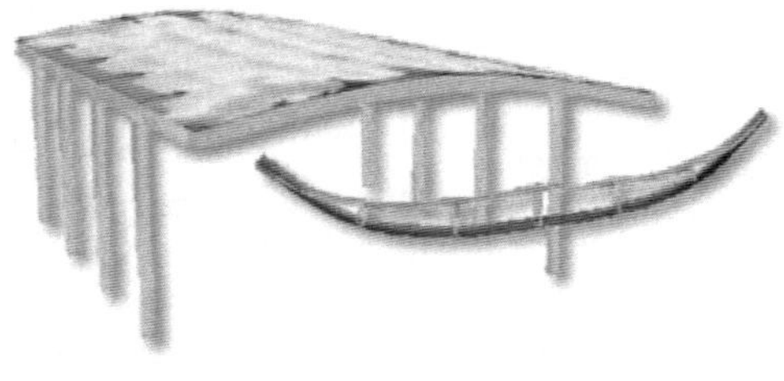

Figure 7. Visualization of structural deformations under uniform snow load.

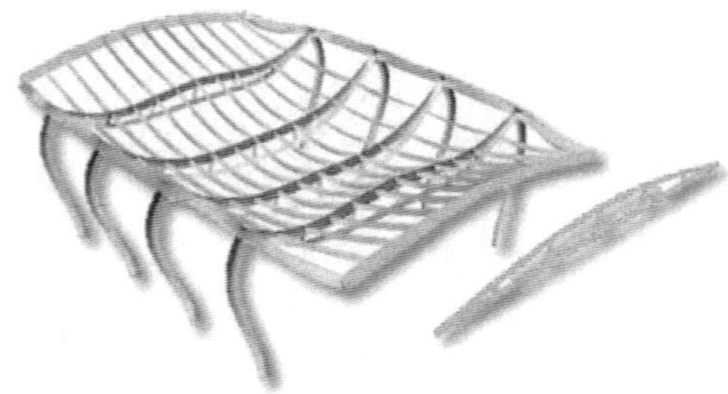

Figure 8. Visualization of structural deformations under uniform temperature load.

4. Conclusion

KBO and KBOS prestressed concrete roof girders were widely used in Poland in the second half of the last century. Due to their long-term operation in difficult, changeable conditions, they require proper observation and control, preferably by continuous, structural health monitoring. Theoretical analysis of such objects will remain imperfect, so *in situ* measurements are always very valuable [8,9], allowing for justified and effective decision making in real time.Observation of structural behavior, especially of prestressed concrete structures using SHM systems, is increasingly used in civil engineering [10,11,12] also including construction process [13,15].

References

[1] Olszak W 1961 *Teoria konstrukcji sprężonych* (Warsow: PWN)

[2] Katalog Projektów Konstrukcji Sprężonych 1956 (Warszawa: Biuro Studiów i Projektów Typowych Budownictwa Przemysłowego)

[3] Sieńko R 2010 *Konstrukcje kablobetonowe* XXV Ogólnopolskie Warsztaty Pracy Projektanta Konstrukcji (Szczyrk: Poland)

[4] Nürnberger U 2002 Corrosion induced failures of prestressing steel *Otto-Graf-Journal* **13**

[5] Dyduch K and Sieńko R 2008 Stany graniczne pasów dolnych kablobetonowych dźwigarów dachowych KBOS *Inż. Budow* **12** 668–671

[6] Zieliński Z 1962 *Prefabrykowane betonowe dźwigary sprężone, wydanie drugie poprawione*, (Warsow: Arkady)

[7] Sieńko R, Howiacki T and Jedliński S 2013 Design errors resulting from the adoption of incorrect structure's model *Tech. Tran.* **2** 126–136

[8] Dyduch K and Sieńko R 2004 Wybrane wyniki weryfikacji doświadczalnej pracy kablobetonowych dźwigarów dachowych KBOS-18 *50 Konferencja Naukowa "Krynica 2004", Warszawa-Krynica*

[9] Sieńko R and Szydłowski R 2004 *Badanie kablobetonowych dźwigarów dachowych KBOS-18 w trakcie symulacji korozyjnego pękania cięgien* Praca zbiorowa Dolnośląskie Wydawnictwo Edukacyjne Wrocław pp 453-460

[10] Lin Y B, Chang K C, Chern J C ,Wang L A 2004 The health monitoring of a prestressed concrete beam by using fiber Bragg grating sensors *Smart Mater. Struct.* **13** 712-718

[11] Matthys S and Taerwe L 2005 *Experimental testing of post-tensioned concrete girders instrumented with optical fibre gratings* 17th International Conference on Optical Fibre Sensors Bellingham WA

[12] Wang M L 2007 Long-term health monitoring of post-tensioning box girder bridges *World Forum on Smart Materials and Smart Structures Technology* China

[13] Uva G, Porco F, Fiore A and Porco G 2014Structural monitoring using fiber optic sensors of a pre-stressed concrete viaduct during construction phases Case Studies in Nondestructive *Test.and Eva.* **2** 27–37

[14] Bednarski Ł and Sieńko R 2013 Pomiary odkształceń konstrukcji czujnikami strunowymi *Inż. Budow.* **11** 615-619

[15] Rybak J and Pieczynska-Kozłowska J M 2014 Vibration monitoring as a tool for a calibration of geotechnical technologies *14th Int. Mul. Sci. Geo. SGEM 2014* (Sofia: STEF92 Technology Ltd.) pp 1043 - 1050

Influence of bacteria on degradation of bioplastics

M Blinková[1] and K Boturová[1]

[1]VŠB – Technical University of Ostrava, Institute of Environmental Engineering, Czech Republic

Email: marieblinkova@gmail.com

Abstract. The degradation rate of bioplastic in soil is closely related to the diversity of soil microbiota. To investigate the effect of soil bacterial on biodegradation, 4 bacterial strains of soil – *Pseudomonas chlororaphis*, *Kocuria rosea*, *Cupriavidus necator* and *Bacillus cereus*, were used to accelerate the decomposition of bioplastics manufactured from Polylactid acid (PLA) by direct action during 250 days. The best results were obtained with bacterial strains *Cupriavidus necator* and *Pseudomonas chlororaphis* that were isolated of lagoons with anthropogenic sediments.

1. Introduction

Bioplastics are products with character of conventional plastics that are made from plant biomass or its parts – organic substances like polysaccharides (starch, cellulose), polyesters produced by organisms and also synthetic biodegradable polymers. Biopolymers are high-molecular organic compounds mostly produced by metabolic reactions of living organisms [1].

All polymers that are subject to change by microorganisms like bacteria, fungi, algae and protozoa, are considered to be biodegradable. These biologically degraded plastics are more environmentally friendly than the products from petroleum because they decompose in shorter time intervals. Not only the process of decomposition but also the final products of bioplastics degradation are significant – only water, carbon dioxide, eventually methane and residual biomass are produced [2]. With regard to the fact that bioplastics serve as a raw material for manufacturing household items, packaging materials and other everyday life needs, they make a big amount of municipal waste and that is the reason why is necessary to focus on faster and more efficient recycling [1].

Degradation capacity of the environment is defined as a maximal level of microbial activity that is possible to expect based on conditions of ecosystem where biodegradation takes place. Resulted activity of microbial consortium depends not only on physical-chemical conditions but also on biological factors like number of microorganisms, their ability to adapt, genetic predispositions and other physiological conditions. Other factors that affect the whole process of biodegradation are the presence of chemical substances, eventually properties of contaminating substances like toxicity, bioavailability and biodegradability [3].

The aim of this study was to compare biodegradation rate and efficiency of conventional and ecological plastics when exposed to bacterial biomass for 250 days. Progress of experiment was continuously monitored visually by taking photos at 15 days interval with binocular microscope. Intermediate result of this experiment was to verify which bacterial species effect the decomposition of bioplastics the most and to use this acquired knowledge for further study of biodegradation processes.

IOP Conf. Series: Earth and Environmental Science **92** (2017) 012004 doi:10.1088/1755-1315/92/1/012004

2. Methods and experimental

Plastic product that is often used as an everyday item was monitored in the experiment. It is a plastic drinking cup with volume of 0.2 L manufactured by American company Nature Works LLC (see figure 1). Since 1989 Nature Works LLC deals with production of environmentally friendly plastics and because of their wide field of activity it is now possible to buy these products in Europe and even in Czech Republic. This plastic cup is made from polylactic acid (PLA). This compound is derived from renewable resources such as starch that is easy to decompose. Lactic acid is obtained by fermentation of the glucose that is followed by polymerization to obtain polylactic acid [1].

For experimental purpose the plastic cup was cut to samples of size 2 x 1 cm. During the process of biodegradation every sample of plastic cup was added in bacterial medium to observe rate and efficiency of reaction and also to monitor changes during experiment.

Four strains of soil bacteria were used for the experiment – *Pseudomonas chlororaphis, Kocuria rosea, Cupriavidus necator* and *Bacillus cereus*. Strains were isolated from metal contaminated soils and some of them were recently used for successful biodegradation of chemical substances [2,6,7]. Bacteria were statically cultivated at 25°C on tryptone-soya agar (M290, HiMedie Laboratories, Mumbai, India). Plastic samples of size 2 x 1 cm were leached in the trypton-soya broth (M011, HiMedie Laboratories, Mumbai, India) with bacterial culture added in the exponential phase of growth with total volume of 100 mL. Progress of experiment was continuously monitored visually by taking photos (zoom 40x and 100x) at 15 days interval with binocular microscope Olympus CX4.

Figure 1. Plastic drinking cup 0.2 L (Nature Works LLC, USA) and its parts used in degradation experiments.

3. Results and discussion

Biodegradable rate and efficiency with selected bacterial strains is summarized in table 1.

Table 1. Biodegradation process of plastic cup sample

Bacterial strain	Number of days 50 d	100 d	150 d	200 d	250 d
Bacillus cereus ZK-27	+	+	++	++	++
Cupriavidus necator POP-31	++	++	++	+++	+++
Kocuria rosea SL-1	+	+	+	+	++
Pseudomonas chlororaphis ZK-1	++	++	+++	+++	+++

As shown in table 1, bacterial strains *Cupriavidus necator* POP-31 and *Pseudomonas chlororaphis* ZK-1 were the most suitable for biodegradation of the sample. After 250 days biodegradation reached its maximal effect and caused distinctive damage of the sample not only on the surface but also in the structure of material. Progress of biodegradation is shown in figures 2 – 4.

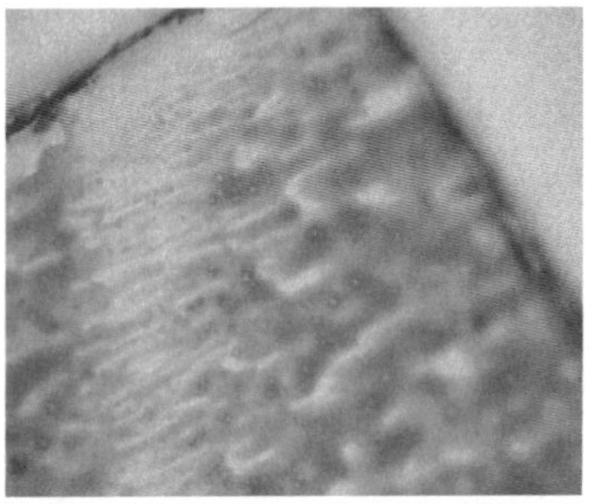

Figure 2. Green color loss in the sample of plastic

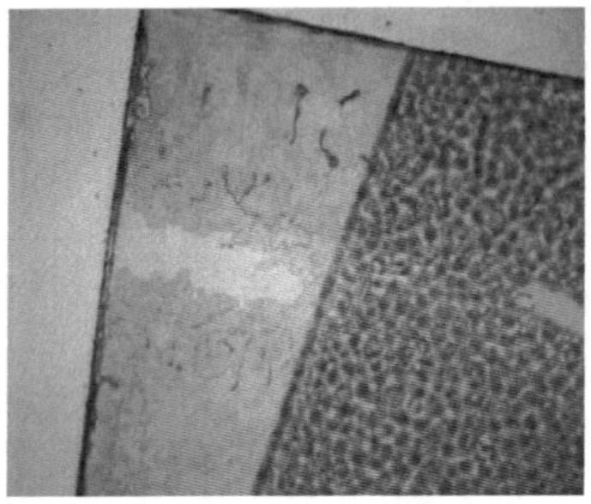

Figure 3. Biodegradation : first holes formation

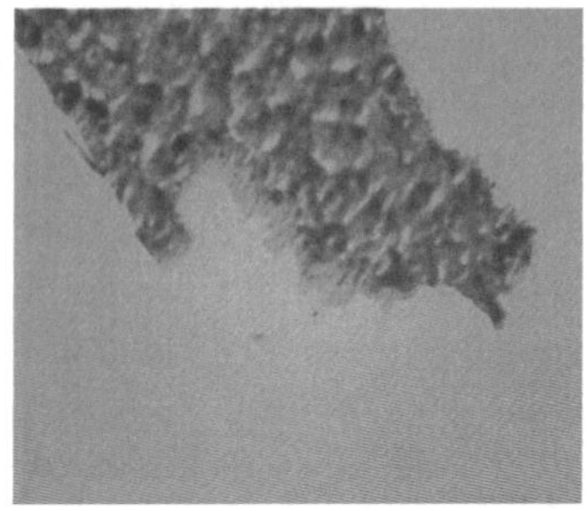

Figure 4. Biodegradation: structure disruption

Bacterium *Pseudomonas chlororaphis* ZK-1 was isolated as an arsenic-tolerant strain from sediments of lagoon in Zemianske Kostoľany [7,8,15]. Bacterium *Cupriavidus necator* POP-31 was identified as a new isolate from lagoon in Poproč and from the soil with high concentrations of antimony and lead [4]. Generally, these bacteria are common soil species that are low nutrient-demanding and due to their adaptability are often used for biodegradation of organic pollutants [7,9,10,11,12,14]. One of the significant factors that determine bacterial behavior in biodegradation process is the characteristics of genome. The fact that bacteria *Cupriavidus* possess wide spectrum of plasmids and megaplasmids has already been proofed by research of the genetic nature of biodegradation [5]. Plasmids and megaplasmids are the reasons why this bacterial strain is resistant to heavy metals and organic toxic substances and is able to degrade them with high rate and efficiency [13,16,17,18]. Biotechnological use of *Pseudomonas chlororaphis*, which contributed on the biodegradation the most, is not fully clarified yet. Bacterial genus *Pseudomonas* is well known for its so called degradable plasmids that are able to metabolized low molecular organic compounds as well as high molecular organic compounds like polymer chain of petroleum substances [14,19].

Study focused on ability of microorganisms to decompose waste plastics on landfill in Bangkok was published in 2015. Lab experiment was based on simulated conditions of landfill environment (lighting conditions with the use of UV-B lamp, presence of indigenous microflora, amount of plastic material in municipal waste). For experimental method were used sample of common plastic products from polyethylene (high density polyethylene HDPE/ low density polyethylene LDPE), polypropylene (PP) and polystyrene (PS), that were cut to pieces of 9 cm^2; from microbial strains were selected bacterial species of *Flavobacterium, Xanthobacter, Methylobacter, Methylococcus, Nitrobacter* and *Nitrosomonas.* Experiment lasted for 210 days and it was concluded that products from HDPE degraded in higher rate with mass loss of 15%, from LDPE of 4, 96%, PP 6, 7% and for products from PS mass loss was 5, 29% [21].

Some studies confirm that ingested polyethylene creates changes in digestive system of fishes, birds and sea animals that cause disruption of function of digestive organs and may cause death. Increasing amount of waste, decrease of landfill capacity and long-term degradation of conventional plastics caused rising interest in biodegradation of PE. Degradation ability of microbial biomass on LDPE was tested in Saudi Arabia [22] and India [20] – especially presence of bacteria able to metabolize LDPE. Bacterial genus *Bacillus, Staphylococcus, Micrococcus* and *Pseudomonas* were tested in this study; the most efficient bacterium was *Brevibacillus borstelensis* isolated from soil. Incubation of polyethylene with bacterium *B. borstelensis* decreased mass weight of 30% and maximal rate of biodegradation was monitored after 90 days of incubation at 50°C [20].

4. Conclusion

The results of experiments show that the biodegradation of PLA materials accelerated the most with effect of gram negative bacteria *Pseudomonas chlororaphis* and *Cupriavidus necator*. These results could be used in processes of controlled biodegradation and in other biotechnological processes that accelerate decomposition of environmentally degradable materials.

Biodegradable materials are becoming a new alternative for plastic manufacturing. Physical-chemical qualities of these substances are improving due to progressive research with the aim of better mechanical properties and resistance to different environmental effects but with preserved easy biodegradation ability.

In these days, biodegradation is considered to be an important part of degradation process when decomposition happens in presence of living microorganisms [23,24,25]. Biological degradation of bioplastics is much more effective and faster than degradation of conventional plastics. Limiting factors of biodegradation process are conditions needed for growth and reproduction of microorganisms but the basic factor that affects the rate of biodegradation is composition on these plastics – from which material are biodegradable plastics made and if this material could be possible source of carbon and energy for their growth and reproduction.

Acknowledgments

This research was financially supported by Project for Specific University Research (SGS) No. SP2017/8 from the Ministry of Education, Youth and Sports of the Czech Republic & Faculty of Mining and Geology of VŠB – Technical University of Ostrava. The authors would like to thank for the project support.

References

[1] Stevens E S 2002 *Green Plastics: An Introduction to the New Science of Biodegradable Plastics* (Princeton: Princeton University Press)

[2] Šimonovičová A, Ferianc P, Vojtková H, Pangallo D, Hanajík P, Kraková L, Feketeová Z, Čerňanský S, Okenicová L, Žemberyová M, Bujdoš M and Pauditšová E 2017 *Chemosphere* **171** pp 89–96

[3] Honzík R 2004 Plasty se zkrácenou životností a způsoby jejich degradace *Biom.cz*

[4] Babičová A and Dlabaja M 2016 *16th Int. Mul. Sc. GeoCon. SGEM 2016 (Albena)* **1** (Sofia: STEF92 Technology Ltd.) pp 445–450

[5] Diels L, Dong Q, Van der Lelie D, Baeyens W and Mergeay M 1995 *J. of Ind. Micr. & Bio.* **14** pp 142–153

[6] Kolenčík M, Vojtková H, Urík M, Čaplovičová M, Pištora J, Cada M, Babičová A, Feng H, Qian Y and Ramakanth I 2017 *Wat., Air, and Soil Pol.* 228-224

[7] Šimonovičová A, Peťková K, Jurkovič Ľ, Ferianc P, Vojtková H, Remenár M, Kraková L, Pangallo D, Hiller E and Čerňanský S 2016 *Wat., Air, and Soil Pol.* 227-336

[8] Peťková K, Vojtková H, Jurkovič Ľ, Ferianc P and Remenár M 2014 *14th Int. Mul. Sci. GeoCon. SGEM 2014 (Albena)* **1** (Sofia: STEF92 Technology Ltd.) pp 399–404

[9] Vojtková H 2013 *13th Int. Mul. Sci. GeoCon. SGEM 2013 (Albena)* (Sofia: STEF92 Technology Ltd.) pp 85–92

[10] Vojtková H 2014 *14th Int. Mul. Sci. GeoCon. SGEM 2014 (Albena)* **1** (Sofia: STEF92 Technology Ltd.) pp 451–458

[11] Vojtková H 2015 *15th Int. Mul. Sci. GeoCon. SGEM 2015 (Albena)* **1** (Sofia: STEF92 Technology Ltd.) pp 291–296

[12] Vojtková H 2016 *16th Int. Mul. Sci. GeoCon. SGEM 2016 (Albena)* **1** (Sofia: STEF92 Technology Ltd.) pp 755–760

[13] Vojtková H, Janulková R and Švanová P 2012 *Inż. Min.* **13** pp 49–54

[14] Vojtková H, Kosina M, Sedláček I, Mašlaňová I, Harwotová M and Molinková V 2015 *Fol. Mic.* **60** pp 411–416

[15] Vojtková H, Peťková K and Jurkovič Ľ 2013 *13th Int. Mul. Sci. GeoCon. SGEM 2013 (Albena)* **1** (Sofia: STEF92 Technology Ltd.) pp 141–146
[16] Janssen P J, Van Houdt R, Moors H, Monsieurs P, Morin N, Michaux A, Benotmane M A, Leys N, Vallaeys T, Lapidus A, Monchy S, Médigue C, Taghavi S, McCorkle S, Dunn J, Van der Lelie D and Mergeay M 2010 *PLoS ONE* 5 e10433
[17] Kašáková H, Vojtková H and Jablonka R 2012 *12th Int. Mul. Sci. GeoCon. SGEM 2012 (Albena)* **2** (Sofia: STEF92 Technology Ltd.) pp 435–440
[18] Monchy S, Benotmane M A, Janssen P, Vallaeys T, Taghavi S, Van der Lelie D and Mergeay M 2007 *J. of Bac.* **189** pp 7417–7425
[19] Vega D, Cooke R and Marty J L 1988 *FEMS Mic. Let.* **49** pp 199–202
[20] Kumar S, Sudip and Raut S 2015 *J. of Env. Chem. Eng.* **3** pp 462–473
[21] Muenmee S, Chiemchaisri W and Chiemchaisri C 2015 *Int. Bio. & Bio.* **102** pp 172–181
[22] Sen S K and Raut S 2015 *J. of Env. Chem. Eng.* **3** pp 462–473
[23] Závada J, Nadkanská H, Bouchal T, Gibesová B, Cibulcová A and Blažek V 2014 *14th Int. Mul. Sci. GeoCon. SGEM 2014 (Albena)* **3** (Sofia: STEF92 Technology Ltd.) pp 79–84
[24] Závada J, Nadkanská H, Smatanová N, Šašek P and Bouchal T 2014 *14th Int. Mul. Sci. GeoCon. SGEM 2014 (Albena)* **1** (Sofia: STEF92 Technology Ltd.) pp 393–400
[25] Závada J, Nadkanská H, Bouchal T, Šašek P and Smatanová N 2014 *14th Int. Mul. Sci. GeoCon. SGEM 2014 (Albena)* **1** (Sofia: STEF92 Technology Ltd.) pp 33–40

IOP Conf. Series: Earth and Environmental Science **92** (2017) 012005 doi:10.1088/1755-1315/92/1/012005

Chemical properties of fly ashes produced in the Arcelor Mittal power station and the Třinec ironworks power station

L Bulikova[1]

[1]Faculty of Civil Engineering, VŠB – Technical University of Ostrava, Ludvíka Podéště 1875/17 708 33 Ostrava-Poruba, Czech Republic

E-mail: lucia.bulikova@vsb.cz

Abstract. Fly ash is an inorganic secondary material coming from combustion of coal in power stations. Tested fly ashes were produced by different processes. Therefore they have different parameters. Their chemical properties were determined using chemical analyzes and compared. Results of laboratory tests show possibility to use of fly ashes in road construction as binder for soil treatment.

1. Introduction and background

Fly ash is an inorganic product of combustion of pulverized coal combustion in power and heating stations. Fly ash consists of small particles of siliceous glass and it is collected by electrostatic precipitation of flue gases. The chemical composition and reactions during combustion are influenced by the presence of water, oxygen, sulphur content, furnace type and burning process. Mineralogical composition of fly ash depends on the geological factors of coal and also on combustion conditions [1,2,5].

There are three types of coal-fired boiler furnaces used in power stations. They are referred to as dry-bottom boilers, wet-bottom boilers (high temperature combustion), and cyclone furnaces (fluid combustion). Traditional combustion of coal in furnaces is carried out at high temperatures in the range of1200-1600°C, while fluid combustion at lower temperatures. Consequently, flue gas is desulfurized with limestone and the final product is gypsum. Produced calcium oxide CaO is low reactive due to high temperature in the furnace and typical product of high temperature combustion is siliceous fly ash. Fluid combustion is a technology, when coal is burned at 850°C along with sorbent (milled limestone), that is added due to the bonding of sulfur content in coal. During combustion, SO_2 is absorbed by particles of CaO and anhydrite ($CaSO_4$) is formed. This process affects considerable reduction of pollutants NO_x in exhalations. Typical product of fluid combustion in Czech republic is calcareous fly ash [2,4,6].

2. Methods and result of tests

A set of methods for chemical and mineralogical analyses of two samples of fly ash from the Morava region was used. It includes Differential Scanning Calorimetry (DSC) and Thermal Gravimetric Analysis (TGA), Fluorescence Spectrometry, X-ray diffraction and Scanning Electron Microscopy, were carried out in the ICT VŠB TU Ostrava - Institute of Clean Technologies for Mining and Utilization of Raw Materials for Energy Use.

First one (SFA), produced by Arcelor Mittal Ostrava power station was collected from the tip near its source. Second one (CFA) was produced in the Třinec ironworks power station and it was collected from dry storage silo. Both fly ash samples were produced by burning of coal from the Ostrava-Karvina coalfield.

Chemical analysis was performed to classify fly ash and determine its reactivity. The analysis showed that SiO_2, Al_2O_3 and Fe_2O_3 represented more than 70% content. The major differences between both samples of fly ashes are in higher content of sulfur trioxide SO_3 7.5% and higher content of CaO 16.7% in case of CFA. The CaO content was only 4.9% in case of SFA (table 1).

Table 1. Chemical composition of fly ash samples.

Element	SFA (%)	CFA (%)
SiO_2	42.7	40.9
Al_2O_3	24.5	32.3
Fe_2O_3	8.6	6.5
CaO	4.9	16.7
MgO	0.7	1.2
SO_3	0.9	7.5
K_2O	2.4	1.7
LOI	11.4	7.8

Based on chemical composition of fly ashes, samples were classified as class F – siliceous and class C - calcareous (table 3) according to the ASTM C-618 classification. Siliceous fly ash (Arcelor Mittal Ostrava power station) is characterized by low volume changes and higher pozzolanic activity. Calcareous fly ash (Třinec ironworks power station) is characterized by higher volume changes and hydraulic properties for higher content of CaO. This fly ash is typical material that hardens in mixture with water without admixture of any binder. The contents of anhydrite (7.45%) and lime (5.66%) as highly reactive substances cause hydraulic behaviour of CFA (table 2). On the other hand SFA fly ash sample contains passive component- mullit (9.77%). X-ray diffraction quantified amorphous phase by method of internal standard ZnO. Amorphous phase is represented by siliceous glass principally.

Table 2. Results of X-ray diffraction.

Element	SFA (%)	CFA (%)
Quartz	12.08	11.65
Hematite	1.43	3.39
Calcite	1.97	1.93
Zincite	9.14	9.08
Periclase	0.73	1.31
Mullite	9.77	-
Magnetite	1.26	-
Gypsum	1.47	-
Muscovite	-	4.63
Lime	-	5.66
Anhydrite	-	7.45
Amor.	62.2	54.83

Table 3. Classification according to ASTM C-618.

Class	Fly ash	Ca, Fe, Sum of oxides (%)	CaO (%)
F	Siliceous	Minimum Ca Considerable amount Fe $(Al_2O_3+SiO_2+Fe_2O_3) > 70$	< 10
C	Calcareous	Considerable amount Ca $(Al_2O_3+SiO_2+Fe_2O_3) > 50$	10 - 20 > 20

Electron microscope micrographs of fly ash samples from the AM power station (SFA) and from the Třinec ironworks power station (CFA) are shown in figures 1 and 2. Spherical particles of SFA microspheres and mica particles were observed.

Microspheres are formed during combustion in the temperature range of 1200-1500 °C and they influence fly ash properties (density, porosity etc.). Particles of calcareous fly ash have shape of irregular grains, due to lower temperature combustion of coal in the Třinec ironworks power station [6,7].

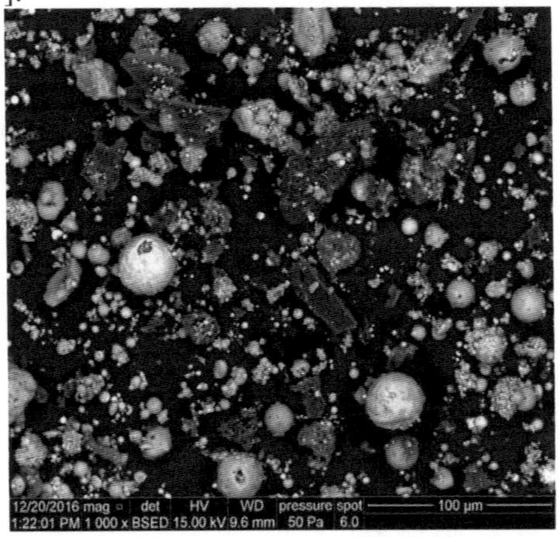

Figure 1. Siliceous fly ash (the SFA samples).

Figure 2. Calcareous fly ash (the CFA samples).

The colour of fly ash is changed from light gray to black depends on the amount of unburned carbon or the presence of pyrite (dark) and shale (light) in coal. Colour is influenced by the amount of lime or iron too [5].

Figure 3. Different colour of fly ash samples (left SFA, right CFA).

3.Application in road construction

Treatment and stabilization of soils change the properties of soils to improve their engineering utilisation. Bulk density, moisture content, plasticity, susceptibility to volume changes, shear strength and bearing capacity are the most improving properties of treated soil. Binders that are used in soil treatment are represented mainly by cement, lime or hydraulic road binder. Fly ash has been used as an alternative binder in last decades [1,3,5,7].

Fly ash is used to mechanical soil treatment. Mechanical treatment of soil improves mechanical properties of treated soil (e.g. shear strength or compressibility), workability. Siliceous and calcareous fly ash is used in geotechnical applications closely connected with road construction to increase strength properties of soils, to control contraction and swelling of soils, to reduce water content in soils and to modify a particle size distribution [1,7,8,10].

Soil treatment by fly ash in the Czech Republic is governed by the European standard EN 14227-15 Hydraulically bound mixtures - Part 15: Hydraulically stabilized soils, Technical Requirement TP 94 - Soil treatment and Technical requirements TP 93 - Design and construction of highway structures using fly ash and ash. Regulations contain technical requirements for materials, material testing, requirements for construction of road layers, effects of soil treatment and environmental aspects [7].

4. Conclusions

Fluid combustion is more effective than high temperature combustion therefore it becomes more useable. However, combustion products have chemical and mineralogical properties that may cause problems in their utilization.

Composition of fly ash depends on type of fuel and furnace. In the past, defects occurred at application of inappropriate materials to road embankments and therefore specific research helps to avoid similar mistakes.

Laboratory tests show the potential of utilisation of fly ashes from the Morava region in road construction. Based on initial results from chemical analyses fly ashes can be used in treatment of soils. In the further research it is recommended to verify mechanical parameters of mixtures with concrete type of soil. The next step in future work would be to check the mixture properties in large-scale tests on construction site using the appropriate machinery.

Acknowledgement

Financial support through the Student Grant Competition VŠB-TUO - Project No. SP2017/155 is gratefully acknowledged.

References

[1] American coal ash association 2003 *Fly ash facts for highway engineers* (Washington DC: FHWA) pp 47-52
[2] Fečko P 2003 *Fly ashes* (Ostrava: Editorial Centre VŠB-TU Ostrava) p 187
[3] Kresta F 2013 *Secondary materials in highway construction* (Ostrava: Editorial Centre VŠB-TUOstrava) pp 54-66
[4] Brandštetr J, Havlica J 1996 Phase composition of solid residues of fluidized bed coal combustion quality tests and application possibilities *Chemical Papers* **50** pp. 188-194
[5] Mezencevová A 2003 Possibilities of utilizing power plant fly ashes *Acta Montanistica Slovaca* **8** pp 146-151
[6] Michalíková F, Sisol M, Krinická I 2010 Chemical and mineralogical properties of ash of coal combustion in thermal power plants *Waste forum* **11** pp 15-16
[7] Bulíková L, Kresta F 2017 Possibility of use of fly ash from the Morava region to soil treatment in road constructions *Conference Fly ash in civil engineering 2017* pp 7-13
[8] Kresta F 2013 Fly ash utilisation in road construction in Czech Republic 5th *symp. on ash, slagand waste landfills in power plants and mines* pp 118-125

[9] Ministry of Transportation Department of Roads 2011 Design and construction of highway structures using fly and ash *Technical Requirement 93*
[10] Ministry of Transportation Department of Roads 2013 Soil treatment *Technical Requirement 94*

An advanced analysis and modelling the air pollutant concentration temporal dynamics in atmosphere of the industrial cities: Odessa city

V V Buyadzhi[1], A V Glushkov[1], O Yu Khetselius[1], V B Ternovsky[1], I N Serga[1] and N Bykowszczenko[2]

[1] Odessa State Environmental University, L'vovskaya str.15, 65016, Odessa-16, Ukraine
[2]Institute of Chemistry and Environmental Protection, Technical University of Szczecin, Piastów 42 av., 71-065, Szczecin, Poland

E-mail: buyadzhivv@gmail.com

Abstract. Results of analysis and modelling the air pollutant (dioxide of nitrogen) concentration temporal dynamics in atmosphere of the industrial city Odessa are presented for the first time and based on computing by nonlinear methods of the chaos and dynamical systems theories. A chaotic behaviour is discovered and investigated. To reconstruct the corresponding strange chaotic attractor, the time delay and embedding dimension are computed. The former is determined by the methods of autocorrelation function and average mutual information, and the latter is calculated by means of correlation dimension method and algorithm of false nearest neighbours. It is shown that low-dimensional chaos exists in the nitrogen dioxide concentration time series under investigation. Further, the Lyapunov's exponents spectrum, Kaplan-Yorke dimension and Kolmogorov entropy are computed.

1. Introduction

The problem of quantitative treating air pollution temporal and spatial dynamics in atmosphere of the industrial cities is remained by one of the most actual and important problem of the modern applied ecology, the environmental protection [1-6]. As a rule, the deterministic models or simplified ones, based on a simple statistical regressions, are usually used to estimate air pollution [1-3]. The problem of any prediction of air pollutants temporal or spatial dynamics is remained practically unsolved hitherto. In the last years a new approach to air pollution problem is provided by using methods of advanced non-linear analysis, chaos, dynamical systems theories (see [5-21] and Refs. therein). The studies concerning non-linear behaviour in the time series of atmospheric constituent concentrations are sparse, and their outcomes are ambiguous.

In ref. [5] there is an analysis of the NO_2, CO, O_3 concentrations time series. Also, it was shown that O_3 concentrations in Cincinnati (Ohio) and Istanbul are evidently chaotic, and non-linear approach provides satisfactory results [4]. In Ref. [6] it has been fulfilled the detailed analysis of the NO_2, CO, CO_2 concentration time series in the Gdansk region (Polland) and it has been definitely obtained the evidence of a chaos. Moreover it has been given a short-range forecast of atmospheric pollutants time evolution using non-linear prediction method. These studies show that a chaos and dynamical system theories methodology can be applied and the short-range forecast by the non-linear prediction method can be satisfactory.

In this paper for the first time we present the results of analysis and numerical modelling the air pollution (dioxide of nitrogen) concentration temporal dynamics in atmosphere of the industrial city Odessa. A chaotic behaviour has been discovered and in details investigated by using nonlinear methods of the chaos and dynamical systems theories [7-22]. To reconstruct the corresponding strange chaotic attractor, the time delay and embedding dimension are computed. The former is determined by the methods of autocorrelation function and average mutual information, and the latter is calculated by means of correlation dimension method and algorithm of false nearest neighbours. All calculations are performed with using "Geomath" and "Quantum Chaos" computational codes [21-35].

2. Data and method of analysis and modelling

In our study, we use the nitrogen dioxide (NO_2) concentration data observed in the atmosphere of the Odessa city from 1976 till 2000 years [2]. The multi-year hourly concentrations (one year total of 20x8760 data points, 1990) are analyzed. The typical temporal series of concentration (in mg/m^3) of the NO_2 (cite 1) is presented in figure 1.

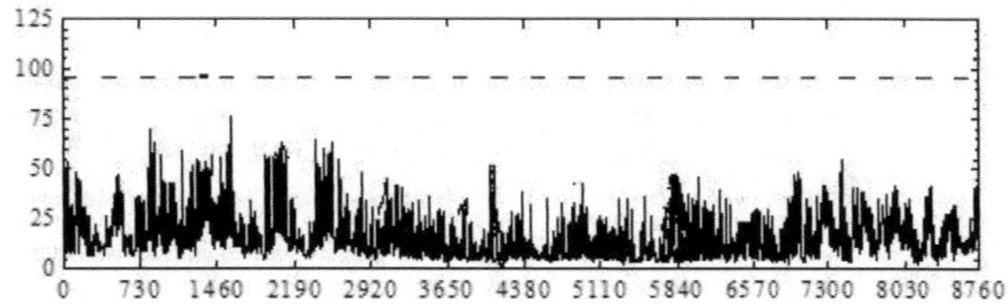

Figure 1. The temporal series of concentrations (in mg/m3) of the of the NO_2 (see text).

In Refs. [7-20] it has been developed computational code for studying chaotic features of the complex non-linear systems and in details described a procedure of testing of the chaos elements in the corresponding time series. Here we are limited only by the key aspects. As usually, we consider scalar measurements $s(n)=s(t_0+ n\Delta t) = s(n)$, where t_0 is a start time, Δt is time step, and n is number of the measurements. In a general case, $s(n)$ is any time series, but here $s(n)$ corresponds to an atmospheric pollutant concentration. The first fundamental step of modelling is in reconstruction of the corresponding phase space using as well as possible information contained in $s(n)$. From the mathematical viewpoint, this procedure results in set of d-dimensional vectors $\mathbf{y}(n)$ replacing scalar measurements. One should further to operate with lagged variables $s(n+\tau)$, where τ is some integer to be defined, results in a coordinate system where a structure of orbits in phase space can be captured. Using a set of the time lags to create a vector in d dimensions, $\mathbf{y}(n)=[s(n), s(n + \tau), s(n + 2\tau),..,s(n +(d-1)\tau)]$, the required coordinates are provided. The dimension d is defined as an embedding dimension, d_E.

In Refs. [6-8,14,15] a few approaches to the choice of proper time lag are presented. This point is important for the subsequent reconstruction of phase space. The first approach is to compute the linear autocorrelation function $C_L(\delta)$ and to look for that time lag where $C_L(\delta)$ first passes through 0. The alternative approach is based on using method of an average mutual information [9]. Let us remind that the mutual information I of two measurements a_i and b_k is symmetric and non-negative, and equals to 0 if only the systems are independent. The average mutual information between any value a_i from system A and b_k from B is the average over all possible measurements of $I_{AB}(a_i, b_k)$. In Ref. [4] it is suggested, as a prescription, that it is necessary to choose that τ where the first minimum of $I(\tau)$ occurs. The fundamental goal of the d_E calculation is in the further reconstruction of the Euclidean space R^d large enough so that the set of points d_A can be unfolded without ambiguity. The embedding dimension, d_E, must be greater, or at least equal, than a dimension of the corresponding chaotic attractor, d_A, i.e. $d_E > d_A$. The correlation integral analysis is one of the widely used techniques to investigate the signatures of chaos in a time series. This method is based on using the correlation integral, $C(r)$. According to Ref. [10], the correlation integral is defined as:

$$C(r) = \lim_{N\to\infty} \frac{2}{N(n-1)} \sum_{\substack{i,j \\ (1\le i<j\le N)}} H\left(r - \| \mathbf{y}_i - \mathbf{y}_j \|\right)$$

where H is the Heaviside step function with $H(u) = 1$ for $u > 0$ and $H(u) = 0$ for $u \le 0$, r is the radius of sphere centered on $\mathbf{y}_i$ or $\mathbf{y}_j$, and N is the number of data measurements. As usually, if the corresponding time series is characterized by an attractor, then the correlation dimension d is defined by a limit of relation of the log $C(r)$ to log of the corresponding radius (look details in Ref. [10]. In a case of the chaotic system the correlation exponent attains saturation with an increase in the embedding dimension. The saturation value of this exponent is defined as the correlation dimension (d_2) of the attractor. The technique of application the correlation integral method (say, the Grassberger-Procaccia algorithm [10]) is presented in Refs. [14,15].
Another method for determining d_E comes from asking the basic question addressed in the embedding theorem: when has one eliminated false crossing of the orbit with itself which arose by virtue of having projected the attractor into a too low dimensional space? This method is called as the method of false nearest neighbours. As a rule, the simultaneous application of two methods provides more exact determination d_E. It is noteworthy that the nearest integer above the saturation value provides the minimum or optimum embedding dimension for reconstructing the phase-space or the number of variables necessary to model the dynamics of the system. This concept can be applied, since the embedding dimension determined by both the correlation dimension method and the algorithm of false nearest neighbours are identical. The further important step in studying the chaotic time series of the dynamical system is determination of predictability, which can be estimated by the Kolmogorov entropy. The Kolmogorov entropy is proportional to a sum of the positive Lyapunov's exponents. Let us remind that the Lyapunov's exponents spectrum is one of the fundamental dynamical invariants for non-linear system with chaotic behaviour. According to definition, the Lyapunov's exponents are related to the eigenvalues of the linearized dynamics across the attractor. These parameters indicate the complexity of dynamics of the studied system. As usually, the positive values the Lyapunov's exponents show local unstable behaviour of the system, and respectively, their negative values show stable behaviour. The largest positive value of the Lyapunov's exponents determines some average prediction limit. Since the Lyapunov's exponents are defined as asymptotic average rates, they are independent of the initial conditions, and hence the choice of trajectory, and they do comprise an invariant measure of the attractor. An estimate of this measure is a sum of the positive Lyapunov's exponents. The estimate of the attractor dimension is provided by the conjecture d_L and the Lyapunov's exponents are taken in descending order. The dimension d_L gives values close to the dimension estimates discussed earlier and is preferable when estimating high dimensions. To compute Lyapunov's exponents, we use a method with linear fitted map (version [14], although the maps with higher order polynomials can be used too.

3. The results and conclusions

Table 1 summarizes the results for the time lag, which is computed for first ~10^3 values of time series. The autocorrelation function crosses 0 only for the NO_2 time series, whereas this statistic for other time series remains positive. The values, where the autocorrelation function first crosses 0.1, can be chosen as τ, but earlier it had been showed that an attractor cannot be adequately reconstructed for very large values of τ. So, before making up final decision we calculate the dimension of attractor for all values in Table 1. If time lags determined by average mutual information are used, then algorithm of false nearest neighbours provides $d_E = 6$ for all air pollutants. Table 2 shows the results of computing a set of the dynamical and topological invariants, namely: correlation dimension (d_2), embedding dimension (d_E), two Lyapunov exponents λ_1,λ_2), Kaplan-York dimension (d_L) and average

Table 1. Time lags (hours) subject to different values of C_L and first minima of average mutual information (I_{min1}) for the time series of NO_2, SO_2 concentrations for two sites of the Odessa city (see text)

	NO_2 (cite 1)	NO_2 (cite 2)
$C_L = 0$	-	-
$C_L = 0.1$	142	156
$C_L = 0.5$	7	9
I_{min1}	10	13

limit of predictability (Pr_{max}, hours) for two the NO_2 concentration time series for the Odessa during the period: Jan.-Dec., 1990. From the data it can be noted that the Kaplan-Yorke dimensions (which are also the attractor dimensions) are smaller than the dimensions obtained by the algorithm of false nearest neighbours. It is very important to pay the attention on the presence of the two (from six) positive Lyapunov's exponents λ_i (i=1,2). One could conclude that the system broadens in the line of two axes and converges along four axes that in the six-dimensional space. Other values of the Lyapunov's exponents λ_i are negative.

Table 2. The correlation dimension (d_2), embedding dimension (d_E), first two Lyapunov exponents, (λ_1,λ_2), Kaplan-Yorke dimension (d_L), and the Kolmogorov entropy, average limit of predictability (Pr_{max}, hours) for the time series of the NO_2 and SO_2 concentrations (Odessa city, 1990)

	Site 1 (Odessa) NO_2	Site 2 (Odessa) NO_2
λ_1	0.0187	0.0191
λ_2	0.0059	0.0049
d_2	5.28	5.26
d_E	6	6
d_L	4.09	3.92
K_{entr}	0.025	0.024
Pr_{max}	41	42

To conclude, for the first time we have presented the results of analysis and modelling the air pollutant (NO_2) concentration temporal dynamics in the Odessa's atmosphere. We have applied a number of nonlinear methods of a modern chaos and dynamical systems, such as an autocorrelation function method and the mutual information approach, a correlation integral analysis and the false nearest neighbours algorithm, the Lyapunov exponent's analysis and surrogate data method etc. To reconstruct the corresponding strange chaotic attractor, the time delay and embedding dimension are computed. The former is determined by the methods of autocorrelation function and average mutual information, and the latter is calculated by means of correlation dimension method and algorithm of false nearest neighbours. Further, the Lyapunov's exponents spectrum, Kaplan-Yorke dimension and Kolmogorov entropy are computed. A chaotic behaviour in the time series of NO_2 concentrations for two sites of the Odessa city was discovered and investigated. It has been shown that the low-dimensional chaos exists in the nitrogen dioxide concentration time series under investigation. The Lyapunov exponent's analysis has supported this conclusion.

References

[1] Bykowszczenko N, Baranowska-Bosiacka I, Bosiacka B, etal 2006 *Pol J Environm. Stud.* **15** 41
[2] Bunyakova Yu Ya, Glushkov A V 2011 *Analysis and estimation of anthropogenic loading influence on industrial city air basin* (Odessa: Ecology)

[3] Boy M, Hellmuth O, Korhonen H, Nilsson E D, ReVelle D, Turnipseed A, Arnold F, Kulmala M 2006 *Atm. Chem. and Phys.* **6** 4499
[4] Chelani A B 2005 *Int. J. Environ. Stud* **62** 181
[5] Dirks K N, Nanni A, Dirks V I 2006. *Atm. Science Lett.* **7** 15
[6] Khokhlov V, Glushkov A, Loboda A, Bunyakova Yu 2008 *Atm. Environm.* **42** 7284
[7] Packard N H, Crutchfield J P, Farmer J D, Shaw R S 1980 *Phys. Rev. Lett.* **45** 712
[8] Abarbanel H D I, Brown R, Sidorowich J J, Tsimring L Sh 1993 *Rev. Mod. Phys* **65** 1331
[9] Gallager R G 1968 *Information theory and reliable communication* (NY: Wiley).
[10] Grassberger P, Procaccia I 1983 *Physica D* **9** 189
[11] Gottwald G A, Melbourne I 2004 *Proc. Roy. Soc. London. Ser. A. Math. Phys. Sci.* **460** 603
[12] Sivakumar B 2004 *Chaos, Solitons & Fractals* **19**(2) 441
[13] Schreiber T 1999 *Phys. Rep.* **1** 308
[14] Glushkov A V, Svinarenko A A, Buyadzhi V V, Zaichko P A, Ternovsky V B 2014 *Adv.in Neural Networks, Fuzzy Systems and Artificial Intelligence, Series: Recent Adv. in Computer Engineering* (Gdansk: WSEAS:) **21** 143
[15] Glushkov A, Vaschenko V, Gerasimenko G, Bunyakova Yu 2013 *Dynamical Systems Applications*, Eds. J Awrejcewicz, M Kazmierczak, P Olejnik, J Mrozowski (Lodz: WPL) **T2** 241
[16] Khetselius O Yu 2013 *Dynamical Systems Theory*, Eds. J Awrejcewicz, M Kazmierczak, P Olejnik, J Mrozowski (Lodz: WPL) **T1** 145
[17] Glushkov A V, Buyadzhi V V, Kvasikova A S, Ignatenko A V, Kuznetsova A A, Prepelitsa G P and Ternovsky V B 2017 *Quantum Systems in Physics, Chemistry and Biology, Series: Progress in Theoretical Chemistry and Physics,* Eds. A Tadjer, R Pavlov, J Maruani, E Brändas and G Delgado-Barrio (Cham: Springer) **30** 169
[18] Glushkov A V, Kuzakon V M, Bunyakova Yu, Buyadzhi V, 2015 *Proc. Int. Geom. Centre.* **8** 91
[19] Glushkov A V, Buyadzhi V V, Ponomarenko E L 2014 *Proc. Int. Geom. Centre.* **7** 24
[20] Prepelitsa G P, Glushkov A V, Lepikh Ya I, Buyadzhi V V, Ternovsky V B and Zaichko P A 2014 *Sensor Electr. and Microsyst. Techn.* **11** 43
[21] Glushkov A V 2012 *Methods of a chaos theory* (Odessa: Astroprint)
[22] Glushkov A V, Bunyakova Yu Ya, Khokhlov V N, Prepelitsa G P and Tsenenko I A 2005 *Sensor Electr. and Microsyst. Tech.* **1** 80
[23] Glushkov A V 2008 *Relativistic Quantum theory. Quantum Mechanics of Atomic Systems* (Odessa: Astroprint)
[24] Khetselius O Yu 2008 *Hyperfine structure of atomic spectra* (Odessa: Astroprint)
[25] Khetselius O Yu 2009 *Int. J. Quant. Chem.* **109** 3330
[26] Khetselius O Yu 2012 *J. Phys.: Conf. Ser.* **397** 012012
[27] Khetselius O Yu 2009 *Phys.Scr.* **T135** 014023
[28] Khetselius O Yu 2008 *AIP Conf. Proceedings* **1058** 363
[29] Khetselius O Yu, Florko T A, Svinarenko A A, Tkach T B 2013 *Phys.Scr.* **T153** 014037
[30] Florko T A, Ambrosov S V, Svinarenko A A, Tkach T B 2012 *J. Phys.: Conf. Ser.* **397** 012037
[31] Svinarenko A A 2014 *J. Phys.: Conf. Ser.* **548** 012039
[32] Svinarenko A A, Ignatenko A V, Ternovsky V B, Nikola V V, Seredenko S S, Tkach T B 2014 *J. Phys.: Conf. Ser.* **548** 012047
[33] Malinovskaya S V, Dubrovskaya Yu V, Vitavetskaya L A 2005 *AIP Conf. Proc.* **796** 201
[34] Sofronkov A N, Vivdenko N, Magdenko E and Chivireva N 1986 *Sov. Materials Sci.* **21** 381
[35] Bunyakova YuYa, Khetselius O Yu 2009 *Proc. 8th Int. Carbon Dioxide Conf.* (Jena, Germany) p T2-098

Acknowledgements
The useful comments of the anonymous AEE 2017 (Ostrava, Czech Republic) referees are acknowledged.

IOP Conf. Series: Earth and Environmental Science **92** (2017) 012007 doi:10.1088/1755-1315/92/1/012007

The application of electrodialysis for the recovery of phosphorus from wastewater sludge liquid discharge

M Červenková[1], J Chromíková[1], S Heviánková[1] and Z Wranová[1]

[1]VSB-Technical University of Ostrava, Faculty of Mining and Geology, Institute of Environmental Engineering, 17.listopadu15, 708 Ostrava – Poruba, Czech Republic

Email: michaela.cervenkova@vsb.cz

Abstract. The paper deals with recovery of phosphorus from the liquid discharge from dewatered wastewater treatment sludge via electrodialysis. Using electrodialysis, phosphorus was transferred into concentrate, where phosphorus concentrated to673.6 mg/l (from the original value 79.87 mg/l). This enriched concentrate may be further processed, e.g. phosphorus can precipitate into the form of struvite.

1. Introduction

This paper focuses on the recovery of phosphorus from wastewater by electrodialysis to verify the efficiency of the electrodialysis process in the removal of phosphorus from the liquid discharge from dewatering the wastewater treatment sludge (further referred to as liquid discharge).The reason forphosphorus recovery from waste products of different operations is that phosphorus reserves in the form of mineral resourcesare decreasing. Extracted phosphorus is used, for example, in agriculture during the production of fertilizers or feeds, in medicine and in the production of detergents. Potential, renewable or recyclable phosphorus sources may be wastewaters, wastewater sludge, manure slurry, yellow, brown and grey waters, food waste, agricultural waste or guano [1,2,3,4].

Wastewaters and wastewater sludge contain sufficient phosphorus, and thus it is advantageous to recover it. Phosphorus is removed from wastewater most often by precipitation in the form ofstruvite, hydroxyapatite or aluminium and ferric phosphate. Other methods for phosphorus removal from wastewater are, for example, biological processes and membrane processes with subsequent precipitation of phosphorus or recycling of phosphorus in the form as fly ash from sewage sludge. Phosphorus is usually removed from wastewater in excess sludge, which can be used as fertilizer after draining, stabilization and hygienisation. The disadvantage is thatsome wastewater sludge may contain toxic elements and such sludge cannot be applied onto agricultural land [1].

2. Materials and Methods

For the recovery of phosphorus from liquid dischargewe selected an electro-membrane separation process – electrodialysis. The aim of the laboratory measurements was the verification of efficiency of this process. Phosphorus was obtained from model waters (salt solution Na_2SO_4 and $NaH_2PO_4 \cdot 2H_2O$) and real samples of liquid discharge obtained from a wastewater treatment plant. Specifically, we used liquid discharge from a wastewater treatment plant with a capacity over 100 000 EP (equivalent population).

2.1.Determined indicators

For the evaluation of experiments in the samples we determined the following indicators – pH, conductivity, turbidity and total phosphorus (P_{total}). The pH value and conductivity were determined by laboratory multimeters pH/Cond 340i, turbidity using a spectrophotometer HACH LANGE DR 2800 and total phosphorus using the method of oxidative decomposition to dissolved inorganic orthophosphates by ČSN EN ISO 6878.

2.2.Used samples

In the first series of tests of electrodialysis, the so called salt tests with 2 % Na_2SO_4 solution were carried out. It is a standard solution for checking the effectiveness and correctness of composition of an electrodialysis unit. In the next series of the tests of electrodialysis we carried out salt tests with 0.1% $NaH_2PO_4 \cdot 2H_2O$ solution. This sample was used for checking the permeability of membranes for phosphate. After salt tests we carried out electrodialysis tests with real samplesof the so called liquid discharge. Table 1 shows the analysis of the real sample.

Table 1. Input analysis of liquid discharge.

Indicator [unit]	Real sample
pH [-]	7.79
Conductivity κ [μS/cm]	5.01
Turbidity [NTU]	> 40
P_{total} [mg/l]	79.87
NH_4^+ [mg/l]	1260
COD_{Cr} [g/l]	318.60

For the correct operation of electrodialysis it is necessary to pre-treat the sample. Herein the pre-treatment method was pressure filtration. Using an ILMVAC PS20-1 vacuum pump, the aim of pre-treatment is, for example, to prevent clogging of membranes and damage in the electrodialysis unit.

2.3.Electrodialysis unit

For the tests we used a laboratory electrodialysis unit P ED(P)-Z/10-1.0 (MEGA a.s.). The unit consists of an electrodialysis module, tank for diluent (D), for concentrate (C) and for electrode solution (E), of separate pumps for D, C and E circle, of rotameters for D, C and E circle and of DC voltage source. Electrodialysis module consists of regularly alternating cation and anionic membranes RALEX which are separated by polyethylene separators. For the tests we used an electrodialysis module with 10 couples of ion exchange membranes of the total area of membranes of 1 344 cm^2 [5].

2.4.Methodology of electrodialysis tests

All the electrodialysis tests were carried out in batch mode with regulated voltage 14 V. The electrodialysis tests with real samples were carried out as below:

- pre-treatment of liquid discharge by pressure filtration;
- dosing of the relevant volumes of the test sample into the D and C circle;
- dosing 2 % Na_2SO_4solution into the E circle;
- setting the regulated voltage 14 V;
- switching on a pump and voltage;
- recording the values of pH, κ, I and U in the time interval 5 minutes;
- measuring volumes of output products at the end of the test;
- analysis of output products.

IOP Conf. Series: Earth and Environmental Science **92** (2017) 012007 doi:10.1088/1755-1315/92/1/012007

2.5.Desalination rate

In all the tests we also calculatedthe desalination rate using the equation below:

$$DR = \frac{\kappa_{input} - \kappa_{output}}{\kappa_{input}} 100,$$

where DR is the percentage of desalination [%]; κ_{input} is the input conductivity of diluent [μS/cm]; κ_{output} is the output conductivity of diluent[μS/cm].

2.6.Mass balance

Next, mass balance was carried out, which is grounded in the fact that substances entering the electrodialysis process must also leave the process – see the equation below:

$$V_{input} \cdot c_{input} = V_D \cdot c_D + V_C \cdot c_C ,$$

where V_{input} is the volume of solution entering the electrodialysis process [l]; c_{input} is the concentration of solution entering the electrodialysis process [mg/l]; V_D is the volume of diluent leaving the electrodialysis process [l]; c_D is the concentration of diluent leaving the electrodialysis process [mg/l];V_C is the volume of concentrate leaving the electrodialysis process [l];c_C is the concentration of concentrate leaving the electrodialysis process [mg/l].

2.7.Carried out tests

As mentioned above, thetotal three series of tests were carried out – salttestwith 2 % Na_2SO_4 solution, salt test with 0.1 % $NaH_2PO_4 \cdot 2H_2O$ solution,and tests with real samples. The aim of these tests was also to reduce the volume of waste product – concentrate. Therefore, different volumes of tests samples were tested.

The first series of tests with 2% Na_2SO_4 solution was carried out to check the efficiency of electrodialysis. Input volumes of salt solution are:

- D:C = 1000 ml : 1000 ml

The second series of tests with 0.1 % $NaH_2PO_4 \cdot 2H_2O$ solution was carried out to check the permeability of the membranes for phosphates. Input volumes of the solution for the tests are:

- D:C = 1000 ml : 1000 ml
- D:C = 1000 ml : 500 ml

The third series of tests was carried out with real samples of liquid discharge. Input volumes of samples for the tests are:

- D:C = 1000 ml : 1000 ml
- D:C = 1000 ml : 500 ml
- D:C = 5000 ml : 500 ml

3. Results and discussion

As the blind test of electrodialysis unit, we selected standard salt test with 2 % Na_2SO_4 solution. Tests lasted for 65 minutes and the percentage of desalination reached 99.5% efficiency. The tests had a standard course. Salt tests with 0.1% $Na_2H_2PO_4 \cdot 2H_2O$ solution were carried out to check the efficiency of phosphate transition by ion exchange membranes. These tests were evaluated as effective.

The following tables summarize the results of electrodialysis tests of real samples of sludge water/liquid discharge. The first tests with lower volumes of input sample were terminated after 40 to 45 minutes when the current dropped to 0.04A. The percentage of desalination was 98.6 – 98.99 %. The results are summarized in tables 2 and 3.

Table 2. Selected analysis indicators V1, D1 and C1.

Indicator [unit]	V1	D1	C1	Balanceinput	Balance output	Deviation [mg/l]	Deviation [%]
V [ml]	2000	1000	1000	-	-	-	-
pH	8.245	4.854	8.382	-	-	-	-
κ [mS/cm]	6.17	0.05	10.70	12.34	10.75	1.59	12.88
COD_{Cr} [mg/l]	269.29	194.98	325.85	538.58	520.83	17.75	3.30
N_{ammon} [mg/l]	1050.00	9.80	1850.00	2100	1859.80	240.20	11.44
P_{total} [mg/l]	198.86	10.432	262.104	397.72	272.536	125.184	31.48

Table 3. Selected analysis indicators V2, D2 and C2.

Indicator [unit]	V2	D2	C2	Balance input	Balance output	Deviation [mg/l]	Deviation [%]
V [ml]	1500	1000	500	-	-	-	-
COD_{Cr} [mg/l]	268.28	128.30	348.75	402.42	302.675	99.745	24.79
N_{ammon} [mg/l]	958.30	8.26	2825.30	1437.45	1420.91	16.54	1.15
P_{total} [mg/l]	68.46	8.48	104.32	102.69	60.64	42.05	40.95

Tests with larger volumes of input sample (5500 ml) lasted for 3 hours and 55 minutes when the current dropped to 0.03 A. The average percentage of desalination was 98.67 %. The results were summarized in table 4.

Table 4. Selected analysis indicators V3, D3 a C3.

Indicator [unit]	V3	D3	C3	Balance input	Balance output	Deviation [mg/l]	Deviation [%]
V [ml]	5680	5180	500	-	-	-	-
COD_{Cr} [mg/l]	268.28	176.193	375.487	1523.83	1100.42	423.41	27.79
N_{ammon} [mg/l]	958.30	6.93	8500	5443.14	4285.90	1157.24	21.26
P_{total} [mg/l]	68.46	10.76	673.60	388.85	392.54	3.69	0.95

The results show the electrodialysis process as an effective method for phosphorus removal from liquid discharge. The phosphorus is concentrated into the concentrate stream. Based on the implementation of mass balances, we may state that higher deviations (tables 3 and 4) are most likely analysiserrors, particularly in the concentrates, where the samples had to be diluted several times. However these differences did not effect for the correct operation of electrodialysis and there were no presented features of membrane fouling. For minimum deviations in the results, it is suitable to use specific analytical methods in the future that would be more appropriate for analyses of concentrates and concentrated samples.

4. Conclusion

The aim of this paper was to verify the possibility of using electrodialysis for phosphorus recovery from liquid discharge rich in phosphorus.The condition for the reduction of phosphorus concentration from wastewater is its reduction to a minimum and its transfer into the concentrate as a product of electrodialysis. To verify electrodialysis as a suitable process for phosphorus recovery from liquid discharge we carried out several tests, with salt solutions and with real waters.

According to the measured results, electrodialysis is suitable for phosphorus recoveryfrom liquid discharge. At all the tests with real samples of liquid discharge achieved the desalination rate from 96.01 to 98.99 %. Phosphorus was transferred from pre-treated samples into the concentrate stream with output phosphorus concentration 673.6 mg/l.This concentrate can be further processed, for example, and phosphorus may precipitate into the form of struvite.

Acknowledgments
This paper was published with the support of the Ministry of Agriculture within the Project QJ1320159 "Research in processing, use and disposal of waste product from biogas stations".

References

[1] Holba M and Došek M 2016 Potřeba recyklace fosforu *Odpadové forum* http://www.odpadoveforum.cz/TVIP2016/prispevky/309.pdf

[2] Cieślik B and Konieczka P 2017 A review of phosphorus recovery methods at various steps of wastewater treatment and sewage sludge management The concept of "no solid waste generation" and analytical methods *J.l of Clean. Prod.* pp 1728-740

[3] Holba M 2011 *Vhodné technologie recyklace fosforu z povrchových a odpadních vod* http://voda.tzb-info.cz/vlastnosti-a-zdroje-vody/7128-vhodne-technologie-recyklace-fosforu-z-povrchovych-a-odpadnich-vod

[4] Holba M, Plotěný M, Maršálková E and Maršálek B 2011 Klady a zápory technologických procesů pro recyklaci a odstranění fosforu z povrchových a odpadních vod *Řešení extrémních požadavků na čištění odpadních vod* J Foller pp 142-147

[5] MEGA a.s. 2009 TECHNICKÁ DOKUMENTACE – ELEKTRODIALYZAČNÍ MODUL ED(R)-Z/10-1.0Stráž pod Ralskem

Study of heavy metals removal from model solutions by wooden materials

S Demcak[1], M Balintova[1] and M Demcakova[2]

[1]Technical University of Kosice, Faculty of Civil Engineering, Institute of Environmental Engineering, Vysokoskolska 4, Kosice, 040 01, Slovak Republic
[2]Slovak Academy of Sciences, Institute of Experimental Physics, Watsonova 47, Kosice, 040 01, Slovak Republic

Email: stefan.demcak@tuke.sk

Abstract. The paper deals with the heavy metals removal by wooden material. The adsorption experiments were carried out using of wood sawdust and bark from poplar tree for the removal of copper and zinc cations from model solutions with initial concentration of 10 mg.L^{-1}. The FTIR was used for functional groups determination not for analysis of hemicelluloses, cellulose and lignin in structure of wood sawdust and bark were studied by infrared spectrometry. Poplar sawdust efficiency for Cu(II) and Zn(II) removal from aquatic model solutions reached approximately 80.0%. Changes of pH values were also observed in model solutions that were probably caused by ion exchange between the ions of sorbents and the ions in solutions.

1. Introduction

The increasing of anthropogenic activities in last decades had the negative impact on all parts of the environment, mainly on aquatic environment. Industrial waters are specific, because contains of contaminants that may destroy waste water treatment process [1]. Heavy metal ions are contained as pollutants in wastewaters which must be removed primarily, due to their mobility in natural aquatic ecosystems and due to their toxicity because they are stable and persistent environmental contaminants since they cannot be degraded and destroyed. These metal ions can be harmful or toxic to living organisms in water and also they can be caused a serious public health problem for human health [2, 3]. Processes of finding of the new and cheap ways for treatment of wastewaters contaminated by heavy metals can increase the quality of the environment in the affected localities and thus prevent adverse effects on living organisms [4].The most commonly used technique for the heavy metals removal from aqueous solutions is chemical precipitation, membrane filtration, electro extraction, ion exchange and sorption [5, 6].

The process of adsorption by sorbents implies that adsorbents bind molecules by physical attractive forces, ion exchange, and chemical binding. It is advisable that the adsorbent is available in large quantities, easily regenerable, and cheap [7]. The application of bio-sorption in environmental treatment of wastewaters contaminated by heavy metals has become a significant research area in the past 10 years [8]. Physic-chemical processes based on adsorption on natural organic materials are cheap and effective techniques for metals removal from wastewater [9]. Adsorbent materials derived from low-cost agricultural wastes or by-products from timber industries can be used for the effective removal and recovery of heavy metal ions from wastewater streams [8, 10-12]. The major advantages of bio-sorption technology are its effectiveness in reducing the concentration of heavy metal ions to very low levels and the use of inexpensive bio-sorbent materials. Bio-sorption of heavy metals using wooden materials from aqueous solutions is a relatively new process which has been proven very promising in the removal of contaminants from aqueous effluents. The sorption capacity of wooden materials with content of lignocelluloses for metal ions is generally described as adsorption. The major advantages of wooden materials as adsorbents over conventional treatment methods include: low-cost, high efficiency, minimization of chemical and/or biological sludge, regeneration of bio-sorbent, no additional nutrient requirement and possibility of metal recovery [13]. For the overall understanding of the sorption process, it is necessary to characterize and determine the properties of the used adsorbents.

In this study the removal of Cu(II) and Zn(II) metal ions by using poplar wood sawdust and bark were investigated. For characterization of functional groups, which can be responsible for metal binding the wood sawdust and bark of poplar were analysed by infrared spectrometry.

2. Material and methods

The sawdust and bark of poplar, species of locally available wood, was dried and sieved, and the fraction with particle size max. 2.0 mm (sawdust) and max. 8.0 mm (bark) was used for adsorption experiments. 1 g of each dry adsorbent was mixed with 100 mL of each model solution prepared by dissolving of their appropriate sulphate salts in deionised water, include initial metal concentrations. Batch adsorption experiments were carried out on static conditions. The sorbent-sorbate interaction time was 24 h. After the end of experiments, poplar wood sawdust and bark were removed by filtration through a laboratory filter paper. Concentrations of appropriate ions were determined by colorimetric method (Colorimeter DR890, Hach Lange, Germany) with appropriate reagents to determine concentration of dissolved cooper, and zinc. The pH values of solutions were also measured by pH meter inoLab pH 730 (WTW, Germany). In both cases, the efficiency of ion removal η (in %) was calculated using the following equation (equation (1)):

$$\eta = \frac{(c_0 - c_e)}{c_0} * 100\%, \qquad (1)$$

where c_0 is the initial concentration of appropriate ions (mg.L^{-1}) and c_e equilibrium concentration of ions (mg.L^{-1}). All adsorption experiments were carried out in triplicate under the batch conditions and results are given as arithmetic mean values.

IR spectra of dried poplar sawdust and bark were studied for characterization of present functional groups, which can be responsible for metal binding. Infrared spectrometry of poplar sawdust was carried out on Bruker Alpha Platinum-ATR spectrometer (Bruker Optics, Ettingen, Germany). A total of 24 scans were performed on sample in the range of 4,000–400 cm^{-1}.

3. Results and discussion

3.1. Infrared spectra of polar sawdust and bark

The important role of FTIR spectroscopy is related to the characterization of sorbents, or to describe of functional groups which are involved to the mechanism of sorption processes. Metal ions adsorption capacity is strongly influenced by the surface structures of C–O and C–OH functional groups which are present in natural wooden organic materials [14]. Functional groups in poplar wood sawdust and bark were determined using FTIR spectroscopy. The IR spectra of poplar sawdust and bark are shown in figure 1. Poplar sawdust was detailed studied and characterized by Demcak et al. [3]. As can be seen from figure 1, the IR spectrum of the bark (the deadwood part of the wood pole) has a similar pattern to sawdust. A strong representation of hydroxyl functional groups is visible in both structures (3,650–3,000 cm^{-1}). There are no significant changes in the characteristic IR absorption bands of bark and sawdust indicating that the functional groups present of poplar on the surface (bark) and in the core (sawdust). The IR spectra of bark showed a strong band at 3,287 cm^{-1} indicating the presence of hydroxyl groups. The peaks at 2,918 and 2,890 cm^{-1} are due to the C–H stretching frequency and the peak at 1,622 cm^{-1} is due to C=O stretching mode of the primary and secondary amides (NH_2–C=O) [15]. The peak at 1,512 cm^{-1} is indicative of the N–H stretching of the primary and secondary amides, and the presence of amide or sulphamide band, respectively. The band at 1,317 cm^{-1} indicates a presence of carboxylic acids [16]. Weak band at 1,512 cm^{-1} is attributed to aromatic CC and two sharp peaks in area from 1,750 to 1,600 cm^{-1} which are characteristic of carbonyl group stretching were also observed. The strong C–O band at 1,022 cm^{-1} confirms the lignin structure of the bark [15].

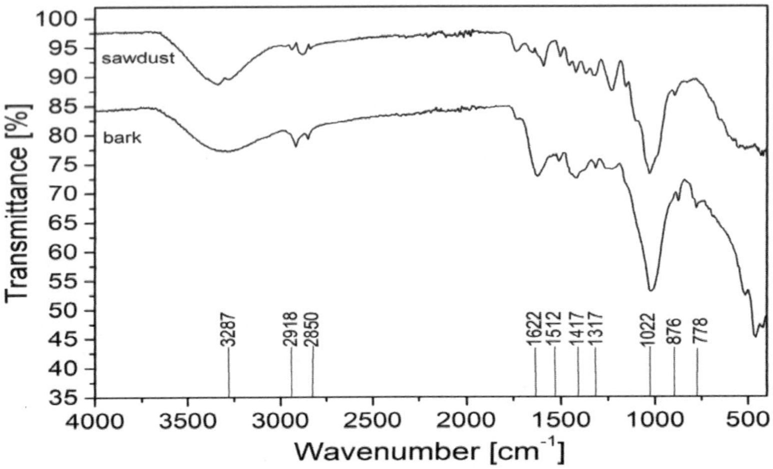

Figure 1. Infrared spectra of poplar wood sawdust and bark.

3.2. Adsorption study of poplar sawdust and bark

The results of static sorption experiments are shown in table 1. Poplar sawdust used as a sorbent has shown a very good efficiency for the removal of studied metal ions. The efficiency which were achieved by sawdust in sorption of Cu(II) and Zn(II) from solutions were over η = 80%. Changes of pH values were also measured in the solutions after adsorption experiments by sawdust. The pH of the solutions is an important parameter for the characterization of adsorption processes and has an influence on the surface charge of the adsorbent, the degree of ionization and the type of used sorbent [17]. The adsorption experiments in the case of the poplar sawdust significantly increased the pH of the solution and it indicating that the ion exchange mechanism was also occurred in the adsorption process [3].

In the case of polar bark, the efficiency Cu(II) and Zn(II) removal from model solutions η = 70.3% and η = 79.5%. It was found that after experiments the pH of the analysed solutions was increased what indicates that also like in case of poplar sawdust adsorption the adsorption processes and ion exchange run simultaneously.

Table 1. Results of sorption experiments with poplar sawdust and bark.

Adsorbent	Initial concentration c_0 (Cu^{2+}) = 10 mg.L^{-1} at pH = 4.6		Initial concentration c_0 (Zn^{2+}) = 10 mg.L^{-1} at pH = 4.7	
	c_e (Cu^{2+}) [mg.L^{-1}]	pH	c_e (Zn^{2+}) [mg.L^{-1}]	pH
sawdust	1.58	5.4	1.92	4.8
bark	2.97	5.2	2.05	4.6

c_0- initial concentration of metal ions in model solution
c_e- residual concentration of metal ions in solution after adsorption

4. Conclusion

Bio-sorption is a relatively new process that has proven very promising in the removal of contaminants from aqueous effluents. Natural organic wooden materials are inexpensive and effective metal ion adsorbents from agro waste materials or by-products from timber industries to offer these adsorbents as replacements for existing commercial materials were investigated.

The higher sorption efficiency was achieved with sawdust. It could be caused due to their higher specific surface area and smaller fraction. For the using of bark as adsorption materials, it is certainly appropriate to milled it to a smaller fraction and thereby to increase the surface area of the absorbent. As a limitation of bark adsorbents, it is leaching of organic-dyes into the solution.

Acknowledgements

This work has been supported by the Slovak Grant Agency for Science (Grant No. 1/0563/15).

References

[1] Igwe J C 2007 A Review of Potentially Low Cost Sorbents for Heavy Metal Removal and Recovery *Terrestrial and Aquatic Environmental Toxicology* **1(2)** 60-69
[2] Tangahu B V, Sheikh Abdullah S R, Basri H, Idris M, Anuar N and Mukhlisin M 2011 A Review on heavy metals (As, Pb, and Hg) uptake by plants through phytoremediation *International Journal of Chemical Engineering*
[3] Demcak S, Balintova M, Hurakova M, Frontasyeva M V, Zinicovscaia I and Yushin N 2017 Utilization of poplar wood sawdust for heavy metals removal from model solutions *Nova Biotechnologica et Chimica* **16(1)** 26-31
[4] Babel S and Kurniawan T A 2003 Low-cost adsorbents for heavy metals uptake from contaminated water: a review *Journal of hazardous materials* **97(1)** 219-243
[5] Fu F and Wang Q 2011 Removal of heavy metal ions from wastewaters: a review *Journal of environmental management* **92(3)** 407-418
[6] Zinicovscaia I and Cepoi L 2016 Cyanobacteria for bioremediation of wastewaters *Springer International Publishing* ISBN: **978-3-319-26749-4**
[7] Hashem M A 2007 Adsorption of lead ions from aqueous solution by okra wastes *International Journal of Physical Sciences* **2(7)** 178-184
[8] Demirbas A 2008 Heavy metal adsorption onto agro-based waste materials: a review *Journal of hazardous materials* **157(2)** 220-229
[9] Barakat M A 2011 New trends in removing heavy metals from industrial wastewater *Arabian Journal of Chemistry* **4(4)** 361-377
[10] Crini G 2006 Non-conventional low-cost adsorbents for dye removal: a review. *Bioresource technology* **97(9)** 1061-1085
[11] Bailey S E, Olin T J, Bricka R M and Adrian D D 1999 A review of potentially low-cost sorbents for heavy metals *Water research* **33(11)** 2469-2479
[12] Keränen A, Leiviskä T, Zinicovscaia I, Frontasyeva M V, Hormi O and Tanskanen J 2016 Quaternized pine sawdust in the treatment of mining wastewater *Environmental technology* **37(11)** 1390-1397
[13] Ahalya N, Ramachandra T V and Kanamadi R D 2003 Biosorption of heavy metals *Res. J. Chem. Environ.* **7(4)** 71-79
[14] Ricordel S, Taha S, Cisse I and Dorange G 2001 Heavy metals removal by adsorption onto peanut husks carbon: characterization, kinetic study and modeling *Separation and purification Technology* **24(3)** 389-401
[15] Reddy D H K, Seshaiah K, Reddy A V R, Rao M M and Wang M C 2010 Biosorption of Pb 2+ from aqueous solutions by Moringa oleifera bark: equilibrium and kinetic studies *Journal of Hazardous Materials* **174(1)** 831-838

[16] Reddy D H K, Seshaiah K and Lee S M 2012 Removal of Cd (II) and Cu (II) from aqueous solution by agro biomass: equilibrium, kinetic and thermodynamic studies. *Environmental Engineering Research* **17(3)** 125-132
[17] Bulut Y and Zeki T E Z 2007 Removal of heavy metals from aqueous solution by sawdust adsorption *Journal of Environmental Sciences* **19(2)** 160-166

Environmental impact of thermal insulations: How do natural insulation products differ from synthetic ones?

M Dovjak[1], M Košir[1], L Pajek[1], N Iglič[1], D Božiček[1] and R Kunič[1]

[1]Chair of Buildings and Constructional Complexes, Faculty of Civil and Geodetic Engineering, University of Ljubljana, Jamova cesta 2, 1000 Ljubljana, Slovenia

E-mail: david.bozicek@fgg.uni-lj.si

Abstract: As the environmental awareness of the public is rising and at the same time contemporary buildings are becoming more and more energy efficient, the focus is shifting towards the usage of environmentally friendly building products. Human decisions are often driven by emotions and perceptions. Consequently, there exists a strong tendency towards preferring "natural" constructional products to the synthetic ones, especially in the case of thermal insulations. Life cycle assessment (LCA) has enabled an opportunity to widen the meaning of the word "environmentally friendly", giving researchers and building designers an objective decision making tool to determine the environmental impact of building products, building components and buildings as a whole. The purpose of this study was to compare the environmental impact of various thermal insulations for the cradle to gate life cycle stages, based on a unified functional unit. Overall, 15 most commonly used thermal insulation products were analysed and classified into natural and synthetic groups. Based on the differentiation, we compared the impact in the selected environmental categories and identified the most influential environmental drivers. The results show that in some environmental categoriesnatural thermal insulations perform better (i.e. global warming potential), whilein others (i.e. eutrophication potential) they underperform. However, environmental impact trends can be identified, specifically for the natural and the synthetic materials.

1. Introduction

The building sector is responsible for a major share of global material consumption and energy use. Consequently, it has a profound impact on the environment [1]. Duringthe last decade, the European Union has been pushing on its member states to implement energy efficiency measures in buildings and promoted the use of renewable energy [2]. Since buildings are getting more and more energy efficient and thereby the environmental impact related to the operational phase of the buildings life cycle is ever smaller, other phases of the buildings life cycle are becoming significant from the environmental perspective [3]. Decisions about choosing the environmentally most appropriate building materials and products are increasing their role in the building design process. Thermal insulation products are among the most detrimental parts of the building envelope, directly affecting the energy use by reducing the heat flow through the opaque building envelope and thereby contributing a major role in providing comfortable living conditions [4]. Despite the fact that a myriad of different insulation types can be considered as equally efficient in reducing the heat flow and providing suitable thermal comfort, when appropriately incorporated in the building envelope, their environmental impact can vary substantially.

IOP Conf. Series: Earth and Environmental Science **92** (2017) 012009 doi:10.1088/1755-1315/92/1/012009

Life cycle assessment (LCA) has been proven asa useful tool in evaluating the impact on the environment and human health of various goods and services [5]. The presented study focuses on the environmental impact of various thermal insulation materials in the cradle to gate life cycle stages. However, the goal of this study was not to determine the type of insulation material with the lowest environmental impact. Rather, the goal wasto elaborate a comparative analysis of thermal insulations and obtain a sense of impact that different insulation materials have on the environment. The analysed insulation materials were separated in three groups; inorganic-synthetic, organic-synthetic and organic-natural. Based on this classification we will try to pinpoint typical drivers that influence the environmental impacts in each group and emphasize the differences between them.

2. Methodology and materials

2.1. Environmental data

The LCA data of the analysed insulation products were collected through EPD-s from the program operators Institut Bauen und Umwelt e.V. and Bau EPD GmbH, members of the ECO Platform (http://www.eco-platform.org/). This approach enabled us to evaluate LCA data of widely used building products that are sold on the EU market.

The LCA results reflect the cradle to gate life cycle stages (A1 – A3). We are focused on the environmental categories defined in EN 15804 [6]: Global Warming Potential (GWP), Ozone Depletion Potential (ODP), Photochemical Ozone Creation Potential (POCP), Acidification Potential (AP), Eutrophication Potential (EP), Abiotic Depletion Potential of Elements (ADPE) and Abiotic Depletion Potential of Fossil resources (ADPF). The defined functional unit is 1 m^2 of the insulation material with its thickness adapted to the thermal transmittance of 0.25 $W/(m^2K)$.

For each insulating material as many as possible EPD-s were collected. Each EPD had to be ISO 14025 [7] andEN 15804 compliant, andthe EPD-s that lacked the results interpretation were excluded. The results interpretation in the EPD-s is of vast importance for us, as it provided information about the most important factors thatdetermine the environmental impact of a certain environmental category. The LCA results used for each insulation in the study are the average values of the insulation products. These valuescan be considered asaverage environmental characteristics of insulation products sold on the EU market.The described methodological approach has it flaws, which need to be accounted for. The life cycle inventory (LCI) of the insulation products isusually a combination of manufacturer provided data from the production plants and generic data from an LCI database. Two databases are commonly used, the GaBi and Eco invent database. Although the first one is used in the majority of the analysed EPD-s, using different databases can lead to disparate LCA results. Another aspect to consider is the fact that EPD-s do not provide information about which midpoint life cycle inventory method (LCIM) was used to perform the LCA study. Using different midpoint LCIM methods (e.g. CML2010, ILCD) can also affect the LCA results [8]. This are the main reasons why we will focus on identifying those drivers thatinfluence the environmental impact, rather than evaluating which insulation materials are less or more preferable from the environmental perspective.

2.2. Insulation materials

Altogether 15 insulation products were analysed. The inorganic-synthetics are glass wool (GW), low and high density stone wool (LdSW, HdSW) and foamed glass (FG). The organic-synthetic group consist of expanded polystyrene (EPS), expanded polystyrene with infrared reflectors (EPSir-r), extruded polystyrene (XPS), polyurethane (PU), polyisocyanurate (PIR) and phenolic boards (PHE). The insulations in the organic-natural group are low and high density wood fiber boards (LdWF, HdWF), cellulose fiber (CF, loose cavity insulation) and straw bales (SB). Also the environmental impact of vacuum insulation panels (VIP) is analysed, but based on its specific characteristics (composite insulation product consisting of the core material, in our case fumed silica, and vapour resistant barrier foil) we observed it separately as it did not fit in any of the defined groups.

3. Results

A brief overview of the results (figures 1–7) shows that VIPs are the most impactful in almost all the categories, except the POCP one. Although for other insulations the impact in each category is influenced by various factors, for VIPs the majority of the environmental burden (over 90%) can be contributed to processes related to extraction and production of the core material, namely fumed silica. In the GWP category (figure 1), we can notice that all the organic-natural insulations show a net negative impact. That is due to the CO_2 sequestration in wood and straw. This means that there is more CO_2 stored in the materials, than CO_2equivalent greenhouse gases are released during the defined life cycle phases.

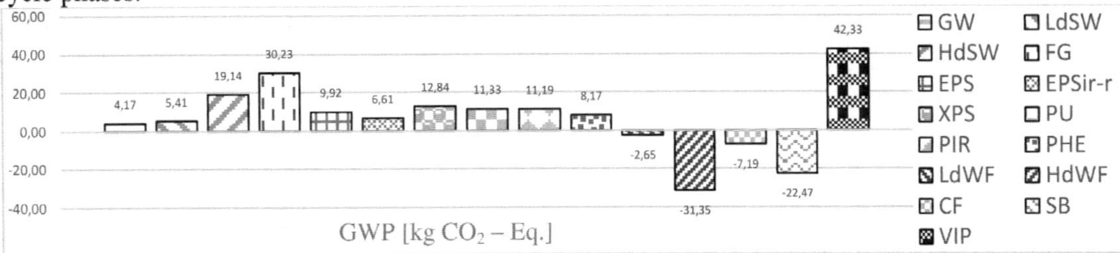

Figure 1. GWP of the analysed insulation materials.

Besides VIP, the PU and PIR insulations stand out in the ODP category (figure 2). This can be assigned to the production processes related to the basic raw materials (A1) methylene diphenyl diisocyanate (MDI) and polyol.

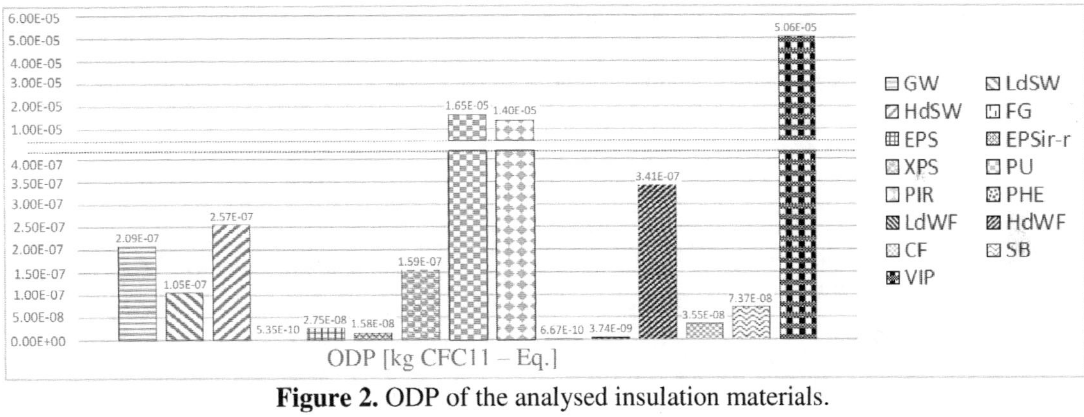

Figure 2. ODP of the analysed insulation materials.

Due to the pentane emission in the production process of EPS and EPSir-r (A3) they show the largest impact in the POCP category (figure 3).

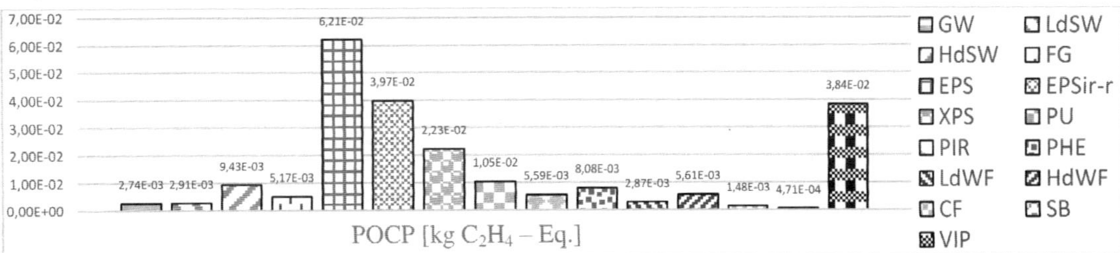

Figure 3. POCP of the analysed insulation materials.

The AP potential (figure 4) is closely related to the emissions during the production process (A3) and the processes of raw material acquisition and processing (A1). Besides VIP, HdSW has the largest AP.

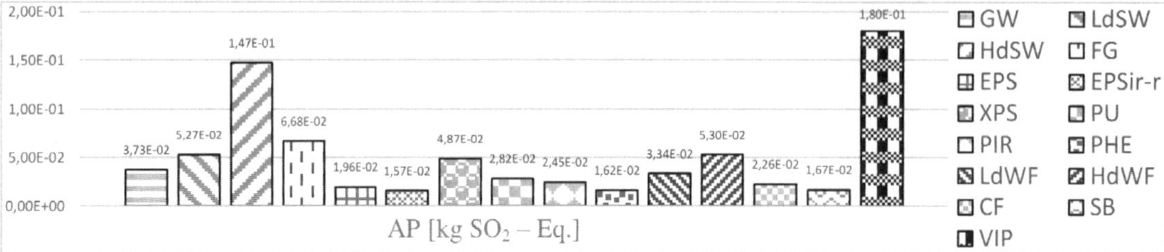

Figure 4. AP of the analysed insulation materials.

Interesting to note is that SB has a profound impact in the EP category compared to other materials (figure 5). The major contributor is the use of farming machinery (diesel) – for sowing, harvesting and fertilising – and the use of fertilisers and pesticides.

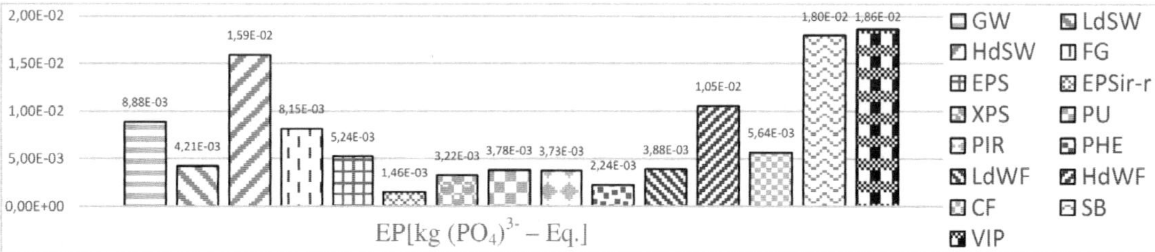

Figure 5. EP of the analysed insulation materials.

Surprising is the high ADPE score (figure 6) for CF and GW insulation. However, they have one thing in common, thatis the use of boron compounds: in GW borax pentahydrate ($Na_2B_4O_7{\cdot}5H_2O$) as part of the basic raw material and in CF boric acid (H_3BO_3) for the protection against fire and insects.

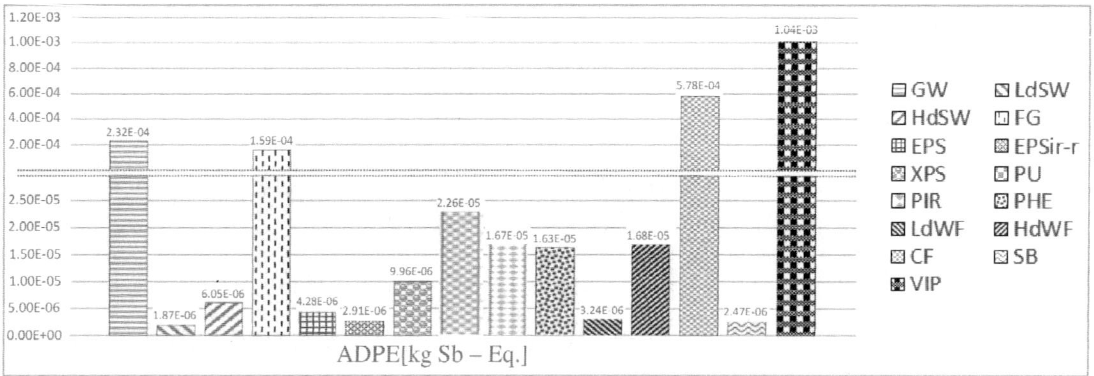

Figure 6. ADPE of the analysed insulation materials.

Besides VIP, the FG insulation boards show the largest contribution in the ADPF category (figure 7) due to the energy usage for melting the raw materials in the electric furnace (up to 1300°C) and the curing in the cellulating oven at 850°C.

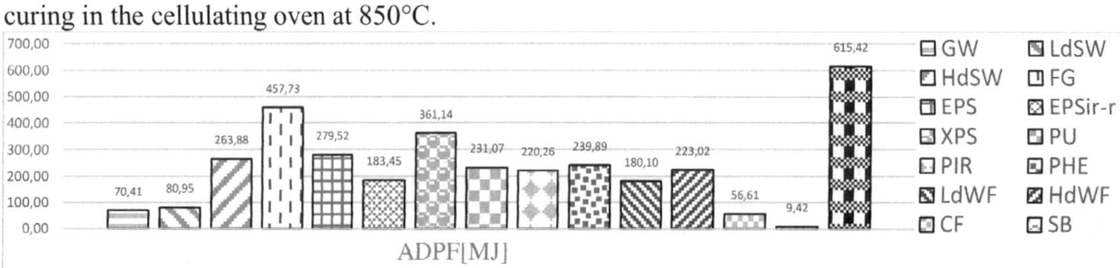

Figure 7. ADPF of the analysed insulation materials.

4. Discussion

Inorganic-synthetic materials: Their production process (A3) strongly determines their environmental impacts, as a great amount of energy is needed and emissions are released when melting the mineral materials (up to 1300°C for GW and FG and approx. 1500°C for SW). In the case of GW and SW the added binders (usually phenol formaldehyde resin) also contribute a notable part to the environmental impact, although they present only a small percent of the mass share (3% –9%).

Organic-synthetic materials: Processes related to the provision of raw materials (polystyrene for EPS, EPSir-r and XPS, MDI and polyol for PU and PIR and phenolic resin for PHE) determine their environmental impact in all the categories except the POCP one. In this category, emissions during the production phase are detrimental. For EPS and EPSir-r it is the pentane emission and for the othersthe emissions of the used blowing agents (e.g. CO_2, HFO 1234 ze, etc.).

Organic-natural materials: Basic raw materials (wood, cellulose, straw) havea minor environmental impact compared to the impact of the production process and the one related to other additives (binders, fire retardants, insecticides). Although different factors determine the environmental impact of WF insulation (production process, additives), the conclusion regarding the environmental impact of CF and SB insulation is straightforward. The environmental impact of the former is defined by the mineral fire retardants and boric acid, and of the later by the activities related to sowing (use of farming machinery – diesel), harvesting, fertilising (fertiliser) and pest control (insecticides), which is most notable bythe high impact of the EP category (figure 5).

5. Conclusion

This study investigated the environmental impact of different insulation materials (products) in the cradle to gate life cycle phases. They were separated in three groups: inorganic-synthetic, organic-synthetic and organic-natural. The drivers which influence the environmental impacts were highlighted and based on the grouping, conclusions regarding their environmental impacts were made. The study showed that there are parallels concerning the factors that influence the environmental categories between the insulations in each group and that for each of the defined groups, specific drivers influence the environmental impact.

References

[1] Ortiz O Castells F and Sonnemann G 2009 Sustainability in the construction industry: A review of recent developments based on LCA *Constr. Build. Mater.* **23** 28–39

[2] European Parliament and European Council 2010 Directive of the European Parliament and of the Council of 19 May 2010 on the energy performance of buildings (2010/31/EC).

[3] Blengini G A and Di Carlo T 2010 The changing role of life cycle phases, subsystems and materials in the LCA of low energy buildings *Ener. Build.* **42** 869–80

[4] Berardi U and Naldi M 2017 The impact of the temperature dependent thermal conductivity of insulating materials on the effective building envelope performanc *Ener. Build.* **144** 262–75

[5] Cabeza L F Rincón L Vilariño V Pérez G Castell A 2014 Life cycle assessment (LCA) and life cycle energy analysis (LCEA) of buildings and the building sector *Renew. Sustain. Ener. Rev.* **29** 394–416

[6] EN 15804: 2012 Sustainability of construction works 2012 Environmental porduct declarations - Corerules for the product category of construction products

[7] ISO 14025: 2006 Environmental label sand declarations 2006 Type III environmental declarations - Principle sand procedures

[8] Gelowitz M D C Mc Arthur J J 2017 Comparison of type III environmental product declarations for construction products: Material sourcing and harmonization evaluation *J. Clean. Prod.* **20** 125–33

Reducing SO_2 Emissions in the Czech Republic in Accordance with Legislation Valid from 2021

P Dvořák[1], O Palička[1], R Fojtů[1], Z Vávrová[1] and M Niedoba[2]

[1] Centrum ENET, Technical University of Ostrava, Czech Republic
[2] ENERGETIKA TŘINEC a.s., Třinec, Czech Republic

E-mail: pavel.dvorak1@vsb.cz

Abstract. This article discusses the impact of new emission limits for sulphur dioxide (SO_2) emissions which will be effective from 2021 for large combustion plants (LCP) in the Czech Republic and as well as possible ways and techniques to achieve these emission limits.

1. Introduction

Reduction of SO_2 emission is essentially a permanent trend of last decades and it is an ongoing task for all employees in operations, maintenance and development of energy equipment. Permanent emissions reductions targets are included in the legislations of all industrialized countries.

This article provides information about current SO_2 emissions of selected combustions plants, which are divided into several performance classes (predominantly large combustion plants). Article also provides an overview of the most common desulphurisation methods for these combustion plants. In article are listed new BAT emission limits (approved by the Parliament on July 31, 2017), which will come into force in 2021. Now, it is therefore possible to compare the status of current SO_2 emissions with new limits and to estimate if it will be necessary to invest in new desulphurisation units.

2. Emission limits of SO_2 for existing and new energy sources valid from 2021

Table 1 compares the daily and yearly average of SO_2 emission limits (mg/Nm3) for individual combustion plants with different heat input (MWt) according to valid and planned legislation since 2021.

Table 1. SO_2 emission limits values for LCP.

IED Direction 2010/75/ES since 2016		BAT-AELs-FD BREF LCP				
		New plant			Existing plant	
Heat input [MW_t]	Monthly avg. [mg/Nm^3]	Heat input [MW_t]	Daily avg. [mg/Nm^3]	Yearly avg.	Daily avg. [mg/Nm^3]	Yearly avg.
50-100	400	<100	170-220	150-200	170-400	150-360
100-300	250	100-300	135-200	80-150	135-220	95-200
>300	200	>300 PC coal	25-110	10-75	25-165	10-130
		>300 FBC coal	25-110	20-75	50-220	20-180

3. Actual situation of emission SO_2 in the Czech Republic

In the Czech Republic 3 desulphurisation methods are used. There are dry desulphurisation method (mostly used in fluidized bed boilers), wet desulphurisation method (mostly in large power plants) and semi-dry desulphurisation method (mostly smaller plants). Nowadays there is no problem in the fossil fuel combustion plants to meet the current emission limit. As it can be seen in the table 2, most desulphurisation devices will easily meet the new emissions limit according BAT. Existing desulphurisation units are not working at their maximum.

Value of emission is inversely proportional to the operating costs of the desulphurisation units. Desulphurisation units based on wet desulphurisation process is characterized among other things, almost linear relationship between consumption of sorbent and desulphurisation efficiency. For desulphurisation units based on a dry and semi-dry method, this dependence is not linear but exponential.

Table 2. SO_2 emission of real combustion plants in the Czech Republic.

Output of Power/ Heat Station	Fuel	Desulphurisation Method	Limit According BAT [mg/Nm^3]	Current Emission SO_2 [mg/Nm^3]
>300 MW_t	Brown Coal	Wet method	10 - 130	110
100 - 300 MW_t	Black Coal	Semi – dry method	95 - 200	190
<100 MW_t	Brown Coal	Wet method	150 - 360	80
<100 MW_t	Black Coal	Dry method	150 - 360	366
<100 MW_t	Black Coal	Dry method	150 - 360	290

4. Flue gas desulphurisation

Flue gas desulphurisation is the most frequently used method of reducing SO_2 emissions. This method can be based on a variety of chemical or physico-chemical principles. The basic principle is the application of absorption substances to the flue gas stream. These substances react with SO_2 and then they are separated from the flue gas. Methods can be systematically divided into several groups according to different criteria. One of the basic criteria is phase of used substance. According this we can desulphurisation methods divide into: wet, semi-dry and dry methods.

4.1. Wet desulphurisation method

Wet method of desulphurisation is the most common method of flue gas desulphurisation on large energy sources in the Czech Republic. The functional principle is capturing acid components of the flue gas in an alkaline medium substance, in this case in calcium carbonate suspension ($CaCO_3$). The basic parts of the wet desulphurisation absorber can be seen in figure 1.

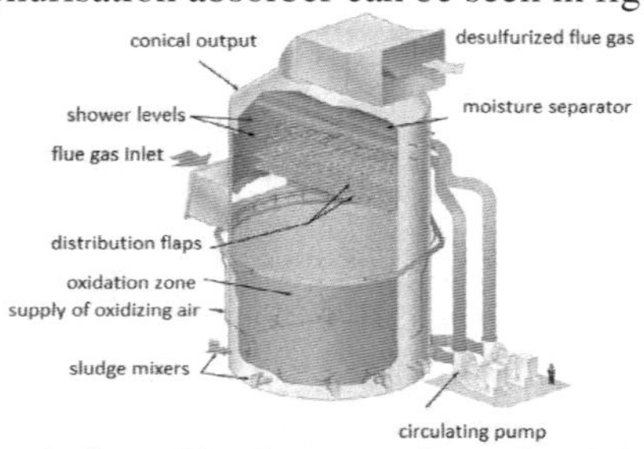

Figure 1. Description of basic parts of wet desulphurisation unit.

Optimal temperature for reaction is between 55 °C to 60 °C. It means that the flue gas must be cooled before entering the absorber. Regenerative or recuperative heat exchangers are most often used. Exiting flue gas is fed into the chimney through the heating part of the heat exchanger. Temperature increases by about 20 - 30 °C with respect to the dew point. Total reaction of the desulphurization process is:

$$2CaCO_3 + 2SO_2 + O_2 + 4H_2O \rightarrow 2CaSO_4 \cdot 2H_2O + 2CO_2 \quad (1)$$

In current embodiment, desulphurisation unit primarily consists of absorber tower, where flue gas is contacted with liquid phase (slurry) and where the desulphurisation process is primarily carried out. Suspension is sputtered in the plane of the nozzles in the absorber tower. Wet limestone scrubbing can be described with the following equations. Dissolving SO_2 in water to form sulfuric acid:

$$SO_2 + H_2O \rightarrow H_2SO_3 \quad (2)$$

This acid dissociates into H^+, HSO_3^- and SO_3^{2-} ions according following equations:

$$H_2SO_3 \leftrightarrow H^+ + HSO_3^- \quad (3)$$

$$HSO_3^- \leftrightarrow H^+ + SO_3^{2-} \quad (4)$$

These ions react with calcium carbonate to form sulphite and calcium bisulphite:

$$CaCO_3 + 2H^+ + 2HSO_3^- \rightarrow Ca(HSO_3)_2 + H_2O + CO_2 \quad (5)$$

$$Ca(HSO_3)_2 + 2H^+ + SO_3^{2-} \rightarrow CaSO_3 + 2H_2O + 2CO_2 \quad (6)$$

The ultimate desulphurisation product is formed at the bottom of absorber, where droplets of the suspension fall and where sulphite and calcium bisulphite are oxidized to calcium sulphate dihydrate according following equations:

$$CaSO_3 + 2H_2O + ½O_2 \rightarrow CaSO_4 \cdot 2H_2O \quad (7)$$

$$Ca(HSO_3)_2 + ½O_2 + 2H_2O \rightarrow CaSO_4 \cdot 2H_2O + H_2SO_3 \quad (8)$$

Resulting sulfuric acid reacts with excess lime also to calcium sulphate dihydrate:

$$H_2SO_4 + CaCO_3 + H_2O \leftrightarrow CaSO_4 \cdot 2H_2O + CO_2 \quad (9)$$

Suspension is continuously pumped into the spray nozzles from bottom of the tank. Spray nozzles are located at the top of the absorber tower. At the bottom of the absorber is suspension aerated with oxidizing air and mixed with circulating pumps and agitators. Fresh water suspension of finely ground limestone is also fed to the sump. Product of the desulphurisation (gypsum) forms crystals which must be separated from suspension. At first suspension is concentrated in hydro cyclones and then is concentrated mixture put to belt filter, where is process water removing.

4.2. Dry additive method of desulphurisation with limestone

Flue gas can be desulphurised also at high temperatures. Dry additive method is based on reaction between solid and gaseous phase. Product of reaction is a solid phase. So, it can be easy separated from flue gas. For example, in an electrostatic precipitator or a fabric filter. This method can be used for the desulphurisation of flue gas from conventional pulverized or fluidized boilers. The sorbent can be limestone, hydrated lime, burnt lime, dolomite and others. Total equation of the process using calcium carbonate is:

$$CaCO_3 + SO_2 + ½O_2 \rightarrow CaSO_4 + CO_2 \quad (10)$$

The whole *desulphurisation* process can be divided into two partial reactions, thermal decomposition of calcium carbonate "calcination" and "sulphation", which is a reaction between calcium oxide and sulphur dioxide.

Calcination: $$CaCO_3 \rightarrow CaO + CO_2 \quad (11)$$

Sulphation: $$CaO + SO_2 + ½O_2 \rightarrow CaSO_4 \quad (12)$$

The individual reaction processes can be seen in Figure 2. During sulphation process is first formed a surface layer of calcium sulphite. In the next phase a barrier to next sulphation of core is formed, so sulphur dioxide must diffuse over the product of the desulphurisation reaction.

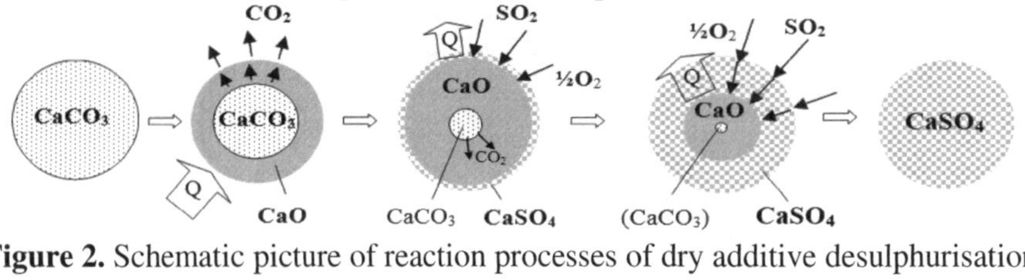

Figure 2. Schematic picture of reaction processes of dry additive desulphurisation.

The ideal reaction conditions are temperature - t < 900°C, excess air - n > 1 and sufficient reaction time. At higher temperatures calcinate is deactivated because of change in porous structure and resulting calcium sulphate is decomposed to the original reactants. This desulphurisation method can be used for granulation and fluid boilers. These types of boilers are characterized by a different combustion process, so the ways of their desulphurisation are also differ.

4.2.1. Dry additive desulphurisation method for granulation boilers

The basic principle of additive desulphurization is absorption of SO_2 on CaO particles. There are several ways to dose CaO. It can be dosed with coal, pneumatically into the area of the burner zone or in the area above the burner zone. One of the main parameters of this process is temperature. Based on experience ideal temperature range is between 750 - 1200 °C. Reaction takes place only on the surface of sorbent particles, because of limited residence time of both phases. Sulphur dioxide doesn´t have enough time to diffuse into the core of sorbent particles. Sorbent is not sulphated more than 15%. The advantage of this method is low investment costs. The disadvantages include low efficiency which is between 20 to 40%.

4.2.2. Dry additive desulphurisation method for fluidized bed boiler

Calcium and calcium-magnesium salts are used for additive desulphurisation in fluidized bed boilers. Sorbent is fed to the fluidized bed. Residence time of both phases is several times longer than in granulating boilers and can take up to tens of minutes. SO_2 molecules have sufficient time to diffuse into sorbent particle cores. Sorbent is sulphated much higher than in the previous desulfurization process. To achieve 80-90% desulphurisation is necessary to supply limestone with an excess of 2 to 3 times corresponding amount of combustible sulphur in the fuel.

4.3. Semi-dry method of desulphurisation

This method is something between dry and wet desulphurisation. Substances, which capture sulphur oxides, enter the process in the liquid phase and stand out in the solid phase.

4.3.1. Absorption of sulphur dioxide in a water suspension of lime

The method is based on injecting lime milk into the flue gas stream at temperature 140-250 ° C. In this reaction sulphur dioxide is absorbed by slurry and then oxidized to calcium sulphate. During this process all water evaporated is, so product of reaction can be separated on fabric or electrostatic filters. Desulphurisation product, unreacted sorbent and ash are collected on the filters. Captured mixture is recirculated for better utilization CaO. The efficiency of this method may reach up to 92%. Description of basic parts of spray dryer can be seen in figure 3.

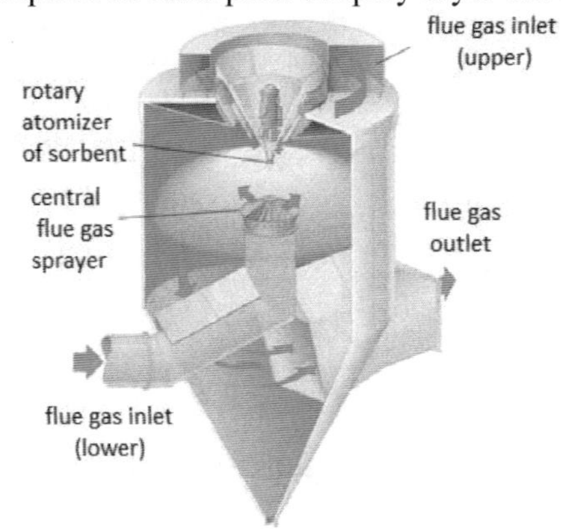

Figure 3. Description of the basic parts of the spray dryer of the semi-dry desulphurisation method.

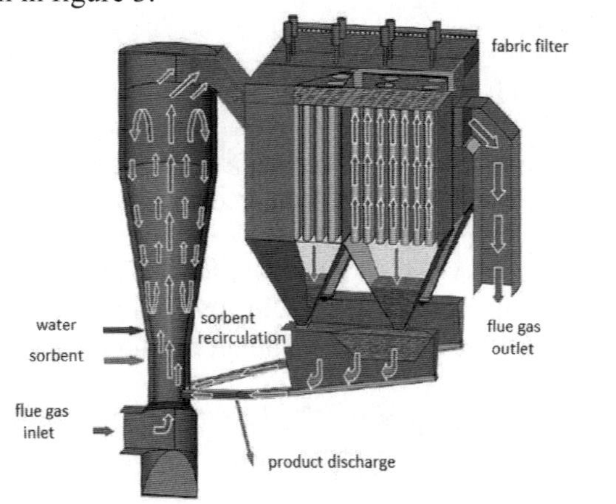

Figure 4. Description of basic parts of fluidized bed desulphurisation reactor.

4.3.2. Absorption of sulphur dioxide in fluidized bed reactor

This technology is based on adsorption of SO_2 in reactor. Description of basic parts of this technology can be seen in figure 4. Flue gas flows through reactor and at the same time is into it injected water and sorbent. As a sorbent is most commonly used CaO and $Ca(OH)_2$. Optimal temperature of desulphurisation process is ensured by adiabatic injection of water into reactor. Chemical reactions of desulphurisation process are:

$$SO_2 + Ca(OH)_2 \rightarrow CaSO_3 + H_2O \tag{13}$$

$$SO_3 + Ca(OH)_2 \rightarrow CaSO_4 + H_2O \tag{14}$$

$$2HCl + Ca(OH)_2 \rightarrow CaCl_2 + 2H_2O \tag{15}$$

$$2HF + Ca(OH)_2 \rightarrow CaF_2 + 2H_2O \tag{16}$$

$$CO_2 + Ca(OH)_2 \rightarrow CaCO_3 + H_2O \tag{17}$$

Desulphurisation of flue gas takes place mainly in a fabric filter which is located just behind the fluidized bed reactor. A major part of separated product is returned to the reactor to better utilization of sorbent. The rest of separated product is taken from the process into product silo.

Advantages of this method include the high efficiency of SO_2 desulphurisation up to 90%. The efficiency of removing HF, HCl and SO_3 is high too. Another advantage is absence of waste water and low consumption of process water. There is also reduction of dust emissions. The technological part of desulphurisation unit is compact, and for flue gas flows up to 800,000 Nm^3/h, it is more investment-friendly than wet desulphurisation method. Disadvantages of this method are high demands on compressed air and the low usability of desulphurisation product, which must be deposited in the landfill.

5. Conclusion

Most of installed desulphurisation units in the Czech Republic fulfil current emission limits without any problems. It can be assumed that new emission limits will be met for most of large energy sources. The necessary increase in efficiency of flue gas desulphurisation will be accompanied only by an increase in the operating costs of the individual desulphurisation units.

Acknowledgments

This paper is supported by the following project: LO1404: Sustainable development of ENET Centre; CZ.1.05/2.1.00/19.0389 Development of the ENET centre research infrastructure; SP2017/96 Students Grant Competition, Czech Republic.

References

[1] Frolka P 2016 *BREF Large combustion plants* XVII Forum for information exchange about BAT: MPO
[2] Dvořák P 2015 *Design of desulphurization plant for K1 to K4 boilers in Heat plant Přívoz* Technical University of Ostrava
[3] Technical reports - Accredited workplace for diagnostics and operation of power generation equipment Centrum ENET
[4] Best Available Techniques (BAT) 2016 Reference Document for Large Combustion Plants: Industrial Emissions Directive 2010/75/EU. Final Draft. JOINT RESEARCH CENTRE Institute for Prospective Technological Studies Sustainable Production and Consumption Unit European IPPC Bureau
[5] Technical reports - Department of Energy, Faculty of Mechanical Engineering VŠB-TUO
[6] Process chemistry in the energy sector 2014 VŠB-TU Ostrava

Evaluation of the damaged depths of slag-based mortars in aggressive sulphate conditions

A Estokova[1], M Smolakova[1] and A Luptakova[2]

[1]Technical University of Kosice, Faculty of Civil Engineering, Institute of Environmental Engineering, Kosice, Slovakia
[2]Institute of Geotechnics, Slovak Academy of Science, Kosice, Slovakia

Email: michaela.smolakova@tuke.sk

Abstract. Resistance of cement composites against various chemical processes such as sulphate attack and corrosion is nowadays one of the most desired property. Mechanism largely involved in sulphate corrosion is the ingress of aggressive ions through diffusion. The result of the ingress of aggressive ions is leaching of the elements from the mortar matrix and the deterioration of the composite. In the experiment, the slag-based cement mortars samples were exposed to an aggressive microbiological sulphate environment, represented by the bacteria *A. thiooxidans,* for the period of 6 months. The paper presents the investigation of damaged depths of cement mortars based on the leaching characteristics of the Ca ions. The damage depths (EDD) of cement composites were much higher (up to 0.2 mm) for samples placed in the biotic environment, which confirms the aggressive action of *A. thiooxidans*.

1. Introduction

In urban environmental engineering, most of the facilities for sewage transportation are made from concrete and mortar structures. They are generally placed underground and are affected by the acid, erosion, microorganisms and other corrosive agents. The growing problem of concrete deterioration and bio-deterioration results in a huge economic losses all around the world [1].

To date, sulphate attack is still a significant problem in soil or solutions containing SO_4^{2-}, because it affects the mechanical performance and durability of mortar and cement concrete infrastructures [2,3]. The mechanism of concrete degradation exposed to external sulphate attack depends on exposure conditions, such as temperature, associated cations and sulphate concentration [4,5,6,7].

Most hydraulic structures have components that are constantly or partially submerged in water. The varying water chemistry of different water bodies affects the surface of these structures in many ways. For example, water can have the ability to dissolve components of the cement, i.e., leaching. The leaching kinetics is generally slow, but the impact can be essential from a long-term standpoint.

It is well known that cementitious materials are not thermodynamically stable, when they are in contact with more or less aggressive water. Portland cement concrete and mortars are easily attacked by water and acidic solutions with low pH. Compounds with elements such as Na and K, are quickly leached out when concentration gradients are present. In this case, the pH of the concrete pore solution may considerably decrease [8,9].

Leaching is a diffusion reaction which takes place when concrete is exposed to poorly mineralized water, acid water or biogenic acids made by microorganisms. Degradation consists in dissolution of calcium and hydroxide ions out of the cement matrix, which causes an increase in porosity and transport properties of concrete surface. Leaching is accelerated in neutral and acid solutions, and it may be coupled with the ingress of aggressive ions such as sulphate, chloride or magnesium [7,10,11, 12]. It is widely known that leaching of calcium ions can have enormous effects on the physical characteristics of the structure. Leaching of Ca leads to loss of mass and mechanical strength and alkalinity. Loss of mass in the cement matrix is linked with an overall increase in porosity [9,13,14,15]. The diffusion kinetics can be described by Fick's law, which relates the altered thickness to the square root of time [16,17]. Leaching is consequently quite rapid at the beginning. As leaching progresses, porosity of the cement paste and ionic diffusivity increases, because ions diffuse through interconnected pores. It has been reported that the leaching rate of cement mortars is lower than that of cement pastes because particles in the corroded layer increase the diffusion pathway. In general, the lower the pH of the external solution is and the higher the concentration gradient is, the faster the transfer kinetics will be [9].

Effects like leaching needs to be taken into consideration in the design of concrete structures, which are in contact with wastewater treatment plants, mortar and concrete products such as sewage pipes and also water tanks [7].

The objective of this work was to evaluate the leaching characteristic of main inorganic components of cement mortars with ground granulated blast furnace slag exposed to microorganisms *Acidithiobacillus thiooxidans* for the period of 6 months and consequently, based on the leach-out amounts, to assess the damaged depths.

2. Material and Methods

2.1. Mortar mixtures

Three mortar mixtures where designed for this experiment. Sample R represents the reference sample made with ordinary Portland cement (OPC) only. Samples M1 and M2 represents the samples where the cement was partially replaced by ground granulated blast furnace slag in 65 % and 95 % by weight of cement, respectively. The mixtures were designed according to standard STN EN 196-1 [18] and were made of three parts of normalized sand (1350 ± 5 g), one part of cement (450 ± 2 g) and half part of water (225 ± 1 g) having water to cement ratio w/c = 0.50. The test samples with dimensions 160 mm × 40 mm × 40 mm have been designed in cooperation with cement factory plant which we cooperate with.

The chemical composition of main components of used Portland cement CEM I 52.5 R was as follows: 57.15 wt.% of CaO, 18.11 wt.% of SiO_2, 4.02 wt.% of Al_2O_3, 2.69 wt.% of Fe_2O_3, 1.49 wt.% of SO_3, 1.37 wt.% of MgO, 1.12 wt.% of K_2O, 0.33 wt.% of P_2O_5, 0.18 wt.% of TiO_2, 0.06 wt.% of Cl. The content of main components of incorporated ground granulated blast furnace slag (GBFS) was as follows: 39.55 wt.% of CaO, 38.95 wt.% of SiO_2, 10.11 wt.% of MgO, 8.33 wt.% of Al_2O_3, 0,74 wt.% of MgO, 0.57 wt.% of SO_3, 0.54 wt.% of Fe_2O_3, 0.48 wt.% of K_2O, 0.37 wt.% of TiO_2, 0.04 wt.% of P_2O_5, 0.02 wt.% of Cl.

2.2. Bacterial exposure

The wastewater-like environment was simulated by sulphur-oxidising bacteria *Acidithiobacillus thiooxidans*, because of their ability to produce sulphuric acid. The nutrient medium by Waksman and Joffe (Waksman, S.A; Joffe, J. S. 1922) was used for the preparation of the active bacterial culture of *A. thiooxidans* in the experiment. The composition of the nutrient medium was as follows: 0.25 g/L $CaCl_2.6H2O$, 0.2 g/L $(NH_4)_2SO_4$, 3.0 g/L K_2HPO_4, 0.5 g/L $MgSO_4.7H_2O$, traces of $FeSO_4.7H_2O$, 10.0 g/L elementary S and up to 1000 mL supplemented by distilled water. Bacteria *Acidithiobacillus thiooxidans* were isolated from an acid mine drainage from Pech shaft (locality of Smolnik, Eastern Slovakia). Elemental sulphur was added to the nutrient medium.

The exposure of cement mortars to bacterial culture has taken place in glass containers under aerobic laboratory conditions over a period of 6 months. Optimal growth temperature of bacterial culture (28 - 30°C) and pH interval 2.0-3.5 were ensured during the experiment [19,20,21].

IOP Conf. Series: Earth and Environmental Science **92** (2017) 012011 doi:10.1088/1755-1315/92/1/012011

The mortar specimens exposed to bacterial medium with *A. thiooxidans* were labelled as R/A, M1/A, M2/A, and the control samples placed in abiotic environment, with no bacteria, were labelled as R/C, M1/C, M2/C, respectively. The volumes of liquid phase for the mortar samples used in experiment are described in table 1.

Table 1. Volume of liquid phases used in the experiment.

Mortar specimen	Volume of inoculum of *A. thiooxidans* (ml)	Volume of nutrient medium (ml)	Volume of sterilized water (ml)	Amount of elemental sulphur (g)
R/A	512	768	1280	6.4
R/C	-	1280		
M1/A	512	768		
M1/C	-	1280		
M2/A	512	768		
M2/C	-	1280		

Chemical composition of concrete samples′ leachates was investigated once a month using X-ray fluorescence analysis. The SPECTRO iQ II (Ametek, Germany) with SDD silicon drift detector with resolution of 145 eV at 10 000 pulses was used for the analysis. The primary beam was polarized by Bragg crystal and Highly Ordered Pyrolytic Graphite - HOPG target. The basic measurements conditions were as follows: time of measurement - 180 s; voltage - 25 kV and 50 kV; current - 0.5 and 1.0 mA; atmosphere - helium. The concentrations of measured elements were calculated using the calibration method of fundamental parameters for liquids.

2.3 Equivalent Damaged Depth

The damaged depths of studied cement mortars after the bacterial exposition were evaluated using equation (1) Equivalent Damaged Depth (EDD) (mm) according to [7]:

$$EDD = \frac{c(Ca)L}{c(Ca)composite} \tag{1}$$

Where $c(Ca)_L$ is total amount of leached calcium (mg/cm^2) and $c(Ca)_{composite}$ is theoretical initial total calcium content in volume unit of composite (mg/cm^3). Equivalent damaged depth (EDD) was selected based on the hypothesis that the calcium was totally leached in damaged depth. Assuming that Ca ions only came from composite and that no portlandite remained in corroded layers, equivalent damaged depths may be used to evaluate the level of deterioration regarding the depth [7].

3. Results

The leaching trends of calcium during the bacterial and non-bacterial exposure are presented in figures 1 and 2. As it can be seen, the leaching trends for calcium ions was proven to be more significant for mortars exposed to *A. thiooxidans* (R/A, M1/A, and M2/A) compared to samples exposed to medium without bacteria (R/C, M1/C, and M2/C). The highest amount of leached Ca ions (923 mg/L) were noticed for reference specimen R/A. The sample with 95 wt. % of blast furnace slag was found to be the most durable after the 6-month experiment, evaluating the Ca^{2+} leaching only.

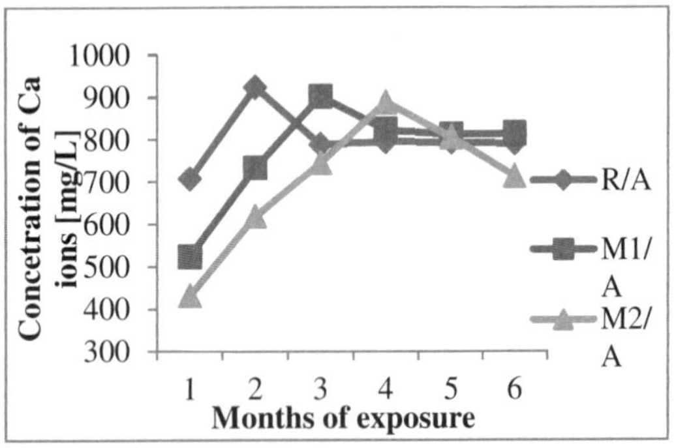

Figure 1. The concentration of Ca^{2+} ions under bacterial exposure.

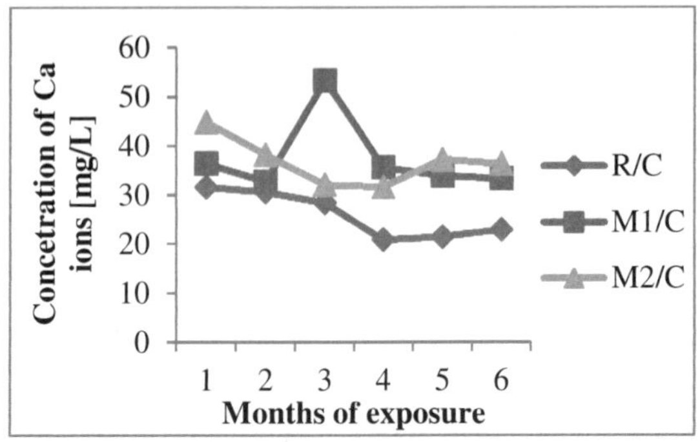

Figure 2. The concentration of Ca^{2+} ions under non-bacterial exposure.

The bacterial exposure was confirmed to be more aggressive also in terms of damaged depths of mortar samples. The calculated equivalent damaged depths (EDD) are given in table 2.

Table 2. Equivalent damaged depth (EDD).

Mortar sample	Total leached calcium (mg/cm^2)	Initial calcium content in binder (mg/cm^3)	Equivalent damaged depth EDD (mm)
R/A	8.012	350.359	0.228
R/C	0.274	357.722	0.007
M1/A	7.822	354.292	0.221
M1/C	0.462	354.609	0.013
M2/A	7.705	341.496	0.225
M2/C	0.388	348.951	0.011

As mentioned above, the most intensive Ca^{2+} leaching was observed for samples R/A, M1/A and M2/A exposed to bacteria and the calculated EDD of these samples reached much more higher values than EDD calculated for the mortars with no bacterial exposure. The calculated indicators of equivalent damaged depths of mortars due to bacterial influence were in a very close interval 0.221 to 0.228 mm.

Therefore, a longer experimental period is necessary to make any clear conclusions regarding the durability of mortars of various compositions. Similarly, the EDD of mortars samples without any bacteria ranged in a very close interval from 0.007 to 0.011 mm. Oddly enough, the reference sample without any blast furnace slag (R/C) had the lowest EDD compared to the other two samples (M1/C and M2/C, respectively).

4. Conclusion

The presented work shows the results of leaching performance of Ca ions from cement mortars with different share of blast furnace slag after 6 months in bacterial and non-bacterial environment and evaluates the equivalent damaged depths of mortars. Samples exposed to the medium with sulphur-oxidizing bacteria *A. thiooxidans* had higher leaching rates and damaged depths were therefore significantly higher.

The mortars with GBFS as admixture can lead to good resistance to leaching and binder incorporating such latent hydraulic material as slag actually produces less portlandite, with compact matrix. Different chemical composition of binder of samples M1 and M2 can lead to noticeable changes in the damaged depths of mortars.

Acknowledgements
This paper has been prepared within the Grant No. 2/0145/15 of the Slovak Grant Agency for Science.

References

[1] Kong L, Zhang B and Fang J 2017 Study on the applicability of bactericides to prevent concrete microbial corrosion *Constr. Build. Mater.* **149** 1-8

[2] Marchand J, Odler I and Skalny J P 2003 *Sulfate attack on concrete* CRC Press

[3] Yu D, Guan B, He R, Xiong R and Liu Z 2016 Sulfate attack of Portland cement concrete under dynamic flexural loading: A coupling function *Constr. Build. Mater.* **115** 478-485

[4] Santhanam M, Cohen M D nad Olek J 2001 Sulfate attack research—whither now? *Cem. Concr. Res.* **31** 845-851

[5] Higgins D D and Crammond N J 2003 Resistance of concrete containing GGBS to the thaumasite form of sulfate attack *Cem. Concr. Comp.* **25** 921-929

[6] Dehwah H A F 2007 Effect of sulfate concentration and associated cation type on concrete deterioration and morphological changes in cement hydrates *Con. Build. Mater.* **21** 29-39

[7] Rozière E, Loukili A, El Hachem R and Grondin F 2009 Durability of concrete exposed to leaching and external sulphate attacks *Cem. Concr. Res.* **39** 1188-1198

[8] Ekström, T 2003 *Leaching of concrete: the leaching process and its effects* Division of Building Materials LTH Lund University

[9] Rosenqvist M, Pham L W, Terzic A, Fridh K and Hassanzadeh M 2017 Effects of interactions between leaching, frost action and abrasion on the surface deterioration of concrete *Constr. Build. Mater.* **149** 849-860

[10] Carde C and Francois R 1999 Modelling the loss of strength and porosity increase due to the leaching of cement pastes *Cem. Concr. Comp*. **21** 181-188

[11] Mainguy M, Tognazzi C, Torrenti J M and Adenot F 2000 Modelling of leaching in pure cement paste and mortar *Cem. Concr. Res.* **30** 83-90

[12] Moranville M, Kamali S and Guillon E 2004 Physicochemical equilibria of cement-based materials in aggressive environments-experiment and modeling *Cem. Concr. Res.* **34** 1569-1578

[13] Duchesne J and Bertron A 2013 Leaching of cementitious materials by pure water and strong acids (HCl and HNO_3) *Performance of Cement-Based Materials in Aggressive Aqueous Environments* Alexander M, Bertron A, De Belie N (Eds.) 91-112 Springer Netherlands

[14] Shi C and Stegemann J A 2000 Acid corrosion resistance of different cementing materials *Cem. Concr. Res.* **30** 803-808

[15] Beddoe R E and Dorner H W 2005 Modelling acid attack on concrete: Part I. The essential mechanisms *Cem. Concr. Res.* **35** 2333-2339

[16] Adenot F and Buil M 1992 Modelling of the corrosion of the cement paste by deionized water *Cem. Concr.Res.* **22** 489-496

[17] Pavlik V 1996 Corrosion of hardened cement paste by acetic and nitric acids Part III: Influence of water/cement ratio *Cem. Concr. Res.* **26** 475-490

[18] STN 73 1316: 1989 *Determination of moisture content, absorptivity and capillarity of concrete*

[19] Kovalcikova M, Estokova A and A. Luptakova 2015 Application of granulated blast furnace slag in cement composites exposed to biogenic acid attack *IOP Conf. Ser.: Mater. Sci. Eng.* **96** (United Kingdom: IOP Publishing Ltd) 7

[20] Estokova A, Smolakova M and Luptakova A 2016 A contribution to sustainable structures by using waste in concrete materials for improving their durability *41st IAHS world congress Sustainability and Innovation for the Future (Algarve)* 9

[21] Strigac J, Martauz P, Estokova A, Stevulova N and Luptakova A 2016 Bio-corrosion resistance of concrete containing antimicrobial ground glanulated blast furnace slag BIOLANOVA and novel hybrid H-CEMENT *Sol St Phen* **244** 57-64

Magnesium Contamination in Soil at a Magnesite Mining Area of Jelšava-Lubeník (Slovakia)

D Fazekašová[1], J Fazekaš[1], O Hronec[1] and M Horňak[1]

[1]Department of Environmental Management, Faculty of Management, University of Prešov,Konštantínova 16, 080 01 Prešov, Slovakia

E-mail: danica.fazekasova@unipo.sk

Abstract. Magnesium contamination in soil at a magnesite mining area of Jelšava-Lubeník (Slovakia) and their effect to the soil characteristics were determined. Soil samples were collected in the area Jelšava – Lubeník with specific alkaline pollutants, it is one of the most devastated regions of Slovakia and with the alarming degree of environmental damage. Magnesite air pollutants are a mixture of MgO and $MgCO_3$ due to which a soil reaction can move above pH 8. Production of clink accompanies the enormous emissions of dust particles of MgO into the air and the leakage of gaseous compounds, mainly SO_2 and NO_x. The total content of heavy metals in soils (Pb, Zn, Cr, Mn, Mg) were determined by atomic absorption spectrometry and X-ray fluorescence spectrometry. Soil reaction was determined in solution of 0.01 M $CaCl_2$. The research showed that the investigated sites are mostly strongly alkaline. Based on the obtained results, it can be stated that the contents of Pb, Zn is below the level of toxicity but for Cr, Mn and Mg, it does not apply. Their significant exceedance points contamination in which we can take into account the harmfulness and toxicity.

1. Introduction

The pollution of the biosphere with toxic heavy metal is a widespread ecological problem resulting from anthropogenic activities like fossil fuel burning, ore mining and smelting, industrial and municipal waste disposal and agricultural activities [1,2]. The soil samples represent an excellent medium to monitor heavy metal pollution because anthropogenic heavy metals are usually deposited in top soils [3]. The result of economic activities of anthropogenic activities occurs in relation with nature to acidification, alkalization and metallization elements of the environment [4]. Soil alkalinity is conditioned by the presence of alkaline salts which are easily hydrolysed and allow formation of alkaline. The fundamental reason of soils alkalinity is exchangeable sodium and content of Na_2CO_3 or $NaHCO_3$ in soil solution and worsens some soil processes like soil acidity. Strong alkaline reaction of pH / KCl> 7.7 is inflicted by air alkaline pollutants from a magnesium factory currently located in the Jelšava and Lubeník. Magnesite air pollutants are a mixture of MgO and $MgCO_3$ due to which a soil reaction can move above pH 8 [4,5,6]. The area Jelšava–Lubeník with specific alkaline pollutants, is one of the most devastated regions of Slovakia and with the alarming degree of environmental damage. The aim of the study was to assess the level of heavy metal pollution in soils from the dust deposition of the magnesium factory Jelšava – Lubeník (Slovakia).

2. Material and methods

2.1. Study area

Jelšava and Lubeník lie in the south-central part of the Slovak Ore Mountains, in the valley Muran, in the district called Jelšava podolie. Jelšava podolie geomorphologically belongs to Revúca Highlands [7].Geologically, the area belongs to the Central Western Carpathians. Twelve research sites in the country agrarian of the problem area Jelšava and Lubeník were monitored (figure 1). Soil samples were sampled on permanent research sites, which are used as permanent grasslands and are in the immission field of the magnesium factory Jelšava-Lubeník (Slovakia), from A horizons the depth of 0.05 m to 0.15 m.

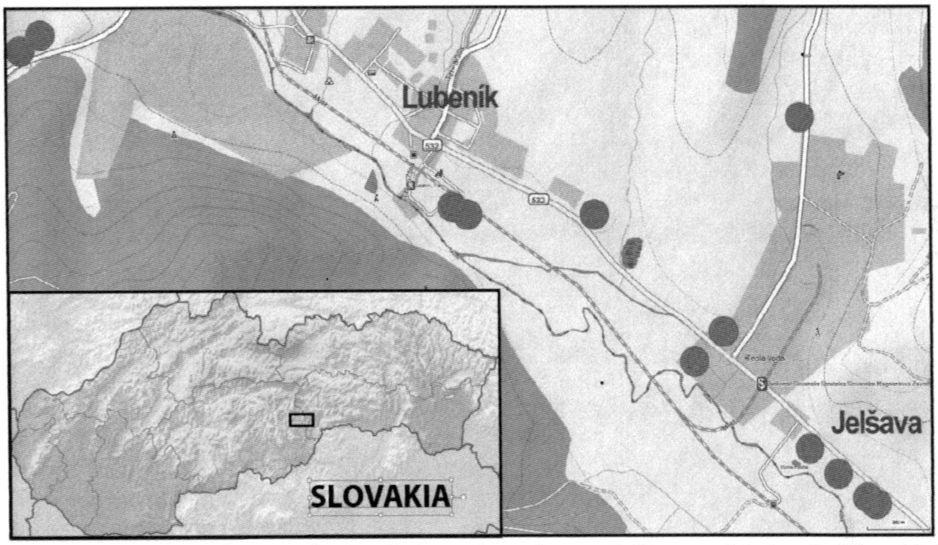

Figure 1. Location of research sites in investigated areas Jelšava and Lubeník (Slovakia).

2.2. Soil assays

After homogenization, the soil samples were manually crumbled, dried at a room temperature, sieved (<2 mm) and stored in polyethylene bags until theanalysis. We studied and evaluated the soil reaction in 1N solution $CaCl_2$ (5 g of soil mixed with 25 ml of 0.01M $CaCl_2$) using a pH meter Mettler Toledo. The total content of Pb, Zn, Cr, Mn and Mg was determined by X-Ray fluorescence spectrometry following the methodology as devised by Fiala [8]. The assessed values of heavy metals in soils were compared to the limit values of Slovak soils (Act No. 220/2004).

2.3. Statistical analysis

STATISTICA 13 software was used for all data analyses. The significance level between the heavy metals and other soil properties was calculated by Spearman's correlation coefficient.

3. Results and discussion

3.1. Soil reaction

The values of soil reaction measured on the Jelšava–Lubeník area are listed in table 1. The research showed that the investigated sites are slightly acidic (pH 6.23-6.36), neutral (pH 6.66– 6.68) and strongly alkaline (pH 7.7 to 8.8). The highest values pH were found at the post Lub9, 10 and Jel 4, 11, 12, 13 and 14 (figure 2). Mg flings with significant reactive caustic magnesite are aggressive in natural environments, already in small quantities, in contact with the soil and crop moisture, form saturated solutions with a high alkaline pH value [9].

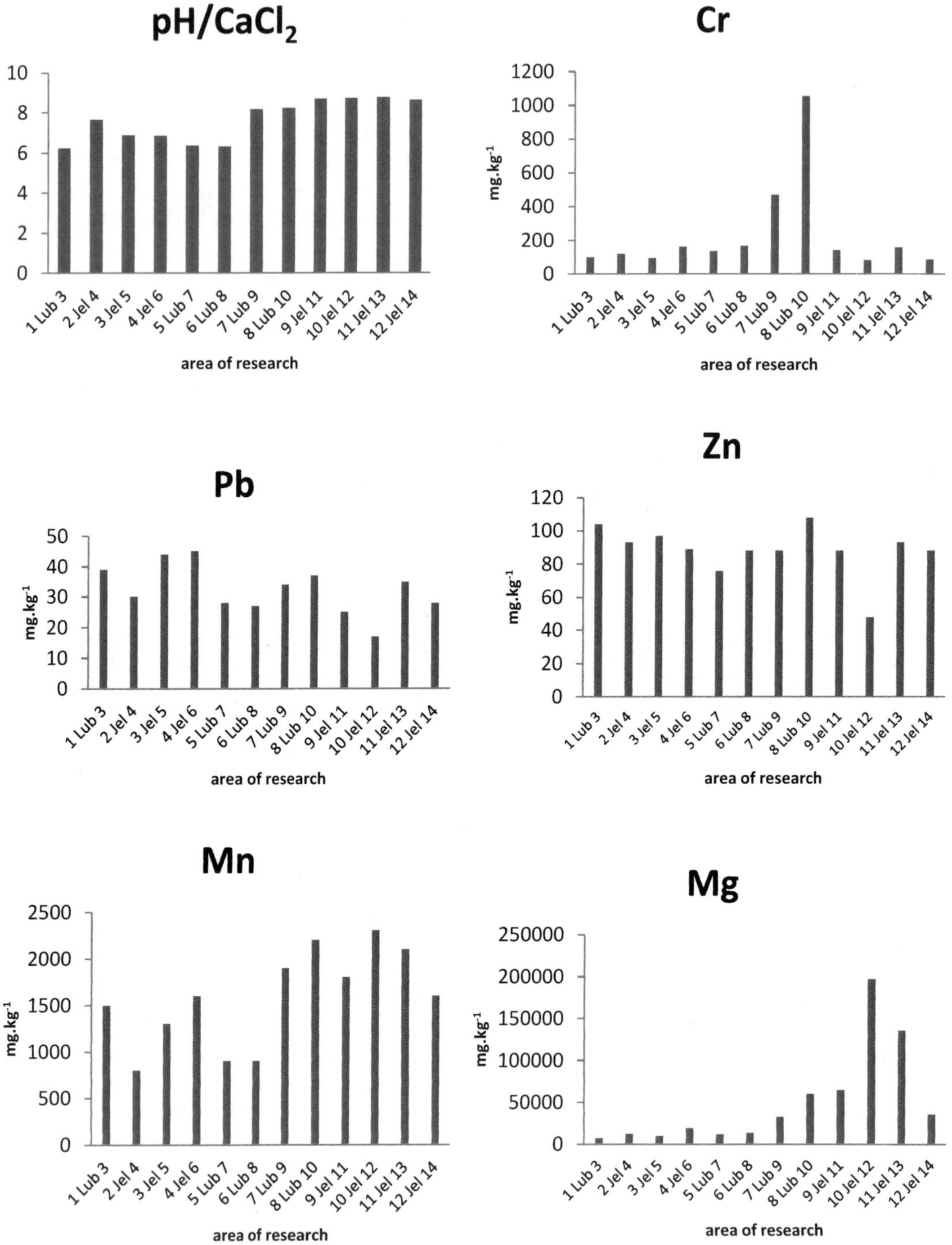

Figure 2. Values of soil reaction and heavy metal content measured of investigated areas in Jelšava and Luberník (Slovakia).

3.2. Heavy metal pollution

The content of heavy metals (Pb, Zn) on the studied area was not exceeding those set out by law (Act No. 220/2004). The chromium contents in the soil on the studied area were within the range 83 -

1055 mg kg^{-1} (table 1), the highest values were found at the post Lub 10 (figure 2).The median level of chromium in the soil of Slovakia was 85 mg kg^{-1} to A-horizon [10]. However, data presented by Čurlík and Šefčik [11] indicate that high Cr levels (up to 6096 mg kg^{-1}) are in both, A and C horizons of soils from Outer Carpathians. Chromium along with cobalt and nickel are considered as metals that come from geogenic load [12].Hexavalent chromium (Cr^{+6}) is classified as one of the most important environmental contaminants [13]. Readily soluble Cr^{6+} in soils is toxic to plants and animals. Therefore, the variability in the oxidation states of Cr in soils is of a great environmental concern [14].

Magnesium is considered the fifth major nutrient in plant nutrition. They are located in several primary and the secondary minerals. Content values of the available magnesium are in the topsoil of agricultural land in Slovakia in the range 200-400 mg kg^{-1} Mg high content of this element in the soil [15]. On the studied area, we found significant contamination of soil by magnesium and its values were within the range from 7000 to 197000 mg kg^{-1}, which is on average 18 to 493 times exceeded. The highest values were found at the post Jel 12 (figure 2).

The measured levels of manganese have a similar pattern and contents in the range 800-2300 mg kg^{-1} (table 1) were found, the highest values were found at the post Jel 12 (figure 2). The average content of manganese in the soil of the Slovak Republic is in the range of 0.85 to 112.90 mg kg^{-1}, indicating a significant spatial heterogeneity of the elements, however the dominates medium supply of this element in soils. Kabata-Pendias [14] referred to the value of 1500 mg kg^{-1}, which show symptoms of the toxicity of manganese. Based on the obtained results, it can be stated that the contents of Pb, Zn is below the level of toxicity but for Cr, Mn and Mg, it does not apply. Their significant exceedance points contamination in which we can take into account the harmfulness and toxicity.

Table 1. Variance analyses of heavy metals [mg kg^{-1}] and soil reaction of investigated areas in Jelšava and Lubeník (Slovakia).

	pH/$CaCl_2$	Pb	Zn	Cr	Mn	Mg
Mean	7.63	32.42	88.33	231.08	1575.00	49841.67
Minimum	6.23	17.00	48.00	83.00	800.00	7000.00
Maximum	8.80	45.00	108.00	1055.00	2300.00	197000.00
Standard deviation	1.03	8.16	15.17	279.49	517.20	59039.25
*Limit value		70	150	70	-	-

*Act No. 220/2004 Coll. of Laws

Relationships between heavy metals are listed in table 2. Spearman's correlation coefficients confirmed negative correlation between Pb–Mg and but only Zn– Mg were significant. A significant positive correlation ($p<0.05$) between soil pH – Mnand pH –Mg was determined in this study. Positive correlations were detected between Mn and Mg (table 2).

Table 2. Correlations between heavy metals and soil reaction (pH) of investigated areas in Jelšava and Lubeník (Slovakia).

	Pb	Zn	Cr	Mn	Mg
pH/$CaCl_2$	0.409	0.239	0.211	0.718*	0.696*
Pb		0.723*	0.210	0.066*	-0.519*
Zn			0.411	0.154	-0.615*
Cr				0.417	0.004
Mn					0.740*

$p<0.05$*

4. Conclusion

In this paper we presented the results of the study carried out in former mining area of Jelšava - Lubeník with specific alkaline pollutants is one of the most devastated regions of Slovakia and with the alarming degree of environmental damage. The major component of environmental pollution in Jelšava - Lubeník is magnesite powder belonging to aerosol particles which is crucial for the deposition process of gravitational sedimentation. The research showed that the investigated sites aremostly strongly alkaline. Based on the obtained results it can be stated that the contents of Pb, Zn is below the level of toxicity but for Cr, Mn and Mg it does not apply. Their significant exceedance points the contamination in which we can take into account the harmfulness and toxicity.

Acknowledgments
The study was supported by VEGA 1/0127/16 and KEGA 011PU-4/2016.

References

[1] Mahmood T 2010 Review Phytoextraction of heavy metals – the process and scope for remediation of contaminated soils *Soil & Environ.* **29** pp 91-109

[2] Adriano D C 2001 *Trace elements in the Terrestrial Environment* (New York: Springer) p 867

[3] Singh R Chavan S L and Sakale P H 2006 Heavy metal concentration in water, sediments andbody tissues of Red Worm (tubifex SPP) collected from Natural Habitats in Mumbai, India *Environ. Monit. Assess.* **129** pp 471–481

[4] Hronec O et al. 2008 Heavy metals in soils and plants in Rudniansko-Gelnická loaded area *Acta regionalia et environmentalica* **1** p 24-28

[5] Fazekašová D Barančíková G Torma S Ivanová M and Manko P 2014 *Chemical and environmental aspect of the component of the environment and landscape* (Prešov: University of Prešov in Prešov, Faculty of Management) p 257

[6] Wang L Tai P Jia Ch Li X Li P and Xiong X 2015 Magnesium Contamination in Soil at a Magnesite Mining Region of Liaoning Province, China *Bull. Environ. Contam. Toxicol.* **95** p 90–96

[7] Mazúr E and Lukniš M 1980 Regional division SSR (*Bratislava: Institute of Geography)*

[8] Fiala K et al. 1999 Partial monitoring system – soil, binding methods (*Bratislava:Research Institute of Soil Science and Conservation)*

[9] Baluchová B Bačík P Fejdi P and Čaplovičová M 2011 Mineralogical research of the mineral dust fallout from the years 2006-2008 in the area of Jelšava (Slovak Republic) *Mineralia Slovaca* **43** pp 327-334

[10] Ševčík P Pramuka S and Gluch A 2008 Assessment of soil contamination in Slovakia according index of geoaccumulation *Agriculture* **54** pp 119-130

[11] Čurlík J and Šefčík P 1999 *Geochemical Atlas of the Slovak Republic* (Bratislava: MŽP SR) p 99

[12] Takáč P Kozáková Ľ Vaľková M and Zeleňák F 2008 Heavy metals in soils in the middle Spiš *Acta Montanistica Slovaca* **13** pp 82–86

[13] Kafka Z and Punčochářová J 2002 Toxicity of heavy metals in nature *Chem. Listy* **96** pp 611-617

[14] Kabata–Pendias A 2011 *Trace Elements in Soil and Plants* (London: CRC Press) p 505

[15] Kobza J Barančíková G Dodok R Hrivňáková K Makovníková J Mališ J Pálka B Styk J and Širáň M 2010 *Soil monitoring of Slovakia* (Bratislava: Research Institute of Soil Science and Conservation) p 38

IOP Conf. Series: Earth and Environmental Science **92** (2017) 012013 doi:10.1088/1755-1315/92/1/012013

Assessment of post-mining damage

B Gibesová[1], J Koudelková[2], J Mudruňka[1] and D Vrublova[1]

[1]VŠB-Technical University of Ostrava - Institute of combined study in Most, Budovatelů 2830/3, Most 434 01, Czech Republic
[2]VŠB-Technical University of Ostrava - Faculty of Mining and Geology, 17. listopadu 15/2172, 708 33 Ostrava-Poruba, Czech Republic

E-mail: beata.gibesova@vsb.cz

Abstract. The article describes characteristic features of post-mining and mining activities. These includes drop basins, settling ponds with flotation tailings, mine water discharge into surface water, coal waste dumps and their thermic activities. In conclusion, the financial demands oi individual remediation and reclamation activities are quantified. The costs of land reclamation measures with respect to the subsequent use of the landscape are also calculated for the Ostrava-Karvina Coal District.

1. Introduction

Currently, a number of organizations worldwide are engaged in the assessment of damage arising during mining and mineral processing. In this paper, we describe the assessment of damage arising in connection with black coal mining in the Ostrava-Karvina Coal District. In the Czech Republic, the Ostrava-Karvina Coal District is an important producer of coal, with an annual production of 12 million tones.

Both the active and closed mines in the Ostrava-Karvina Coal District have negative impacts on the environment. The paper describes how the mining activities have contributed both to underground and surface damage, and enumerates the manifestations of mining and some related problems, such as manifestations of mining on the surface and underground, problems related to settling ponds with flotation tailings, coal waste dumps, mine water discharge into surface water and mine gas emissions[6].

To be specific, a more detailed attention will be paid to coal waste dumps. After the termination of mining it is necessary to take certain measures to bring the territory designated to storing waste rock and coal slurries into an acceptable state. This leads to a disturbed ecosystem representing considerable costs due to landscaping.

The success of the land reclamation measures is also influenced by the subsequent use of the exploited rock in dumps and coal sludge from the settling ponds. At present, the waste rock from mining may be used in the building industry. Coal slurries in their different forms are used as fuel.
The paper compares the costs of land reclamation with respect to the subsequent use of the landscape. It also discusses the legislation related to the assessment of post-mining damage.

2. Manifestation of mining on the surface and underground

Free underground space is created by underground mining. Surrounding rock fills this space and the rock decrease continues through the immediate roof to the surface then the subsidence trough is formed on the surface according to the worked out seam depth, the depth of deposit occurrence and other influences. Its extent corresponds to the worked out underground space. The largest decrease in the surface will be reflected at the center of the basin and at the edges of the basin. Depression surface is slower with increasing distance from the center of the basin. This situation is on the figure 1.

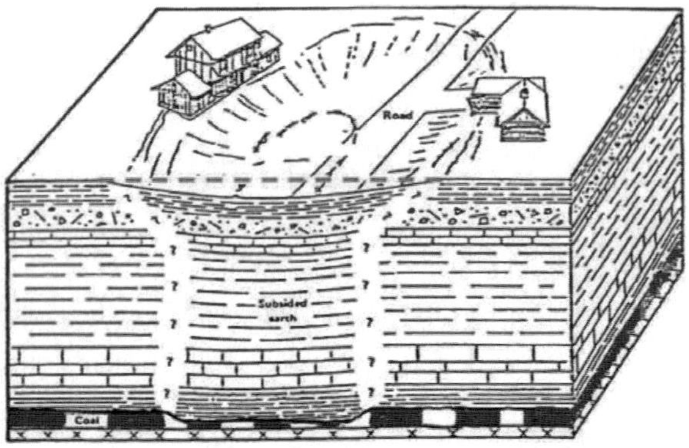

Figure 1. The scheme of the drop basin.

Figure 2. An example of the influence of undermining from the Karviná region [2].

The surface deformations in the Karvina and Orlova part of OKR can be demonstrated by figure 2, this is a so-called oblique church of St. Peter of Alcantara in Karvina-Doly. Church foundations has been deformed according to the shape of the subsidence trough and the whole building has been tilted. Mining technologies to minimize the impact of mining on the surface is chosen rarely in present time [3].

2.1. Settling ponds with flotation tailings

The coal slurries from coal preparation are a mixture of very fine particles of coal dispersed in water. Currently, coal slurries are not produced due to qualitative changes in mineral processing. Coal slurries in old settling ponds are currently used for other purposes or energy generation. The largest increase in slurry production occurred during the introduction of mining combines (when higher percentages of small dust particles were formed) in the late 1950s and 1960s. A dramatic decline in the production of coal slurries occurred in the 1990s in connections with the mining phase-out and the use of new technologies in slurry processing [1].

2.2. Mine water discharge into surface water

The water that has accumulated in underground mines all closed, was drained and pumped to drown low-lying active region after 1990 when the decline of mining. This condition should be achieved by maintaining the level of mine water Ostrava Mining Basin below about -388.5 m. This state is provided by special submersible pumps for central pumping of mine water on Jeremenko pit. The retention space of around 1.3 million cubic meters is formed in mines of Ostrava Mining Basin between height levels about -389.5 m and -371.5 m below the surface more.

The pumping and treating process of mining waters must be performed so that the discharge of surface water flow did not cause their contamination, and that represents a significant financial clause in billion CZK per year. Therefore, it would seem a good idea utilization of mine water temperature (about 26 °C) for such heat pumps.

2.3. Coal waste dumps

The catalogue from year 1989 presents that the former district of Ostrava has 116 waste dump disposal areas. At the Karvina district are located 143 of these and at Frýdek-Místek district there are 22 waste dump disposal areas. The condition of this areas is very various. Some waste dump disposal areas are still using for the waste dump storage and they hold their original character. The remaining areas are stabilized and partially reclaimed especially the waste dump disposal areas from the former mine activity. At some waste dump disposal areas got to the spontaneous combustion of a coal substance – the fire bank.

The thermic active Heřmanice waste dump ranks among the widest waste dump complex at the Ostrava Region, this is on the figure 3.

Figure 3. The thermic active Heřmanice waste dump [4].

Then next thermic active waste dump is The Ema Heap (figure 4), it is the unique technical monuments and it is the dominant feature of the Ostrava City.

Figure 4. The view from the active waste dump – The Ema Heap [5].

2.4. Thermic process redevelopment method

The relatively common method of the redevelopment of waste dump thermic activity is grouting of inert materials. The inert material is forces in or gravity flow supplies into drill holes at the thermic active part of the waste dump. This inert material is the mixture of a flue ash and various materials. This method causes the free space sealing into the waste dump whereby some combustion gas is transmitted. This is not the single redevelopment method.

The cost overview of the waste dump forming and subsequent reclamation for one billion tones is in the table 1. There is introduced what expensive costs should be.

Table 1. The rough costs for a rock fill or a waste dump storage.

1	2	3	4
The low dumping costs (CZK)	A land buying up (CZK)	The waste dump storage costs (CZK)	The reclamation costs (CZK)
600 mil.	82.5 mil	100 mil	5.1 mil

2.4.1. Costs for individual redevelopment and reclamation actions.

The caused damages can be divide to five groups. At the figure 5 there are costs for individual action which this environment damages are fixed. The expensive costs for the river landscaping entail to consider if the mining profit will be sufficient for cover the costs.

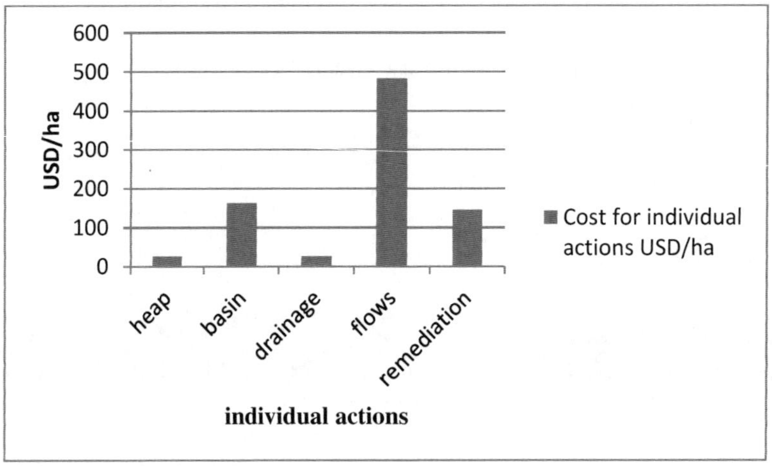

Figure 5. Costs for individual action USD/ha.

3. Conclusion

In the assessment of the damage caused by mining and treatment of minerals, the analysis in this article has been done, which is not self-serving. Ways to effectively remedy the damage was found and, if possible, exploit the economic value of the waste generated by the extraction and treatment of mineral resources.

The following analysis is based on this:

- Prior to mining, an economic and social analysis needs to be carried out in the area of interest. Conclusion is decision if the extraction or setting of the safety pillar is preferable.
- When closing mines and degassing boreholes, degassing and gas extraction for subsequent energy use must be used as much as possible.
- The using of the temperature of the mine water or the temperature of mine workings for the efficiency of the heat pump.

- In selected wasted dumps, there will be the technology of preparation plant or progressive sorting equipment to obtain valuable aggregate or coal using.
- For the decontamination and a recovery of sludge and settling tanks of coal sludge will make use the co-financing of municipalities, towns, regions, ministries (MIT, ME) and especially European funds.
- Obtain the European subsidies for projects that eliminate the impact of mining damages.

Acknowledgement
This paper was compiled within the Project of Specific University Research (SGS) no. SP2017/30 Evaluation of dustiness on surface mine near Bilina town.

References

[1] Trčková H 2015 *Analysis of damage on the environment by the extraction and management of mineral resources* (Ostrava: VŠB - TU Ostrava)
[2] Statutární město Karviná 2012 *Šikmý kostel sv. Petra z Alkantary – rarita v celém Česku*
[3] Hummel M, Hummelová I, Koudelková J and Černá K 2015 Mining of Protection Pillars without Subsidence *Inter.J. of Min. Sci. and Tech.* **5** 335-341
[4] Hylská M 2015 Rekultivace jižního plata odvalu Heřmanice
[5] HORNICKÝ KLUB 2009 *Aktivní odval Ema*
[6] Hudeček V, Urban P, Zapletal P and Košňovský V Elimination of safety risks at mined Coal faces in the Paskov mine Staříč plant – OKD, a.s. Czech Republic *Acta montanistica Slovaca* **18** 172-179

Modelling dynamics of atmosphere ventilation and industrial city's air pollution analysis: New approach

A V Glushkov[1], O Yu Khetselius[1], E V Agayar[1], V V Buyadzhi[1], A V Romanova[1] and V F Mansarliysky[1]

[1]Odessa State Environmental University, L'vovskaya str.15, Odessa, 65016, Ukraine

E-mail: florkota@gmail.com

Abstract. We present a new effective approach to analysis and modelling the natural air ventilation in an atmosphere of the industrial city, which is based on the Arakawa-Schubert and Glushkov models, modified to calculate the current involvement of the ensemble of clouds, and advanced mathematical methods of modelling an unsteady turbulence in the urban area. For the first time the methods of a plane complex field and spectral expansion algorithms are applied to calculate the air circulation for the cloud layer arrays, penetrating into the territory of the industrial city. We have also taken into account for the mechanisms of transformation of the cloud system advection over the territory of the urban area. The results of test computing the air ventilation characteristics are presented for the Odessa city. All above cited methods and models together with the standard monitoring and management systems can be considered as a basis for comprehensive "Green City" construction technology.

1. Introduction

The aim to ensure a steady pace of economic development, which has the known organizational forms of the globalization and industrialization, cannot be realized except through an increase in the already excessive burden on the environment with the further increasing atmospheric pollution. Although the development of global industrial, transport and communication systems rules out the possibility of localization of human influence on the environment, the large industrial cities are in the most difficult situation. As a rule, the indicated systems are the main source of environmental (atmospheric) pollution [1-3]. The composition of atmosphere in the industrial city is influenced by many factors such as an effect of pollution sources, climatic, meteorological factors, the city's topography, the energy and transport factors, processes of dissipation, relaxation, self-cleaning, regeneration etc [1]. At present time a number of different models is carried out to estimate the temporal and spatial structure of the atmosphere pollution [1-12]. As a rule, these models base on the simplified molecular diffusion or statistical regression equations algorithms (see [1-25] and Refs. therein). It is worth also to mention the known computational fluid dynamics models such as "Weather Research & Forecasting–Chem" and others [5-7]. Now it is clear that the most of models especially for urban zones should be advanced. The known torch-like model of molecular diffusion does not work if the atmosphere contains elements of convective instability and a vortices diffusion plays the dominant role.

The aim of our research is to develop a new "Green City" construction technology that includes as monitoring, management measures as a group of the physical, chemical, ecological blocks which allow to provide quantitative reliable modelling an atmospheric ventilation dynamics and precise analysis of the city air pollution. In this paper we briefly present a new generalized approach to an atmospheric turbulence and natural air ventilation for the industrial city. It is based on the Arakawa-

Schubert and Glushkov models [12,13], modified to calculate the current involvement of the ensemble of clouds. For the first time methods of a plane complex field theory and spectral expansion algorithms are applied to calculate the air circulation characteristics for the cloud layer arrays, penetrating into the territory of the city. Some numerical test data are listed for the Odessa city.

2. Models of atmospheric turbulence and air ventilation for the industrial city

Here we briefly consider generalized models of atmospheric turbulence and air ventilation for the industrial city, modified to calculate the current involvement of the ensemble of clouds. To calculate the involving streams (*the real involving mass effect is emerged due to a disbalance of vertical and down-running streams*), reaching the territory of city, the modified Arakawa-Schubert equations system is solved [1,12,13]. Sketch for air ventilation between a city and its periphery in a presence of the atmosphere clouds convection is presented in figure 1.

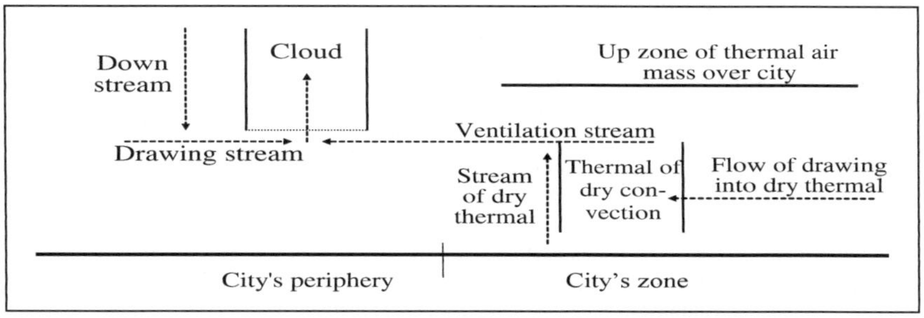

Figure 1. Sketch for the air ventilation between a city and its periphery.

It is important to note that if a square of the cloud base is substantially larger than a square of cross-section of the dry thermal top, a ventilation current captures several dry thermals or else compensates a mass-balance current by an involvement current from the periphery of the city. A signal of the destruction of thermal air mass over a city is an appearance of the convective cloudiness over the city's periphery. The corresponding convective clouds are usually formed by ridges on the secondary fronts or in the lines of convective instability arising in the synoptic processes. It is well known that the city area has a fairly complex geometric relief, so the application of any method in its pure form is possible only to a flat surface. Indeed, the air flows in the city area are far from isotropic picture. The horizontal turbulent vortices within the city are involving the circulation currents. More consistent description of city's turbulence should reflect a complex picture of the anisotropic vortex flow structures. *The turbulent vortices over the urban area must be in the resonant interaction contact with the turbulent vortices of the cloud-based arrays in order to obtain an effective air ventilation.* It is possible if the front convection currents coincide with currents of the thermal convection over the city zone in the phase setting. The antiphase state of the urban convective vertical currents with the cloud convection currents would mean mutual compensating the two mechanisms. These physical features of air ventilation predetermine the necessary modification of the well-known Arakawa-Schubert model. The model (equation (1)) includes the budget equations for mass, moist static energy, total water content plus the equations of motion (look details in [12]). In [12] it is also defined a cloud work function which is an integral measure of the buoyancy force in the clouds. If A is a work of the convective cloud then it consists of the convection work and work of down falling streams in the neighbourhood of a cloud:

$$dA/dt = d\,A/dt_{conv} + dA/dt_{downstr}, dA/dt_{downstr} = \int_0^{\lambda_{\max}} m_B(\lambda')K(\lambda,\lambda')d\lambda', \qquad (1)$$

Here λ is a velocity of involvement, $m_B(\lambda)$ is an air mass flux, $K(\lambda, \lambda')$ is the Arakawa-Schubert integral equation kernel, which determines the dynamical interaction between the neighbours clouds. The actual form of $K(\lambda, \lambda')$ and $F(\lambda)$ as well as their derivations are given in Appendix B [12]. In the case of air ventilation emergence, mass balance equation (2) in the convective thermals is as follows [13]:

$$m_B(\lambda) = F(\lambda) + \beta \int_0^{\lambda_{\max}} m_B(\lambda')K(\lambda,\lambda')d\lambda' \quad (2)$$

Here β is parameter which determines disbalance of cloud work due to the return of part of the cloud energy to the organization of a wind field in their vicinity, and balance regulating its contribution to the synoptic processes. The solution of the equation (2) with accounting for air stream superposition of synoptic processes can be determined by a resolvent method (equation (3)):

$$m_B(\lambda) = F(\lambda) + \beta \int_0^{\lambda_{\max}} F(s)\Gamma(\lambda,s;\beta)ds, \quad \Gamma(\lambda,s;\beta) = \sum_{i=1}^{\infty} \beta^{i-1} \cdot K_i(\lambda,s) \quad (3)$$

The key idea [13] is to determine the resolvent as an expansion to the Laurent series in a complex plane ζ. Its centre coincides with the centre of the city's "heating" island and the internal cycle with the city's periphery. The external cycle can be moved beyond limits of the urban recreation zone. The Laurent representation for resolvent is provided by the standard expansion (see equation (4)):

$$\Gamma = \sum_{n=-\infty}^{\infty} c_n(\zeta - a)^n, \quad c_n = \frac{1}{2\pi i} \oint_{|\zeta|=1} \frac{\Gamma(\zeta)d\zeta}{(\zeta - a)^{n+1}} = \frac{1}{2\pi i} \int_0^{2\pi} \Gamma(e^{it})e^{-int}dt, \quad (4)$$

where a is center of the series convergence ring.

The method for calculating a turbulence spectra inside the urban zone should be based on the standard tensor equations of turbulent tensions (look details in [1,11]). As usually, it is convenient to partition velocity $\boldsymbol{u}(v_x,v_y,w)$, pressure p, temperature θ etc into equilibrium and departures from equilibrium values (for example: $p=p_o+p'$etc). One could write the system of equations for the Reynolds tensions, moments of connection of the velocity pulsations with entropy ones and the corresponding closure equations. The important parameter of the turbulent processes is the kinetic energy of turbulent vortices $b^2 = \overline{u_k' u_k'}$, which can be found from the equation (5) (with physical explanations of any term):

$$\frac{\partial b}{\partial t} + \frac{\partial u_k b^2}{\partial x_k} + \frac{\partial}{\partial x_k}\left(\overline{u_k' u_i' u_j'} + 2\overline{u_k' p'}\right) = -2\overline{u_k' u_i'}\frac{\partial u_i}{\partial x_k} - 2\frac{g}{\theta_0}\overline{w'\theta'} \quad (5)$$

Advection | Turbulent diffusion | Effect of forces of the tension | Interaction: Reynolds tension-averaged motion | Accounting for swimming forces

Here g is the magnitude of the acceleration vector due to the planet's gravity, θ_0 is the equilibrium potential temperature, θ', p' are departures from equilibrium values.

The velocity components, say, v_x,v_y, of an air flux over the city area are determined in an approximation of "shallow water" [3,11]. In contrast to the standard difference methods of solution, here we use the spectral expansion algorithm [1,27]. The necessary solution, for example, for the v_x-iv_y component for the city's heat island has the form of expansion into series on the Bessel functions. From the other side, the velocity of an air flux over the city's periphery in a case of convective instability can be found by method of a plane complex field theory in a full analogy with the known Karman vortices chain model (equation (6)) [10,11]:

$$v_x - iv_y = \frac{df}{d\xi} = \frac{\Gamma}{2\pi i}\left[\frac{1}{\zeta-\zeta_0} + \sum_{k=1}^{\infty}\left(\frac{1}{\zeta-\zeta_0-kl} + \frac{1}{\zeta-\zeta_0+kl}\right)\right] + \frac{d}{d\zeta}\left[\sum_{k=1}^{n}\Gamma_k \ln(\zeta-b_k)\right] \quad (6)$$

Here Γ_k –circulation on the vortex elements, created by clouds, b_k – co-ordinates of these elements, Γ – circulation on the standard Karman chain vortices of, l – distance between standard vortices of the Karman chain, ζ - co-ordinate of the convective perturbations line (or front divider) centre, ζ_0 - kl – co-ordinate of beginning of the convective perturbation line, ζ_0+kl – co-ordinate of end of this line.

Equating the velocity components determined in the shallow water model and model (6), we find the spectral matching between the wave numbers that define the functional elements in the Fourier-Bessel series with the source element of a plane field theory. It is also worth to remind that any vector field u can be separated into rotational and divergent parts, i.e., $u = \nabla\psi + u_\chi$ (the Helmholtz's theorem). If the vector field is a horizontal wind, one can define a current function ψ, to express the rotational part, and a velocity potential χ, to express the divergent part. Namely these parameters are of a great interest in applied analysis of an air ventilation in the urban zone. Below we present the results of test computing the air ventilation parameters for a few synoptic situations in the Odessa city. All calculations are performed with using of the "Geomath" (blocks of numerical differentiation and integration, solution of the "shallow water" equations etc) and "Quantum Chaos" (blocks of expansion of the Reynolds tensions into series on the vector-tensor spherical functions, velocity components into the Bessel functions series etc) PC codes [1,13-20,27-34].

3. The numerical test results for air ventilation in the Odessa city and conclusions

The test computing is performed with using natural and model data on a cloudiness, convection intensities, the Odessa city topography parameters, including the parameters in (4),(6) etc (all data are taken from Refs. [1,8,13]). Basically, it is assumed that the clouds masses are coming to the city by lines of convective instability. The distance between the convective clouds is assumed to be 300-700 m. In figure 2a (left part) the Odessa city area (Google Map) is presented. In figures 2b and 2c we list the data on ventilation potential (b) (it is equivalent to a field of potential in the complex velocity potential function) and a current function (c). Figure 2b shows the results for the synoptic situation in Odessa, when the clouds run from the sea by two lines of convective disturbances and penetrate deeply into the Gulf of Odessa and city respectively. The clouds are marked as the black squares. The contours of a complex potential reflect time variation of the velocity field, namely 0.5 m/s per hour.

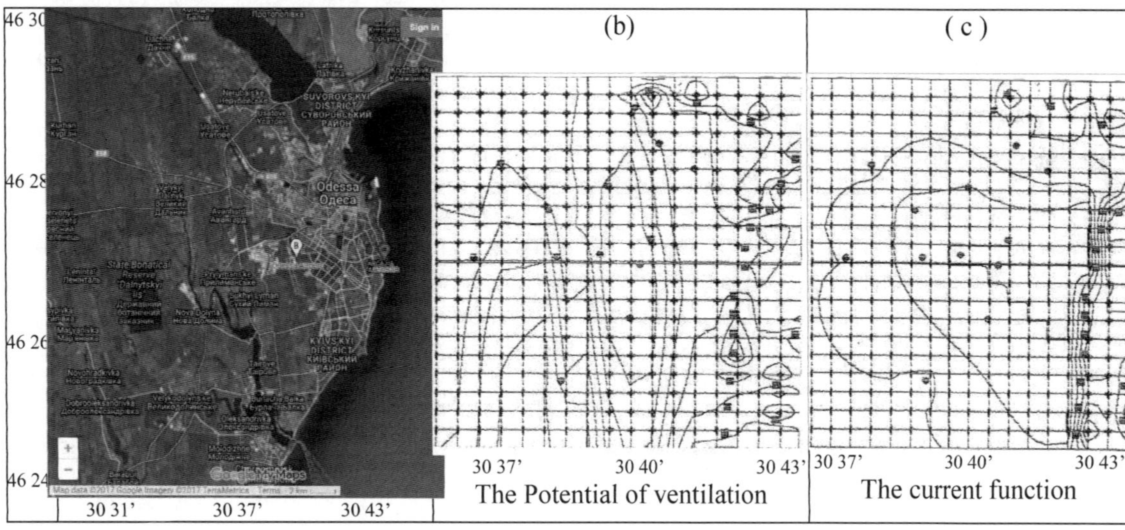

Figure 2. (a) Odessa city area map, (b) Potential of ventilation, (c) Current function for the situation b.

Density of current lines is adequate to ventilation flow speed, ~ 1 m/s to 0.5 cm of gradient in Fig 2. If $v_x>0$, the velocity increases in the direction of positive foci (and similarly on y). This means that the potential function draws flow in positive foci. Compression of the current function isolines means increasing a velocity. The direction of flow is obtained from the definition of the current function, i.e., $v_x>0$, if $\partial\varphi/\partial y$. The isolines are not signed, because modular values depend on many factors, notably intensity of convection, which determines the involvement currents power and density of the cloud arrays. The dry thermals (marked by black circles in figure 2) are located in the city area. They create their involvement currents and increase the intensity of the annular heat circulation. The picture in figure 2b of the air ventilation is typical for most of the city. However, in the current function field (figure 2c) penetration of air ventilation is expressed more weakly. To conclude, we present a new approach to modelling an air ventilation and turbulence in the urban area and list the test data on air ventilation for the typical synoptic situation in the Odessa area.

References

[1] Bunyakova Yu Ya, Glushkov A V 2011 *Analysis and estimation of anthropogenic loading influence on industrial city air basin* (Odessa: Ecology)
[2] Vardoulakis S, Dimitroulopoulou C, Thornes J et al 2015 *Env. Int.* **85** 299
[3] Vulfson A N and Borodin O O 2016 *Phys. Uspekhi.* **59** 109
[4] Aakriti Grover A, Singh R B 2015 *Environments.* **2** 125
[5] Argyropoulos C D and Markatos N C 2015 *Appl. Math. Modelling* **39** 693
[6] Argyropoulos C, Sideris G, Christolis M, Nivolianitou Z, Markatos N 2010 Atm. Env. **44** 803
[7] Kashia E, Mirzaeib F and Mirzaeib F 2015 *J. of Loss Prevention on the Proc. Industr.* **36** 125
[8] Lenschow D H, Wyngaard J C, Pennell WT 1980 *J. Atmos. Sci.* **37** 1313
[9] Dirks K N, Nanni A, Dirks V I 2006 *Atm. Science Lett.* **7** 15
[10] Ogura Y, Phillips N A 1962 *J. Atmos. Sci.* **32** 173
[11] Faber T E 1995 *Fluid dynamics for physicists* (Cambridge: Univ. Press)
[12] Arakava A, Shubert W H 1974 *J. of Atm. Sci.* **31** 674
[13] Glushkov A V 1999 *Met., Clim. and Hyd.* **38** 77
[14] Glushkov A V 2012 *Methods of a chaos theory* (Odessa: Astroprint)
[15] Buyadzhi V V, Glushkov A V, Mansarliysky V F, Ignatenko A V and Svinarenko A A 2015 *Sensor Electr. and Microsyst. Techn.* **12** 27
[16] Rusov V D, Glushkov A V, Vaschenko V N etal 2010 *J. Atm. and Solar-Terrestrial Phys.* **72** 498
[17] Khokhlov V, Glushkov A, Tsenenko I 2004 *Nonlin. Processes in Geophys.* **11** 285
[18] Glushkov A, Khokhlov V, Prepelitsa G and Tsenenko I 2004 *Optics of atm. and ocean* **14** 219
[19] Glushkov A V, Safranov T A, Khetselius O Yu, Ignatenko A V, Buyadzhi V V and Svinarenko A A 2016 *Env. Prob.* **1** 115
[20] Glushkov A V, Svinarenko A A, Buyadzhi V V, Zaichko P A, Ternovsky V B 2014 *Adv.in Neural Networks, Fuzzy Systems and Artificial Intelligence, Series: Rec. Adv. in Comp. Eng.* (Gdansk) **21** 143
[21] Glushkov A, Khetselius O, Brusentseva S, Zaichko P, Ternovsky V 2014 Ibid. **21** 69
[22] Glushkov A V, Kuzakon V M, Ternovsky V B and Buyadzhi V V 2013 *Dynamical Systems Theory*, Eds. J. Awrejcewicz, M Kazmierczak, P Olejnik and J Mrozowski (Łódz) **T1** 461
[23] Glushkov A V, Prepelitsa G P, Svinarenko A A and Zaichko P A 2013 *Dynamical Systems Theory*, Eds. J. Awrejcewicz, M Kazmierczak, P Olejnik and J Mrozowski (Łódz) **T1** 467
[24] Khetselius O Yu 2013 *Dynamical Systems Applications*, Eds. J. Awrejcewicz, M. Kazmierczak, P. Olejnik, J, Mrozowski (Łódz) **T2** 145
[25] Khetselius O Yu, Brusentseva S V and Tkach T B 2013 *Dynamical Systems Applications*, Eds. J. Awrejcewicz, M. Kazmierczak, P. Olejnik, J, Mrozowski (Lodz, Polland) **T2** 251
[26] Glushkov A V, Kuzakon V, Bunyakova Yu and Buyadzhi V 2015 *Proc. Int. Geom. Centre.* **8** 91
[27] Glushkov A V 2008 *Relativistic Quantum theory. Quantum Mechanics of Atomic Systems* (Odessa: Astroprint)
[28] Khetselius O Yu 2008 *Hyperfine structure of atomic spectra* (Odessa: Astroprint)
[29] Glushkov A V and Ivanov L N 1993 *J. Phys B: At Mol Opt Phys.* **26** L379
[30] Khetselius O Yu 2009 *Int. J. Quant. Chem.* **109** 3330
[31] Khetselius O Yu 2009 *Phys.Scr.* **T135** 014023
[32] Khetselius O Yu 2012 *J. Phys.: Conf. Ser.* **397** 012012
[33] Florko T A, Ambrosov S V, Svinarenko A A, Tkach T B 2012 *J.of Phys.:Conf. Ser.* **397** 012037
[34] Svinarenko A A 2014 *J. Phys.: Conf. Ser.* **548** 012039

IOP Conf. Series: Earth and Environmental Science **92** (2017) 012015 doi:10.1088/1755-1315/92/1/012015

Isotropic assessment of clay soil in terms of volume changes

M Gomboš[1], A Tall[1], B Kandra[1] and D Pavelková[1]

[1]Institute of Hydrology, Slovak Acagemy of Sciences SAV, Hollého 42, 071 01 Michalovce, Slovakia

E-mail: gombos@uh.savba.sk

Abstract. One of the basic characteristics of heavy soils is their capacity to change their volume in horizontal plane they are represented by formation of cracks and in vertical plane by vertical movement of soil surface. Non-dimensional geometric factor r_s is the ratio of participation in soil volumetric changes of both crack formation and vertical movements. In theory, r_s can acquire values in the interval: $r_s \in \langle 1, \infty \rangle$.

1. Introduction

Heavy soil is characterized by a high content of clay particles. These particles in soil moisture change cause shrinking and swelling processes. These processes occur in three dimensions. In horizontal plane they are represented by formation of cracks and in vertical plane by vertical movement of soil surface. With the formation of cracks, soil environment becomes a two-domain structure, cracks being one domain and soil matrix being the other one. Shrinkage cracks are distributed through the unsaturated zone of a soil profile.

The aim of this chapter is to evaluation of isotropy of soils from the view of volume changes. Evaluating criteria for isotropy was value of geometric factor - r_s.

2. Materials and methods

Soils volumetric change is a three-dimensional process. In nature, drying of soils is partly reflected in cracks formation and partly in the soil surface subsidence. Therefore soil changes are horizontal, caused by opening and closure of cracks, and vertical as soil surface movement. In laboratory, soil volumetric changes are visible as changes in geometric dimensions of an undisturbed specimen of soil [1]. Calculations are based on the following equations (1), (2), (3) (figure 1):

$$\Delta V = \Delta V_v + \Delta V_h \qquad [\mathrm{m^3}] \tag{1}$$

where

$$\Delta V_v = z_s^2 \times \Delta z \qquad [\mathrm{m^3}] \tag{2}$$

$$1 - \frac{\Delta V}{V} = \left[1 - \frac{\Delta \mathrm{z}}{\mathrm{z}}\right]^{\mathrm{r_s}} \tag{3}$$

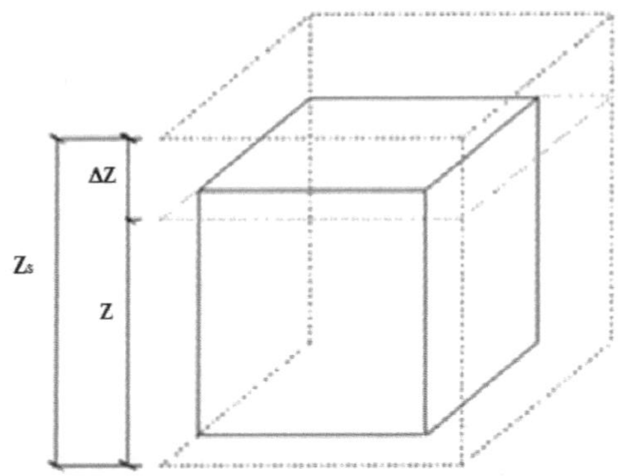

Figure 1. Volume change of isotropic soil sample in shape of cube during drying process, (dashed line express the sample in saturated state (volume Vs), continues line express the sample after shrinkage (volume V)).

From equations (1), (2), (3) and figure 1 apparent a mathematical expression of the relationship between soil volumetric change and vertical subsidence is as follows [1] in the equation (4) :

$$\Delta z = z_S - \left[\left(\frac{V}{V_S}\right)^{\frac{1}{r_S}}\right] \times z_S \quad [\mathrm{m}] \tag{4}$$

where ΔV - total volumetric change of a soil specimen, [m^3], ΔV_v - vertical volumetric change, [m^3], ΔV_h - horizontal volumetric change of a soil specimen, [m^3], V - soil volume after shrinkage, [m^3], V_s - volume of a saturated soil specimen, [m^3], Δz -change in height of a soil specimen, [m], z_s - height of a saturated soil specimen, [m], z - height of a soil specimen after shrinking [m], r_s - geometric factor [-].

Horizontal volumetric change can be expressed by the following combination of equations (1), (2), (4):

$$\Delta V_h = V_S \times \left[\left(\frac{V}{V_S}\right)^{\frac{1}{r_S}} - \frac{V}{V_S}\right] [\mathrm{m}^3] \tag{5}$$

Non-dimensional geometric factor r_s is the ratio of participation in soil volumetric changes of both crack formation and vertical movements. It can be influenced by external load and possibly by terrain settling process, which occurs when the clay sheet particles are orientated in one predominant direction. r_s can reach the following values:

$r_s = 1$	no cracking process, all soil volumetric changes are vertical;
$1 < r_s < 3$	vertical movement predominates over crack formation;
$r_s = 3$	isotropic shrinking;
$r_s > 3$	crack formation predominated over vertical movement;
$r_s \rightarrow \infty$	all soil volumetric changes are horizontal; i.e. only cracks are formed.

In nature, isotropic shrinking with $r_s = 3$ can occur in most soils. Provided that during drying the vertical change in height of a soil specimen and the volume of water saturated soil are measured, the equation (4) can be used to calculate the geometric factor r_s. This factor can be calculated from the equation mentioned previously in its analytical form:

$$r_S = \frac{log\left(\frac{V}{V_S}\right)}{\log\left[\frac{-(\Delta z)+z_S}{z_S}\right]} \quad [-] \tag{6}$$

Geometric factor r_s has been studied in selected areas of ESL table 1) on a set of 160 samples [2,3].

3. Results and discussion
Table 1 shows that max r_s was in the locality of Vysoká (13.0) and the smallest in locality Zatín and Pribeník (0.5). These are sites with high and almost the same content of clay. The highest average r_s was in the high (5.7) area and the smallest in the Somotor 2 site (2.3). From the results shown in table 1 indicates that the profile Senné 2000, on average, identified isotropic shrinkage (r_s = 3).

Table 1. Rating profiles the average value of r_s.

Locality	Coordinates	rs in layer				I. fr.	clay	profile rating
		max	min	max-min	avg	< 0,001 mm	< 0,002 mm	
Michalovce	N48° 44,255′ E21° 56,664′	5,3	2,9	2,3	3,9	23,54	26,90	dominated by the formation of cracks
Milhostov	N48° 40,185′ E21° 44,248′	5,9	2,8	3,1	3,9	27,28	29,15	dominated by the formation of cracks
Pribeník	N48° 23,688′ E21° 59,547′	3,4	2,9	0,5	3,1	29,70	31,90	dominated by the formation of cracks
Senné	N48° 39,802′ E22° 02,892′	4,4	2,6	1,8	3,0	51,03	54,84	isotropic shrinkage
Sírnik	N48° 30,538′ E21° 48,830′	4,8	2,7	2,0	3,3	30,55	35,18	dominated by the formation of cracks
Somotor 1	N48° 23,748′ E21° 48,471′	10,0	3,8	6,2	5,3	23,72	25,89	dominated by the formation of cracks
Somotor 2	N48° 23,173′ E21° 48,237′	3,0	1,7	1,3	2,3	18,30	20,15	dominated by vertical movement
Horeš	N48° 22,540′ E21° 53,907′	3,9	2,7	1,2	3,1	41,19	43,74	dominated by the formation of cracks
Kamenec	N48° 21,048′ E21° 48,877′	3,4	2,2	1,2	2,8	29,81	32,11	dominated by vertical movement
Vysoká	N48° 36,796′ E22° 06,898′	13,0	3,0	10,0	5,7	11,27	12,88	dominated by the formation of cracks
Zatín	N48° 28,725′ E21° 54,918′	3,4	2,9	0,5	3,4	28,45	31,55	dominated by the formation of cracks

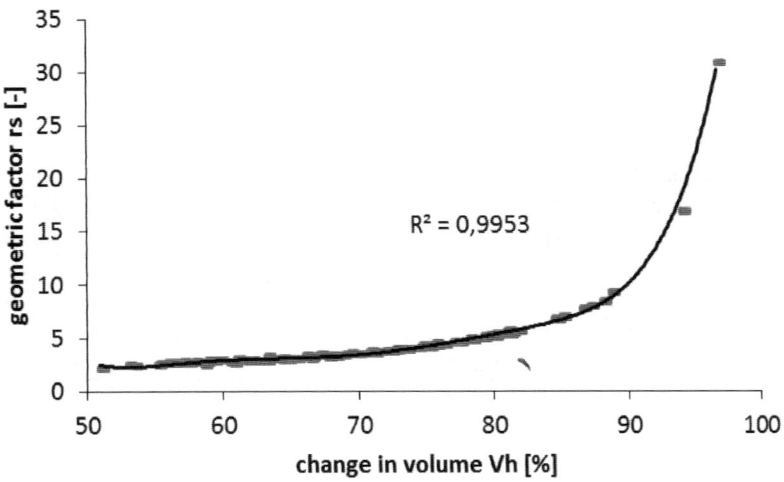

Figure 2. The dependence on the geometric factor r_s from volume changes V_h in the horizontal direction. Volume change V_h is expressed in % of the total measured volume changes.

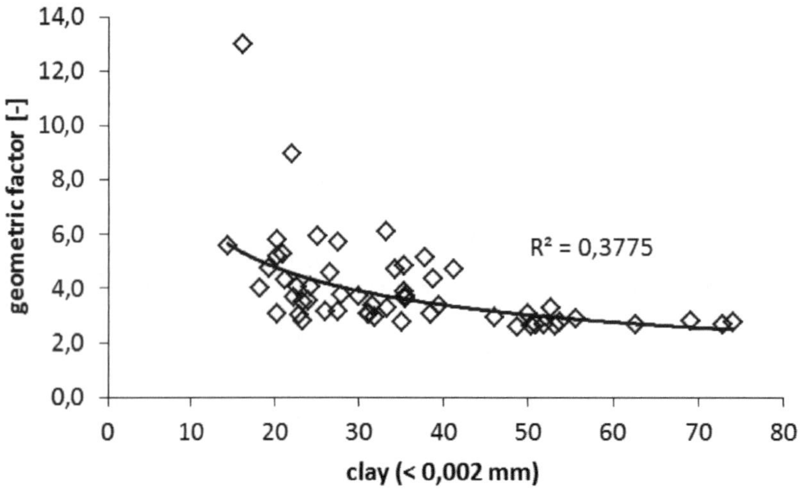

Figure 3. The dependence on the geometric factor r_s from volume changes V_h from the clay content. Volume change V_h is expressed in % of the total measured volume changes.

The dependence of the geometric factor r_s on volume changes V_h in the horizontal direction (figure 2) and on the clay content (figure 3) is shown. Volume change V_h is expressed in % of the total measured volume changes. It is clear from the figure 3 that with the increase in volume corresponding to the horizontal direction, the geometric factor increases. It is clear from the figure 2 that with the increase in volume corresponding to the horizontal direction, the geometric factor increases. It is also clear that with the growth of clay in the soil the variability of r_s is reduced. The geometric factor is approaching 3. This means that the increase in clay content in the soil causes isotropic shrinkage, respectively vertical movement begins to predispose isotropic shrinkage to crack formation.

Conclusions

In the current contribution, the isotopic properties of heavy soils were analyzed in VSN in selected soil profiles in terms of spatial realization of their volume changes. The basis of the analysis was the results of experimental measurements in the field and in the laboratory. The assessment of soil isotropy in relation to their volume changes and based on the geometric factor r_s showed that its values ranged from 0.5 to 13.0. The effect of the horizontal component of volume changes and clay content in the soil on the geometric factor r_s was documented. The results of this analysis will be used in the numerical simulation of the water regime and its prognosis under the conditions of heavy soils under VSN.

Acknowledgments

This study is the result of the project implementation: Centre of excellence for the integrated river basin management in the changing environmental conditions, ITMS code 26220120062; supported by the Research & Development Operational Programme funded by the ERDF (50%).

Authors are also grateful to Scientific Grant Agency of the Ministry of education, science, research and sport of the Slovak Republic: (project VEGA: 2/0062/16).

References

[1] Bronswijk J J B 1989 Prediction of actual cracking and subsidence in clay soils *Soil Science* **148** p 87-93
[2] Velísková Y 2010 Changes of water resources and soils as components of agro-ecosystem in Slovakia *Növénytermelés* **59** p 203-206
[3] Šoltész A and Baroková D 2011 Impact of landscape and water management in Slovak part of the Medzibodrožie region on groundwater level regime *J. Landsc. Manag.* **2** p 41-45

[4] Koltai G, Hegedős Mikéné, F Végh, K R, Orfánus T and Rajkai K 2010 Soil moisture monitoring as resilience indicator on the Danube lowlands *Növénytermelés* **59** p 291-294
[5] Orfánus T, Stojkovová D, Nagy V and Nemeth T 2016 Variability of soil water content controlled by evapotranspiration and groundwater-root zone interaction *Archives of agronomy and soil science* **62** p 1602-1613
[6] Tarnik A and Igaz D 2015 Quantification of soil water storage available to plants in the Nitra river basin *Acta scientiarum polonorum-formatio circumiectus* **14** p 209-216

IOP Conf. Series: Earth and Environmental Science **92** (2017) 012016 doi:10.1088/1755-1315/92/1/012016

Evaluation of eutrophication of Ostravice river depending on the chemical and physical parameters

A Hlavac[1], I Melcakova[1], J Novakova[1], H Svehlakova[1], L Slavikova[1], L Klimsa[1] and M Bartkova[1]

[1]VSB – Technical University Ostrava, 17. listopadu 15/2172, 708 33 Ostrava-Poruba, Czech republic

Abstract: The main objective of this study was to evaluate which selected environmental parameters in rivers affect the concentration of chlorophyll a and the distribution of macrozoobenthos. The data were collected on selected profiles of the Ostravice mountain river in the Moravian-Silesian Region. The examined chemical and physical parameters include dissolved oxygen (DO), flow rate, oxidation-reduction potential (ORP), conductivity, temperature, pH, total nitrogen and phosphorus concentration.

1. Introduction

One of the important elements of the assessment of the ecological status of surface water under the Water Framework Directive (WFD 2000/60 / EC) is the quantity and condition of phytoplankton [1]. The level of nutrients in rivers has increased dramatically over the last 50 years. The result is the excessive growth of algae (phytoplankton) called eutrophication. Eutrophication mainly affects the stagnating water with a low flow rate. Due to human activity, eutrophication is spreading to the lower parts of the rivers of surface running waters, near large cities with high atmospheric deposition and around agricultural areas due to the spillage of fertilizers [2]. Last but not least, the nutrient addition to the recipient by discharging water from small domestic treatment plants that cannot effectively remove pollution [3].

This research was inspired by a study carried out in 2002-2007 in the Czech Republic (at the lower reaches of the Berounka river), where annual changes in phytoplankton biomass values were observed. Significant influence of temperature and flow on the development of phytoplankton biomass was found. The significant effect of nutrient concentration has not been demonstrated [4]. One of the objectives of this research is to verify the hypothesis whether these effects are reflected to the same extent on the gravel-flowing parts of the river. Also, the abundance and composition of the macrozoobenthos community (ie, organisms> 1 mm) is considered to be a reliable indicator of our water quality [5]. These organisms are usually bound in flowing waters to the surface of stones (epibenthos), where they inhabit the top or bottom of the stones [6].

2. Matherial and methods

2.1. Locality selection

Three locations (sampling points) on the Ostravice river were selected. The sampling points were chosen so as not to be affected by point contamination (e.g. effluent from the sewage treatment plant), which led to a distortion of the concentration of phytoplankton concentration on the measured parameters. Other conditions were good site availability, good hydraulic conditions for discharge measurement and taking samples. It was also found that the sampling point was not under the bridge or in its immediate vicinity, possibly before or under the weir.

2.1.1. Locality 1 – Pržno

This site is near Pržno near Frýdlant nad Ostravicí. Sampling was done approximately 100 meters from the bridge across the river before the effluent from the wastewater treatment plant. In the river were seen boulders (12-25%), stones (25-50%), rough gravel (12-25%), gravel (6-12%), small gravel (6-12%), submerged vegetation (6-12%), emerald vegetation (12-25%), marginal vegetation (> 50%). The river width was 12.3 m and depth was about 0.5 m. The width and the depth of the river fluctuated depending on the rainfall. The flow in this section of the river was slightly turbulent. In this area, the banks were covered with trees and grasses. The cover the river of vegetation was less than 10%. On the shores were the stands of *Salix alba, Alnus glutinosa, Impatiens glandulifera* and grasses (eg *Phragmites spp.*)

2.1.2. Locality 2 – Staré město near Frýdek Místek

Sampling was collected out approximately 200 m above the mouth of river Morávka. In the river were boulders (6-12%), stones (25-50%), rough gravel (12-25%), gravel (6-12%), small gravel (12-25%), submerged vegetation (<6 %), emerald vegetation (6-12%), marginal vegetation (25-50%). At this site, the width of the river was 7.2 m, and the depth was approximately 1-1.5 m. The flow was slightly turbulent in this section. The cover the river of vegetation was less than 10%. On the banks were determined these species: *Corylus avellana, Swida spp., Populus spp., Acer spp., Dipsacus fullonum, Impatiens glandulifera, Solidago virgaurea.*

2.1.3. Locality 3 – Frýdek Místek – Lískovec

The locality is located in Frýdek Místek in local area Lískovec, about 200 m from the railway station Frýdek-Lískovec. In the river were boulders (12-25%), stones (> 50%), rough gravel (12-25%), gravel (6-12%), small gravel (12-25%), submerged vegetation (6-12%) and marginal vegetation (> 50%). The river width was about 12 m and the depth was about 1.5 m, both values vary depending on the rainfall. The flow in this section of the river was slightly turbulent. The riverbanks were overgrown with grasses, sometimes *Swida spp and Sambucus nigra*. Larger distance from the shore was a deciduous forest that did not reach the river. The cover the river of vegetation was less than 10%.

2.2. Methodology of measurement of physical and chemical parameters

Selected parameters (dissolved oxygen (DO), temperature, pH, conductivity, oxidation-reduction potential (ORP) and flow rate) were measured 1 x monthly from November 2016 at all locations, except for flow rate measurements that were measured only at Locality 1 and 2. At location 3 the flow was taken over from the Czech hydrometeorological institute. Dissolved oxygen and temperature values were measured by the Oxi 3310 SET 1 (incl. CellOx® 325), pH and ORP values were measured with the pH 3310 SET 2 (incl. SenTix® 41) and conductivity was measured by a pocket conductivity meter HQ 30d with electrode CDC401, from HACH company. Flow rate was determined using the hydrometric method. For this measurement was used hydro propeller STS 005 (range 5 cm – 5 $m \cdot s^{-1}$) from Greisinger company. Total phosphorus and total nitrogen were measured using the cuvette tests (LCK 138, range 1-16 mg/L TN and LCK 349, range 0.05-1.50 mg/L PO_4-P) and spectrophotometer DR 6000 from HACH company. The methodology of chlorophyll a determination is defined by the legislation ČSN ISO 10260 [7].

2.3. Method of sampling of macrozoobenthos

Macrozoobenthos was taken 1 x in November 2016. Four sites were selected for each of the selected localities - macrozoobenthos sampling microhabitats: (upper side of stones, bottom of stones, side of stones and sampling in surplus using Surber sampler). Macrozoobenthos was collected using the

Surber sampling method using the semi-quantitative method as a standard three-minute multihabitat sampling using kick sampling methods by ČSN 757703 [8].

To evaluate statistical dependencies, regression and correlation analysis and Principal Component Analysis (PCA) were selected. For data processing were used MS Excel and Canoco for Windows 4.0 [9].

3. Results and discussion

The data comes from research that has not yet been completed and is a subset of data. The data was obtained by measuring 1 x monthly at each locality from November 2016 to August 2017. At each locality, the measured data was subjected to a correlation analysis to determine the dependence between the individual parameters. The results are expressed using the determination coefficient R^2, see table 1.

Table 1. Expression of dependence on individual localities between the selected parameters using the determinative coefficient.

	R^2		
DEPENDENCE	Locality 1	Locality 2	Locality 3
Chlorophyll a and flow rate	0.012	0.112	0.002
Chlorophyll a and total N	0.003	0.301	0.001
Chlorophyll a and total P	0.573	0.385	0.420
Chlorophyll a and temperature	0.043	0.060	0.301
Chlorophyll a and DO	0.016	0.010	0.329

It can be seen from table 1 that there is a significant dependence between chlorophyll a and phosphorus only at locality 1 (in Pržno), where the correlation coefficient R^2 reached 0.573 (see figure 1).

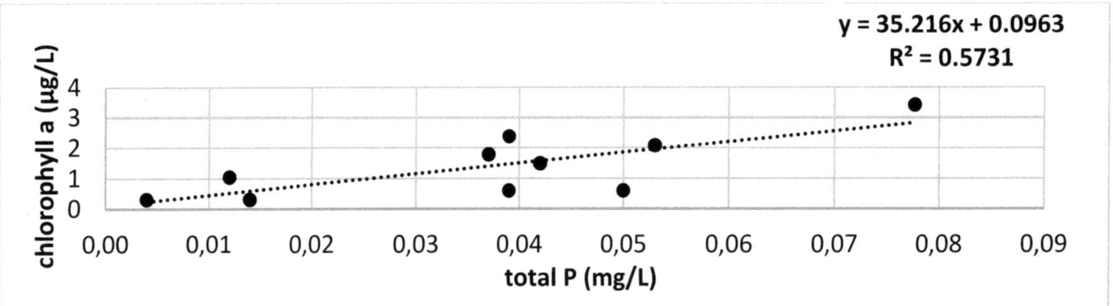

Figure 1. Dependence of chlorophyll a concentration and concentration of total phosphorus - Locality 1.

The data on the frequency of individuals and species of macrozoobenthos found in November 2016 were subjected to a PCA analysis in Canoco for Windows 4.0. The output is the ordination graph shown in figure 2. Thus, two axes are apparent from the intensity of the correlation relationship between the variables and the major axes. Both of these axes (axis 1 (x) and axis 2 (y)) show dependencies of the measured parameters on the occurrence of certain species at localities and microhabitats. Numbers 1-10 in the ordination chart represent a particular habitat on a particular site, for clarity the numbers are assigned to the individual microhabitats and localities in table 2.

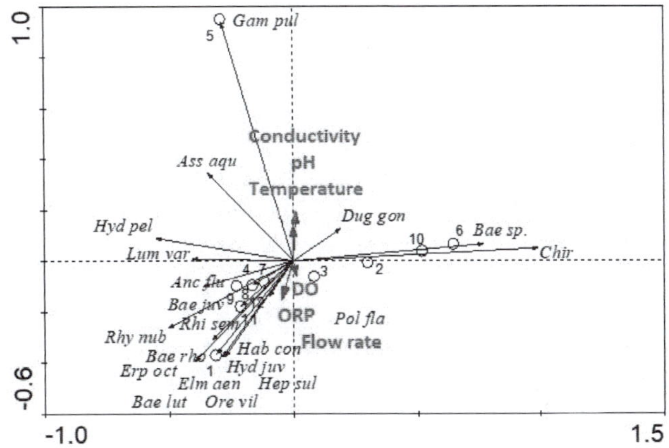

Figure 2. Graph of PCA analysis.

Table 2. The numbers of individual microhabitats in individual localities.

	Number of locality		
MICROHABITATS	Locality 1	Locality 2	Locality 3
The upper side of the stones	2	6	10
The underside of stones	3	7	11
The side of the stones	4	8	12
Surber sampler	1	5	9

From the PCA analysis graph, the variability of the environmental parameters, which corresponds to the time of measurement, can be observed. From the point of view of the biotopes, the similarity of the habitats forming the cluster in the 3rd quadrant and the biotopes corresponding to the side and bottom of the stones, eventually the gravel deposits (abstraction through the binary network) of the localities 1, 2, 3 with the most frequent grain fractions of the stones and the rough gravel. Locality 1 shows slightly different properties due to its strong torrential character, which corresponds to the composition of the composition of *Heptagenia sulphurea, Hydropsyche juvenile, Elmis aenea, Orectochilus villosus, Baetis lutheri, Rhyacophila nubila.* Other biotopes corresponding to the top and bottom sides of the stones (2, 3) at locality 1 differ (less oxygen on the underside, strong stream on the upper side), lower dependence on these biotopes occurred in *Polycentropus flavomaculatus*, whose larvae form nets and are capable with the stones to snap. Taxa of biotopes 6, 10 (upper sides of stones) at locations 2 and 3 show a negative correlation with ORP and flow, including species representation on the given biotopes. The occurrence of *Baetis sp.* and the family *Chironomidae,* slightly *Dugesia gonocephala. Baetis sp.* and *Chironomidae*, however, exhibit higher variability of occurrence within given samplings. Biotope 5 (Surber sampler) at locality 3 is significant for the presence of representatives of the highly variable species *Gammarus pulex.* The occurrence of this species is negatively correlated with the oxygen content. Negative correlations with oxygen are also found in *Hydropsyche pellucidula, Lumbriculus variegatus* (both benthic) and *Asellus aquaticus* (bottom of stones) on locality 3.

4. Conclusion

When evaluating the data in this work, a low dependence between chlorophyll a̲ and temperature was found by regression and correlation analysis at all localities. The relationship between chlorophyll a̲ and pH was not established. At site 2, there was a weak relationship between chlorophyll a̲ and total nitrogen. The localities showed a weak (locality 1 and 2) to no (location 3) dependence between chlorophyll a and flow rate. Hypothesis, which results from a study carried out in 2002-2007 on the river Berounka (Desortová, Punčochář, 2010) was not confirmed at the Ostravice mountain river. Our research shows that chlorophyll a concentration is dependent on temperature and total phosphorus concentration, the effect of flow rate has no significant effect on chlorophyll a concentrations. It should be noted that the data were obtained from November 2016 to August 2017 and the results presented here may change in the future, as research is still ongoing.

In terms of macrozoobenthos data, *Dugesia gonocephala, Baetis sp., Gammarus pulex, Asselus aquaticus, Hydropsyche pellucidula* and *Lumbriculus variegatus* and *Chironomidae* were found to be dependent on temperature, pH and conductivity. The narrow dependence of species on temperature, pH, flow and conductivity can be found in *Lumbriculus variegatus* and *Leuctra fusca*. On the other hand, dependence on these parameters has not been demonstrated for these species: *Radix peregra, Baetis vernus, Simulium sp., Ameletus inopinatus, Psychomyia pusilla, Habroleptoides confusa, Ephemera danica, Dicranota sp.* and *Chironomidae.* The following species depended on ORP and oxygen *Habroleptoides confusa, Hydropsyche juvenile, Heptagenia sulphurea, Elmis aenea, Orectochilus villosus, Baetis lutheri, Erpobdella octoculata and Rhyacophila nubila, Rhithrogena semicolorata, Baetis juv., Baetis sp., Ancylus fluviatilis, Polycentropus flavomaculatus, Heptagenia sulphurea, Atherix ibis* and *Isoperla oxylepis.* These species are closely related to pH value: *Baetis alpinus, Radix peregra, Dugesia gonocephala, Asellus aquaticus, Habroleptoides confusa, Erpobdella octoculata* and *Hydropsyche juv.* Temperatures, oxygen and flow are dependent species *Hydropsyche fulvipes, Paraleptophlebia sp.* and *Ephemera ignites.* Conductivity and ORP are dependent species of *Psychomyia pusilla, Ameletus inopinatus, Ancylus fluviatilis, Ephemera danica* and *Dicranota sp.* The species *Simulium sp., Baetis sp., Baetis scambus, Baetis juv., Atherix ibis, Isoperla oxylepis, Rhithrogena semicolorata, Baetis lutheri* and *Baetis rhodani* do not show any dependence on any of the measured parameters. Except for pH value, the taxa always depended on more parameters.

References

[1] WFD Water Framework Directive 2000 2000/60/EC 1–72

[2] Kočí V, Burkhard J, Maršálek B. 2000 Eutrofizace na přelomu tisíciletí *Eutrofizace* 3-13

[3] Klimša L. 2015 The influence of small domestic treatment plant with activated sludge on water quality of stream in Hradiste, Ostrava Diploma thesis VSB – Technical University of Ostrava

[4] Desortová B and Punčochář P 2011 Variability of phytoplankton biomass in a lowland river: Response to climate conditions *Limn. – Eco. and Man. of Inl. Wat. Berlin: Akademie-Verlag,* **41** 160-166

[5] Smith V H, Tilman G D, Nekola J C 1999 Eutrophication: impacts of excess nutrient inputs on freshwater, marine, and terrestrial ecosystems *Env. Pol.* **100** (1-3) 179-196

[6] Lambert A and Roux C 2014 Eutrophication: causes, economic implications and future challenges (New York: Nova Science Publishers, Inc.)

[7] ČSN ISO 10260: (757575) *Water quality. Measurement of biochemical indicators. Spectrophotometric determination of chlorophyll-a concentration*

[8] ČSN EN ISO 10870 (757703) *Water quality - Instruction for sampling methods and equipment for sampling freshwater macrozoobenthos*

[9] Slavíková L 2017 Seasonal preferences of microhabitats of selected orders of macrozoobenthos in Ostravice river, Ostrava Diploma thesis. VSB – Technical University of Ostrava

Estimating effectiveness of crop management for reduction of soil erosion and runoff

K Hlavcova[1], Z Studvova[1], S Kohnova[1] and J Szolgay[1]

[1]Dept. of Land and Water Resources Management, Faculty of Civil Engineering, Slovak University of Technology, Radlinského 11, 810 05 Bratislava, Slovakia

E-mail: kamila.hlavcova@stuba.sk

Abstract. The paper focuses on erosion processes in the Svacenický Creek catchment which is a small sub-catchment of the Myjava River basin. To simulate soil loss and sediment transport the USLE/SDR and WaTEM/SEDEM models were applied. The models were validated by comparing the simulated results with the actual bathymetry of a polder at the catchment outlet. Methods of crop management based on rotation and strip cropping were applied for the reduction of soil loss and sediment transport. The comparison shows that the greatest intensities of soil loss were achieved by the bare soil without vegetation and from the planting of maize for corn. The lowest values were achieved from the planting of winter wheat. At the end the effectiveness of row crops and strip cropping for decreasing design floods from the catchment was estimated.

1. Introduction

Soil erosion by water is one of the most widespread forms of soil degradation [1] which can lead not only to soil loss but also to other soil threats. It causes the greatest loss of soil in Europe compared to other soil degradation processes and it is listed in the Soil Thematic Strategy of the European Commission in 2006 as one of the eight threats to soil. Recent policy formulated by the European Commission have called for quantitative assessments of soil loss intensity at the European level [2]. The analysis and quantification of soil erosion processes and the assessment of their impact on soil loss on slope, catchment or regional scales are required by water managers and catchment stakeholders. Various erosion models have been developed to predict soil erosion intensities by water. These models differ in terms of their complexity, inputs, and spatial and temporal scales, so that different modelling approaches can lead to significantly different soil erosion rates [3].

Approaches to soil conservation on cultivated lands are based on agronomic measures, soil management or mechanical methods. Agronomic measures for soil conservation use the protective effect of plant covers to reduce soil erosion. The simplest way to combine different crops is to grow them consecutively in rotation or in strips. With strip-cropping, row crops and protection-effective crops are grown in alternating strips aligned on the contour. Erosion is limited to the row-crop strips and soil removed from these is trapped within and behind the next strip downslope [4].

This paper presents an engineering approach for the quantification of soil erosion and sediment transport on a catchment scale. Methods based on crop management are applied for the reduction of soil loss and sediment transport from the catchment and their effectiveness for decreasing soil erosion and runoff is estimated. The paper is structured in the following way: A brief introduction of the problems studied is presented in Chapter 1. In Chapter 2 the methodology used is described. We characterise the USLE, SDR and WaTEM/SEDEM models for estimating the soil erosion and sediment transport on a catchment scale. Chapter 3 describes the study area of the Svacenický Creek catchment and the Svacenický Creek polder. Chapter 4 contains the results of modelling the soil erosion and sediment transport by the USLE/SDR and WaTEM/SEDEM models under selected crops and crop management. Chapter 5 discusses the results and states our conclusions.

2. Methods

2.1. Erosion models

Two modelling approaches were applied for quantifying the soil erosion by water on a catchment scale. First, the mean annual soil loss and sediment transport from the catchment were estimated by the Universal Soil Loss Equation (USLE) and the Sediment Delivery Ratio (SDR) model. The second approach was presented by the WaTEM/SEDEM spatially- distributed soil erosion and sediment delivery model.

The Universal Soil Loss Equation (USLE) is an empirical model which can be used to estimate soil loss with an emphasis on sheet and rill erosion. In the modified version of USLE the USLE 2D methodology was applied to calculate the LS topographical factor. The LS factor is derived for closed eroded units based on a raster digital elevation model and increasing slope's length in the direction of the surface runoff is expressed by a unit contributing area, which is defined by several algorithms.

The USLE model does not take into account the sediment transport and deposition. Therefore, the sediment transport was estimated using the Sediment Delivery Ratio (SDR) model by [5]. The SDR calculates the percentage of total soil loss that is delivered to a catchment outlet by the equation (1):

$$SDR = 1{,}366.10^{-11}.A^{-0{,}0998}.S_r^{0{,}3629}.CN^{5{,}444} \tag{1}$$

where A is the catchment area [km^2], Sr is the relief ratio [$m.km^{-1}$], CN is the average SCS curve number of the catchment.

The second modelling approach is represented by the WaTEM/SEDEM spatially distributed soil erosion and sediment delivery model developed at the Physical and Regional Geography Research Group of KU Leuven [6]. The model consists of two submodels that calculate water and tillage erosion, including sediment delivery to rivers. The soil loss is estimated by the RUSLE equation [7] using a 2D approach to the topographical factor. The second submodel calculates the amount of sediments that is exported towards surface water. This is done by routing the sediments towards the surface water and taking into account the possible deposition of sediments. This deposition is controlled by the transport capacity (Tc), which is calculated for each pixel. The transport capacity is proportional to the volume of the potential gully erosion (see equation (2)):

$$Tc = ktc.Eprg = ktc.R.K.(LS - 4.12.Sg.0.8) \tag{2}$$

Where ktc is the transport capacity coefficient, $Eprg$ is potential gully erosion, Sg is the local slope (meter per meter), and R, K and LS are the factors from the RUSLE equation.

2.2. The SCS-CN methodology

The SCS-CN method is a rainfall-runoff model developed for United States by the U.S. Soil Protection now NRCS-Natural Resources Conservation Service in 1956. In this method a relationship between watershed characteristics and antecedent rainfall meets in the Curve number parameter. With this simple parameter rainfall depth is transformed to runoff depth. Tables and figures for estimating the CN parameter for soil cover complexes of USA are given e.g. in the publication of USDA [8]. The method of SCS - CN is based on the formula (equation (3)):

$$Q = \frac{(P-I_a)^2}{S+(P-I_a)} \tag{3}$$

where *Ia* is initial abstraction [mm], *S* is potential maximum retention [mm], *Q* is direct runoff depth [mm], *P* is rainfall depth [mm], *CN* is curve number [-]. The relation between *S* [mm] and *CN* [-] has the form (see equation (4)):

$$S = 25{,}4\left(\frac{1000}{\mathrm{CN}} - 10\right) \tag{4}$$

While *Ia* can be expressed as $\lambda.S$ (in the original form $\lambda = 0.2$, by estimation of design 100-year discharge in our study, the parameter $\lambda = 0$) the original formula can be in the form (see equation (5)):

$$Q = \frac{(P-\lambda S)^2}{S+(P-\lambda S)} \tag{5}$$

3. Pilot area

The catchment of the Svacenický Creek with its area of 8.61 km^2 is located in the Myjava River basin in Western Slovakia (figure 1). The creek has a length of 5.4 km and is a right-hand tributary of the Myjava River. The Svacenicky Creek polder is a part of the flood protection measures in Turá Lúka and is located near the outlet of the Svacenický Creek. The Svacenicky Creek polder, which was built in 2012, can reduce a 100-year maximum design flood of 16.00 $m^3.s^{-1}$ to 0.21 $m^3.s^{-1}$. The projected volume of the polder´s storage capacity is 215,808 m^3, and it is able to capture a flood wave of 207,330 m^3.

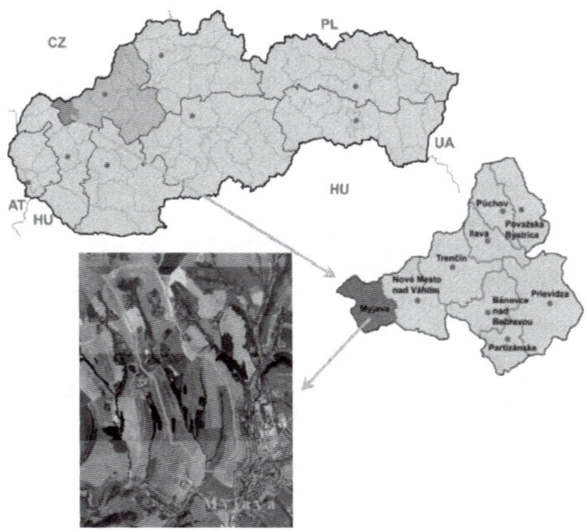

Figure 1. Location of the Svacenický Creek basin.

For modelling the erosion and sediment transport, the catchment's outlet at the Svacenický Creek polder was considered. The area of the catchment to the polder is 626.32 ha, with 10.76 ha of bare soil (1.72 % of the area), 513.32 ha of arable land (81.96 %), 46.45 ha of forest (7.42 %), 46.78 ha of urban areas (7.47 %) and 9.01 ha of roads (1.44 %).

4. Results

First, the modelling was provided for two different crops (winter wheat and maize for corn) grown on arable lands; and bare soil. The results of modelling the erosion and sediment transport from the Svacenický Creek catchment are presented in the form of mean annual soil loss in (t ha^{-1} $year^{-1}$) and the total mean annual sediment transport to the catchment outlet (polder) in (t $year^{-1}$).

Table 1 presents the results of the mean annual soil loss for the selected crops and the bare soil. The comparison shows that the greatest intensities of soil loss were achieved by the bare soil without vegetation and from the planting of maize for corn. The lowest values were achieved from the planting of winter wheat. Table 2 presents the total mean annual sediment transport to the catchment outlet.

In all tables "WS" is the LS algorithm developed by Wischmeier- Smith; "Nearing-WS" is the LS algorithm developed by Nearing and Wischmeier Smith; and "Nearing-Mc" is the LS algorithm developed by Nearing and Mc Cool.

Table 1. Comparison of the mean annual soil loss (t ha^{-1} $year^{-1}$) according to the USLE and WaTEM/SEDEM models and various LS algorithms.

Model	LS algorithm	Soil loss (t ha^{-1} $year^{-1}$)		
		Winter wheat	Maize for corn	Bare soil
USLE/SDR	WS	8.93	45.41	74.45
USLE/SDR	Mc Cool	9.74	49.49	81.13
USLE/SDR	Nearing-WS	7.86	39.97	65.52
USLE/SDR	Nearing-Mc	9.43	47.95	78.61
WaTEM/SEDEM	WS	11.15	54.77	84.95
WaTEM/SEDEM	Mc Cool	12.44	61.35	95.66
WaTEM/SEDEM	Nearing-WS	9.78	48.03	74.05
WaTEM/SEDEM	Nearing-Mc	12.06	59.44	92.64

The strip cropping was based on alternating row crops (winter wheat or maize for corn) with protection effective crops (grass) aligned on the contour. Table 2 presents the comparison between the mean annual sediment transport for the row crops (winter wheat or maize for corn) and strip cropping (winter wheat/grass and maize for corn/grass).

Table 2. The mean annual sediment transport (t $year^{-1}$) according to the USLE/SDR and various LS algorithms.

Model	LS algorithm	Sediment yield (t $year^{-1}$)			
		Winter wheat (row crops)	Maize for corn (row crops)	Winter wheat/grass (strips)	Maize/grass (strips)
USLE/SDR	WS	2684.36	18374.76	1149.33	6822.75
USLE/SDR	Mc Cool	2925.02	20022.09	1252.37	7434.43
USLE/SDR	Nearing-WS	2362.42	16171.00	1011.49	6004.47
USLE/SDR	Nearing-Mc	2834.38	19401.67	1213.57	7204.06

Table 3. Design floods for various types of crop management.

	CN	Hz [mm]	Runoff [mm]	V (m^3)	Q_{100} ($m^3.s^{-1}$)
Winter wheat (row crops)	72.94	32.51	8.34	52237.78	11.81
Maize for corn (row crops)	77.04	32.51	9.17	57426.38	12.98
Winter wheat / grass	70.89	32.51	7.73	48394.67	10.94
Maize for corn / grass	72.94	32.51	8.34	52237.78	11.81
Bare soil	82.78	32.51	9.77	61179.20	13.83

Evaluation of effectiveness of the strip cropping for the reduction of direct runoff form the catchment was provided by the SCS-CN methodology. The 100-year design rainfall (Hz) of the duration equal to the time of concentration was derived by simple scaling methodology from the daily design rainfall at the Myjava station. The flood wave volume (V) of a triangular shape was estimated and used for estimation of the maximum peak discharge. The values of estimated 100-year design maximum discharges (Q_{100}) for various types of the crop management are presented in table 3. We can observe that the peak discharge has lowered when combining crop with grass strips in comparison to row crops and the highest runoff was achieved for bare soil. In the comparison with bare soil, the row crops (winter wheat) decreased design discharges Q_{100} from the catchment by 14 %, maize for corn by 6 %, strip cropping (wheat/grass) by 20 % and strip cropping (maize for corn/grass) by 14 %.

5. Conclusions

The main objective of this study was to quantify both the soil loss from agriculturally arable lands and the transport of sediments to the dry water reservoir (polder) of the Svacenicky Creek. For estimating the mean annual soil loss and sediment transport in the Svacenický Creek catchment an empirical (USLE/SDR) and conceptual (WaTEM/SEDEM) approaches were applied. The results of various modelling approaches were consistent and comparable. The AUV EcoMapper was used to gather the data on the Svacenicky Creek polder in September 2015. From the filed measurement, we determined that during the last four years the clogging of the reservoir by sediments was over 10.474 m^3. The sediment transport was estimated to be 4994.4 $t.year^{-1}$. The modelled sediment transport from the Svacenický Creek catchment to the polder proved that the applied rotation of winter wheat and maize /grass on the filed gave comparable result of annual soil erosion to the field measurements. Next the erosion modelling was provided for various crop management on arable lands based on row crops and strip cropping. The comparison shows that the greatest intensities of soil loss were achieved by the bare soil without vegetation and from the planting of maize for corn. The lowest values were achieved from the planting of winter wheat. The strip cropping consisted of alternating row crops with protection effective crops aligned on the contour. The agronomic measures based on strip cropping reduced the sediment transport from the catchment area by 50 – 60 % compared to the row crops. We can observe that the peak discharge has lowered when combining crop with grass strips in comparison to row crops and the highest runoff was achieved for bare soil.

Acknowledgements
This work was supported by the EU-FP7 RECARE project under the 603498 project ID, by the Slovak Research and Development Agency under Contract No. APVV-15-0425 and VEGA Grant Agency No. 1/0710/15.

References

[1] Bosco C, de Rigo D, Dewitte O, Poesen J and Panagos P 2015 Modelling soil erosion at European scale: towards harmonization and reproducibility *Nat. Hazards Earth Syst. Sci.* **15** pp 225–245

[2] Panagos P, Borrelli P, Poesen J, Ballabio C, Lugato E, Meusburger K, Montanarella L, Alewell C 2015 The new assessment of soil loss by water erosion in Europe *Environmental Science & Policy* **54** pp 438–447

[3] Shen Z Y, Gong Y W, Li Y H, Hong Q, Xu L, Liu R M 2009 A comparison of WEPP and SWAT for modeling soil erosion of the Zhangjiachong watershed in the Three Gorges reservoir area *Agr. Water Manage* **96** pp 1435–1442

[4] Morgan R P C Soil erosion and conservation 2005 *National Soil Resources Institute, Cranfield University, Blackwell publishing* pp 304

[5] Williams J R 1977 Sediment delivery ratios determined with sediment and runoff models. In: *Erosion and solid matter transport in inland waters IAHS-AISH publication* **122** pp 168-179

[6] Van Rompaey A, Verstraeten G, Van Oost K, Govers G and Poesen J 2001 Modelling mean annual sediment yield using a distributed approach *Earth Surface Processes and Landforms* **26** (11) pp 1221-1236

[7] Renard K G, Foster G R, Yoder D C, McCool, D K 1994 RUSLE revisited: status, questions, answers, and the future *J. Soil Water Conserv.* **49** (3) pp 213– 220

[8] http://www.nrcs.usda.gov/Internet/FSE_DOCUMENTS/stelprdb1043088.

Practical assessment of the SWMM programme

P Hlustik[1]

[1]Institute of Municipal Water Management, Faculty of Civil Engineering, Brno University of Technology, AdMaS centre, Purkyňova 139, 612 00 Brno, Czech Republic

E-mail: hlustik.p@fce.vutbr.cz

Abstract: The article describes the advantages and disadvantages of the SWMM programme user environment when working with it. The Storm Water Management Model (SWMM) is a programme developed by the U.S. EPA (United States Environmental Protection Agency). The SWMM programme is used worldwide to plan, analyse and design rainfall-runoff, combined and separate sanitary sewage systems and other drainage systems in urban areas [1]. The programme is freely available to download from the U.S. EPA website [2].

1. Introduction

Introduction: It is a dynamic rainfall-runoff simulation model used for a single rainfall event or for long-term simulation of runoff quantity and quality of water in predominantly urbanised areas.
The model monitors the quantity and quality of runoff, flow-rate, water depth and quality in each sub-catchment area. The simulation time in individual channels and pipelines consists of multiple time steps. The results can be displayed as river basin maps, time series charts and tables and as statistical frequency analyses. The run-off process is based on sub-catchments with rainfalls with resulting surface runoff and its pollution. Furthermore, the runoff is conveyed into a system of *conduits, manholes* and *stormwater tanks, pumping stations, and treatment controls* can also be included. The SWMM monitors the quantity and quality of runoff within each area. It also monitors the flow-rate, depth, and water quality in the individual pipe and channel sections during the simulation which consists of several time steps [3].

2. Materilas and Methods

It is programmed in C language and is only compatible with Windows operating systems (XP, Vista, 7, 8.1). Drawings covering the area of interest are not required; however, they streamline and accelerate working with the programme [4].

The SWMM programme work is based on the hydro-technical situation derived from the ideal roof method in AutoCAD. The SWMM programme can read only data in standard image formats (JPG, JPEG, BMP, etc.). PDF or DWG files are not supported in the workspace. The drawing format size is not important, the scale is not restricted. The best drawing format as the basis for the hydro-technical situation in SWMM is BMP. The format size is not important. The drawing can be easily zoomed in and out in the programme. The A1 format is also a suitable input data. The BMP image extension has proven to be the best in terms of readability in the SWMM programme. A minor disadvantage of this image format is its size. The input data in A1 format and resolution of 300-DPI has a size of 200MB.

2.1. Data import into the SWMM programme Soil reaction

Data imports into the SWMM programme may be primarily addressed in the form of an image with BMP extension; data can also be imported by entering coordinates of individual objects (e.g. manholes, outlets, CSOs). Rainfall data can also be entered as a text file, which must be entered into: Rain Gage → File Name and then the path to the text file containing data for the required simulation is entered. If no data is available (coordinates, map data in image format), the situation can be plotted directly into the SWMM programme because the visual form of the situation is irrelevant. All required lengths, areas, etc. are entered into tables of the respective objects and sections.

2.2. Data export from the SWMM programme
Data exports from SWMM can be addresses in two ways offered by the programme. All outputs are addressed via File → Export. The first export option is Map Export, where the programme offers three types of output extensions, the most useful being the DXF extension file, which can be opened in AutoCAD. AutoCAD plots individual manholes and sections between manholes. This gives the sewerage system in the relevant area. Another most common form of output is the Status and Summary Report, where the programme generates results of the entire simulation in an RPT text file that can be opened in Notepad.

2.3. Measured data/input data
The created project is a set of pre-defined objects where corresponding data needs to be filled in. The quantity and type of input data required by the programme depends on outputs required by the investor. If this only concerns assessment of the capacity of the sewerage system, the following data is required: geodetic survey of the sewerage system (manhole cover altitude and manhole depth), hydrological data of the area (total rainfall) and hydraulic data of the sewerage sewer system and estimated roughness coefficient.

Units of the input and output (calculated) quantities can be entered in SI metric system or in US formats.

3. Results
The programme environment is well-organised and working in the programme is not demanding. Given its features, the environment is similar, for example, to EPANET 2.0 [3] and therefore designers working in similar programme should have no difficulties running this programme.

Each object is specified using data that is entered into the relevant object. Some of the data in the table must be filled in and are necessary to start the simulation. Other data serves to define the simulation results more accurately. In order for the required values to be filled in correctly, the table of the object also contains a note field, in which help is displayed, indicating data required to be entered in the respective box and unit.

3.1. Rain Gage
The design rainfall for a given location is determined using traditional methods. The input parameters are: Rain format can be entered as intensity [l/s ha], volume [mm]; Time interval [h:m]; The data source can be entered directly in the programme or a file and Time series can be attached.

The programme enables rain data input in formats provided by US meteorological institutions (NWS, NCDC) and in customised text files which must contain the following consecutive data: station ID, year, month, day, hours, minutes, rainfall values. This gives us the real rainfall. There are no rainfall databases or predefined diagrams here. However, the rainfall time behaviour can be entered with relevant intensities.

3.2. Subcatchment
This defines the area where surface runoff will be simulated. The runoff is discharged into manholes. The runoff coefficient in % must be entered for each catchment. The runoff coefficient value was determined on the basis of a sample unit hectare.

Input parameters: Rain Gage; Outlet; Area [ha]; Width [m]; % Imperv: runoff coefficient Ψ (if Ψ= 0,4, 40 is entered). To complete other data, a detailed land survey is needed. However, such data is not essential for preliminary simulation results.

3.3. Junction

Junctions in the SWMM are understood as manholes, the definition of which requires geodetic data only. Input parameters: Invert El. [m a.s.l.] and Max. Depth [m].

3.4. Conduit

The individual manholes are simply connected by conduits. Select the *Conduit* from the object menu and by clicking on the manhole sign enter the beginning of the section, and by clicking on the sign for the next manhole upstream draw the junction. Hydrological data of the network is required to define conduit sections.

Input parameters: Opening and Closing Manhole, Shape, Max. Depth [m], Length [m] and Roughness: Manning's roughness coefficient *n* must be entered.

When entering a section cross-section, use can be made of a predefined offer in the programme. There are 6 open (Rectangular, Trapezoidal, Triangular, Parabolic, Power, Irregular), and 17 commonly used closed cross-sections (Circular, Force Main, Filled Circular, Closed Rectangular, Horizontal Elliptical, Vertical Elliptical, Gothic, Catenary, Semi-Elliptical, Arch, Rectangular Triangular, Rectangular Round, Basket handle, Semi-Circular, Custom, Modified Basket handle, Egg, Horseshoe). If a cross-section other than those offered by the programme exists on the network, you can create you own cross-section by entering the coordinates X (height), Y (width). Dimensions (for circular), depth (for open) of the profiles are given in meters and are unlimited. Besides these basic parameters, which are necessary for the simulation, it is also possible to enter other ones so as to make the results more accurate. Extension of the entered parameters depends on the supporting data. These include: Inlet, Outlet offset: inlet height, outlet height over the respective manhole bottom; Initial flow; Energy loss: at the conduit inlet, conduit outlet, Seepagelossrate and check valve.

3.5. Outfall

The final junction on the network is determined by predefined boundary conditions. The outlet object is taken as the end junction of the entire network. During simulation, this junction is opened for the flow-rate from the sewer system and the conduit does not get filled which would be alerted by the programme as a calculation error. The outlet object is most often placed as the last manhole upstream the WWTP. Input parameters: Invert. [m a.s.l.] and Type.

3.6. Storage

Any size can be defined. Volume properties can be determined by the function or table (flood curve) where each water table in the reservoir determines the flooded area. Input parameters: bottom level [m. a.s.l.], maximum depth [m], evaporation [mm/d] and infiltration [mm/h].

3.7. Pump

It is entered as a section between junctions. The pump is defined by a curve. There are 4 types of curves used to select the desired parameters. The input parameters are the curve and the on/off level [m]. Description of curves for setting up the pump:

- Off-line pump in a wet sump where the flow is increased gradually.
- In-line pump, the flow-rate increases gradually with the water table at the inlet junction.
- In-line pump, the flow-rate changes continuously with the difference between the levels of inlet and outlet junctions (centrifugal pump).
- In-line pump, the flow-rate changes continuously with the water table in the inlet manhole.

4. Discussion

Once all parts of the project are entered, the simulation can start. Wastewater runoff and water quality transport can be simulated. The outputs of the program are fully with the legislation of the Czech Republic [5,6].

4.1. Wastewater outflow

After the simulation, the programme can show the course of filling and emptying in the selected profile. Another option is to suspend the ongoing simulation at a certain time step to assess the condition at a given moment (time step). We recommend setting the duration of the simulation as at least 4 times longer than the rainfall event itself as the programme also considers the time of delay and the later rainwater runoff from the catchment area. The programme also offers a spreadsheet form of the output where the flow-rates in l/s, time, flow velocity, conduit filling percentage in terms of the flow-rate and water level are evaluated. This output enables comparison of the maximum values in the above parameters only.

4.2. Programme outputs

After the simulation, the outputs can be viewed. The outputs can be either in the form of text or graphics in the form of a longitudinal section profile or depending on the flow-rate over time. Simulation of outputs can be performed between individual manholes or selected sections.

5. Conclusion

The applicability of SWMM programme is suited for small and medium-sized municipalities with a not-too-branched sewer system. When working with larger sewer systems, the situation may become unclear and the method of calculating the surface runoff using the ideal roof method is less suitable for larger municipalities.

The advantages and disadvantages of the SWMM model are summarised in the following points:

Advantages of the SWMM programme:

- Freely available programme.
- Programme stability.
- Variability of inputs and outputs.
- Miscellaneous tools to facilitate work with the programme.
- Simulation of water quality behaviour and pollution transport (Pollutant Editor).

Disadvantages of the SWMM programme:

- Inability to enter data in dwg format.
- Does not work with GIS compared to completive programmes.
- Cannot make calculations if the slope of the pipeline is in reverse slope.
- Absence of the "Return" button.

Acknowledgements

“The research report was drawn up in the framework of project No. LO1408 "AdMaS UP - Advanced Building Materials, Structures and Technologies" supported by the Ministry of Education, Youth and Sports in the framework of the purpose-oriented support of the “National Sustainability Programme I“.

References

[1] Rossman L 2010 *U.s.epa. Storm water management model user's manual*
[2] http://www2.epa.gov/water-research/storm-water-management-model-swmm
[3] https://www.epa.gov/water-research/epanet
[4] CZN EN 2012: 01 3463 Drawings of engineering structures - Drawings of sewerage

[5] Law no.274/2001 code: Law about water pipelines and sewerage system for public usage and about modification of some laws (law about water pipelines and sewerage systems.) (The Czech Republic: AION) 2010-2016
[6] Law no.183/2006 code: Law about city planning and building code (The Czech Republic: AION) 2010-2016

Implementation of recycled cellulosic fibres into cement based composites and testing their influence on resulting properties

V Hospodarova[1], N Stevulova[1], V Vaclavik[2] and T Dvorsky[2]

[1]Technical University of Kosice, Faculty of Civil Engineering, Institute of Environmental Engineering, Kosice, Slovakia
[2]VSB – Technical University of Ostrava, Faculty of Mining and Geology, Institute of Environmental Engineering, Ostrava-Poruba, Czech Republic

E-mail: viola.hospodarova@tuke.sk

Abstract. Nowadays, the application of raw materials from renewable sources such as wood, plants and waste paper to building materials preparing has gained a significant interest in this research area. The aim of this paper is to investigate the impact of the selected plasticizer on properties of fibres composites made of cellulosic fibres coming from recycled waste paper and cement. Investigations were performed on specimens with 0.5 wt. % of fibre addition without and with plasticizer. A comparative study did not show positive influence of plasticizer on the density and thermal conductivity of 28 days hardened composite. The specimens after 1, 3 and 7 days of hardening with plasticizer exhibited the highest impact on compressive strength in comparison to composite without plasticizer but 28 days hardened specimens reached the same value of strength characteristic (41 MPa).

1. Introduction

Nowadays, the development of natural fibre reinforced composite based products to substitute traditional reinforcing materials (steel, glass, and polymer fibres) is becoming a trend in many fields of their application such as building construction, automotive, engineering etc. The main advantages of using natural fibres are the sustainability, low cost, low density as well as competitive specific mechanical properties [1]. There has been observed increase in the use of natural fibres for making new environmental friendly and biodegradable composite materials [2].The increasing interest is in the utilization of recycled fibre materials as reinforcing element for building application. One of this potential material is waste cellulosic fibres that are obtained from wood, local annual plants, agricultural crops, industrial residues and waste paper, as well [1,3,4,5].

There are many factors that can affect the performance of natural fibre reinforced composites with inorganic or organic matrix. The properties of the natural fibre composites depend on the properties of matrix and fibres, adhesion of fibres to matrix at interface and their content and distribution in matrix. Another important factor influencing the composites properties is the processing parameters. Therefore, suitable processing techniques and parameters must be carefully selected in order to yield the optimum composite materials [2,6]. Cellulosic fibres are the most commonly used for composites reinforcement in organic or inorganic matrix to develop a natural fibre composite with appealing mechanical properties [2,7]. Composites based on natural cellulosic fibres offer environmental advantages such as reduced dependence on non-renewable energy and material sources, lower greenhouse gas emissions, enhanced energy recovery, and the end of life biodegradability of components [7].

The objective of this work is to investigate the impact of plasticizer implementation on properties of cement composites based on 0.5 wt. % addition of recycled cellulosic waste paper fibres. Density, thermal conductivity after 28 days of hardening and compressive strength after 1, 3, 7 and 28 hardening days were tested. As a reference sample, the cement composite without fibres and plasticizer was prepared.

2. Materials and methods

2.1. Materials

The cellulosic fibres used in this study coming from different sources of recycled waste papers. These unbleached recycled cellulosic fibres (G-3/00T) were provided by Slovak company Greencel Ltd (Hencovce, Slovakia). Properties of grey recycled waste paper fibres G-3/00T with 80 % of cellulose content are shown in table 1.

Portland cement type CEM I 42.5 N, from Cement Factory Ltd. (Povazska cementaren Ladce, Slovakia), was used for preparing cement composites.

Standard silica sand was supplied by company Filtracni pisky Ltd (Chlum, Czech Republic) conforming with European standard STN EN 196-1 [8].

Water for the cement mixtures preparation was used conforming to standard STN EN 1008 [9].

Superplasticizer Glenium Sky 665 acquired from BASF The Chemical Company Ltd (Praha, Czech Republic) was used to adjust the workability of fibre cement mixtures in accordance with European standard STN EN 934-2+A1 [10].

Table 1. Properties of waste paper fibre Greencel.

Characteristic of cellulosic fibre	Max. length [μm]	Bulk density [kg/m^3]	Dry matter [%]	Ash [%]	Thermal conductivity λ [W/m.K]	Volume heat capacity x10^{-6} [J/m^3.K]	Thermal diffusivity x10^6 [m^2/s]
G-3/00T	1200	30-50	93	20	0.0595	0.1709	0.3478

2.2. Preparation of cellulosic fibres reinforced cement plasters

The fibre cement mixtures were prepared in accordance with European standard STN EN 196-1 [8]. In the mix design, the weight proportion of cement:sand:water was kept 1:3:0.55. The first set of cement mixture (RPC) was prepared with 0.5 wt. % recycled waste paper fibre addition. The second set (RPCP) had the same content of components, however there was added superplasticizer in amount of 0.5 wt. % from cement weight. Reference specimen (RF) (mixture without fibre and superplasticizer) was also prepared.

The soaking of recycled fibres and manual mixing in approximately 50 wt. % of water was the first step in the preparation of mixture. Next, remaining water, superplasticizer (only for the second set), the required amount of sand and the cement were mixed by mechanical stirring in a mixer. The fresh fibre cement mixture was cast into standard steel prism in the dimension of 40x40x160 mm. Consequently, consolidation was made for the mixture with paper fibres by a jolting apparatus and for the mixture with paper fibres and superplasticizer by a vibrating table, respectively. The samples were demoulded after 24 hours and cured for 3, 7 and 28 days in water bath (laboratory conditions + 20 °C). Each set of composite samples consisted of 3 prismatic bodies.

2.3. Methods of testing
Physical property, densities of fibre cement composites were calculated after 28 days of hardening (STN EN 1015-10/A1) [11].

The coefficients of thermal conductivity of hardened fibre cement composites were measured on surface specimens by using the commercial device ISOMET 2114 with surface probe.

The compressive strength of fibre cement composites was also tested according to the European standard STN EN 1015-11/A1 [12] by using the compression test machine (FORM+TEST Seidner & Co. GmbH, Riedlingen, Germany) after 1, 3, 7 and 28 days of hardening.

3. Results and discussion

3.1. Density and thermal conductivity of fibre cement composites
Table 2 reports the values of important parameters such as density and thermal conductivity of fibre cement composites as well as reference composite. While density of reference sample after 28 days of hardening reached 2244 kg/m^3, specimens (RPC and RPCP) with the addition of recycled waste paper fibres in amount of 0.5 wt. % showed the decrease in density. The lowest value of density and its reduction around 6.8 % in comparison with reference sample had the RPC composite. Although the RPCP specimen containing superplasticizer and the same amount of cellulosic fibres than composite RPC, there was observed only slightly decrease (0.8 %) in comparison with the reference sample.

The lowest value of thermal conductivity (1.78 W/m.K) and its maximal reduction (34 %) of 28 days hardened fibre cement composites was found in the RPC specimen in comparison with reference sample (2.70 W/m.K). This trend was also observed in RPCP specimen with superplasticizer, however this reduction, compared with control specimen RF, was 11.4 %.

Generally, the fibre additions into matrix have positive impact on density and thermal conductivity of fibre cement composites, as well. Values of density and thermal conductivity decrease with increasing the fibre content in composites [13,14]. The incorporation of cellulosic fibres into matrix creates the voids and the heat-insulating properties are enhanced when density decreases [14].

Table 2. Density and thermal conductivity values of cement composites based on recycled waste paper fibre G-3/00T Greencel.

Sample	Density [kg/m^3]	Thermal conductivity [W/m.K]
RF	2244	2.70
RPCP	2227	2.39
RPC	2091	1.78

3.2. Compressive strength of fibre cement composites
The experimental results of compressive strength of cement composites based on recycled waste paper fibres after 1, 3, 7 and 28 days of hardening are shown in figure 1. From figure 1, it is obvious that there is observed increasing trend in the development of compressive strength values of cement composites containing recycled cellulosic fibres in the hardening time. The lower value of compressive strength after 28 days of hardening was recorded in the RPC fibre cement composite when compared with reference specimen RF. It was approximately decrease about 26 %. However, using superplasticizer into fibre cement matrix led to higher value of compressive strength of RPCP composite in comparison to reference sample. After 28 days of hardening, the composite with superplasticizer (RPCP) and reference sample reached slightly similar value of tested mechanical parameter.

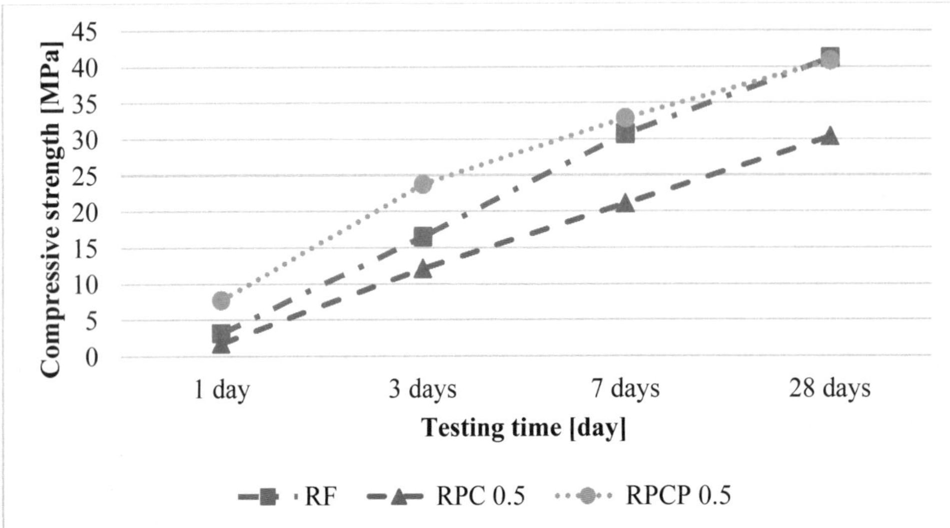

Figure 1. Compressive strength values of reference sample (RF) and cement composites based on recycled waste paper fibres tested after 1, 3, 7 and 28 days of hardening.

4. Conclusion

In this work, the influence of recycled waste paper fibres addition and superplasticizer on density, thermal conductivity and compressive strength of hardened cement composites based on recycled waste paper fibres was investigated. Two sets of fibre cement composites with the same amount of recycled waste paper fibres 0.5 % with and without superplasticizer were prepared.

Cellulosic fibres in the mixture have significant impact on resulting density and coefficient of thermal conductivity. In this case, incorporation of cellulosic fibres into the mixture shifts the results of tested parameters to better values of density (lighter in weight) as well as thermal conductivity of these composites. However, the composite with fibres and superplasticizer after 1,3 and 7 days of hardening showed higher compressive strength values than fibre composite without plasticizer, while the same strength parameter value of both of 28 days hardened composites was observed.

Acknowledgments

The authors are grateful to the Slovak VEGA Grant Agency for financial support of the project VEGA 1/0277/15 and Project Institute of Clean Technologies for Mining and Utilization of Raw Materials for Energy Use. Reg. No. LO1406.

References

[1] Shawia N B, Jabber M A and Mamouri A F 2014 Mechanical and physical properties of natural fiber cement board for building partitions *Phys. Sci. Res. Int.* **2** 49-53

[2] Ramesh M 2016 Kenaf (Hibiscus cannabinus L.) fibre based bio-materials: A review on processing and properties *Prog. Mater. Sci.* **78** 1-92

[3] Savastano H, Warden P G and Coutts R S P 2005 Microstructure and mechanical properties of waste fibre–cement composites*Cem.and Concr. Comp.* **27** 583-592

[4] Carvalho M A, Calil Júnior C, Savastano Junior H, Tubino R and Carvalho M T 2008 Microstructure and mechanical properties of gypsum composites reinforced with recycled cellulose pulp*Mat. Res.* **11** 391-397

[5] Hospodarova V, Stevulova N and Sicakova A 2015 Possibilities of using cellulose fibres in building materials *2015 IOP Conf. Ser.: Mater. Sci. Eng.* **96** 1-7

[6] Nishino T, Hirao K and Kotera M 2006 X-ray diffraction studies on stress transfer of kenaf

reinforced poly (L-lactic acid) composite *Compos. Part A Appl. Sci. Manuf.* **37** 2269-73
[7] Joshi S V, Drzal L T, Mohanty A K and Arora S 2004 Are natural fiber composites environmentally superior to glass fiber reinforced composites? *Compos. Part A Appl. Sci. Manuf.* **35** 371-376
[8] STN EN 196-1: 2016 *Methods of testing cement* Part 1: Determination of strength
[9] STN EN 1008: 2003 *Mixing water concrete* Specification for sampling, testing and assessing the suitability of water, including water recovered from processes in the concrete industry, as mixing water for concrete
[10] STN EN 934-2+A1: 2013 *Admixtures for concrete, mortar and grout* Part 2: Concrete admixtures Definitions, requirements, conformity, marking and labelling
[11] STN EN 1015-10/A1: 2007 *Methods of test for mortar for masonry* Part 10: Determination of dry bulk density of hardened mortar
[12] STN EN 1015-11/A1: 2007 *Methods of test for mortar for masonry* Part 11: Determination of flexural and compressive strength of hardened mortar
[13] Asasutjarit C, Hirunlabh J, Khedari J, Charoenvai S, Zeghmati B and Shin U C 2007 Development of coconut coir-based lightweight cement board *Constr.and Build. Mat.* **21** 277-288
[14] Onésippe C, Passe-Coutrin N, Toro F, Delvasto S, Bilba K and Arsène M A 2010 Sugar cane bagasse fibres reinforced cement composites: thermal considerations *Compos. Part A Appl. Sci. Manuf.* **41** 549-556

The Evaluation of Basal Respiration for Various Soil Textures in Ecologically Sensitive Area

P Huličová [1], D Kotorová [2], D Fazekašová [1] and J Hynšt [3]

[1]Department of Environmental Management, Faculty of Management, University of Prešov in Prešov,Konštantínova 16, 080 01 Prešov, Slovakia
[2]National Agriculture and Food Centre, Agroecology Research Institute, Špitálska 1273, 071 01 Michalovce, Slovakia
[3]Central Control and Testing in Agriculture, Hroznová 63/2, 603 00 Brno, Czech Republic

E-mail: petra.hulicova@unipo.sk

Abstract. The present contribution was focused on monitoring changes in the soil basal respiration in different textures of soil in the dry polder Beša. The research was conducted between 2012 and 2014 on soil type Fluvisol locations on three soil textures: clay – loam soil, clayey soil and clay soil in three soil depths. The basal respiration (BR) has been determine by soil CO_2 production measuring from incubated soil samples in serum bottles in laboratory condition. Release Co_2 has been analysed by gas chromatography. Content of clay particles were in the range 52.18 % to 81.31%, indicating the high difference between the minimum and maximum content. By using of multiple LSD-test we recorded statistically significant impact of clay on basal respiration. Results confirm the values of basal respiration with the depth of the soil profile decreased.

1. Introduction

Ecological stability of landscape area is dynamic ability of ecosystems permanently to Ecological stability of landscape is dynamics ability of ecosystems for permanent keeping and renovate of conditions of its existence by auto-regulation mechanisms [7]. Keeping of landscape ecological stability is basic and very important conditions for its permanent sustainable development and have long-time strategy importance for society development in this region. Ecologically stabile are ecosystems near to climax stadium of vegetation, and ecosystems with very high portion of biota (mainly forests, natural meadows and pastures, water ecosystems).

The problem of flooding occurs in the country long ago. Floods of agricultural land are dangerous and harmful, especially if lasting longer. In this causes no only water and airy regime of soil are changed, but also structure and properties of soil.

In recent decades on the ESL floods are increasingly. It is in connection with course of meteorological conditions in spring months, but also with very high ability of heavy soils to keep water. In this context knowledge of changes of physical and hydro-physical soil properties by effect of flooding of soil surface is important. It is known that flooding of territory causes negative changes of soil properties and create un-favourable share between water and air in soil profile. Dry polders on the East-Slovakian Lowland (ESL) were constructed for retention of flood water in Medzibodrožie.

IOP Conf. Series: Earth and Environmental Science **92** (2017) 012020 doi:10.1088/1755-1315/92/1/012020

Area of dry polder Beša is landscape compound from various ecosystems (forests, natural meadows and pastures, water ecosystems, agro-ecosystems). For these ecosystems is typical high degree of biodiversity. Ecological stability of this area is in long and no-regularly time intervals disturbed by human activities, that by mechanical saturation of polders in time of extremely floods in region of ESL. The aim of the present study was to investigate the variability of release CO_2 in selected soil types and different depths in the soil profile in an ecologically sensitive area, which is disrupted by irregular flooding.

2. Materials and methods

In years 2012 – 2014 investigation of spatial variability of soil profile was realised. The experimental site dry polder Beša was situated in central part of the East Slovak Lowland and has continental climate. The study area is the dominant soil type Fluvisols with gleyic subtypes. Smaller area covers Haplic Gleysols, Haplic Luvisols, Eutric Regosols. The soils in this area can be characterized as deep and heavy soils. Widespread types of soil are clayey soil and clay less represented are clay-loam and sandy-loam soil types.

For the determination of granulometric composition, we collected samples from each experimental profile. Soil samples were taken from three soil profiles to depth 0.0 – 0.6 m from each 0.2 m (with three replications). Sampling was realized in spring time (2012 – 2014). By pipette method, we determined grain size fractions as follows [8]1st fraction – clay, <0.001 mm; 2nd fraction – medium and fine silt, 0.001 to 0.01 mm; 3rd fraction – coarse silt, 0.01 to 0.05 mm; 4th fraction – medium and fine sand, 0.05 to 0.25 mm; 5th fraction – coarse sand, 0.25 to 2.00 mm.

The basal respiration (BR) has been determine by soil CO_2 production measuring from incubated soil samples in serum bottles for 24 h. 15 g of field moist soil samples have been weighed into each of three 120-ml serum bottles that were sealed with butyl rubber stoppers and incubated at 25 °C. After 3 and 24 h 0.5 ml sample of the internal atmosphere in each bottle has been analysed by gas chromatography (Agilent Technologies 7890A GC System equipped with a thermal conductivity detector).

Respiration has been calculated from the CO_2 increase during the 21 h incubation period (24–3 h). At the end of measurements, the total headspace volume for each replicate bottle has been determined by measuring the volume of water required to fill the bottle. The measured CO_2 amounts have been corrected for the gas solved in the liquid phase. The results are expressed per gram of dry soil and hour [4].

For total assessment of relations between the parameters and factors under, we used mathematical-statistical methods of package STATGRAPIHCS. Obtained data were tested by statistical methods using LSD-multiple test at significant level $\alpha = 0.05$.

3. Results and discussion

Content of clay particles in soil profile is important information about possibilities of use land in concrete area and also water storage. Polder Beša is no-regularly overflowing area and so on majority of lands the perennial grass stands are cultivated. From table 1 it becomes clear high variability of clay particles content (<0.01 mm), when for all depth the lowest content of clay particles (in average 56.29 %) was determined in profile 1st profile, higher content (in average 68.89 %) was in profile 2nd and the highest content (in average 77.05 %) was in profile 3rd.

The contents of clay particles varied in experimental years and individual depths of soil profile. Shown results (table 1.) confirmed high spatial – horizontal and vertical – heterogeneity of soils in this landscape area. By content of clay particles in experimental soil profiles, soil in 1st profile is clay-loamy soil, in 2nd profile it is clayey soil and 3rd soil profile is clay. These results are similar as published [5] and [2] for heavy soil on the East Slovak Lowland.

Table 1. The average content of I. category grains [%] for investigated area.

Depth [m]	Year	Profile 1st	2nd	3rd
0.0–0.2	2012	53.03	66.35	77.13
	2013	53.87	66.24	67.58
	2014	52.18	65.01	76.59
	$\bar{x}$ Y	*53.03*	*65.87*	*73.77*
0.2–0.4	2012	54.11	67.97	78.13
	2013	61.11	67.50	78.21
	2014	54.77	68.05	76.05
	$\bar{x}$ Y	*56.66*	*67.84*	*77.46*
0.4 – 0.6	2012	54.59	68.94	78.89
	2013	61.58	78.12	81.31
	2014	61.33	71.88	79.60
	$\bar{x}$ Y	*59.17*	*72.98*	*79.93*
$\bar{x}$ profile		56.29	68.89	77.05
soil texture		clay-loamy soil	clayey soil	clay

Soil respiration is considered to represent the overall microbial activity reflecting mineralisation of organic matter in soil. It is the most commonly used biological variable in soil studies. Accurate prediction of the amount of CO_2 respired by soil is of great importance in evaluating the carbon balance of ecosystems. The intensity of respiration depends on the amount and activity of soil microflora, the quantity and quality of the organic substrate, the chemical and physical properties of the soil [6].

The values of soil basal respiration were in the range 1.2–2.3 μg C-CO_2 g^{-1} h^{-1} on clay-loam soil, 0.8 – 4.2 μg C-CO_2 g^{-1} h^{-1} in clayey soil and 3.9 – 5.9 μg C-CO_2 g^{-1} h^{-1} on clay soil. The highest values of basal respiration we found on soil textures of clay and lowest values we found on soil textures of loamy clay (figure 1).

The each soil texture has diverse quantity of microorganisms and these have also various activity. These properties primary caused the variability of basal respiration. From this fact resulting different intensity of basal respiration for individual soil types.

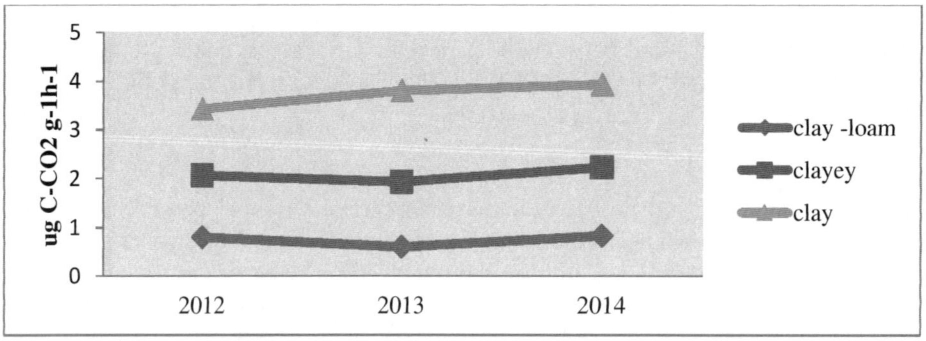

Figure 1. The average values of basal respiration of the investigated area.

We know about basal respiration that its values are increased at increasing temperature. The basal respiration is influenced also by other factors, as are the soil moisture, storage and quality of nutrients in the soil, soil texture, soil aeration and used methods of soil management [1]. The amount of CO_2-release also depends on the depth of soil profile. [3] found that with soil depth the amount of released CO_2 significantly decrease. Our results, obtained from investigated area, were similar and confirmed these tendency. The course of basal respiration for three experimental soil profiles is shown on figures 2–4 Data obtained from dry polder Beša were tested by means of statistical methods. Analysis of variance and LSD-multiple test were used.

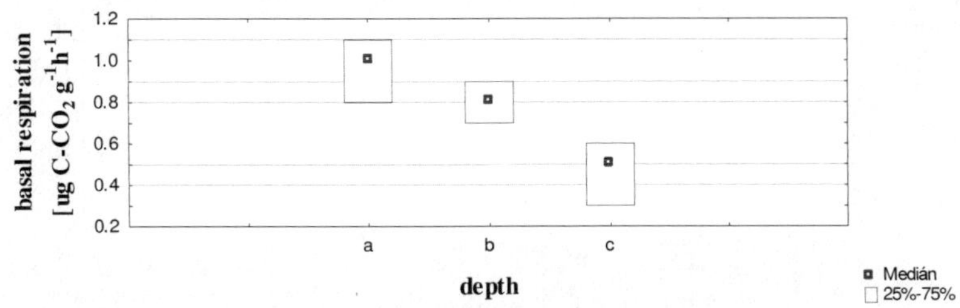

Figure 2. Basal respiration at different depth of clay- loamy soil (a: 0.0 – 0.2 m; b: 0.2 – 0.4 m; c: 0.4 – 0.6 m).

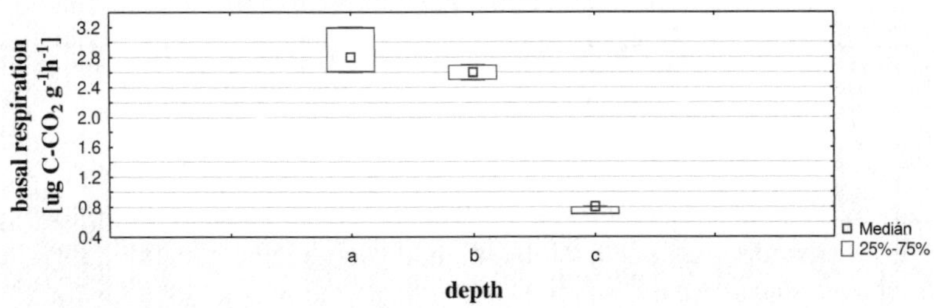

Figure 3. Basal respiration at different depth of clayey soil (a: 0.0 – 0.2 m; b: 0.2 – 0.4 m; c: 0.4 – 0.6 m).

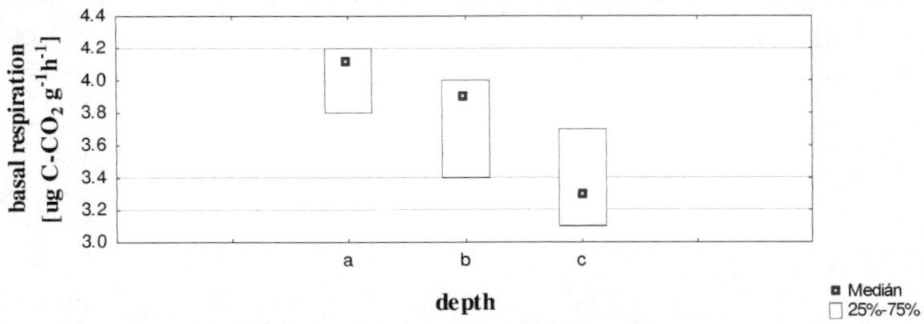

Figure 4. Basal respiration at different depth of clay soil (a: 0.0 – 0.2 m; b: 0.2 – 0.4 m; c: 0.4 – 0.6 m).

Table 2. Analysis of variance for basal respiration of the service area.

Source of variability	Degree of freedom	Calculated F-value	P significance
soil texture	2	815.05	++
Year	2	15.16	++
Depth	2	54.38	++
Duplicating	3	0.09	-
Residual	98		
Total	107		

++$P< 0.01$ +$P<0.05$

From table 2 it becomes clear statistically significant effect of soil textures, experimental year and depth on basal respiration. The strongest effect on basal respiration had soil texture, hence content of clay particles in soil profile. From testing of obtained data by using of multiple LSD-test resulted statistically significant impact of 3rd soil texture (clay) on amount of CO_2-release. Similarly, the basal respiration was statistically significant effected by year 2014 and depth of soil 0.0 – 0.2 m (table 3).

Table 3. Statistical evaluation of observed basal respiration values ($\alpha = 0.05$).

Source of variability	Factor	Basal respiration [μg C-CO_2 g^{-1} h^{-1}]
soil texture	1st	2.0333a
	2nd	2.16a
	3rd	4.8778b
Year	2012	2.8922a
	2013	2.9022a
	2014	3.2767b
Depth	0.0 – 0.2 m	3.4589c
	0.2 – 0.4 m	2.98b
	0.4 – 0.6 m	2.6322a

Where: 1st – clay-loamy soil, 2nd – clayey soil, 3rd – clay, ascenders (a, b, c) between factors suggestive of statistically significant references ($\alpha = 0.05$) – LSD test

4.Conclusion

Based on the research results for observed period 2012 – 2014, we came to the following conclusions: During the research period, soil respiration of texturally-different soils showed considerable variability. The highest values of basal respiration we noticed on clay profile and the lowest in the clay-loamy soil. The differences between experimental soil profiles were statistically significant.

The differences of soil respiration values between experimental years weren't statistically significant. The effect of year 2014, with higher sum of precipitation, was though remarkable.

The highest values of CO_2-release we noticed in the upper layers of the soil profile. In depth of profiles the basal respiration was decreased.

Acknowledgements

This work was supported by APVV-0163-11, APVV-15-0489, VEGA 1/0127/16 and KEGA 011PU4/2016

References

[1] Jenkins M and Adams M A 2010 Vegetation type determines heterotrophic respiration in subalpine Australian ecosystems *Global Change Biol.* **16** pp 209–219

[2] Kotorová D, Šoltysová B and Mati R 2010 *Properties of Fluvisols on the East Slovak Lowland at their different tillage* (Piešťany Agroecology Research Institute) p160

[3] Martin J G and Bolstad P V 2009 Variation of soil respiration at three spatial scales: Components within measurements, intra-site variation and patterns on the landscape *Soil Biol. Biochem.* **41** pp 530-543

[4] Šimek M, Virtanen S, Krištůfek V, Simojoki A and Yli-Halla M 2011 Evidence of rich microbial communities in the subsoil of boreal acid sulphate soil conductive to greenhouse gas emissions *Agric. Ecosyst. Environ.* **140** pp 113-122

[5] Šútor J, Gomboš M, Mati R, Tall A and Ivančo J 2007 *Water in the aeration zone of soils in the East Slovak Lowland.* (Bratislava ÚH SAV, Michalovce Agroecology Research Institute) p 279

[6] Vanhala P, Tamminen P and Fritze H 2005 Relationship between basal soil respiration rate, tree stand and soil characteristics in boreal forests *Environ. Monit. Assess.* **101** pp 85-92

[7] Wall D H 2005 Biodiversity *Encyclopedia of soils in the environment* (Oxford Elsevier Academic Press) pp 136-141

[8] Zaujec A J, Chlpík J, Nádašský J, Szombathová N and Tobiašová E 2009 *Pedology and base of geology.* (Slovak University of Agriculture in Nitra) p 400

Removal of bromates from water

D Barlokova[1], J Ilavsky[1], I Marko[1] and J Tkacova[2]

[1]Department of Sanitary & Environmental Engineering, Faculty of Civil Engineering STU, Radlinského 11, 810 05 Bratislava, Slovakia
[2]Water Research Institute, Nábr. arm. gen. L. Svobodu 5, 812 49 Bratislava, Slovakia

E-mail: jan.ilavsky@stuba.sk

Abstract. Bromates are substances that are usually not present in drinking water. They are obtained by ozone disinfection in the presence of bromine ions in water, as an impurity of sodium hypochlorite, respectively. Because of their specific properties, bromates are classified as vary dangers substances, that can cause serious illnesses in humans. There are several technological processes that have been used to the removal of bromates from water at present. In this article, the removal of the bromates from water by the adsorption using various sorbent materials (activated carbon, zeolite, Klinopur-Mn, Bayoxide E33, GEH, Read-As and Activated alumina) are presented. The effectiveness of selected sorbent materials in the removal of bromates from drinking water moves in the interval from 10 to 40%. Based on laboratory results, the zeolite can be used to reduce the concentration of bromates in water.

1. Introduction

Bromates are substances that are usually not present in natural waters; they can be present in sea and mineral waters (up to 100 µg/L). Presence of bromates in drinking water is usually relating to water disinfection. The water source Dvorníky containing 1 to 40 µg/L of natural bromates is the exception. Bromates (BrO_3^-) are formed as by-products of water disinfection by ozone in the presence of bromine ions (BrO^-). Under certain conditions, bromates may be present in concentrated solutions of hypochlorite used for water disinfection (e.g. at the NaCl electrolysis containing bromine ions or at the solution ageing). Bromates are also formed by oxidation of natural bromine in the presence of oxidisers [1,2].

Pursuant to SR Government Regulation No. 354/2006 Coll. as amended, the limit value of bromates in drinking water in Slovakia is 0.01 mg/L; this value complies with the recommendations of the World Health Organization (WHO), and Council Directive 98/83/EC [3].

The formation of bromates during ozonisation consists of a complex multistage process including bromine ions, molecular ozone, hydroxide radicals, and the water quality as such. In the case of bromide oxidation, reactive bromine is formed to react with water with the formation of hypobromous acid (HBrO) and a bromine ion (BrO^-). Formation of $HBrO/BrO^-$ compounds depends primarily on the pH value of water. The higher pH of water, the more bromine ions (BrO^-) formed, and vice versa, the lower pH of water, the more amount of hypobromous acid (HBrO) formed.

There are also other parameters except for the pH value that can affect the formation of bromates in drinking water such as concentration of bromine in water, the concentration of organic substances, ammonia, alkalinity, water temperature, ozone concentration and dose [4].

The level of concentration of bromates in water has been monitored in Canada and England in particular; in Slovakia, we do not have too many records of their presence in waters. The Canadians have been monitoring presence of bromine in water for many years. In 1993, there were 53 samples of drinking water taken there and the presence of bromates in the range from 0.2 to 0.5 mg/L was detected in two of them. The concentration of bromates in water has been increasing continuously and, in accordance with the results of 2013 published, the average level of concentration of 0.62 mg/L was detected in 22 of 23 samples [4].

The presence of bromates was also detected in European waters, where the concentration approximately 2 μg/L was detected in two of 36 samples of raw water taken including rivers and groundwater [4]. In the public water-mains in the Czech Republic, the concentration of bromates around 2.7 μg/L was detected [2].

As for bottled water, the concentration of bromates in the range from 0.2 to 37.3 μg/L was detected with the average concentration of 3.72 μg/L for water that is not subject to ozonisation, and 18.14 μg/L for water subject to ozonisation [2,4].

Bromates substance is restricted for its negative impacts on the human health. The Environmental Protection Agency (EPA) considers bromates potential human carcinogen, but its effects were assessed on the laboratory animals only [5].

Toxic symptoms occurred due to bromine salts consumed include vomiting, abdominal pain, diarrhoea, various degrees of central nervous system depressions, pulmonary oedema, and haemolytic anaemia. Majority of these effects can be cured. Failure of kidneys and loss of hearing remain as lasting consequences at the use of potassium bromate [5].

In accordance with SR Government Regulation No. 354/2006 Coll., bromates belong among the indicators of the drinking water quality as a part of the complete water analysis [3]. Methods used for defining dissolved bromates shall meet the requirements of Council Directive 98/83/EC. Currently, we have seven methods approved by the Environmental Protection Agency - US EPA used for defining concentration of bromates in drinking water. All the methods are based on the principle of ion chromatography using various detection methods (such as conductivity detector, inductively coupled plasma and mass spectrophotometer, post-column reagent and UV/VIS spectrophotometer, suppressed conductivity detection, and so on). They differ from one another by the detection definition limit to be from 0.02 to 1.3 μg/L [4].

When defining bromates, one can also use isotachophoresis or a combination of capillary isotachophoresis and capillary zone electrophoresis Coulometry and spectrophotometric methods are using reactions of bromates with bromides in a strongly acidic environment with the formation of bromine; methyl orange or fuchsine is used as a colour indicator. These are less selective (detection limit of 1 to 5 μg/L) [6].

An important factor for reducing concentration of a bromate is knowing basic characteristics of a water source such as water temperature, pH, content of bromine ions, water chemistry and its changes during seasons, ozone dose, and so on.

Currently, we have several technological methods for providing bromates removal from drinking water: coagulation, UV radiation, membrane processes, ion exchangers and adsorption [7, 8]. Reduced generation of bromates can also be achieved by controlling the pH value, by adding ammonia or H_2O_2 [4].

Adsorption processes are usually used for the water treatment, and they ensure removal of undesired substances causing water colouring, changed taste and odour properties, substances formed at the water disinfection, for removal of various inorganic and organic substances.

There are several types of sorption materials classified into groups upon the purpose of removing substances from water. In the water management, removal of undesired substances from drinking water is most frequently performed using sorption (filtration) materials such as granulated active coal, zeolite (clinoptilolite), Activated alumina, Bayoxide E33, Kemira CFH 12 and CFH 18, and materials with layers of MnO_2, TiO_2, CeO_2 on their surfaces.

2. Material and methods

Compared to the limit value (according to SR Government Regulation No. 354/2006 Coll.), some natural sources of drinking water can contain higher concentrations of bromates. In the territory of

Slovakia, there is a water source Dvorníky with the concentration of bromates oscillating from 0.001 to 0.04 mg/L. Due to this reason, the experimental part was focused on removing bromates from water with using selected sorption materials.

The chemical analysis of raw water is stated in table 1. Regarding low concentration of bromates as well as for increasing the accuracy of measurements, standard sodium bromate ($NaBrO_3$) has been added into raw water samples from Dvorníky with consequent increase of the concentration of bromates in model water up to the value of 91.3 μg/L.

Table 1. Chemical analysis of groundwater from the water source Dvorníky (2 March 2017).

Parameter	Unit	Value	Parameter	Unit	Value
pH		7.39	Ca^{2+}	mg/L	102
conductivity	mS/m	65	Mg^{2+}	mg/L	25.84
TDS (105°C)	mg/L	570	Ca + Mg	mmol/L	3.60
alkality	mmol/L	5.71	Cl^-	mg/L	24.35
acidity	mmol/L	0.55	NO_3^-	mg/L	29.21
COD-Mn	mg/L	0.68	SO_4^{2-}	mg/L	98.46
Fe total	mg/L	0.02	F^-	mg/L	0.26
Mn	mg/L	0.024	PO_4^{3-}	mg/L	0.63

The following sorption materials were used there:

- GEH (0.6 – 1.6 mm), (ferric hydroxide, GEH Wasserchemie, Germany);
- Bayoxide E33 (0.5 – 2.0 mm), (ferric oxide, Severn Trend, England) ;
- Klinopur Mn (0.6 – 1.6 mm), (zeolite with a layer of MnO_2 on the surface, Zeocem, a.s.;
- Read – As (0.3 – 1.0 mm), (ceric oxide, Nihon Kaisui Co Ltd, Japan);
- Activated alumina (1.5 – 2.5 mm), (aluminium oxide, China) ;
- GAU - Filtrasorb F 400 (0.42 – 1.68 mm), (Chemivron Carbon, Belgium) ;
- Natural zeolite (0.5 – 1.5 mm), Zeocem, a.s. Bystré.

The experiment was performed by means of a static test in the laboratory. 2.0 g of the sorbent was added into 200 ml of model water with pH of 7.38. The samples were mixed continuously, and 100 ml of water was taken off after one hour, four hours, and eight hours. In order to stabilize bromates, we have added EDA – ethylenediamine (0.1 ml into 100 ml of the sample).

The concentration of bromates was defined using the device EcaFlow 150 GLP made by ISTRAN s.r.o. Slovakia, which works on the basis of flow-through coulometry and in-electrode coulometric titrations. Using this method, bromates in the acidic environment are reduced to bromine.

The area of the signal – chronopotentiogram – recognized is directly proportional to the concentration of bromates in water. In order to remove distorting impacts of chlorites and other oxidation reagents, Fe^{2+} is added to the sample [9].

3. Results and discussion

The efficiency of removal of bromates from water using sorption materials was observed at pH of 7.38, input concentration of bromates of 91.3 μg/L, and the contact of water with the sorbent after one hour, four hours, and eight hours. Results of the static test were always obtained from two parallel sample measurements. At every set of analyses, the accuracy of measurements was checked by a reference material with the concentration of 100 μg/L and 75 μg/L of bromates (BrO_3^-). Results are listed in table 2.

Table 2. Values measured at the experiment (pH = 7.38).

No.	Material	0 hour	1 hour	4 hour	8 hour
		bromates [μg/L]			
1	Zeolite - SK (0.5 – 1.5 mm)	91.3	65.8	57.5	55.2
2	Filtrasorb F400 (0.42 - 1.68 mm)	91.3	84.6	77.1	72.0
3	Act. Alumina (1.5 - 2.5mm)	91.3	85.5	80.3	76.4
4	Read - As (0.3 - 1.0 mm)	91.3	80.2	70.5	63.6
5	Klinopur Mn (0.6 - 1.6 mm)	91.3	87.2	85.8	85.2
6	GEH (0.6 - 1.6 mm)	91.3	79.0	72.5	66.0
7	Bayoxide E33 (0.5 - 2.0 mm)	91.3	82.3	78.8	76.1

Based on the defined values, the efficiency of bromates removal - η [%] (equation (2)) and immediate adsorption capacity of selected sorption materials – a_t [mg.g^{-1}] (equation (1)) were calculated:

$$a_t = \frac{(c_0 - c_m)V}{m} \text{ [mg/g]} \quad (1)$$

$$\eta = \frac{(c_0 - c_m)100}{c_0} \text{ [\%]} \quad (2)$$

where a_t is the immediate adsorption capacity in μg.g^{-1}, η is the adsorption efficiency [%], c_0 is the concentration of bromates before the adsorption, 91.3 μg/L, c_m is the concentration of bromates after the adsorption at the time t [μg/L], V is the volume of water solution of 0.2 L, m is the weight of sorption material, 2.0 g.

Results of the adsorption efficiency (in %) and calculated values of the adsorption capacity for particular sorption materials at the concentration of bromates of 91.3 μg/L and pH of 7.38 are shown in figure 1.

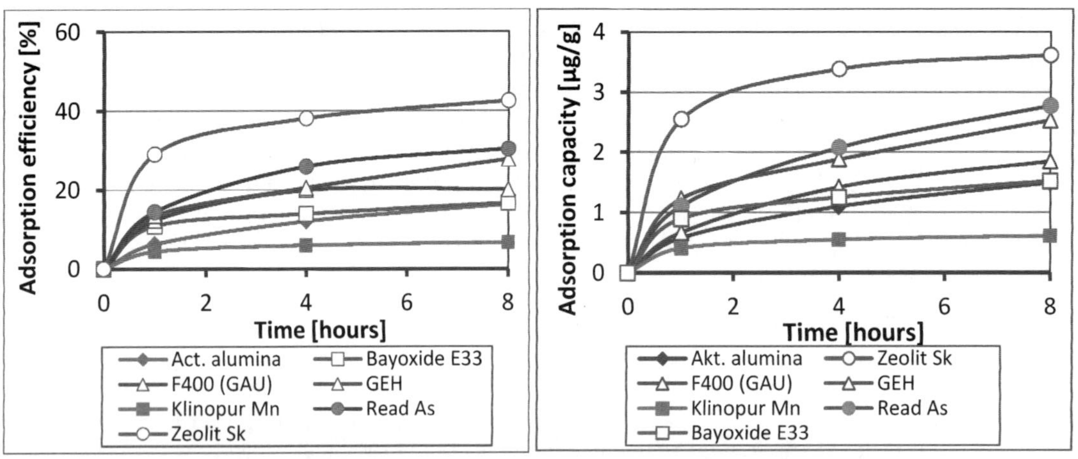

Figure 1. Course of the adsorption efficiency of bromates removal from water (left) and adsorption capacity of bromates removal from water (right).

Based on the given results we can state that, at pH of 7.38 and at the initial concentration of bromates before the adsorption (91.3 μg/L), none of the materials used was able to reduce the concentration of bromates under the limit value defined by GR No. 354/2006 Coll. (10.0 μg/L). Zeolite (clinoptilolite) was proven the most efficient sorption material able to remove 27.93% of bromates from the model water after the first hour and 39.54% of the same after eight hours. The adsorption capacity of zeolite was changing upon the duration of the contact and oscillated from 2.55 $\mu g.g^{-1}$ after the first hour up to 3.61 $\mu g.g^{-1}$ after eight hours. The efficiency of other sorption materials was relatively low, it did not exceed 30.0% after eight hours. In the literature, it is frequently recommended to use GAU for bromates removal, but not every active coal is suitable. In some works, the efficiency of 50-80% is stated, some of them refer to the GAU material specification, some of them even state that GAU is inefficient for bromates [8].

4. Conclusion

The experiment monitored the efficiency of selected sorption (filtration) materials at the removal of bromates from the model sample under laboratory conditions, and the resulting concentration of bromates was 91.3 μg/L. Based on the static tests, the adsorption efficiency of particular materials observed was compared, and their adsorption capacity was calculated.

Based on the results, the possibility of using zeolite from Nižný Hrabovec at the removal of bromates from water comes into account, and this can be caused by sorption, but also by the ion exchange capacity of zeolite. In the case of granulated active coal Filtrasorb F400, the efficiency of bromates removal after the contact with water lasting for eight hours was only 20.2% of the total amount of bromates in water. From among the rest of the materials observed, a good sorption efficiency is shown by GEH and Read-As (30% after eight hours of the contact with water).

Acknowledgements

The experimental measurements were performed with the financial support of the APVV-15-0379 and VEGA 01/0400/15 projects.

References

[1] Marko I 2017 *Removal of Bromates from Water* Thesis (Bratislava: Faculty of Civil Engineering Slovak University of Technology)
[2] Pitter P 2009 *Hydrochemistry* (Prague: UCT Prague)
[3] SR Government Regulation No. 354/2006 Coll. laying down requirements for water intended for human consumption and quality control of water intended for human consumption
[4] Federal-Provincial-Territorial Committee on Drinking Water Health Canada 2017 Bromate in drinking water https://www.canada.ca/en/health-canada/programs/bromate-drinking-water/bromate-drinking-water.html#purpose
[5] WHO 2009 *Bromate in Drinking-water background document for development of WHO guidelines for drinkingwater quality* World Health Organization, Geneva, Switzerland WHO/SDE/WSH/09.01/6
[6] Moricová K, Súkeníková M and Pajtášová M 2014 Methods of Defining Bromates in Disinfected Water *Indust. Toxicol. Coll.* 81-84
[7] Amy G L and Siddiqui M S 1999 Strategies to control bromate and bromide *AWWA Research Foundation and American Water Works Association* (Denver Colorado)
[8] Chen R, Yang Q, Zhong Y, Li X, Liu Y, Li X M, Du W X and Zeng G M 2014 Sorption of trace levels of bromate by macroporous strong base anion exchange resin: Influencing factors, equilibrium isotherms and thermodynamic studies *Desalination* **344** 306-312
[9] Tkáčová J, Horváth M, Kassai A, Kútniková D and Božíková J 2011 Defining Bromates by Flow-Through Coulometry in Waters Intended for Human Consumption *Hydroanalytics* (Hradec Králové)

Water disinfection agents and disinfection by-products

J Ilavský[1], D Barloková[1], O. Kapusta[1] and M Kunštek[1]

[1]Department of Sanitary & Environmental Engineering, Faculty of Civil Engineering STU, Radlinského 11, 810 05 Bratislava, Slovakia

Email: jan.ilavsky@stuba.sk

Abstract. The aim of this work is to describe factors of water quality change in the distribution network and legislative requirements in Slovakia for disinfectants and disinfection byproducts (DBPs). In the experimental part, the time dependence of the application of the chlorine dioxide and sodium hypochlorite on the formation of some by-products of disinfection for drinking water from WTP Hriňová is studied. We monitored trihalomethanes, free chlorine, chlorine dioxide and chlorites.

1. Introduction

Disinfection of drinking water, and also ensure of safety water quality is a major challenge 21 century. This challenge is not only about drinking water, but also about utility, industrial, cooling water and water in swimming pools. The aim is always such a disinfection that is acceptable in terms of environmental impact, the formation of disinfection by-products, but at the same time sufficiently effective and affordable.

Raw water, whether surface or underground, used to treatment of drinking water, contains microbiological and biological pollutants. Therefore, it is absolutely necessary to carry out its disinfection. Disinfection uses a wide range of disinfection techniques, each of which has its advantages and disadvantages.

Emphasis should be put on the best quality of the input water which is in equivalence in terms of physio-chemical properties of the water and has the lowest potential to enable the microorganisms to get overpopulated during the water transport in the distribution system. Therefore the selection of water source represents the first step in meeting this requirement.

Disinfection is conditional upon the doses of disinfection agent (oxidizing agent) not only but also by a well performed and efficient water treatment. To remove the soluble organic substances (so called a natural organic matter), the turbidity respectively means to avoid the bacteria to get increased in number in the distribution system during water transport and formation of the disinfection by-products (DBPs). Natural organic pollution is considered as the main precursor of the formation of disinfection by-products. Possible points of oxidizing (disinfection) agent input in the water supply system are schematically presented in the figure 1.

Most frequently used disinfection agents (oxidant) for ensuring the sanitary safety and used in water treatment are:

- Chlorine gas (Cl_2), or sodium hypochlorite (NaClO);
- Chloramines (NH_2Cl, RNHCl);
- Chlorine dioxide (ClO_2);
- UV radiation;
- Ozone O_3;

and various combinations of them. Combination of oxidants and their impact on formation of the disinfection by-products is a matter of many research projects [1]. In general, when the UV, ClO_2 or O_3 is applied in pre-oxidation as a oxidizing agent, a need for additional disinfection by a chlorine or chloramines resp. appears to ensure the disinfection in the distribution system. Some systems that use the chloramines to reach the disinfection require sufficient amount of free chlorine during the treatment to form chloramines in the distribution system after the NH_4^+ is added into the system.

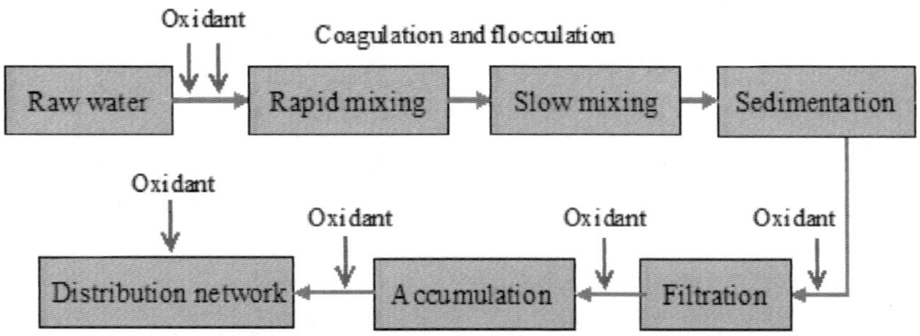

Figure 1. Drinking water treatment scheme with a possible point of addition of oxidizing agent.

Majority of public water supply systems use chlorine compounds for drinking water disinfection. Addition of chlorine can react directly or after the hydrolysis by chlorine acid (HOCl) with organic matter that is common to be presented in the system (precursors such as humine acids, fulvo acids, microorganisms and algae, COOH, phenols, nitrogen organic compounds, etc.) to form halogenated inorganic and organic compounds (disinfection by-products) such as trihalomethans (THMs), halogenated acidic acid (HAA), halogenated acetonitriles (HAN, and etc.), and other substances that has not yet been identified. DBPs of various oxidizing agents are mentioned in the table 1 [2].

Table 1. Disinfectants and Associated Disinfection By-products.

Disinfectant	Disinfectant by-products
Chlorine (e.g. gas, sodium hypochlorite, tablets, OSEC)	Trihalomethanes, Haloacetic Acid, Halogenated Acetonitriles, Chloramines, Chlorinated Acetic Acit, Chloral Hydrate, Chloropicrin, Chlorophenols, dichloropropanone, trichloropropanone, bromate, trichloroacetaldehyd, halofurans, bromohydrins
Chlorine dioxide	Chlorite, Chlorate and Chloride
Chloramines	Dichloramines, Trichloramines, Cyanogen Chloride, Chloral Hydrate,
Ozone	Bromate, Formaldehyd, Aldehydes, Hydrogen Peroxides, Bromoform,

Increase in the concentration of disinfection by-products depends on drinking water quality (ground, surface water resp.), way of the treatment, kind and dose of used oxidant, time of contact of the oxidant with water, concentration of residual oxidant (chlorine), retention time in the distribution system (length of the system), water temperature, pH (formation of THM is more common at the higher pH values), bromides content, concentration nature of organic substances often expressed as total organic carbon (TOC), resp. as UV_{254} : TOC ratio [3].

It is being generally considered that the high concentrations of chloroform and other THMs appear when the pre-chloration of raw water is used. Concentrations of THM are especially high when the retention time of pre-chlorinated raw water in the distribution system is long. When there is a sufficient concentration of residual chlorine in the distribution system and the retention time is long enough all the organic precursors are changed to form the THMs.

Impact of retention time of the water in the distribution system on increase in disinfection by-products concentration is presented in the work of Chen at al. [4]. Concentration of disinfection by-products was monitored in the central distribution system in New Jersey over one year and in 4 localities while the water retention time was drafted as 0, 1, 2 and at least 3 days. Water was disinfected with chloramine to contain free residual chlorine at the concentration 0.5 mg/L. By increasing the retention time in the distribution system the concentration of THMs was increasing but the concentration of halogenated acetonitriles, kethons, chlorpicrine and halogenated acetic acid was decreasing. Changes in the concentrations were more rapid over the summer months in comparison with changes in the winter months.

In the England, an impact of retention time of water in the distribution system was monitored for several years while a THMs concentration increase was found and was up to 40-60% which corresponded to increasing of the distance from the drinking water treatment plant. It was determined within this monitoring that the content of free residual chlorine was decreased from 0,4 mg/l below the detection limit while the retention time was set to be 2-4 days. Authors [5] used a limit temperature of the water 10°C to categorize the water samples into winter category (< 10 °C) and summer category (> 10 °C) and they found that the average of summer concentrations of THMs in the model distribution system is higher than the average of THMs winter concentrations.

Changes in disinfection by-products in the model distribution system were monitored within the work [6] while samples were taken from two distribution systems. Dose of chlorine, temperature and incubation time were adapted to the conditions in drinking water treatment plant. Free residual chlorine was utilized completely in three days of water retention in the system. Concentration of THMs and halogenated acetic acid was increasing while the halogenated acetonitriles were increasing at the beginning and decreasing thereafter.

Laboratory studies aimed to measure the changes in THMs concentration in the samples of water depending on the retention time give a proof that the concentration was increased in the supplying system but the results are not consentient [7,8,9]. Different values are for instance caused by the use of various disinfection agents, used sampling method and the nature of sampled water. For example, the THMs concentration in the sample that was taken every 4 hours from the constantly flowing water tap varies within the range 40% [10].

Strategy of removal of the precursors can be divided into three groups:

- selection of proper source;
- physico-chemical removal of precursors (classic coagulation, accelerated coagulation on granulated activated carbon – GAC, membrane separation);
- oxidation (processes which lead to a change of precursors and that increase the biodegrability).

Pursuant to Government Regulation No. 354/2006 Coll. as amended, the limit values of disinfection agents and disinfection by-products in drinking water in Slovakia shows in table 2; these values complies with the recommendations of the World Health Organization (WHO) and Council Directive 98/83/EC [11].

Table 2. Disinfectants and their by-products (according to the GR No. 354/2006 for drinking water).

Parameter	Symbol	Unit	Limit	Parameter	Symbol	Unit	Limit
chlorine	Cl_2	mg/L	0.3	chlorine dioxide	ClO_2	mg/L	0.2
bromodichloromethane	BDCM	mg/L	0.015	chlorite	ClO_2^-	mg/L	0.2
trihalometanes	THMs	mg/L	0.15	trichloromethane	$CHCl_3$	mg/L	0.04
2,4-dichlorophenol	DCF	mg/L	0.002	tribromomethane	$CHBr_3$	mg/L	-
2,4,6-trichlorophenol	TCP	mg/L	0.2	ozone	O_3	mg/L	0.05

2. Experimental part

The time dependence of the action of the disinfectant on the formation of disinfection by-products for treated water from WTP Hriňová was study. The chemical composition of drinking water before adding a chlorine disinfectant is shown in table 3.

Table 3. The chemical composition of drinking water from WTP Hriňová (21.5.2017).

Parameter	Unit	Limit	Hriňová	Parameter	Unit	Limit	Hriňová
Temperature	^{o}C	8 - 12	7,8	COD-Mn	mg/L	3,0	1,36
pH		6.5– 8.5	7.57	TOC	mg/L	5.0	1.2
conductivity	mS/m	125	12.3	Humic substances	mg/L	2.5	0.98
alkalinity	mmol/L	–	0.04	absorbance A^{254}		0.08	0.042
acidity	mmol/L	–	0.42	turbidity	FTU	5	1
Ca + Mg	mmol/L	1.1–5.0	0.46	oxygen	%	> 50	82.69

Sodium hypochlorite and chlorine dioxide as the disinfectant were used. To 500 ml of bottles filled with water from WTP Hriňová was added an exact amount of disinfectant at certain time intervals, i.e., a concentration of 0.3 mg/L so that there is no air in the sample. These samples were analyzed in an accredited laboratory of the Water Research Institute in Bratislava and the following parameters were observed: trihalomethanes and some other chlorinated organic substances, free (residual) chlorine, chlorine dioxide and chlorites.

3. Results and discussion

The results are shown in table 4 and figure 2. The values of the analyzed parameters are very low, mostly below the detection limit of the analytical method. This can be explained by the low value of humic substances and by the fact that laboratory experiments can not replace the actual operation of the group water supply.

Table 4. Experiment results (values in µg/L).

Compounds	Sodium hypochlorite			Chlorine dioxide			Limit
	24 hours	'2 hours	120 hours	24 hours	'2 hours	120 hours	
Chloroform	< 5.0	6.6	8.5	< 5.0	< 5.0	< 5.0	40
Bromodichlormethane	< 2.0	< 2.0	< 2.0	< 2.0	< 2.0	< 2.0	15
Dibromochloromethane	< 13.0	< 13.0	< 13.0	< 13.0	< 13.0	< 13.0	-
Bromoform	< 12.0	< 12.0	< 12.0	< 12.0	< 12.0	< 12.0	-
Trihalomethanes	< 13.0	< 13.0	< 13.0	< 13.0	< 13.0	< 13.0	100
Free chlorine [mg/L]	< 0.10	< 0.10	< 0.10	< 0.10	< 0.10	< 0.10	0.3
Chlorine dioxide [mg/L]	-	-	-	0.02	0.02	0.02	0.2
Chlorites [mg/L]	-	-	-	0.14	0.15	0.16	0.2

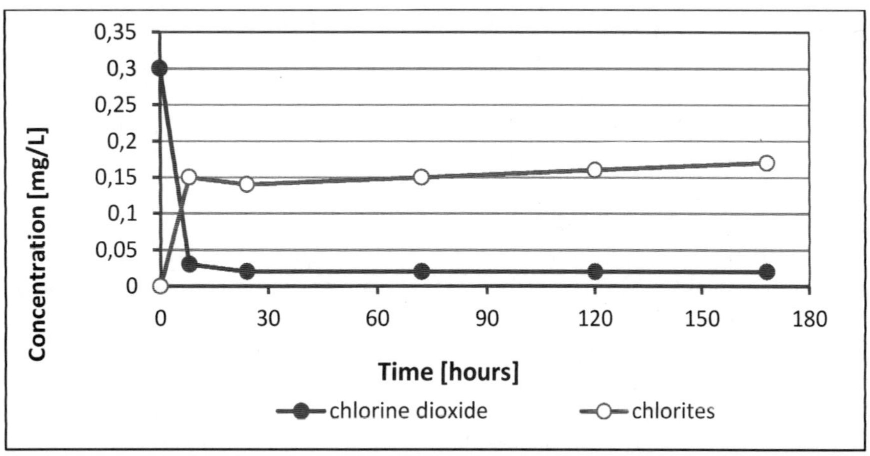

Figure 2. Course of the decomposition of ClO_2 and the formation of ClO_2^- in water.

4. Conclusion

Most of the disinfection methods produce by-products, which depend mainly on the quality of water, the method of water treatment, the type and amount of oxidant used, the oxidant contact time with water, residence time in the distribution system, but also the temperature and pH of the treated water. Concentration of disinfection by-products are regulated by permitted maximum values in drinking water.

The application of the chlorine dioxide and sodium hypochlorite on the formation of some by-products of disinfection for drinking water from WTP Hriňová was studied. The results confirmed formation of chloroform after 72 hours contact time of sodium hypochlorite with water. When chlorine dioxide (ClO_2) was used, chlorite compounds immediately forms in the beginning of water treatment (0.14 to 0.16 mg chlorite per 0.3 mg of chlorine dioxide used). ClO_2 was decomposed after 8 hours to form chlorites (0.15 mg/l), concentration of chlorine dioxide are decreased to 0.03 mg/L. An advantage of using chlorine dioxide is that it does not form halogenated by-products (THMs) as it is the case with chlorine or sodium hypochlorite.

The basic prerequisite for introducing arbitrary disinfectant in water treatment is the perfect knowledge of the water quality as well as knowledge of the technological properties - chemical stability in the distribution system. In order to avoid by-products, it is necessary to remove the precursors from the water and also to use a disinfectant to prevent the formation of halogenated by-products (THMs). An important role in choice of disinfectant plays a residual chlorine concentration in the distribution network.

Acknowledgements

The experimental measurements were performed with the financial support of the APVV-15-0379 and VEGA 01/0400/15 projects.

References

[1] Rook J J 1974 Formation of Haloforms During Chlorination of Natural Waters Water Treatment Exam **23** 234
[2] Krasner S W et al. 1989 The Occurrence of Disinfection By-products in US Drinking Water J AWWA **81** 41
[3] McGraw-Hill 1990 Water Quality and Treatment A Handbook of Community Water Supplies (AWWA)
[4] Chen W J and Weisel C P 1998 Halogenated DBP Concentrations in a Distribution System J AWWA **90** 151
[5] Brett R W and Calverley R A A 1979 One-year Survey of Trihalomethane Concentration Changes Within a Distribution System Jour AWWA **71** 515
[6] Koch B et al. 1991 Predicting the Formation of DBPs by the Simulated Distribution System J AWWA **83** 62
[7] Schreiber J S 1981 The Occurrence of Trihalomethanes in Public Water Supply Systems of New York State J AWWA **73** 154
[8] Kasso W B and Wells M R A 1981 Survey of Trihalomethanes in the Drinking Water System of Murfreesboro *Tenessee. Bull. Environ. Contaminants Toxicol.* **27** 295
[9] Otson R, Williams D T and Bothwell P D 1981 Comparison of Trihalomethane Levels and Other Water Quality Parameters for Three Treatment Plants on the Ottawa River *Environ. Sci. Technol.* **15** 1075
[10] Smith V L et al. 1980 Temporal Variations in Trihalomethane Content of Drinking *Water. Environ. Sci. Technol.* **14** 190
[11] WHO 2005 Trihalomethanes in Drinking-water Background document for development of WHO *Guidelines for Drinking-water Quality* WHO/SDE/WSH/05.08/64

IOP Conf. Series: Earth and Environmental Science **92** (2017) 012023 doi:10.1088/1755-1315/92/1/012023

Benefits of aggregates surface modification in concrete production

J Junak[1] and A Sicakova[1]

[1]Institute of Environmental Engineering, Faculty of Civil Engineering Technical University of Kosice, Vysokoskolska 4, Kosice 04200, Slovakia

Abstract. In our study, recycled concrete aggregates (RCA), which surfaces had been modified by geopolymer material based on coal fly ash, were used to produce the concrete samples. In these samples, fraction 4/8 mm was replaced by recycled concrete aggregate with a range of 100%. To modify the surface of RCA was "Solo" and "Triple stage" modification used. On these samples real density, total water absorption and compressive strength were examined after 28, 90, 180 and 365 days of hardening. The highest compressive strength 56.8 MPa, after 365 days hardening, reached sample which had improved RCA surface by "Triple stage mixing".

1. Introduction

At present, sustainable development is a priority for all spheres of society. Important role in the sustainable development of the built environment, reduction of pollution, conservation of natural resources and energy savings certainly has the entire civil engineering, especially construction materials industry [1].

The manufacture of traditional concrete is one of the most pollutants industries. About 80–90% of GHG (Greenhouse gas) emissions and energy used in the fabrication of concretes derives from the high temperatures required for cement production and the decomposition of the calcium carbonate [2]. One of the main reasons to use Recycled concrete aggregate (RCA) in structural concrete is to make construction more "green" and environmentally friendly. Some major environmental issues associated with construction, are that construction "takes 50 % of raw materials from nature, consumes 40 % of total energy, creates 50 % of total waste." The use of RCA on a large scale may help to reduce the effects of the construction on these factors by reusing waste materials and preventing more natural aggregates from being harvested [3].

Many studies have focused on the physical, mechanical and durability properties of concrete employing recycled aggregates [4] and their findings have concluded that when compared with natural aggregates the lower properties of the recycled aggregates have a detrimental effect on the physical, mechanical and durability properties of the Recycled aggregate concrete (RAC) [5].

It is important to improve the interface between cement matrix and coarse aggregate of low-quality recycled aggregates to promote their application in concrete mixes. Qualitative improvement will in turn enhance the mechanical performance such as the modulus of elasticity and strength, and it will be possible to manufacture concrete with improved durability in terms of drying shrinkage, neutralization, and freezing-thawing. One possibility is a surface modification through the use of an inorganic material on crushed concrete aggregates [5,6].

However, the use of RCA has successfully developed since its initial use via the strict following of minimum qualities, maximum replacement ratios, particular mixing methods or mixing designs using mineral admixtures [5,7].

In our study, construction and demolition wastes (C&DW), namely RCA with modified surface were used to produce the experimental concrete cubes. On these samples real density, total water absorption and compressive strength were after 28, 90, 180 and 365 days tested.

2. Material and methods

Cement, natural aggregates, recycled concrete aggregates, coal fly ash and water were used in our research.

The binder used in the study were Portland cement CEM I 42.5 (according standard STN EN 197-1 Cement. Part 1: Composition, specifications and conformity criteria for common cements, while its specific weight is 3100 kg/m^3.

Two types of aggregates were used in our experiment, natural aggregates - NA and recycled concrete aggregates - RCA. Two different fractions, 0/4 mm and 4/8 mm of natural aggregates for concrete samples preparing were used. Natural aggregates were evaluated according to the standard STN EN 12 620 Aggregates for concrete.

The recycled concrete aggregates in a platform of recycling by crushing concrete waste of demolished buildings were produced. They are composed of old paste and natural aggregates and are free of other materials like bitumen. Separated concrete blocks were fed into crushing unit, then sieved to achieve required sizes. We used only fraction 4/8 mm, in our experiment.

Original coal fly ash (OFA) from Heating plant, Kosice – Slovakia, were used as the main component of geopolymer slurry, in our experiment. The results of chemical analysis and original particle size of the investigated coal fly ash are shown in table 1. The results of this analyses are expressed by following parameters - grain diameters: d (0.1) - indicates that 10% of the whole amount of investigated sample is under measured value of the particle size (μm) and the like parameters d (0.5) – 50% and d (0.9) – 90%. The calcium oxide content of fly ashes is less than 10%, hence they can be classified as Class F according to ASTM 618 standard.

Table 1. The results of chemical and granulometric analysis of coal fly ash.

	Oxide Composition (%)						Grain size (μm)		
Materials	MgO	Al_2O_3	SiO_2	CaO	Fe_2O_3	CaO/SiO_2	d(0.1)	d(0.5)	d(0.9)
OFA	1.18	23.21	51.11	2.58	6.4	0.051	3.97	20.44	84.73

To prepare concrete mixtures drinking water (STN EN 1008 Mixing water concrete) and superplasticizer additive Stachement 2353 was used. Stachement 2353 (Stachema Company) is a high-activity water reducing/superplasticising additive based on polycarboxylates. According to the technical data, this additive can reduce the water content of the fresh concrete while maintaining workability and improving its strength, durability and shrinkage. Table 2 shows basic composition of concrete to 1m^3.

Table 2. Reference concrete mix proportion per cubic meter.

Composition	1 m^3
Cement (kg)	370
Aggregates 0/4 mm (kg)	1100
Aggregates 4/8 mm (kg)	695
Water (l)	205
Plasticizer (%)	0.5

The amount of water in the mixture was indicative. For mixture resulting consistency required slump grade was followed (S2). The total amount of water was composed form effective water and 10% absorbed water of the aggregates and its shown in table 3.
In this experiment four series of mixes were considered (table 3). The first series was the control mix containing 100 % natural aggregates in both fractions. The second mixes were concretes containing 100% RCA as full replacement for natural coarse aggregates in fraction 4/8 mm. The third and fourth mixes were similar to the second series in every respect except the type of used RCA. Recycled concrete aggregates used in these samples had improved surface by coating.

Table 3. Details of experimental mixtures converted to $1m^3$.

Sample	Aggregate 0/4		Aggregate 4/8		RCA aggregate surface modification		Plasticizer (%)	Total Water (l)
	NA	RCA	NA	RCA	Solo	"Triple stage" mixing		
B0	•	-	•	-	-	-	0.5	210
B1	•	-	-	•	-	-	0.5	220
B2	•	-	-	•	-	•	0.5	210
B3	•	-	-	•	•	-	1.5	220

To RCA surface modification geopolymer slurry based on coal fly ash in two ways was used. Geopolymer slurry consisted of coal fly ash, liquid glass, 10M NaOH solution and water. In "Solo modification" firstly was the coating of RCA by geopolymer slurry. After coating the aggregates at ambient conditions for 28 days were cured. Then, like this RCA aggregates into concrete mixture were applied. In the second way ("Triple stage mixing") RCA aggregates directly during mixing of concrete were coated. First, fraction 4/8 of RCA was coated by geopolymer slurry and it was curing. After 30 minutes the concrete mixing continued – gradually the other components were added. After a careful mixing of the all components concrete samples were prepared.

All the samples were elaborated under laboratory conditions and demolded after 24 h. Then, they were immersed in water under constant temperature at 20 ± 3 °C during the corresponding curing time. Real density, total water absorption and compressive strength were tested after 28, 90 and 180 days. The compressive strength was tested using a hydraulic press ELE International and was determined by using the standard (STN EN 12390-3 Testing hardened concrete. Part 3:Compressive strength of test specimens) on specimens 150x150x150 mm. Same specimen dimensions were used for determining the real density in accordance with STN EN 12390-7. Total water absorption by following the standard (STN EN 73 1316 Determination of moisture content, absorptivity and capillarity of concrete) specifications on prism 40x40x160 mm was measured.

3. Results and discussion

Table 4 shows results of real density of tested samples. The highest values of real density were presented in reference sample B0 prepared with natural aggregates only, from 2290 kg/m^3 to 2340 kg/m^3. The lowest density reached sample B1, contained staring RCA with specific gravity 1960 kg/m^3.

Table 4. Average values of density after 28, 90, 180 and 365 days of curing.

Sample	Real density (kg/m^3)			
	28 days	90 days	180 days	360 days
B0	2340	2290	2310	2300
B1	2160	2130	2190	2180
B2	2290	2280	2290	2280
B3	2190	2140	2200	2190

Values observed in samples with surface modified RCA (samples B2 and B3) were quite similar, and observed density values are caused by low density of RCA. Average density of all samples is in the range from 2130 kg/m^3 to 2340 kg/m^3.

Table 5 shows the results obtained in total water absorption test analysis for each mix according STN EN 73 1316. Water absorption of reference sample was about 3% during all periods. Total water absorption other samples with time gradually decreased, firstly values ranged about 9%, after 28 days and got to the level of 3%, after 365 days curing. This value of 3% is at the level of the comparison sample. In the sample prepared with starting RCA total absorption was reduced from 9% to 6%, only.

Table 5. Total water absorption of tested samples after 28, 90, 180 and 365 days of hardening.

Sample	Total water absorption (%)			
	28 days	90 days	180 days	365 days
B0	3.1	2.9	2.7	2.8
B1	9.1	7.8	6.0	6.3
B2	8.8	7.8	3.3	3.4
B3	8.3	7.5	3.0	3.1

Based on the results, we can conclude that both tested methods of aggregates surface modification ("Solo" and "Triple stage mixing") have brought significant results. The values of the measured total water absorption were different at least for both methods, but namely about 0.3%, afterr 180 and 365 days hardening in the same way.

Compressive strength development of concrete mixes which were produced by substitution of natural aggregates with recycled concrete aggregates is shown in table 6.

Table 6. Compressive strength samples prepared with recycled concrete aggregate after 28, 90, 180 and 365 days of curing.

Sample	Compressive strength (MPa)			
	28 days	90 days	180 days	360 days
B0	38.7	45.1	53.1	55.1
B1	29.6	35.5	36.5	40.0
B2	33.5	41.3	55.7	56.8
B3	32.8	43.4	52.8	54.9

The values in the table are the average values of minimum three cubes 150x150x150 mm. No differences in type of failure were witnessed. From the results is clear, that in all samples the compressive strength values increased over time. However, the largest increase can be observed in the sample B2, which contained aggregates with improved surface by "Triple stage" mixing. In this sample, compressive strength increased from 33.5 MPa (after 28 days) to 56.8 MPa (after 365 days hardening). This is also the highest compressive strength reached among the tested samples. It should be noted, however, that high compressive strengths even using "Solo modification" have been achieved (54.9 MPa). The lowest compressive strength, 29.6 MPa, after 28 days reached sample B1, which contained RCA with unmodified surface.

4. Conclusion

To reduce the absorption of the RCA fraction 4/8 mm, the surface of aggregate was activated by a multi-stage mixing and using of geopolymer phases, in our experiment. Experimental concrete samples with using these activated RCA were prepared, while w/c ratio was from 0.52 to 0.56. For mixture resulting consistency required slump grade S2 was followed.

Average density of all samples were in the range from 2160 kg/m^3 to 2340 kg/m^3. The highest values of real density were presented in reference sample with natural aggregates prepared, 2340 kg/m^3. The lowest density reached sample, which contained unmodified aggregates with specific gravity 1960 kg/m^3. Values observed in samples with surface modified RCA were quite similar, about 2200 kg/m^3.

Total water absorption of reference sample was about 3% during all periods, but water absorption samples prepared with improved surface RCA with time gradually decreased, from 9%, after 28 days to 3%, after 365 days hardening. This is a similar value like reached a reference sample. Only in the sample containing starting RCA total absorption was reduced at least, and thus from 9% to 6%.

Compressive strength testing intervals for experimental samples with RCA were 28, 90, 180 and 365 days. Based on the results, we can conclude that all of the tested samples had the same compressive strength development - evenly increasing. However, the largest increase can be observed in the sample which had improved RCA surface by "Triple stage mixing", from 33.5 MPa, after 28 days to 56.8 MPa, after 365 days hardening.

Acknowledgments
This research has been carried out within the Slovak Cultural and Education Grant Agency (contract No. 073TUKE-4/2015).

References

[1] Radonjanin V et al. 2013 Green recycled aggregate concrete *Constr. Build. Mater.* **47** 1503
[2] Letelier V et al. 2016 Assessment of the mechanical properties of a concrete made by reusing both: Brewery spent diatomite and recycled aggregates **114** 492
[3] McNeil K and Kang TH-K 2013 Recycled Concrete Aggregates: A Review *IJCSM* **7** 61
[4] Junak J and Sicakova A 2017 Concrete Containing Recycled Concrete Aggregate with Modified Surface *Procedia Engineering* **180** 1284
[5] Gonzalez-Corominas A and Etxeberria M 2016 Effects of using recycled concrete aggregates on the shrinkage of high performance concrete *Constr. Build. Mater* **115** 32
[6] Junakova N and Junak J 2017 Sustainable use of reservoir sediment through partial application in building material *Sustainability* **9** 852
[7] Junak J 2015 Green concrete: Environmentally friendly concrete made from alternative materials International *Proc. Multi. Scient. GeoConf. Survey. Geo. and Mining Eco. Manag.* p (Sofia: STEF92 Technology Ltd.) p 147

A mathematical model of reservoir sediment quality prediction based on land-use and erosion processes in watershed

N Junakova[1], M Balintova[1] and J Junak[1]

[1]Institute of Environmental Engineering, Faculty of Civil Engineering Technical University of Košice, Vysokoškolska 4, Košice 04200, Slovakia

Email: magdalena.balintova@tuke.sk

Abstract. The aim of this paper is to propose a mathematical model for determining of total nitrogen (N) and phosphorus (P) content in eroded soil particles with emphasis on prediction of bottom sediment quality in reservoirs. The adsorbed nutrient concentrations are calculated using the Universal Soil Loss Equation (USLE) extended by the determination of the average soil nutrient concentration in top soils. The average annual vegetation and management factor is divided into five periods of the cropping cycle. For selected plants, the average plant nutrient uptake divided into five cropping periods is also proposed. The average nutrient concentrations in eroded soil particles in adsorbed form are modified by sediment enrichment ratio to obtain the total nutrient content in transported soil particles. The model was designed for the conditions of north-eastern Slovakia. The study was carried out in the agricultural basin of the small water reservoir Klusov.

1. Introduction

Soil erosion associated with non-point source pollution is a considerable environmental problem with respect to the off-site transport of sediment from agricultural land and nutrient-rich pollutants adsorbed to the eroded sediment into nearby waterbodies and the neighbouring environment.

Persistently high rates of water erosion affect an estimated 1,094 million hectares of land worldwide [1]. A recent new model of soil erosion by water has estimated the surface area affected in the EU-27 at 130 million ha. Almost 20 % is subjected to soil loss in excess of 10 tonnes per ha per year [2]. According to World Commission on Dams [3] it is estimated, that the world annual loss in storage capacity of the world's reservoirs due to sediment deposition is around 0.5-1%, and for individual reservoirs these values can be as high as 4-5% [4].

Together with small fraction of sediment, pollutants including nutrients are transported via surface runoff from arable land in the catchment and they are deposited in reservoirs [5]. Quantity of nutrients concentration in water reservoir sediments is regarded as the environmental pollution index and is often attributed to nonpoint source pollution from agricultural production areas.

In order to protect surface water resources and optimise their use, soil and nutrient loss must be controlled and minimised [6]. Field studies for prediction and assessment of soil erosion and sediment-associated chemical runoff are expensive, time-consuming and need to be collected over many years [7]. Nowadays, it should be rather a preference to predict of agricultural non-point source runoff of sediment and nutrients and their control using mathematical models.

The objective of this study is to propose a mathematical model for predicting the nitrogen and phosphorus concentrations in sediments from small water reservoirs and its application in agricultural watershed of the Tisovec River, Slovakia.

2. Material and methods

2.1. Study area

A mathematical model for prediction of total nitrogen (N) and phosphorus (P) content in eroded soil particles and reservoir bottom sediments has been proposed on the basis of the nutrient transport study realized in the Tisovec River catchment, located in the east of Slovakia, in Bardejov district. This catchment covers about 6.0 km^2 and falls in the Topla partial river basin. This catchment is affected particularly by non-point sources of pollution from agricultural production areas and no significant point source of pollution is located there. Land use within the catchment is found to be mixed type. The upstream part and middle part is an area mainly covered with forest and pastures (together 61 %), while the lower part is an arable land (21 %). The rest of the land area is for other uses.

In the narrowest part of the Tisovec River, the Klusov small water reservoir is located with its surface area of 2.2 ha and the total capacity of 72,188 m^3. During 19 years of the reservoir's operation it trapped approximately 24,500 m^3 of sediments delivered from the upper catchment. This situation resulted in the reduction of the total storage capacity about 33 % and thus reservoir was drained completely and desilted. Therefore the need to monitor and predict the transport of sediments and their quality is essential.

2.2. Sampling procedure

From the drained reservoir, 21 composite sediment samples were collected for determining of total N, P content in bottom sediments. Composite sediment samples were taken in nearness of dam location, especially due to deposition of fine-grained particles (fractions below 63 μm). Other sediments were sampled along the reservoir and also in various sampling depths from one locality because of detailed analysis of the stratigraphy of sediments in a reservoir. Sample S9 was taken from the Tisovec River flowing through the reservoir. Simultaneously, for determining of total N, P content in eroded soil particles, soil samples were collected from 2 units (plots 1004/1 and 2001/1) of arable land situated in the neighbourhood of the reservoir (figure 1). Within each sampling unit, at minimum 30 soil subsamples were taken from several different locations and they were mixed into one composite sample. Sampling depth was 0.30 m. In laboratory conditions, the samples were air dried at room temperature, any coarse lumps were crushed and samples were homogenized.

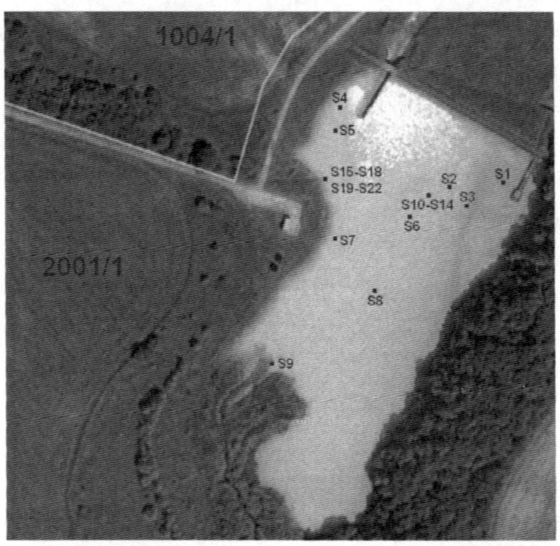

Figure 1. Location of monitored plots in the neighbourhood of reservoir and sediment sampling sites.

Adsorbed form of nutrients in sediment and soil samples were determined for total nitrogen by elemental analysis (EA) and for total phosphorus using inductively coupled plasma-atomic emission spectrometry (ICP-AES) technique.

2.3. Model design

Methodology of the proposed model (figure 2) works on the determination of total N and P concentrations in eroded soil particles in adsorbed form of followed elements. It consists of the soil loss calculation using the Universal Soil Loss Equation (USLE) [8] and the determination of the average soil nutrient concentration in top soils.

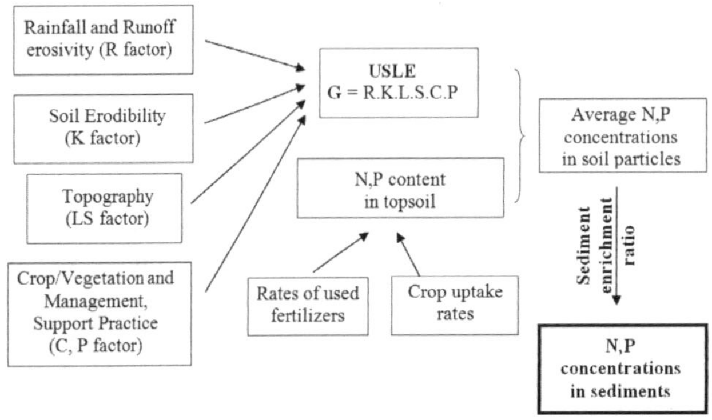

Figure 2. The scheme of suggested model for predicting the content of total nitrogen and phosphorus in reservoir bottom sediments.

The average annual vegetation and management factor in USLE equation (C factor) is specified and divided into five cropping cycle periods (seedbed preparation, establishment, development, maturing crop, stubble field). Average nutrient concentrations in top soils, divided into five cropstage periods, are computed on the basis of subtracting the nutrient input from fertiliser use (provided by the agricultural cooperative) and plant nutrient uptake (calculated according to the average yield in the studied catchment's area and the mean plant nutrient uptake values given in the [9].

The total N, P content in transported soil particles from arable land to reservoir is calculated on the basis of the average N, P concentrations in eroded soil particles in adsorbed form detached through water erosion modified by sediment enrichment ratio [10] using this proposed equation (1):

$$C_X = \frac{\sum_{i=1}^{5} X_i . G_i}{G_{r.i}} . SER = \frac{\sum_{i=1}^{5} X_i . G_i}{G_{r.i}} . e^{2-0.2\ln(G_{r,i}.1000)} \tag{1}$$

where C_X is an average annual concentration of total N, P in transported soil particles from monitored plot in kg N,P ha^{-1} or mg N,P kg soil^{-1}; X_i is residual concentration of total N, P in soil in *i*-period in kg N,P ha^{-1} determined as the subtraction of the total nutrient concentration in soil and plant available nutrients in soil; G_i is soil loss in individual cropping periods in t ha^{-1} and $G_{r,i}$ means an average soil loss in *i*-year from monitored plot in t ha^{-1} year^{-1}. The total concentrations of N, P in reservoir bottom sediments are then computed as a weighted mean of the calculated average annual concentration of total N, P in transported soil particles from monitored plots surrounding the reservoir.

3. Results and discussion

The total nutrient concentrations in eroded soil particles in dissolved form of followed elements weren't considered in this proposed model, because the portion of dissolved phase represented 0.22 – 0.43 % of total N and 0.45 – 0.86 % of total P in the soil sample.

Calculations of total N and P concentrations in eroded soil particles in adsorbed form were realized for two units (plots 1004/1 and 2001/1) of arable land situated in the neighbourhood of the Klusov reservoir. Calculated total N and P concentrations in bottom sediments were determined during the 10-years crop rotation system in the Klusov small water reservoir.

The average annual soil loss ranged from 1.9 to 17.7 t ha^{-1} year^{-1} for plot 1004/1 and from 5.34 to 27.7 t ha^{-1} year^{-1} for plot 2001/1 depending on the crop. Long term average annual soil loss is 8.3/14.3 t ha^{-1} year^{-1} for plots 1004/1; 2001/1 depending on the crop, plot length and slope gradient.
The most widely used fertilisers in the watershed were NPK fertilisers (8.5% or 15% N), ammonium nitrate (27% N). For crops common grown in Slovakian conditions, an average plant nutrient uptake divided into five cropstage periods was devised (table 1).

Table 1. Proposed average plant nutrient uptake divided into five cropping periods.

Crop	(kg ha^{-1})	Cropping period					$\sum_1^5$
		1.	2.	3.	4.	5.	
	N	0.00	15.90	17.75	20.00	0.00	53.65
spring barley	P	0.00	2.05	2.50	3.70	0.00	8.25
	K	0.00	18.65	21.80	19.55	0.00	60.00
winter wheat,	N	0.00	0.00	19.10	71.35	0.00	90.45
triticale,	P	0.00	0.00	1.25	10.75	0.00	12.00
winter rye	K	0.00	0.00	9.55	67.50	0.00	77.05
	N	0.00	9.90	68.90	30.00	0.00	108.80
winter oilseed rape	P	0.00	1.40	15.90	5.00	0.00	22.30
	K	0.00	6.20	88.10	7.70	0.00	102.00
	N	0.00	18.30	17.70	26.30	0.00	62.30
pea	P	0.00	1.90	1.80	2.60	0.00	6.30
	K	0.00	8.50	8.20	12.10	0.00	28.80
	N	0.00	18.27	17.66	53.00	0.00	88.93
potatoes	P	0.00	2.61	2.52	7.58	0.00	12.71
	K	0.00	23.49	22.71	68.15	0.00	114.35
	N	0.00	16.35	16.89	50.18	0.00	83.42
corn silage	P	0.00	2.18	2.19	6.75	0.00	11.12
	K	0.00	13.62	13.67	42.22	0.00	69.51

The average nutrient concentrations in eroded soil particles in adsorbed form were modified by sediment enrichment ratio to obtain the total nutrient content in transported soil particles. Sediment enrichment ratios *SER* computed based on the average soil loss in *i*-year from monitored plots were in range from 0.97 to 1.63.

The total concentrations of N, P in reservoir sediments (table 2) were then computed as a weighted mean of the calculated average annual concentration of total N, P in transported soil particles from monitored plots surrounding the reservoir.

Table 2. Calculated concentrations of total N, P (%) in reservoir sediments.

Year	N (%)	P (%)	K (%)
1.	0.188	0.067	1.956
2.	0.225	0.078	2.307
3.	0.188	0.066	1.944
4.	0.225	0.082	2.387
5.	0.203	0.072	2.131
6.	0.193	0.068	2.032
7.	0.189	0.065	1.954
8.	0.236	0.083	2.469
9.	0.199	0.069	2.071
10.	0.202	0.070	2.096

Calculated concentrations of total N, P in reservoir sediments correspond to the measured nitrogen (0.16 to 0.26 %) and phosphorus (0.05 to 0.11 %) concentrations in collected sediments.

4. Conclusion

Sediment, soil particles that end up in the water, reduces the basin's accumulation capacity. Sediment runoff can also pick up and transport additional pollutants such as nutrients that threaten the good ecological status of waterways and water basins. In order to protect surface water resources and optimise their use, soil and nutrient loss must be controlled and minimised. In this way, mathematical models provide a useful tool for assessing and forecasting the nutrient content in reservoir bottom sediments. This paper summarizes the results of the study aimed at the proposal of a mathematical model to predict the nitrogen and phosphorus concentrations in eroded soil particles and reservoir bottom sediments. The results showed that calculated concentrations of total nitrogen and phosphorus in reservoir sediments correspond to the measured N, P concentrations in collected sediments.

Acknowledgments
This research has been supported by the Slovak Cultural and Education Grant Agency (contract No. 073TUKE-4/2015) and Slovak Grant Agency for Science (Grant No. 1/0563/15).

5. References

[1] Berc J, Bruce J, Easterling D, Groisman P Y, Hatfield J, Hughey B, Johnson G, Kellogg B, Lawford R, Mearns L, Moss R, Nearing M, O'Neal M, Spaeth K 2003 Conservation Implications of Climate Change: Soil Erosion and Runoff from Cropland *A report* (USA, Iowa: Soil and Water Conservation Society)

[2] European Environment Agency 2012 Climate change, impacts and vulnerability in Europe 2012 *An indicator-based report No 12/2012* (Denmark, Copenhagen: European Environment Agency)

[3] World Commission on Dams 2000 Dams and development A new framework for decision making Report of the world commision on dams (UK, London: Earthscan publications)

[4] Verstraeten G, Bazzoffi P, Lajczak A, Rădoane M, Rey F, Poesen J, Vente J 2006 Reservoir and Pond Sedimentation in Europe *Soil Erosion in Europe* eds J Boardman, J Poesen (UK, Chichester: Wiley) pp 757-774

[5] Liu X, Li Z, Li P, Zhu B, Long F, Cheng Y, Wang T, Lu K 2015 Changes in carbon and nitrogen with particle size in bottom sediments in the Dan River, China *Quat. Int.* **380** 305-313

[6] Ongley M 1996 Chapter 13 - Sediment measurements. Water Quality Monitoring: *A practical guide to the design and implementation of freshwater quality studies and monitoring programmes* eds J Bartram, R Balance (UK, London: UNEP/WHO)

[7] Saha S K 2004 Water and wind induced soil erosion assessment and monitoring using remote sensing and GIS *Sat. Rem. Sen. and GIS App. in Agri. Met.* eds M V K Sivakumar, P S Roy, K Harmsen, S K Saha (Switzerland, Geneva: World Meteorological Organization) pp 315-330

[8] Wischmeier W H, Smith D D 1978 Predicting Rainfall Erosion Losses *A Guide to Conservation Planning Agriculture Handbook No. 537* (USA, Maryland: U.S. Department of Agriculture)

[9] Ministry of Agriculture of the Slovak Republic 2005 Decree No. 338/2005 Coll. laying down the Details on the procedure for soil sampling, manner and extent of implementation of agrochemical soil testing, detecting soil properties of forest land, and on keeping the records of soil fertilization and nutritional status of plants on agricultural soil and forest lands (Bratislava)

[10] Menzel R G 1980 Enrichment ratios for water quality modelling *CREAMS, A field scale model for chemicals, runoff, and erosion from agricultural management systems (vol III) Supp. Doc.Cons. Rep. No. 26* ed W Knisel (USA, Arizona: US Department of Agriculture) pp 486-492

Analysis of heavy soils water retention curves with respect to volume changes

B Kandra[1], A Tall[1], M Gomboš[1] and D Pavelková[1]

[1]Institute of Hydrology SAS, Dúbravská cesta 9, 841 04 Bratislava, Slovak Republic

Email: kandra@uh.savba.sk

Abstract. This work analyses the problem of measuring water retention curves in heavy soils. The results present the differences between soil water retention curves measured in soil samples collected from the selected localities of the Czech and Slovak area. In the drying process, the results showed an increased rate of soil shrinkage depending on clay content and water content. The rate of shrinkage affected the results of the water retention curves points measurement.

1. Introduction

Reliable determination of the soil water retention curves (WRC) is a basic assumption in the analysis of the soil water regime. WRC is an important hydrophysical characteristic of soil which has a high level of informative value. It has a sigmoidal course and expresses soil moisture (θ) dependence on its moisture potential (h_w) $h_w = f(\theta)$ [1]. On the precise determination of WRC depends the accuracy of soil retention determination for water and the amount of available water for plants. In addition, WRC is a part of inputs into mathematical models describing the soil water regime in the soil - plant - atmosphere system. Measurement of individual WRC points is time-consuming process, especially in heavy and extremely heavy soils. A specific feature of such soils is the ability to change their volume depending on moisture [2]. The rate of volume changes also depends on the soil texture and, mainly, on the clay fraction content. Decreasing or increasing the soil volume during drying or wetting affects a wide range of soil hydrophysical properties (hydraulic conductivity, porosity, bulk density and retention). Under the conditions of the East Slovakian Lowland (ESL), total volumes of shrinkage up to 40 % of the saturated state were demonstrated. Under natural conditions, volumetric changes are reflected by vertical movement of the soil surface (ESL potential up to 0.13 m, real measured 0.056 m) and crack creation. Many authors have dealt with various ways of expressing soil volume changes and describing their geometry [3, 4, 5]. The hydrophysical properties measurement of the expansive soils is considerably complicated because the samples shrink when moisture decreases during the WRC measurement. The calculation of the water movement in such soil is difficult and, therefore, the influence of shrinking (swelling) is often neglected. During the WRC measurement, the appropriate volumetric changes of the soil samples should be measured. Neglecting this physical phenomenon leads to an inaccurate WRC estimation in heavy soils. This error is further transferred in the subsequent WRC application in calculations.

The aim of the paper is to compare the measured WRC courses of the samples obtained from the heavy soils of the Czech and Slovak Republic in neglecting the volume changes during the measurement and in taking them into account too. Further, to quantify the differences in the retention properties of the investigated soils resulting from WRC comparison.

2. Materials and methods

Within the international cooperation between Czech and Slovak Republic research centres, the 18 soil samples were collected and evaluated for the purpose of WRC measurement. Within the Czech Republic, sampling was realised in the Chomutovsko region at localities near Údlice and Březno. In the Slovak Republic was selected the Milhostov locality in the district of Trebišov, situated in the central part of ESL (Table 1). Intact soil samples were collected by metal cylinders of different depths. The number of samples and sampling depths in individual locations are shown in table 1.

Table 1. Labelling and the sampling depth of soil samples in localities.

Údlice	U1	U2	U3	U4	U5	U6
Soil depth [m]	0.10 – 0.15	0.10 – 0.15	0.10 – 0.15	0.32 – 0.35	0.32 – 0.35	0.32 – 0.35
Březno	B1	B2	B3	B4	B5	B6
Soil depth [m]	0.15 – 0.20	0.15 – 0.20	0.15 – 0.20	0.40 – 0.45	0.40 – 0.45	0.40 – 0.45
Milhostov	M1	M2	M3	M4	M5	M6
Soil depth [m]	0.00 – 0.10	0.20 – 0.30	0.40 – 0.50	0.60 – 0.70	0.80 – 0.90	0.90 – 1.00

Selection of sampling points was based on the laboratory measurements of soil texture, with an emphasis on the high content of clay fraction where significant volume changes were assumed. Soil samples were saturated with water in the laboratory and subsequently placed in overpressure device for the measurement of pF curves. The difference in the measurement methodology between the two workplaces lies in the fact, that the Milhostov samples had been shortened to the volume of the metal sampling cylinder before being placed in the overpressure device. Samples from Údlice and Březno were placed to the device without shortening. Methodology for measuring was based on the ISO standard [6]. Selection of pressures for drainage of samples followed the rules of international ring test laboratories [7] which was aimed at assessing such hydrophysical characteristics. The aim of the ring test was the harmonization of measurement methods for all the participating countries. Samples were normally weighted upon completion of draining under a given pressure and, in addition, geometric parameters were measured using a digital Vernier calliper. At the end of the measurement in the overpressure device, the samples were dried in the oven at 105 °C. From the obtained moistures the volume percentages were expressed and the volumes of soil samples were calculated from the geometric parameters. For each sample, two obtained courses of the measured points were evaluated, the reduced soil water retention curve (WRC_r) and the real soil water retention curve (WRC). The WRC_r did not take into account volume changes and the measurement procedure was the same as for nonexpansive soils. Volumetric moistures θ_r (reduced) were calculated from the volume of metal sampling cylinders V_r. The WRC included the volume changes and the soil volumetric moistures θ_p were expressed to the actual sample volumes V_p in equation (1). The second approach represents a more accurate WRC measurement methodology for heavy (expansive) soils.

$$\theta_p(\theta_r) = \frac{m_p - m_d}{\rho_w \times V_p(V_r)} \times 100 \quad (1)$$

Where

θ_p is the soil water content, expressed as a volume fraction, at the final pressure applied [% Vol.];
θ_r is the reduced soil water content, expressed as a volume fraction, at the final pressure applied [% Vol.];
m_p is the mass of soil sample at the final pressure applied [g];
m_d is the mass of the oven-dried soil sample [g];
ρ_w is the density of water [= 1,00 g.cm^{-3}];
V_p is the volume of the soil sample at the final pressure applied [cm^3];
V_r is the volume of the metallic sampling ring [cm^3].

3. Results and discussion

The basis for the exact calculation of the soil samples volumetric moistures was obtaining the correct sample volumes during draining process. From the measurement of geometric parameters it was clear that the degree of soil shrinkage increased along with the increase in pressure. Figure 1 shows the average volume of samples, depending on the pressure, individually for each locality. The most significant curve decrease in average sample volume was at Údlice locality. The difference between the maximum and minimum V_p value for individual localities was following, Údlice 20.40 cm^3, Březno 14.66 cm^3 and Milhostov 13.75 cm^3. The average change in sample volume was following, Údlice 2.27 cm^3, Březno 1.63 cm^3 and Milhostov 1.53 cm^3. The result of the WRC own measurement is shown in figure 2. In each locality, six samples were collected from a range of different depth (table 1). The total amount of examined samples was 18. The pictures show differences between the real WRC and the reduced WRC_r. In case of the Údlice and Březno locality, a moisture decrease of θ_p compared to θ_r under the influence of initial pressures can be observed. This is due to the fact that samples had not been shortened at the end of saturation process and the sample volumes V_p at θ_p were higher than the volumes of metal sampling cylinders V_r at θ_r. Here was also the difference due to swelling of samples. During the higher pressures, the situation was the opposite and θ_p was higher than θ_r. There was a shrinking of the samples in relation to the volume of the metal sampling cylinder. In the Milhostov locality, only the differences under the influence of shrinkage could have been observed due to shortening of samples after saturation.

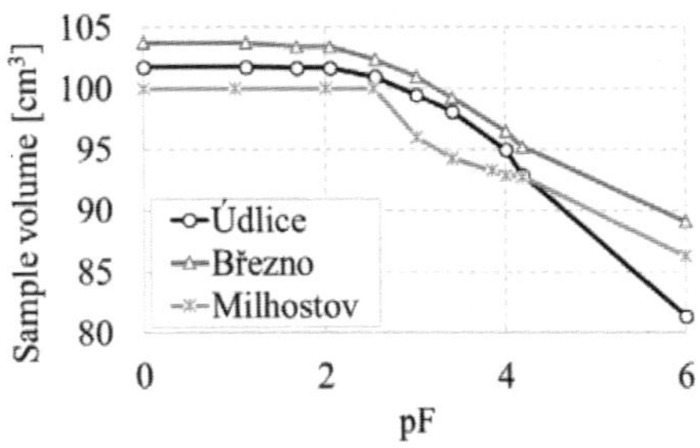

Figure 1. Volume of soil samples according to soil moisture potential.

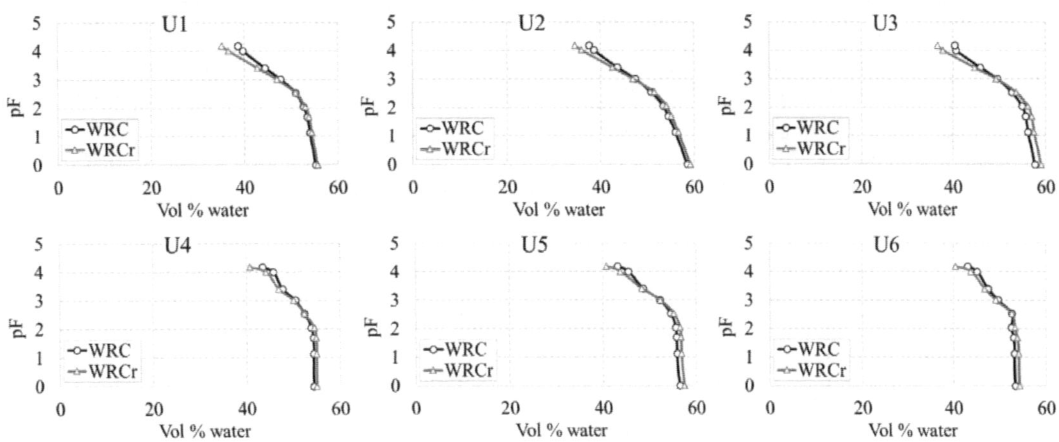

IOP Conf. Series: Earth and Environmental Science **92** (2017) 012025 doi:10.1088/1755-1315/92/1/012025

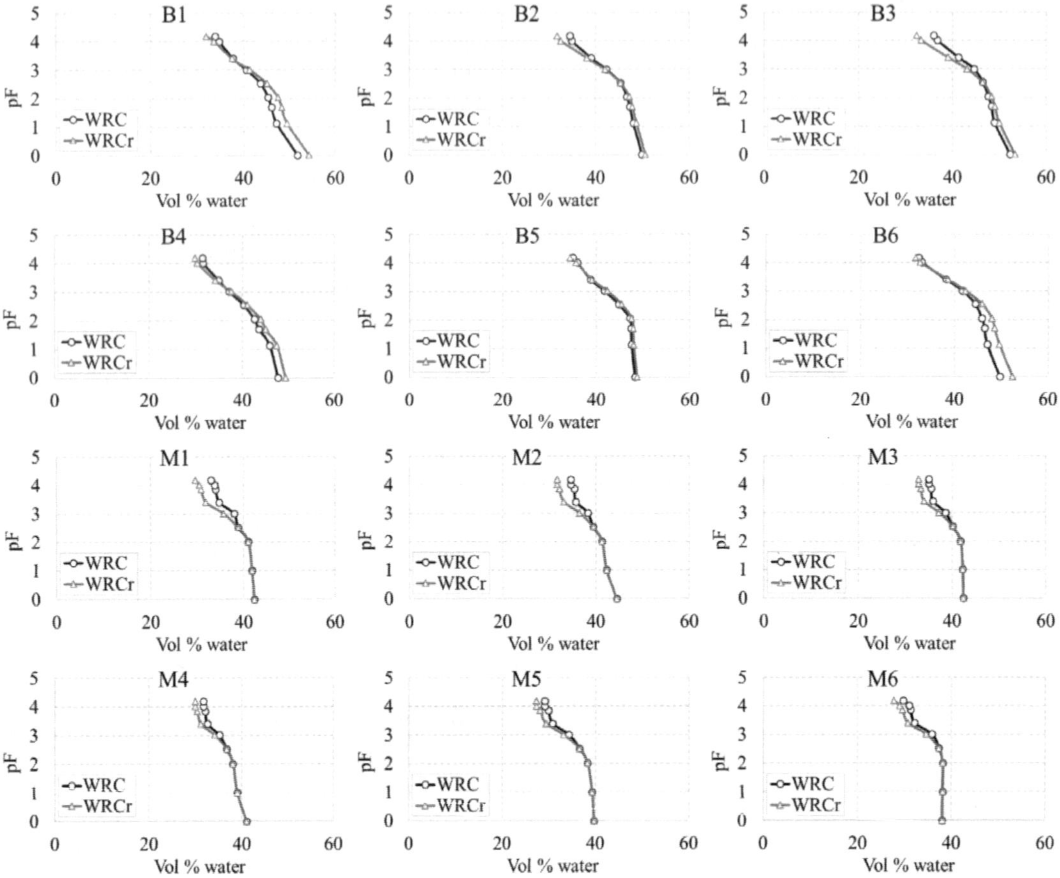

Figure 2. The courses of measured points of real WRC and reduced WRC_r in localities.

Table 2 shows the maximum, minimum and average deviations between θ_p and θ_r for the evaluated localities. The highest difference was in samples U3, B3 and M1 and the lowest in samples U3 and B6. Based on the average deviations, it can be stated that in case of Březno locality, the WRC_r overestimates the WRC measurement results and in case of Údlice and Milhostov localities, the WRC_r underestimate the WRC measurement results. In most samples, the correlation coefficients between the values of the soil moisture potential and the soil volumetric moisture in table 3 show slightly higher dependencies in WRC than WRC_r. On the other hand, the correlation values in WRC are lower than WRC_r only in samples U6, B6 and M6.

Table 2. Deviations of volumetric moistures of soil samples in localities.

$\Delta\theta$ ($\Delta\theta = \theta_p - \theta_r$)	Údlice	Březno	Milhostov
$\Delta\theta_{MAX}$	3.65	3.59	3.38
$\Delta\theta_{MIN}$	-1.35	-2.75	0.00
$\Delta\theta_{AV}$	0.32	-0.27	1.15

The achieved results allowed to optimize the methodology of the soil water retention curves measurement and information about the heavy soils retention properties. In future, the results will enable the elimination of errors resulting from numerical simulations of soil water regime by mathematical models which were caused by incorrectly determined courses of soil retention curves because the volume changes were disregarded. Under this approach, it will be possible to specify and characterize those soils in which the volume changes affect their retention characteristics and dynamics of water regime.

Table 3. The correlation between soil volumetric moisture and soil moisture potential.

CORREL	U1	U2	U3	U4	U5	U6
WRC	-0.927	-0.960	-0.931	-0.892	-0.882	-0.874
WRC_r	-0.916	-0.947	-0.918	-0.888	-0.877	-0.884
CORREL	B1	B2	B3	B4	B5	B6
WRC	-0.984	-0.943	-0.953	-0.979	-0.922	-0.940
WRC_r	-0.974	-0.935	-0.950	-0.971	-0.914	-0.944
CORREL	M1	M2	M3	M4	M5	M6
WRC	-0.947	-0.982	-0.932	-0.976	-0.939	-0.870
WRC_r	-0.936	-0.965	-0.923	-0.964	-0.933	-0.877

4. Conclusion

The paper summarizes the partial results of the water retention curves measurement in soil samples collected from the territory of the Czech and Slovak Republic. The aim of the measurements was to increase precision of the WRC laboratory determination in heavy and extremely heavy soils. With this intention, the sampling was focused on the soils selection with high content of clay particles, where significant volume changes were assumed during saturation and subsequent draining of the samples in the overpressure device. Two WRC point courses were compared. The first WRC_r, was based on the standard approach which followed the same procedure as in nonexpansive soils and the volume changes were not taken into account. The second was WRC which took into account the volume changes. The results of the comparisons showed significant differences between the two courses, pointing to the importance of measuring and involving volume changes in WRC determination.

Acknowledgments

This study is the result of the project implementation: Centre of excellence for the integrated river basin management in the changing environmental conditions, ITMS code 26220120062; supported by the Research & Development Operational Programme funded by the ERDF (90%).
Authors are also grateful to Scientific Grant Agency of the Ministry of education, science, research and sport of the Slovak Republic: (project VEGA: 2/0062/16) (10%).

References

[1] Genuchten van M TH 1980 A closed – form equation for predicting the hydraulic conductivity of unsaturated soils *Soil Sci. Soc. of Ame. J.* **44** 987
[2] Bronswijk J J B and Evers-Vermeer J J 1990 Shrinkage of Dutch clay soil aggregates *Neth. J. of Agr. Sci.* **38** 175
[3] Bronswijk J J B 1988 Modelling of water balance, cracking and subsidence of clay soils *J. of Hyd.* (Amsterdam) **97** 199
[4] Chertkov V Y 2005 The shrinkage geometry factor of a soil layer *Soil Sci. Soc. of Ame. J.* **69** 1671
[5] Peng X R, Horn and Smucker A 2007 Pore shrinkage dependency of inorganic and organic soils on wetting and drying cycles *Soil Sci. Soc. of Ame. J* **71** 1095
[6] ISO 11274: 1998 Soil quality – Determination of the water retention characteristics – Laboratory methods
[7] Cools N, De Vos B 2010 1 st FSCC soil physical ring test 2009 Brusel: Instituut voor Natuur- in Bosonderzoek (INBO) p 47

On the development of a new methodology in sub-surface parameterisation on the calibration of groundwater models

D K S Y Klaas[1,2,3], M A Imteaz[1], I Sudiayem[3], E M E Klaas[3] and E C M Klaas[3]

[1]Department of Civil & Construction Engineering, Swinburne University of Technology, Melbourne, VIC, Australia.
[2] Department of Civil Engineering, Politeknik Negeri Kupang, Indonesia.
[3] Department of Health, Science and Technology, Klinik Garuda, Malang, Indonesia

E-mail: duaklaas@yahoo.com, dklaas@swin.edu.au

Abstract. In groundwater modelling, robust parameterisation of sub-surface parameters is crucial towards obtaining an agreeable model performance. Pilot point is an alternative in parameterisation step to correctly configure the distribution of parameters into a model. However, the methodology given by the current studies are considered less practical to be applied on real catchment conditions. In this study, a practical approach of using geometric features of pilot point and distribution of hydraulic gradient over the catchment area is proposed to efficiently configure pilot point distribution in the calibration step of a groundwater model. A development of new pilot point distribution, Head Zonation-based (HZB) technique, which is based on the hydraulic gradient distribution of groundwater flow, is presented. Seven models of seven zone ratios (1, 5, 10, 15, 20, 25 and 30) using HZB technique were constructed on an eogenetic karst catchment in Rote Island, Indonesia and their performances were assessed. This study also concludes some insights into the trade-off between restricting and maximising the number of pilot points and offers a new methodology for selecting pilot point properties and distribution method in the development of a physically-based groundwater model.

1. Introduction

In groundwater modelling, accurate estimation of groundwater discharge over a catchment requires a proper approach in the parameterisation step [1,2]. In the parameterisation step, parameters are conditioned until the simulated values match the observed measurements. Proper distribution of spatial variations of parameters that could represent intrinsic heterogeneity of the geologic formation, in the absence of adequate pre-measured parameter values, is a challenge for modellers [1,3]. To cope with data limitation, modellers often use numerically aided geostatistical techniques with the aid of high-speed processors. An alternative to the classical zonation method, pilot-point method (PPM) [4] is a geostatistical technique that has been widely used in different hydrogeological conditions in many parts of the world [5,6,7]. However, organising pilot points into a model to achieve a good fit is still considered subjective and less user-friendly for real case examples [8]. Thus, we introduce a practical head zonation-based (HZB) method to assign pilot points in each head guided zone of the model domain by incorporating the practical zone ratio. Therefore, the main objectives of this paper were to present a modelling-oriented analysis of groundwater flow to:

- investigate the impact of pilot-point properties (number, distance and distribution method) on the calibration of groundwater models using HZB method; and

- provide recommendations on the selection of pilot-point properties (zone ratio) to obtain a better calibration result of a groundwater model assessed by statistical measures.

2. Materials and methodology

2.1. Study area

This study exercised the catchment of Oemau spring located in Rote island, Indonesia, geographically located between latitudes 10°46'42.17"S ~ 10 ° 43'36.91"S and longitudes 123 ° 3'14.84"E ~ 123 ° 9'17.64"E. Having a topographically bounded surface-drainage basin area of 20.11 km^2 (figure 1a), the spring is located around 3 km from Ba'a, the capital of the Rote Island.

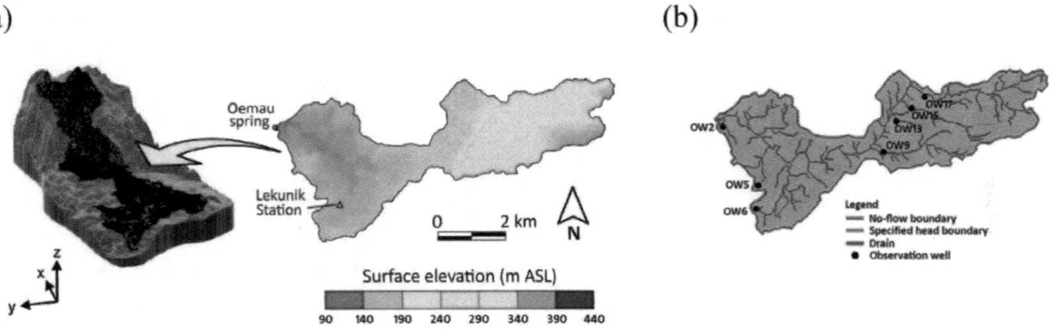

Figure 1. (a) Topography of Oemau spring catchment; (b) conceptual model of the modelled area.

Amoonsonal climate characterised by two distinct seasons: dry (May-November) and wet (December-April) dominates the study area [9]. Humidity is 75-92% and annual rainfall is between 1000 and 2300 mm. Karst landscape characterised by carbonate formations dominates the area. Typified by the absence of preferential flow paths and conduits and geomorphologicaly characterised by low karstification degree, the area is grouped as an eogenetic karst [10].

2.2. Groundwater model

Seven models were developed using HZB method and used to simulate the groundwater flow. The pilot-point distribution is based on the hydraulic-gradient distribution of groundwater flow. Several studies have demonstrated that the groundwater elevation often follows the spatial distribution of hydraulic conductivity [11,12]. It was found that the gradient of groundwater contours relatively corresponds to the hydraulic conductivity of the region. In areas with low conductivity, water travels slower implying a high hydraulic gradient of groundwater. Similarly, the hydraulic gradient would be lower when travelling through a high conductivity area. Therefore, the hydraulic gradient can be used as an initial graphical indication of hydraulic conductivity distribution. In this method, the catchment was divided into four zones following a 40-m head difference between hydraulic head contours [13,14].

The interval is considered sufficient to allow ample space for placing numerous pilot points and to spatially incorporate the subsurface heterogeneity over the model domain. The pilot points were then evenly distributed into the four zones by manual placement. Each zone contained the same number of pilot points. In this study, seven models were developed employing seven zone ratios (i.e., 1, 5, 10, 15, 20, 25 and 30; figure 2) to investigate the model performance. The simulations of groundwater flow under both steady-state and transient conditions were carried out using MODFLOW code [15], operated and visualized within the groundwater modelling system, GMS [16]. The model was discretised in the Layer Property Flow package, LPF [17], using a homogenous horizontal model of 50 m by 50 m grid cells. The catchment boundary was set as a no-flow boundary assuming no flux flowing through the boundary (figure1b). Downstream, a small stream after the spring was set as a specified head boundary using time-variant specified head, CHD [18].A constant value of 0.3 was set for the porosity to represent the carbonate formation [19]. The drain package, DRN [18] was used to model the surface drainage network of the recharge area. The streams' surface drain conductance was set between 2,700 and 8,125 m/d. To simulate recharge to the aquifer, the recharge package RCH [18] was used. The SCS-CN [20] and Penman-Monteith [21] methods were applied to calculate surface runoff and evapotranspiration respectively.

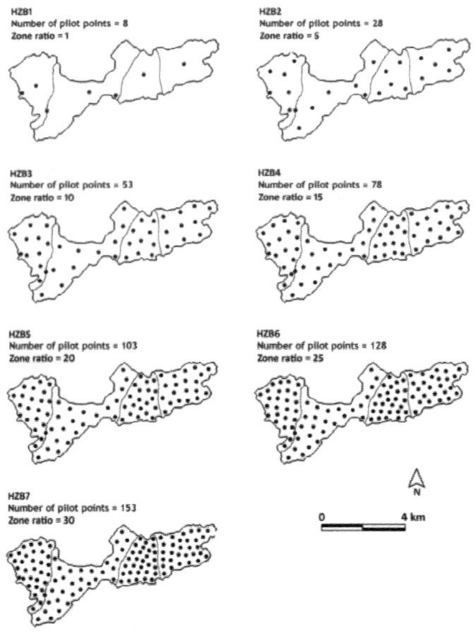

Figure 2. Spatial distributions of pilot points in seven models constructed using HZB distribution.

2.3. Model calibration

The K_h values are set to range between 0.1 and 250 m/d to represent karst limestone [22,23]. The initial K_h values then feed the transient simulation which aims to calibrate the specific yield (Sy) and K_h values. Transient simulation covers the period from January 2011 to August 2011, divided into 8 monthly stress periods, with each comprising of 10time steps. In both steady–state and transient simulations, PEST [24] was used. A calibration criterion for both the steady-state and transient simulations was employed to match simulated heads with observed heads. Daily data collected from 7 observation wells in the catchment for the duration of January, 2011 ~ April, 2012 were used as observed heads. The model calibration was accomplished by analysing the models' performance specified by statistical goodness-of–fit measures – the mean absolute error (MAE_h), root mean squared error of head ($RMSE_h$) and Nash-Sutcliffe coefficient (NSE_h)– as objective functions.

3. Results and discussion

3.1. Model calibration

Generally, the error statistics of all models using the two pilot points distribution methods are small in regards to $RMSE_h$ values (between 0.35 m and 0.71 m), confirming that all models are capable of reproducing hydraulic heads to a satisfactory level.The model performances in transient simulation are overall considered satisfactory in regards to $RMSE_h$ values, which vary from 0.36 m to 0.44 m.

3.2. Effect of pilot-point properties on model performance

Figure 4a show the model performances of the seven models developed using the HZB method. The model performances, represented by $RMSE_h$, increase by 13.49% when the zone ratio increases from 1 to 15, resulting in the increase of pilot-point number from 8 to 78. Conversely, the model performance

IOP Conf. Series: Earth and Environmental Science **92** (2017) 012026 doi:10.1088/1755-1315/92/1/012026

drops by 20.04% for subsequent increase of pilot-point number from 78 to 153 as a result of the zone ratio increase from 15 to 30. Although assigning more pilot points in the parameterisation step helps to increase the model performances, represented by decreased $RMSE_h$, too many pilot points prevent the model from achieving better result during calibration. This condition might be attributed to improper parameterisation of pilot-point properties, which causes over-parameterisation, thus overfitting. Processing speed is principally essential when dealing with a large number of pilot points. As the complexity arises due to the added number of points, large numbers of iterations take place in the optimisation process. Figure 4b shows the processing time incurred by the HZB methods.

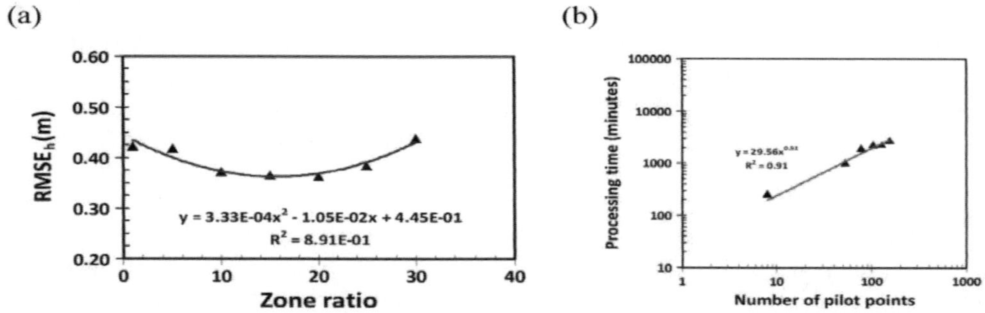

Figure 3. The result of model calibration showing: (**a**) model performance ($RMSE_h$) vs. zone ratio, which shows the impact of pilot-point density on the performances of the HZB models; and (**b**) Processing time incurred by the HZB models.

4. Conclusions

In this study, new approach of using geometric features of pilot point and distribution of hydraulic gradient over the catchment area is presented to efficiently configure pilot point distribution in the calibration step of a groundwater model. The Head Zonation-based (HZB) technique was applied on seven zone ratios (1, 5, 10, 15, 20, 25 and 30) on an eogenetic karst catchment in Rote Island, Indonesia and their performances were assessed. Overall, the observed-versus-simulated scatterplots of the seven models represented by considerably low $RMSE_h$ values in the transient simulation, indicate that a satisfactory level of model-to-observation fitting is obtained. The results show that the model performance increases by 13.49% when the zone ratio increases from 1 to 15, resulting in the increase of pilot-point number from 8 to 78. Conversely, the model performances drop by 20.04% after the increase of pilot-point number from 78 to 153 as a result of increased zone ratio from 15 to 30. It appears that although the use of many pilot points' results in significant increase of model performance, the excessive pilot-point number obviously increases the computational time, and to some extent could lead to a decrease of model performance. Therefore, choosing a reasonable pilot-point number and an appropriate distribution method using a practical guideline is important in the parameterisation step of groundwater modelling. Hence, appropriate selection of pilot-points properties is critical in developing a physically-based groundwater model. Therefore, based on the model simulations developed in this study, the recommended zone ratio is between 10 and 20 when using the HZB distribution method.

Acknowledgments
The authors wish to acknowledge the Australian governmentfor supporting the research through the Australia AwardsScholarship (AAS) and Allison Sudrajat Prize Award in the Swinburne University ofTechnology, Australia.

References

[1] Klaas D K S Y and Imteaz M A 2017 Investigating the impact of the properties of pilot points on calibration of groundwater models: case study of a karst catchment in Rote Island, Indonesia *Hydrogeol. J.* **25** 1703–1719

[2] Klaas D K S Y, Imteaz M A and Arulrajah A 2017 Development of groundwater vulnerability zones in a data-scarce eogenetic karst area using Head-Guided Zonation and particle-tracking simulation methods *Water Res* **122**

[3] Klaas D K S Y, ImteazM A, Sudiayem I, KlaasE M E and Klaas E C M 2017 Novel approaches in sub-surface parameterisation to calibrate groundwater models *IOP Conf. Series: Earth and Environmental Science* **82** 1-8

[4] de Marsily G 1984 *Spatial Variability of Properties* in Porous Media: A Stochastic Approach ed J Bear and M Y Corapcioglu (Boston, USA: NATO ASI Series E) 719-69

[5] Carniato L, Schoups G, van de Giesen N, Seuntjens P, Bastiaens L, Sapion H 2015 Highly parameterized inversion of groundwater reactive transport for a complex field site *J. Contam. Hydrol.* **173** 38-58

[6] Klaas D K S Y and Imteaz M A 2017 Development of a groundwater model for a tropical eogenetic karst aquifer in Rote Island, Indonesia *Proc.37^th^Hydrol. and Water Res.Symp.* 228-235

[7] Klaas D K S Y, Imteaz M A and Arulrajah A 2016 Evaluating the impact of grid cell propertiesin spatial discretization of groundwater model for a tropical karst catchment in Rote Island, Indonesia *Hydrol. Res.*

[8] Certes C and de Marsily G 1991 Application of the pilot point method to the identification ofaquifer transmissivities *Adv. Water Resour.* **14** 284-300

[9] Klaas D K S Y 2008 *Indigenous Water Management: Sustainable Water Conservation Strategies in a Karstic Dominated Area in Rote Island, NTT Province, Indonesia* (Monash University: Melbourne, Australia)

[10] Vacher H L, Mylroie J E 2002 Eogenetic karst from the perspective of an equivalent porous medium *Carbonates Evaporites* **17** 182-196

[11] Worthington S R H, Schindel G M and Alexander E C J 2001 Aquifer-scale Properties for Hydraulic Characterization of Carbonate Aquifers *Proc. Geol. Soc. of Am. Meet.* (Boston, USA)

[12] Scanlon B R, Mace R E, Barrett M E and Smith B 2003 Can we simulate regional groundwater flow in a karst system using equivalent porous media models? *J Hydrol* **276** 137–158

[13] Fetter CW 2001 *Applied hydrogeology* (Upper Saddle River: Prentice-Hall)

[14] Hudak PF 2010 *Principles of hydrogeology* (Boca Raton: CRC)

[15] Mc Donald M and Harbaugh A 1988 *A Modular Three-Dimensional Finite-Difference Groundwater Flow Model (USGS)*

[16] BYU 2014 *Groundwater Modeling System* (GMS) version 9.2.9. (Provo: Environmental Modeling Research Laboratory, Brigham Young University)

[17] Harbaugh A W 2005 *MODFLOW-2005* the U.S. Geological Survey Modular Ground-Water Model, the Ground-Water Flow Process (USGS Techniques and Methods 6-A16, USGS)

[18] Harbaugh A W, Hill M C and Mc Donald M G 2000 *MODFLOW-2000* the U.S. Geological Survey Modular Ground-Water Model: User Guide to Modularization Concepts and the Ground-Water Flow Process

[19] Johnson A I 1967. *Specific Yield – Compilation of Specific Yields for Various Materials* pp 1-74

[20] Viessman W and Lewis G L 2003 *Introduction to Hydrology* (Upper Saddle River: Pearson Education)

[21] Allen R G, Pereira L S, Raes D and Smith M 1998 *Crop evapotranspiration guidelines for computing crop water requirements* FAO Irrigation and Drainage Paper 56

[22] Davis S N 1969 *Porosity and Permeability of Natural Materials in Flow Through Porous Media* (New York: Academic Press)

[23] Freeze R A and Cherry J A 1979 *Groundwater* (Englewood Cliffs: Prentice-Hall)

IOP Conf. Series: Earth and Environmental Science **92** (2017) 012027 doi:10.1088/1755-1315/92/1/012027

Isoconcentration principle – an effective tool of (radio) analysis in environmental protection

J Klas†, J Lesny[1,2], M Hornik[1], I Matusikova[1] and V Adamcova[1]

†With memory of our departed colleague, RNDr. Ing. Ján Klas, DrSc. Such an honourable professor and a faithful friend will be incredibly missed, especially his exceptional scientific inventiveness and his warm heart
[1]Department of Ecochemistry and Radioecology, Faculty of Natural Sciences, University of SS. Cyril and Methodius in Trnava, Nám. J. Herdu 2, Trnava, SK-917 01, Slovak Republic

E-mail: lesny@ucm.sk

Abstract. The paper brings two determination methods for non-radioactive analytes based on isoconcentration principle. The first method represents a determination procedure for a non-radioactive substance by use of its radio-labelled form of known concentration by help of a (non-quantitative, but selective) separation method. The main advantage of this method consists in the fact, that the determination is executable only by the detection of radioactivity. The second described procedure enables the determination of the chosen analyte without requirement of radioactive form of the substance to be determinate.

1. Introduction

An effective protection of the environment puts extremely high demands on the implementation of analytical methods for an extremely wide range of analytes in an extremely wide range of matrices. This is due to many reasons. For humans, the environment presents a crucial part of the material world. According to the definition, the environment is everything which creates natural conditions for the existence of organisms, including humans, and is the precondition of their further development [1]. Components of the environment are mainly water, soil, air and organisms. Therefore, the list of analytes in the environment is particularly long. Analogically, the list of components of the respective matrices, which can naturally contain many potential interferences, is also extremely long.

Analytes of the environment are mostly present in concentrations which are several orders of magnitude lower than the concentrations of the components in the relevant matrices. All the stated facts contribute to the occurrence of a wide range of sources of incorrect results mainly in the application of relative methods requiring calibration.

A series of publications appeared in the 1970s and 1980s developing the isotope dilution analysis by the application of an isoconcentration principle [2–7]. These publications focused mainly on the radioactive analytes. The aim of our paper is to contribute to the possible methodological solutions of the stated problems by proposing and modifying the isoconcentration principle for determination of non radioactive analytes.

2. Methods of determination

2.1. Determination process by use of radio-labeled analyte of known chemical concentration

The modified process of the isotope dilution analysis for the determination of non radioactive analytes (provided that selective separation procedures (reagents) are available, and provided that the radio-labelled analyte of known chemical concentration is also available) consists of the following steps.

2.1.1. Step 1

Creation of two series (I, II) of aliquot volumes and an addition of known equal volumes of unknown concentration, i.e. the same unknown amounts (**x**) of the non radioactive substance, which is be determined, to the individual elements of the first series.

2.1.2. Step 2

Addition of known increasing volumes of known concentration, i.e. known increasing amounts (y_1,...y_Φ...y_j) of the radio-labeled analyte to the elements of the first series and an addition of Φ-multiple of these amounts (Φ.y_1,...Φ.y_Φ...Φ.y_j), to the individual elements of the second series.

2.1.3. Step 3

The same adjustment of reaction conditions (pH, temperature…), the addition of the same amounts of the separation reagent to all the elements in both series and refilling of all elements from both sets to the same volume.

2.1.4. Step 4

Implementation of the separation process for all elements of both sets. Execution of radioactivity detection of separation products. Theoretically, for one of the elements of the first, and second set a condition can occur for which the following applies:

$$\frac{(\mathbf{x}+\mathbf{y}_\phi)}{\mathbf{V}} = \frac{(\mathbf{\Phi}.\mathbf{y}_\phi)}{\mathbf{V}} \tag{1}$$

Separated masses:	$M_{1;1}$	$M_{1;2}$	$M_{1;3}$	$M_{2;1}$	$M_{2;2}$	$M_{2;3}$
Separated activities:	$A_{1;1}$	$A_{1;2}$	$A_{1;3}$	$A_{2;1}$	$A_{2;2}$	$A_{2;3}$
General symbols:		$(M_{1j}\,;A_{1j})$			$(M_{2j}\,;A_{2j})$	

2.1.5. Step 5, Construction of the isotope dilution function

The ratio of separated activities of the individual elements in the second set and the respective separated activities of the individual elements in the first set (which is the function of the known increasing amounts $\mathbf{y_1,...y_\Phi...y_j}$ of radio-labelled analyte) is identified by the symbol $\mathbf{i_j}$ (equation (2)):

$$\frac{\mathbf{A}_{2j}}{\mathbf{A}_{1j}} = i_j = f(y_j) \tag{2}$$

The total activity of the separated products of individual elements in both sets can be expressed as the product of the total mass and its relevant specific activity. We can write (see equations (3)(4)(5)):

$$\boldsymbol{\Phi}(x+y_j)\,(A_{1j}/M_{1j}) = \boldsymbol{\Phi} y_j\,(A_{2j}/M_{2j}) \tag{3}$$

$$(x+y_j)\,(A_{1j}/M_{1j}) = y_j\,(A_{2j}/M_{2j}) \tag{4}$$

$$i_j = \frac{x+y_j}{y_j}\,\frac{M_{2j}}{M_{1j}} = \frac{M_{2j}}{M_{1j}}\left(1+\frac{x}{y_j}\right) \tag{5}$$

Theoretically, two extreme situations may happen:

- separation of the whole quantities of analyte from all elements of both sets;
- separation of the same quantities of analyte from all elements of both sets.

In case of separation of the whole quantities of analyte from all elements of both sets the following equation (6) applies:

$$M_{1j} = x + y_j \quad \text{and} \quad M_{2j} = \phi . y_j \tag{6}$$

Under these circumstances (equation (7)):

$$i_j = \frac{x + y_j}{y_j}; \; i_j = x \tag{7}$$
$$di/dy = 0$$

In case of separation of equal quantities of analyte of all the elements of both sets the following equations (8) (9) applies:

$$M_{1j} = M_{2j} \tag{8}$$

$$i_j = \frac{x + y_j}{y_j} = \frac{x}{y_j} + 1 \tag{9}$$

If we introduce a substitution:

$$z = \frac{1}{y_j}$$
$$i_j = z . x + 1$$
$$di/dy = x$$

2.1.6 Step 6

Determination of the unknown amount of non-radioactive analyte (**x**) from the curve $\mathbf{i_j = f(z)}$. In the intersection of all isotope dilution curves $i_j = f(z)$ at the same time applies the separation of the whole quantities as well as equal quantities of analyte from all the elements in both sets. In this situation equation (10) is:

$$i_j = \phi; \qquad z_\phi = \frac{(\phi - 1)}{x} \tag{10}$$

Thus, according to equation (1) we can write (see equations (11)(12)):

$$x + y = \phi . y \tag{11}$$

$$x = \frac{(\phi - 1)}{z_\phi} \tag{12}$$

Thus, according to the equation (12), the determination of the unknown quantity x is in this case executable without the determination of separated quantities, i.e. only by the radioactivity detection (figure 1).

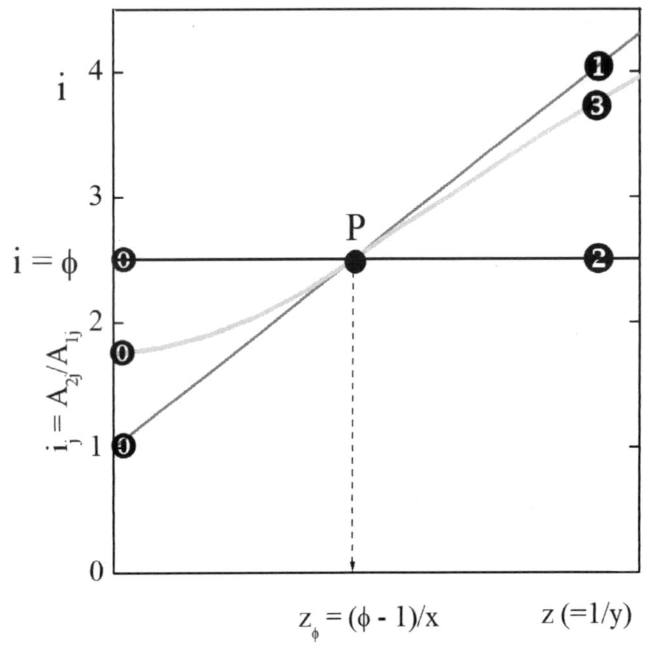

❶ $\frac{di}{dy} = x = constant$

The same amounts of separation product are isolated from all elements in both sets

❷ $\frac{di}{dy} = 0$

Full amount of analyte is isolated from all the elements in both sets.

❸ $0 < \frac{di}{dy} \leq constant$

Useful area of the subject method.

Figure 1. Construction of the isotope dilution function.

2.2. Determination process by use of non radio-labeled analyte of known chemical concentration

The procedure for the determination of the substance ***M*** with the use of isoconcentration principle without the use of (radio) isotopic indicator is as follows:

From the solution I with the substance ***M*** to be determined, two sets of aliquot volumes will be taken. The first set (FS) will contain equal aliquot volumes $\text{ɒ}.V_I$, where ɒ represents an arbitrary selected number **(ɒ > 1)**. Individual elements in this set contain the amount $\text{ɒ}.x$ of the substance to be determined. At the same time, from the solution I we take a second set (SS) of equal aliquot volumes $1.V_I$, which contain the amount x of the substance to be determined. From an other solution (II) of the same studied substance ***M***, we take variable quantities of volumes $\alpha.V_{II}$, where α represents the variable sequence of numbers (α = 0; 0,1; 0,2; ... 1; 1,1; 1,2;...); V_{II} is the volume of the solution II containing a quantity of the substance ***y*** to be determined (at α = 1). To aliquots $\alpha.V_I$ we add the volumes of the second (SS) set of aliquots $1.V_I$, which creates a mixed set of aliquots (MS). This results in two sets of aliquots with various volumes of the studied substance: FS with equal $\text{ɒ}.x$ content of the substance ***M***, and MS with variable $x + \alpha.y$ content of the substance ***M***. All aliquots shall be completed to the same volume ***V***. The stated two sets of aliquots are in mathematical equations indicated by subscript $_\text{ɒ}$ or $_\alpha$.

If we mark the function of a measurable quantity related to the amount of the substance ***M*** (or its concentration), for FS with $I_\text{ɒ}$ and for MS with I_α, then we can state (see equations (13)(14)):

$$I_\text{ɒ} = k_\text{ɒ}.\text{ɒ}.x \qquad (13)$$

$$I_\alpha = k_\alpha.(x + \alpha.y) \qquad (14)$$

Where ***k*** is a function between the measurable amount (or concentration) of the substance ***M*** in the first and mixed set of aliquots. This quantity can be optical, electrochemical, physical or nuclear properties associated with the quantity (concentration) of the studied substance ***M***.

Quantities, which can be determined by measurements for the stated sets $I_\text{ɒ}$ and I_α can be formulated as a function:

$$i = \text{ɒ}.I_\alpha/I_\text{ɒ} = (k_\alpha/k_\text{ɒ})\,(1 + \alpha.(y/x)) = f(k_\alpha, k_\text{ɒ}, \alpha) \qquad (15)$$

If in isoconcentration point $\boldsymbol{P}$ applies $k_{\alpha P} = k_{ɒP}$, then the isoconcentration point $\boldsymbol{P}$ with coordinates (α_P, $ɒ$) is the intersection of the equation (15) with the a priori selected constant value $i = I_P = ɒ$ (see equation (16)):

$$I_P \;=\; ɒ.I_{\alpha P}/I_{ɒP} \;=\; (1 + \alpha_P\,(y/x)) \;=\; ɒ \tag{16}$$

of which

$$x \;=\; y.\alpha_P/(ɒ - 1) \quad or \quad y \;=\; x.(ɒ - 1)/\alpha_P \tag{17}$$

With the use of equations (17), if the value $\boldsymbol{y}$ is known, it is possible to calculate the value $\boldsymbol{x}$, or if the value $\boldsymbol{x}$ is known, the value $\boldsymbol{y}$ can be calculated. If both these values are known, it is possible to calculate their ratio (see equation (18)):

$$x/y \;=\; \alpha_P/(ɒ - 1) \tag{18}$$

In case of determination of the quantity $\boldsymbol{x}$ there is a partial compensation of the effect of present substances in the matrix on the substance $\boldsymbol{M}$, because their quantities in the set of aliquots FS and MS are in the ratio $ɒ : \alpha$, while by determination of $\boldsymbol{y}$ there is no such compensation ($0 : \alpha$).

The uncertainty of this method is, in general, consist in the fact, that between the measurable quantity ($I_{ɒ}$, I_{α}) and a variable amount of the substance $\boldsymbol{M}$ (or its variable concentration) the functions k_{α} and $k_{ɒ}$ are not generally equal ($k_{\alpha} \neq k_{ɒ}$).

The advantage is an unquestionable fact, that in the isoconcentration point $\boldsymbol{P}$ (α_P, $ɒ$) the concentrations of aliquots of the substance $\boldsymbol{M}$ in the set FS and in the set MS are equal and the measured quantities (unless there is an interference of the substance $\boldsymbol{M}$ with other present substances in the matrix, or if the measured quantity is not affected by other factors) are also the same. In this case $k_{\alpha} = k_{ɒ}$ and the result of determination can be perceived as reliable.

References

[1] Ďurčo P et al. 2007 *Bezpečnostnoprávna terminológia* (Bratislava: Akadémia PZ) p 176
[2] Garten R P H and Tölgyessy J 2001 *Radionuclides in analytical chemistry. In: Ullman's Encyclopedia of Industrial Chemistry* (Weinheim: Wiley-VCH)
[3] Klas J, Tölgyessy J and Klehr E H 1974 *Radiochem. Radioanal. Lett.* **18** p 83
[4] Klas J, Tölgyessy J and Lesný J 1985 *Sub-superekvivalentová izotopová zrieďovacia analýza* (Bratislava: Veda) p 156
[5] Klas J, Tölgyessy J and Lesný J 1977 *Radiochem. Radioanal. Lett.* **31** p 171
[6] Lesný J, Tölgyessy J and Klas J 1976 *Radiochem. Radioanal. Lett.* **26** p 363
[7] Rao V R S, Rao CH P and Tataiah G 1977 *Radiochem. Radioanal. Lett.* **30** p 365

IOP Conf. Series: Earth and Environmental Science **92** (2017) 012028 doi:10.1088/1755-1315/92/1/012028

The properties of waste water from a concrete plant

L Klus[1], V Václavík[1,2], T Dvorský[1], J Svoboda[1] and R Papesch[1]

[1]Institute of Environmental Engineering, Faculty of Mining and Geology, VSB-Technical University of Ostrava, Czech Republic
[2]Institute of Clean Technologies for Mining and Utilization of Raw Materials for Energy Use Faculty of Mining and Geology, VSB-Technical University of Ostrava, Czech Republic

Email: lukas.klus@vsb.cz

Abstract. This paper presents the results of a research dealing with the testing of the properties of wastewater forming during the production of concrete mixture in a concrete plant. The test results are compared with the limit values of concrete mixing water according to ČSN EN 1008 in order to use this waste water as a partial replacement of mixing water during the production of concrete.

1. Introduction

The cost-effective management of raw materials and recycling materials by returning them into the production processes is a trend that is nowadays present in almost all fields. This trend is also evident in the building industry, where recycled materials are widely used as substitutes for the primary raw materials of newly developed building elements. One of the main reasons is reducing the environmental loads and cutting the manufacturing costs of products. This article deals with the issue of wastewater from concrete mixture production. It describes its properties according to ČSN EN 1008 [1] with an emphasis on the possible use as a substitute for mixing water in the production of concrete composites.

At present, we already know the results of a research dealing with the utilization of wastewater from a car wash as a substitute for the mixing water in the production of concrete [2, 3], household waste water used as the mixing water in the production of concrete [4]. The use of sludge water from a concrete plant as a substitute for mixing water in the interval of 10 - 50% and its impact on the concrete strength parameters are described in [5].

A concrete plant in Ostrava, which has the production capacity of 85 m^3/h of concrete mixture, was chosen for the experimental research of the properties of waste water from the production of concrete mixture. Sedimentation tests and tests of the following properties were performed using the samples of waste water: pH, temperature, conductivity and the concentrations of chlorides, sulphates, and nitrates. The determined concentrations were compared with the concentrations of pure mixing water (surface water from a water tank) and with the contamination limit values of the mixing water, which are defined in the ČSN EN 1008 standard [1].

2. Methods

2.1. Sampling
Two types of samples (sludge water from a concrete agitator rinse and pure mixing water) were taken within the scope of the experimental research. The samples were taken at five time intervals during one working day. The samples were stored and subjected to laboratory tests over the next 14 days. 5 samples of sludge water and 5 samples of pure mixing water were taken during one day of sampling. The samples of sludge water had to be modified for further tests by means of a filtration process using a ChS Vaccspace vacuum pump.

2.2. Determination of the pH value, temperature and conductivity
The pH values of the samples of sludge water and pure mixing water were measured potentiometrically by means of an electrochemical analytical method based on the measurement of the equilibrium cell voltage, which is composed of indicative and reference electrode. The temperature and conductivity determination was performed using Merci pH/Cond 3401 instrument during each sampling.

2.3. Sedimentation test
A sedimentation test of sludge water was performed on 5 samples from 1 day of sampling in 2000 ml graduated cylinders for 30 minutes. The sludge sediment was measured at 5-minute intervals.

2.4. Determination of the mineralogical composition of the sludge
The determination of the mineralogical composition of the sludge sample from the concrete plant waste water was carried out using Bruker Advance D8 powder diffractometer equipped with Lynx Eye linear semiconductor detector and SOL-XE energy dispersion detector.

2.5. Determination of the chloride concentration
A methodology in accordance with ČSN ISO 9297 [6] was used to determine the concentration of chlorides in the samples of sludge water and pure mixing water.

2.6. *Determination of the sulphate concentration*
A methodology in accordance with ČSN 75 7477 [7] was used to determine the concentration of sulphates in the samples of sludge water and pure mixing water.

2.7. *Determination of the nitrate concentration*
A methodology in accordance with ČSN ISO 7890-3 [8] was used to determine the concentration of nitrates in the samples of sludge water and pure mixing water.

3. Results and discussion
The table 1 presents the average values of 5 samples used to determine pH, temperature and conductivity. Significant differences in the values of pH and conductivity are clearly visible when the values of sludge and pure mixing water are compared. The increase in the pH value to approximately double value in case of sludge is caused by the binder component of the concrete mixture entering the sludge water. The dispersed particles of the binder component and the fine share of the filler in the concrete mixture cause a significant reduction of the conductivity in sludge water, by app. 26 times, when compared with the conductivity value of pure mixing water.

Table 1. Results of the values of pH, temperature and conductivity.

pH	t [°C]	Conductivity [μS/cm]
Sludge water		
12.56	20.76	13.00
Pure mixing water		
6.77	18.78	336.12

The figure 1 presents the settling curves of the wastewater samples from the concrete plant taken at different time intervals during one day. It is clear that the sludge settling values show a linear decrease in all samples over a time interval of 5 to 30 minutes. However, they vary in the settling values in the individual points of the time interval. The value of the variation range from five samples in the time interval of 5, 10, 15, 20, 25 and 30 min. reaches the values of 22, 29, 30, 31, 32, 34. It is therefore apparent that the particles that are capable of settling in the form of sludge are unevenly distributed in the waste water from the concrete plant. The average settling values of 5 samples of sludge water in the time interval of 5 - 30 min. are shown clearly in tables 2 and 3 that the average density of the sludge water from the concrete plant has reached the value of 1098 kg / m^3 and the sludge content in 1 m^3 represents about 60 kg (5.5%).

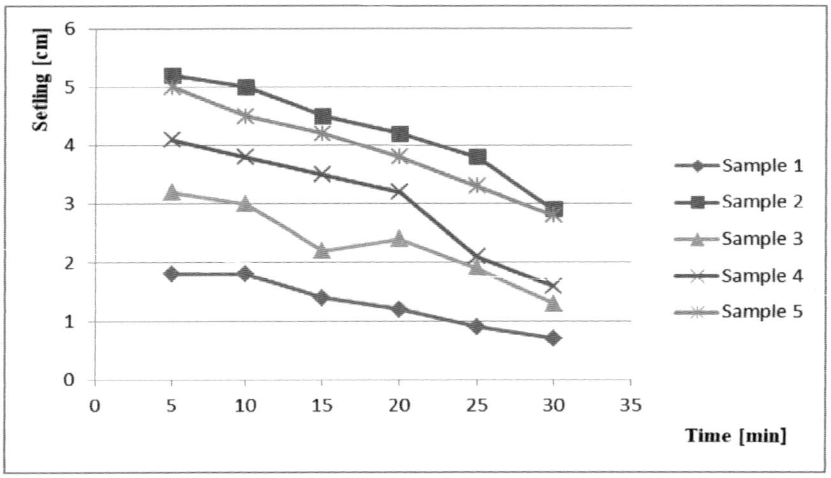

Figure 1. Sedimentation curves of sludge water samples from the concrete plant.

Table 2. Average settling values of the concrete plant sludge water from 5 samples.

Time [min.]	5	10	15	20	25	30
Sludge settling [cm]	1.86	2.40	2.96	3.26	3.62	3.86

Table 3. Sedimentation test result.

Sample	1	2	3	4	5	Average
Weight of sludge water in 1 litre [g]	1126.20	1067.80	1109.60	1099.60	1087.80	1098.20
Weight of solid share in 1 litre of sl. water [g]	19.6	45.6	125	93.8	15.6	59.92

The figure 2 presents the average values of mineralogical composition of the sludge acquired from the waste water produced by the concrete plant. Limestone, which is calcium carbonate $CaCO_3$ with a hardness of 3 and a density of 2.7 g/cm^3 and which decomposes in acids, has the highest share among the identified minerals. Other minerals with the lower percentage share are:

- bassanite - $CaSO_4.0,5H_2O$, hydrates to gypsum, density is 2.5-2.7 g/cm^3,

- quartz – SiO_2, hardness 7, density 2.6 g/cm^3, soluble in hydrofluoric acid,
- portlandite – $Ca(OH)_2$, hardness 2.5-3, density 2.26 g/cm^3,
- oligoclase An16 – $(Ca,Na)(Al,Si)_4O_8$, hardness 6-6.5, density 2.65 g/cm^3,
- chlorite IIb – $(Mg,Fe)_6\,(Si,Al)_4O_{10}\,(OH)_8$, hardness -, density 2.5 g/cm^3,
- muscovite – $KA_2(AlSi_3O_{10})(F,OH)_2$, hardness 4, density 2.76-3 g/cm^3,
- orthoclase - $KAlSi_3O_8$, hardness 6, density g/cm^3,
- anhydrite – $CaSO_4$, hardness 3.5, density 2.95 g/cm^3, the highest content was found in the last sample,
- sylvite – KCl, hardness 2, density 1.49 g/cm^3.

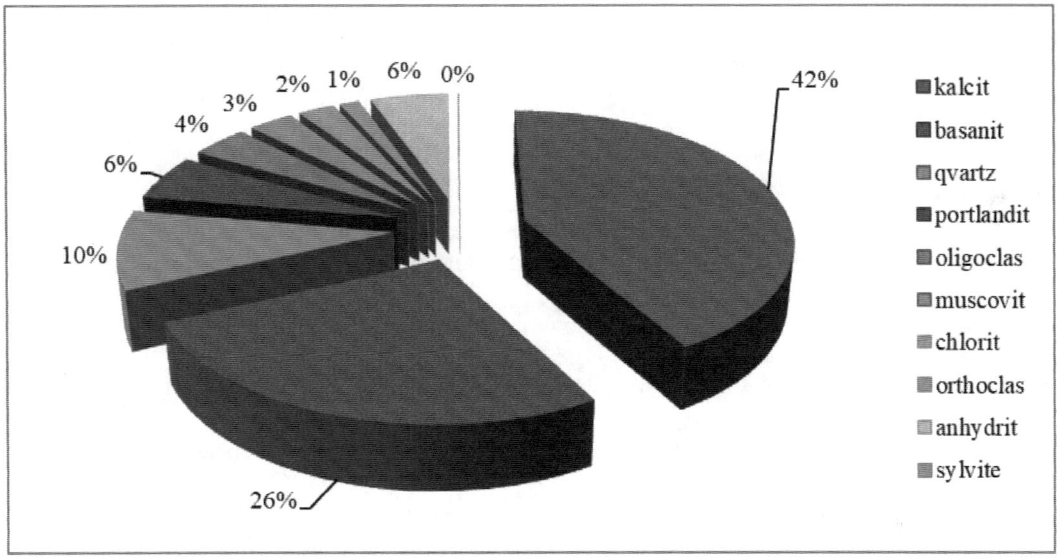

Figure 2. Mineralogical composition of waste water sludge from a concrete plant.

Table 4. Measured concentration values and limit concentrations according to ČSN EN 1008.

Properties	Measured concentrations of sludge water [mg/l]	Measured concentrations of pure mixing water [mg/l]	Limit concentrations according to ČSN EN 1008 [mg/l]
Chlorides	251.7	14.2	-
Plain concrete			< 4500
Reinforced concrete			< 2000
Pre-stressed concrete			< 600
Sulphates	1702.1	40,3	< 2000
Nitrates	-	9.4	< 500

The table 4 shows a comparison of the tested properties of the waste water from the concrete plant and the pure mixing water with the recommended concentrations according to ČSN EN 1008. At first glance, it is evident that the measured average concentrations of chlorides, sulphates and nitrates meet the limit concentrations. The concentration of chlorides in sludge water compared to the limit value according to ČSN EN 1008 reaches the percentage value of 5.6% for plain concrete, 12.6% for reinforced concrete, and 42% for pre-stressed concrete. The sulphate concentration is represented by the percentage share of 85.1% when compared to the limit value.

4. Conclusion

This article presents the results of the tests of sludge water from the concrete plant and pure mixing water. It presents the average values of five samples, because the composition of the waste water from the concrete plant varies during the day and directly depends on the type of concrete being produced at the concrete plant. The tested properties included pH, temperature, conductivity, the mineralogical composition of waste water sludge, and the content of chlorides, sulphates and nitrates. The results meet the requirements for concrete mixing water as set out in ČSN EN 1008. The direction of further research will be focused on the replacement of pure mixing water with 20% and 50% of sludge water from the concrete plant and its impact on the strength characteristics of cement composites.

Acknowledgement
This article was written in connection with project Institute of clean technologies for mining and utilization of raw materials for energy use - Sustainability program. Identification code: LO1406. Project is supported by the National Programme for Sustainability I (2013-2020) financed by the state budget of the Czech Republic.

References

[1] ČSN EN 1008: 2003 Mixing water for concrete. Specification for sampling, testing and assessing the suitability of water, including water recovered from processes in the concrete industry, as mixing water for concrete
[2] Shahidan, Shahiron, Mohamad Syamir Senin, Aeslina Binti, Abdul Kadir et all 2017 Properties of Concrete Mixes with Carwash Wastewater (MATEC Web of Con.87)
[3] Al-Jabri K S, A H Al-Saidy, R Taha and A J Al-Kemyani 2011 Effect of using Wastewater on the Properties of High Strength *Concrete. Proc. Eng.* **14** 370-376
[4] Cebeci Oz and Saatci AM 1989 Domestic sewage as mixing water in concrete ACI *Mat. J.* **86** p 503-506
[5] Kadir, Aeslina Abdul, Shahiron SHAHIDAN, Lau Hai Yee, Mohd Ikhmal Haqeem Hassan and Mohd Mustafa Al Bakri Abdullah 2016 The effect on slurry water as a fresh water replacement in concrete properties. *IOP Con. Ser.: Mat. Sci. and Eng.* **133**
[6] ČSN ISO 9297: 1996 Water quality. Determination of chloride. Silver nitrate titration with chromate indicator (Mohr's method)
[7] ČSN 75 7477: 2010 Water quality. Determination of dissolved sulphate. Lead nitrate titrimetric method
[8] ČSN ISO 7890-3: 1995 Water quality. Determination of nitrate. Part 3: Spectrometric method using sulfosalicylic acid

Changes of strength characteristics of pervious concrete due to variations in water to cement ratio

M Kovac[1] and A Sicakova[1]

[1]Technical University of Košice, Faculty of Civil Engineering, 042 00, Košice, Slovakia

Email: marek.kovac@tuke.sk

Abstract. Pervious concrete is considered to be a sustainable pavement material due to high water permeability. The experiment presented in this paper was aimed at study the influence of water to cement ratio on both the compressive and splitting tensile strength of pervious concrete. Typically, less water content in concrete mixture leads to less porosity of cement paste and thus it provides desirable mechanical properties. In case of conventional dense concrete, the lower is the water to cement ratio, the higher or better is the strength, density and durability of concrete. This behaviour is not quite clear in case of pervious concrete because of low amount of cement paste present. Results of compressive and splitting tensile strength of pervious concrete are discussed in the paper while taking into account values measured after 2 and 28 days of hardening and variations in water to cement ratio. The results showed that changes of water to cement ratio from 0.25 to 0.35 caused only slight differences in strength characteristics, and this applied to both types of tested strength.

1. Introduction

Portland cement pervious concrete (PCPC) allows water to penetrate its structure through a series of interconnected voids. This characteristic goes against all design criteria for conventional concrete, whose typical focus is on remaining impermeable. If designed, compacted, and maintained correctly, many sustainability benefits can be seen through the use of pervious concrete. Various other types of pervious/porous pavements show similar benefits. Permeable pavements can be divided into nine different categories, including porous aggregate, porous turf, plastic geo-cells, open-jointed paving blocks, open-celled paving grids, pervious concrete, porous asphalt, soft paving materials and decks [1]. All of these permeable systems are similar in nature, typically consisting of a granular base that not only acts as storage for excess storm water, but also as structural support [2]. The use of these types of pavements has been increasing in recent years due to their many environmental benefits. Like pervious concrete, the long-term performance of these pavements is still a concern and further research is needed to understand these pavements to a depth that traditional pavements are understood [3].

Pervious concrete mixtures have little to no slump [4]. This inherent property is due to the low cement paste content, which only allows for a thin film coating of cement paste around the aggregates [5]. High viscosity paste is needed so that the aggregates can be coated, while resisting the drain-down of the paste. The mitigation of drain-down is pertinent so that the matrix porosity is maintained throughout the width of the concrete section. The structure relies on the stone-to-stone contact achieved through compaction, which allows for the cement paste coated aggregates to bond with one another. It is recommended that the appropriate cement paste, to achieve the proper void structure,

should possess a low water to cement ratio (w/c), about 0.20 to 0.25, in addition to superplasticizer and adequate mixing [6]. Experiment given in [7] shows decrease of compressive strength along with decrease of w/c ratio what is contrary with conventional dense concrete behaviour.
W/c ratio in published experimental works varies in wide range from 0.2 to 0.4 usable for pervious concretes. In principal, pervious concrete mixtures with w/c ratios under 0.3 requires usage of water reducing admixtures while w/c above 0.3 can be mixed without plasticizer [5,7]. In general, w/c in range from 0.27 to 0.34 can be assumed as the most common and wide range applicable for pervious concrete mixtures [8].

This article presents a study that has been conducted to confirm the applicability of the w/c ratio in the range of 0.25 to 0.35 for locally available concrete components. Influence on both the compressive and splitting tensile strength of pervious concrete is evaluated, while taking into account values measured after 2 and 28 days of hardening.

2. Materials and methods

The experiment was focused on the utilization of locally available materials (Eastern Slovakia), given in following list:

- cement CEM II/A-S 42.5 R in accordance with [9];
- natural aggregate (siliceous gravel type) in accordance with [10]. To create open structure of concrete, single size coarse aggregate of fraction 4/8 was used. To maintain desirable performance of concrete, 7% of sand was added to aggregate mixture. Properties of aggregate are given in table 1.

Table 1. Properties of aggregate used in experimental work.

Properties	Aggregate fraction	
	0/4	4/8
Density [kg/m^3]	2600	2580
Dry rodded unit weight [kg/m^3]	1830	1580
Water absorption [%]	1.2	1.4
Void content [%]	30	39

The sieve analysis of aggregate was conducted to get the information about granulometric composition of used aggregate (table 2).

Table 2. Sieve analysis of aggregate used in experiment.

Sieve mesh size [mm]	Cumulative percent passing [%]		
	0/4	4/8	7% 0/4 + 4/8
0	0	0	0
0.063	0.9	0	0.1
0.125	3.3	0	0.3
0.25	10.4	0.1	0.8
0.5	31.8	0.1	2.3
1	54.0	0.2	4.0
2	75.8	0.3	5.6
4	98.2	8.4	14.4
8	100.0	94.1	93.3
16	-	100	100

Proportions of materials were calculated by method given in [11]. Three recipes of pervious concrete mixtures were designed and mixed in this experiment. The first step in calculation was determination of voids content. Voids content was intended as constant in each of mixtures and the value was selected using following criteria: voids content for pervious concrete in accordance with [12] is within 18% to 23% and in accordance with [8] it should range between 15% and 25%.

Assuming these boundaries, 20% voids content was considered in further calculations. Hence, volume of aggregate (0.575 m^3) and volume of cement paste (0.225 m^3) were constant for all mixtures. Mixtures differed by each other only in w/c ratio (0.25, 0.30 and 0.35). The final proportions of mixtures for 1 m^3 of fresh pervious concrete are given in table 3.

Table 3. Proportions of pervious concrete mixtures.

Material	Units	Mixture		
		W35	W30	W25
Cement	[kg/m^3]	327	354	384
Water		115	106	96
Aggregate 0/4		96	96	96
Aggregate 4/8		1375	1375	1375
Plasticizer		0.3	1.4	4.2
w/c	[-]	0.35	0.30	0.25
Calculated unit weight	[kg/m^3]	1913	1931	1951

Samples of pervious concrete were mixed in the forced action concrete mixer. The first step was mixing aggregate for about 15 seconds then, cement was added and mixed another 15 seconds and finally the water with superplasticizer were added and mixed thoroughly. Pervious concrete specimens of 100 mm in diameter and 200 mm height were made by constant compacting (15 hits). After the curing in water, specimens were tested in 2 and 28 days of setting and hardening to compressive and splitting tensile strength.

3. Results and discussion

During the mixing, the nature of fresh mixture workability was controlled visually. Mixture W35 (w/c = 0.35) showed the slightly runny character immediately after mixing however after few next minutes the excessive water has been soaked by aggregate. On the other hand, mixture W25 (w/c = 0.25) showed very stiff nature and despite of higher dosage of plasticizer (lignosulphonate base), it was difficult to process.

3.1. Compressive strength

The results of compressive strength after 2 and 28 days of setting and hardening are given in figure 1. After 2 days of curing, mixtures W35 and W30 reached 9.5 MPa and 9.0 MPa respectively. Mixture W25 with the lowest w/c (0.25) achieved the highest compressive strength 10.5 MPa. After 28 days of curing, mixtures W35 and W30 reached 14.5 MPa and 16 MPa respectively. Mixture W25 with the lowest w/c (0.25) achieved highest compressive strength 17.5 MPa. Difference between the compressive strength values of samples after 2 days of curing (1.5 MPa) and after 28 days of curing (3 MPa) was very slight despite significant w/c variation (from 0.35 to 0.25).

3.2. Splitting tensile strength

The results of splitting tensile strength after 2 and 28 days of setting and hardening are given in figure 2. After 2 days of curing, mixtures W35 and W25 reached 1.3 MPa and 1.4 MPa respectively. Mixture W30 with w/c = 0.30 achieved the highest splitting tensile strength 1.5 MPa. After 28 days of curing, mixtures W25 and W30 reached 1.6 MPa and 1.7 MPa respectively. Mixture W35 with w/c ratio 0.35 achieved the highest splitting tensile strength 2 MPa. Difference between the splitting tensile strength values of samples after 2 days of curing (0.2 MPa) and after 28 days of curing (0.4 MPa) was very slight despite significant water to cement ratio variation (from 0.35 to 0.25).

Mixture with the highest w/c (0.35) obtained the highest splitting tensile strength; it can be explained by better workability and thus better bond between the aggregate and cement paste. In case of mixtures W30 and W25, stiffer cement paste showed weaker adhesion to aggregate grains what resulted in lower splitting tensile strength. Similar behavior was observed and discussed in [13] and [14]. However, due to very slight difference in splitting tensile strength between mixtures (difference was smaller than measurement accuracy), it can be assumed that present variation in w/c ratio had not significant influence on splitting tensile strength.

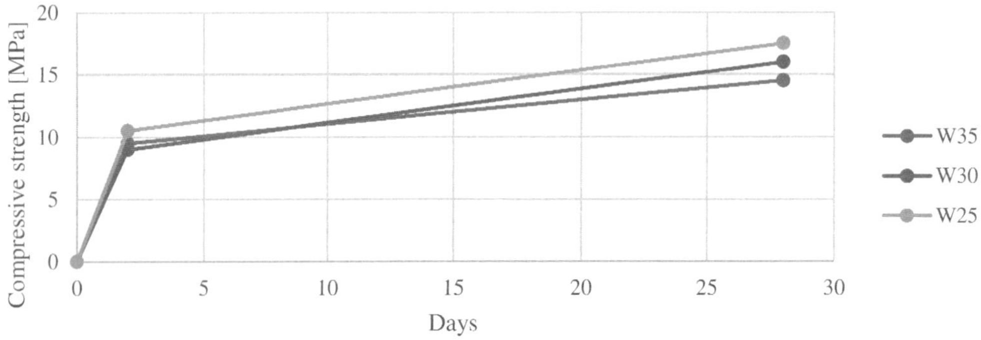

Figure 1. Compressive strength development of tested mixtures.

The nature of failure should also be described since it was clearly visible after the splitting tensile strength test. The crack came through most of the aggregate grains (figure 3); this referred to strong cement stone. Therefore next increasing of the cement paste strength by lowering w/c ratio needn't lead to better strength characteristics because the strength of aggregate can be a limiting factor [14].

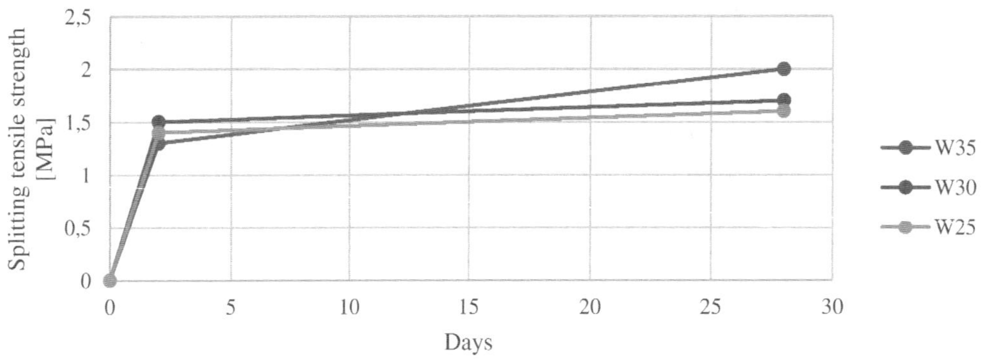

Figure 2. Splitting tensile strength development of tested mixtures

Figure 3. View of the inner structure of pervious concrete mixtures W25, W30 and W35 (left to right) after the splitting tensile strength test.

Conclusion

In the case of conventional dense concrete, the lower is the water to cement ratio, the higher or better are the strength, density and durability of concrete. This article presented results of study that has been conducted to confirm the applicability of the w/c ratio in the range of 0.25 to 0.35 for production of pervious concrete made of locally available components. The following conclusions can be formulated:

- very slight difference in strength characteristics between mixtures allows for suggestion that 0.25 to 0.35 range of w/c ratio has no significant influence on strength characteristics of pervious concrete;
- the nature of failure of all samples points to strong cement stone for each of w/c ratio while the strength of aggregate seems to be a limiting factor. This opinion comes from the crack formation that occurs at yield strength of samples;
- analysing the relationship between the compressive and splitting tensile strength, linear correlation with R value 0.63 was found;
- despite the high dose of plasticizer, w/c lower than 0.35 causes difficult working with pervious concrete. Here the next optimization of the kind and dose of chemical admixtures is necessary.

Results of strength characteristics being found within the presented experiment are promising and open for next experimental work focused on various locally available materials.

Acknowledgments

This work was supported by the project ITMS Center of excellent integrated research of progressive building constructions, materials and technologies [grant number 26220120037].

References

[1] Ferguson B K 2005 Porous Pavement *Integr. Stud. Water Manag.* L. Dev
[2] Hunt P D, W F PE, L S Laura 2006 Urban Waterways, Permeable Pavements, Green Roofs and Cisterns, Stormwater Treatment Practices for Low-Impact Development
[3] Yang J, G Jiang 2003 Experimental study on properties of pervious concrete pavement materials *Cem. Concr. Res.* **33** 381–386
[4] Haselbach L, R Freeman 2006 Vertical Porosity Distribution in Pervious Concrete Pavement ACI J **103**
[5] Wang K, V R Schaefer, J T Kevern, M T Suleiman 2006 Development of Mix Proportion for Functional and Durable Pervious Concrete
[6] Chindaprasirt P, S Hatanaka, T Chareerat, N Mishima, Y Yuasa 2008 Cement Paste Characteristics and Porous Concrete Properties *Constr. Build. Mater.* **22** 894–901
[7] Joshi T, U Dave 2016 Evaluation of strength, permeability and void ratio of pervious concrete with changing w/c ratio and aggregate size *Int. J. Civ. Eng. Technol.* **7** 276–284
[8] Tennis P,D, M L Leming, D J Akers 2004 Pervious Concrete Pavements
[9] STN EN 197 - 1:2011 Cement.Part 1:Composition, specifications and conformity criteria for common cements Slovak Office of Standards, Metrology and Testing Bratislava
[10] STN EN 12620:2008 Aggregates for concrete Slovak Office of Standards, Metrology and Testing Bratislava
[11] Nguyen D H, N Sebaibi, M Boutouil, L Leleyter, F Baurd 2014 A modified method for the design of pervious concrete mix *Constr. Build. Mater.* **73** 271–282
[12] STN 73 6124-2:2013 Road Construction Part 2: Concrete drainage layers Slovak Office of Standards, Metrology and Testing Bratislava
[13] Neamitha M, T M Supraja 2017 Influence of Water Cement Ratio and The Size of Aggregate on The Properties Of Pervious Concrete *Int. Ref. J. Eng. Sci.* **6** 9–16
[14] Chopra M, J Wanielista, A M Mulligan 2007 Compressive Strength of Pervious Concrete Pavements

Water-use in the context of the approaching climate change

S Krocova[1] and V Vaclavik[2]

[1]Department of Fire Protection, Faculty of Safety Engineering, VSB-Technical University of Ostrava, Czech Republic,
[2]Institute of Environmental Engineering, Faculty of Mining and Geology, VSB-Technical University of Ostrava, Czech Republic,

Email: sarka.krocova@vsb.cz

Abstract. Optimal use of surface and groundwater will be a decisive factor in maintaining the lives of people on the Earth and the functioning of infrastructure. A climate change calls for a new way of the management and use of water. What means and methods are used to achieve the required balanced state is presented in the basic scope by the following article.

1. Introduction
Water is essential for life on the Earth. Everyone knows this fact, but not everyone treats water with sufficient respect when using it. Very often, especially in the second half of the last century, water was treated with certain contempt in all countries of the world. Although it was perceived as a significant natural factor, water sources and in particular their long-term renewability, were not adequately protected. The broader spectrum of surface and groundwater use did not comply with the natural laws of their renewability, which is why the overall balance of these waters in the nature gradually deteriorated. Nowadays, a number of states have already reached a deficit in terms of water management due to the coming climate change. Lack of water is one of the main causes of extensive human migration, changes in the type of flora, and a negative impact on the species represented in the fauna as well. It can be assumed that the current situation will be significantly more strained at least in the medium term horizon. The current analyses already indicate that the global reserves of water that can be used for water supply needs as well as to maintain agricultural production has been decreasing [1]. In many countries in Africa and the Middle East, the scarcity of fresh water has been the cause of migration of people to places with a relatively sufficient amount of water. However, these places will quickly become overloaded and will make people move to new locations again.

The countries of the European Union must also be prepared for this situation. Even the current migration events clearly indicate the development in the future. A preventive adoption of the technical and operational measures in the area of surface and groundwater use can significantly reduce the future negative impact in each region of the world.

2. Climate change and its impact on water ecosystems
A climate change is not a modern phenomenon. Written history of mankind and scientific research have documented with sufficient accuracy that this phenomenon is repeated in certain periods. The only difference is represented by the accompanying negative impacts on the flora, fauna and, during the last millennium, on the human population as well [2]. It is necessary to realize that the increasing cultural development of the human society goes hand in hand with rapidly growing consumption of drinking and process water for this group of inhabitants of the world. At the same time, fresh water

consumption in agriculture and in infrastructure of countries increases as well, and these facts are associated with the growing risk of a global lack of water. Whether and to what extent this threat becomes reality will result from the current climate change during this century. The effects of the climate change will most likely become visible in the sphere of aquatic ecosystems in the following areas:

- long-term, further decrease of the reserves of shallow, but also deep groundwater,
- significantly fluctuating capacity of flowing and accumulated surface water in different basins,
- fundamental change of the quality of surface and groundwater due to inorganic and organic material load,
- fluctuating lack of raw water suitable for treatment into drinking water in large regions of the world,
- secondary induced change of cultivation of agricultural crops depending on the intensity of the effect of hydrological drought in the region,
- extinction of indigenous flora species in the areas critically affected by droughts accompanied by significant change of fauna.

The above presented and other impacts show that the long-term decrease of groundwater in various water-bearing beds will be the greatest threat to the existing natural environment and aquatic ecosystems, see figure 1.

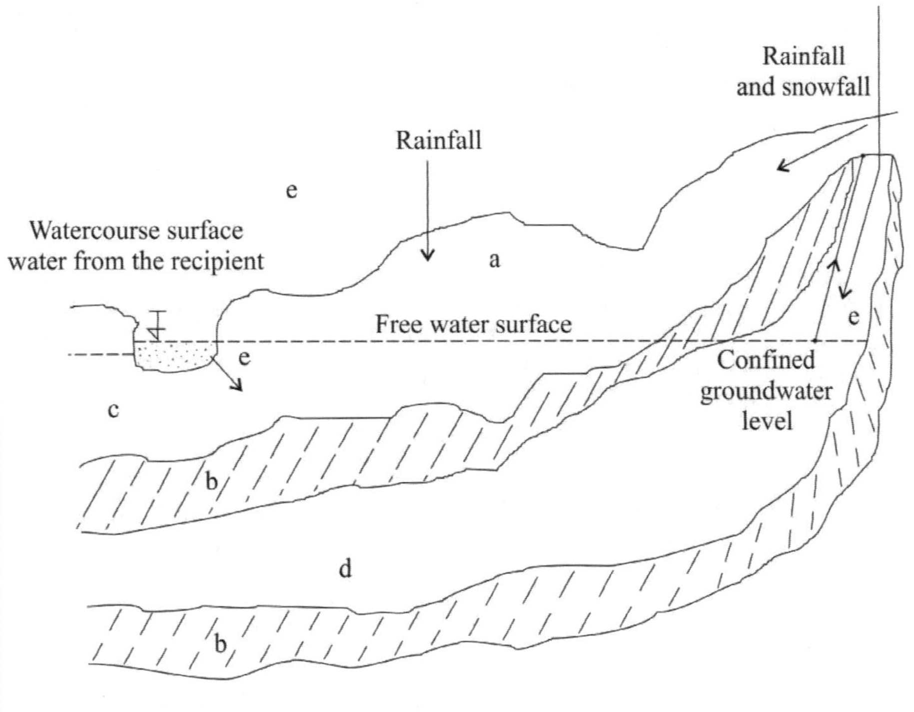

Legend:
a permeable subbase
b impermeable subbase
c shallow water-bearing beds
d deep water-bearing beds
e direction of water infiltration

Figure 1. Scheme of groundwater levels in the original environment.

The lack or a significant reduction in the volume of water in the water-bearing beds shown in the picture will certainly be the biggest threat to the natural environment. It will gradually disrupt the current concept of balance in the landscape and it will subsequently lead to an overall change in the existing natural environment of large regions of the world. Apart from this, the negative changes will probably also be reflected in the water-supply environment of the use of water for drinking purposes by the human population.

3. Water-supply use of surface and groundwater in critical situations

The existence of the majority of human population depends on sufficient amount of water to be used, especially for its treatment into drinking water. At present, more than a third of the people are threatened by the lack of drinking water. This deficit will increase with the rapidly growing number of people in the world and the declining reserves of this raw material. Various countries of the world are and will be affected with different intensity. Most notably, it will affect the countries of Asia and Africa. However, the countries of Europe or the European Union will not be spared either. It will always depend on how the state operates its water-supply systems and how it is able to use the strengths and weaknesses of the aquatic ecosystems for the water-supply purposes.

For example, in the Czech Republic, surface water and groundwater are used for the treatment into drinking water at a ratio of about 50/50 % [3]. Large cities and industrial densely populated areas in the Czech Republic are supplied from surface water sources; see the model diagram in Figure 1.
The advantage of this method of the supply of drinking water to the consumer system is the relative sufficiency and rapid renewability of the source of water during the season. Medium-sized towns and smaller municipalities, if they are not connected to water-supply systems, use treated or at least hygienically safe groundwater to supply the inhabitants with drinking water.

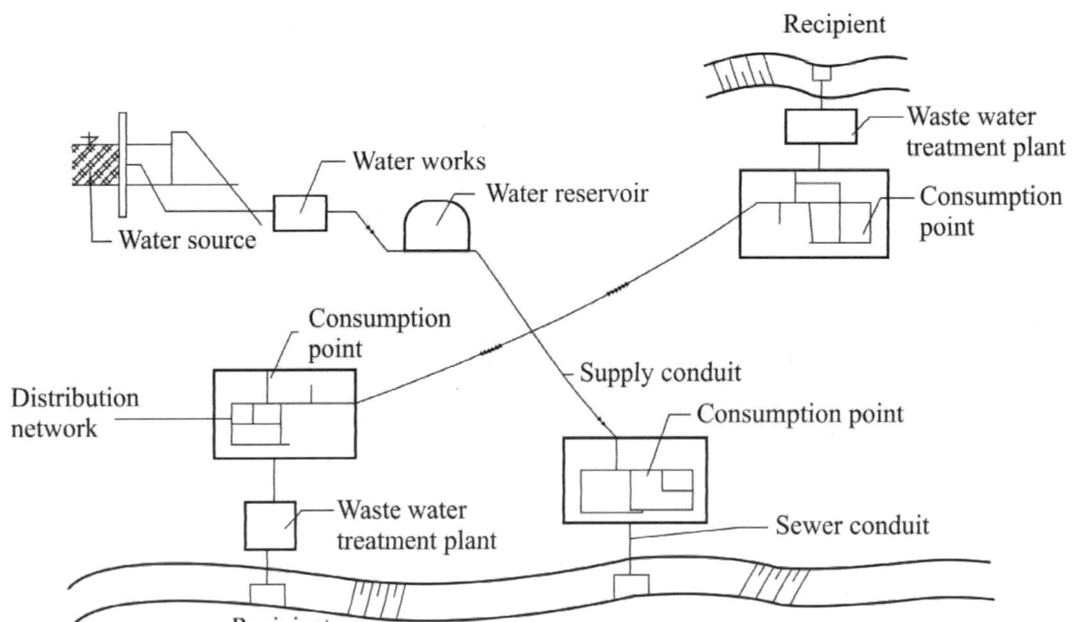

Figure 2. Water-supply system with one source of drinking and fire water.

The overall trend of ensuring the supply of drinking water for its consumers, with significant signs of a climate change in the Czech Republic as well as in other countries of the world, will probably be based on a substantial expansion of the water-supply systems to provide the supply of drinking water for built-up areas from surface, rapidly renewable sources of water.

4. The possible ways of how to reduce the threat of water scarcity in the 21st century

Taking into account the above-mentioned arguments in this article dealing with the issue of water-use in the context of the approaching climate change, there is an increasing need for a global approach to aquatic ecosystems with the international protection of these waters.

However, this water-use process focused on increasing its real efficiency (see figure 3) must always be complemented by a number of specific measures [4]. The basic and quickly evident measures in the field of water protection and the increase of the capacity of aquatic ecosystems in the natural environment may include especially the following steps [5]:

- speed up the process of the construction of new waste water treatment plants and increase the intensity of the treatment of waste water discharged especially into small watercourses above the current limits,
- deactivation of the original drainage systems of the so-called waterlogged lands, with a simultaneous use of these water works to increase the infiltration of rainfall in the area under consideration,
- deregulation, especially of small watercourses, to reduce the water current velocity of surface water in the recipient (creating meanders),
- execution of technical water works in places where the sources of deep water-bearing beds of soil profile are created,
- change the possibility of announcing and operating the protective zones (PZ) of underground, or even surface sources of drinking water in a legal way,
- restore the function of water managers in the legislation in order to increase the water protection and the management process of its use in the natural environment.

Their possible execution, depending on the local conditions of the individual states, has the potential not only to stabilize water in the landscape even in the new climate conditions, but also its quality and to reduce the safety anthropogenic threats of its quality change, especially in industrial densely populated areas.

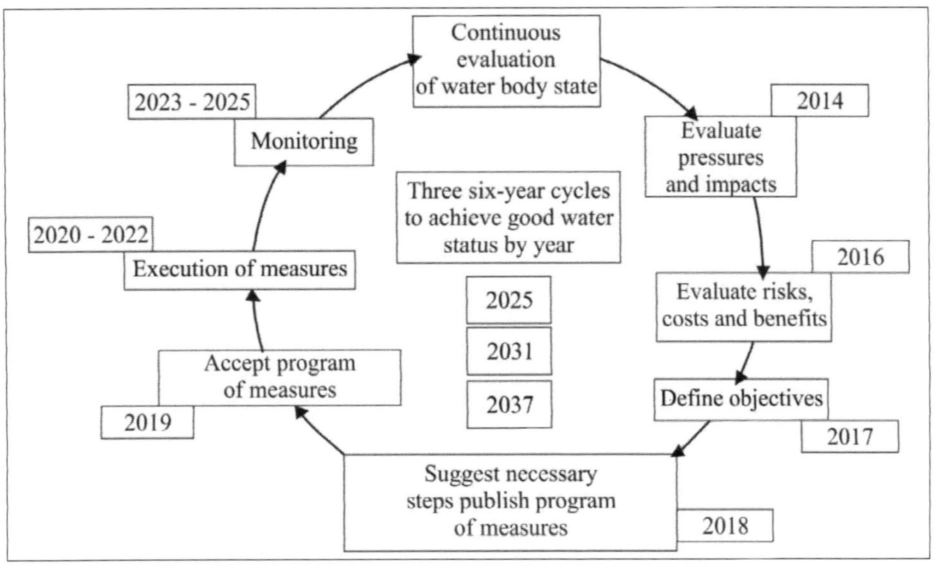

Figure 3. The process of significant improvement of the protection and use of European waters [3].

5. Conclusion

The solution of the negative signs of rapid climate changes cannot be postponed. Any delay in the response to the newly created natural environment has the potential not only to cause material damage in the given state, but, in particular, to cause serious and even irreversible damage to the current living conditions of people, fauna, and flora.

This article was drawn up in order to reduce the extent of the danger through education, by highlighting the real threats and, at the same time, by indicating their elimination. Perhaps, it will trigger a discussion and a new reflection on these threats and contribute to their solution.

Acknowledgement
This work was supported by the research project VI20152019049 „RESILIENCE 2015: Dynamic Resilience Evaluation of Interrelated Critical Infrastructure Subsystems", supported by the Ministry of the Interior of the Czech Republic in the years 2015-2019.

References:

[1] Potop V, Turkott L and Kožnarová V 2009 Drought impact on variability crop yields in Central Bohemia. *Cerl. Res. Commun.* **37** 295-304
[2] Vlnas R et al. 2010 Časová a plošná variabilita hydrologického sucha v podmínkách klimatické změny na území české Republiky (Praha: Výzkumný ústav vodohospodářský T. G. Masaryka)
[3] *Svaz vodního hospodářství v ČR* <http://www.svh.cz/>
[4] Adamec V, Maléřová L and Adamec M 2016 How to assess teritory vulnerability *Sci. for Popul. Protect.* **1** 35-40
[5] Kročová S 2015 Threats of water contamination in an industrial lanscape *Inzyn. Miner.* **36** 125-130

Fuels and their influence on the properties of ash in boilers for local heating

K Kryštofová[1, 2], H Raclavská[1, 3], H Škrobánková[1, 4] and V Sassmanová[2]

[1]ENET Centre-Energy Units for Utilization of Non-traditional Energy Sources, VSB-Technical University of Ostrava, 17. listopadu 15, 708 33 Ostrava, Czech Republic
[2]Department of Power Engineering, Faculty of Mechanical Engineering, VSB-Technical University of Ostrava, 17. listopadu 15, 708 33 Ostrava, Czech Republic
[3]Institute of Geological Engineering, Faculty of Mining and Geology, VSB-Technical University of Ostrava, 17. listopadu 15, 708 33 Ostrava, Czech Republic
[4]Institute of Environmental Engineering, Faculty of Mining and Geology, VSB-Technical University of Ostrava, 17. listopadu 15, 708 33 Ostrava, Czech Republic

Email: helena.raclavska@vsb.cz

Abstract. Local household boilers significantly contribute to the air contamination. The quality of combusted fuel can be monitored by an ash analysis. Cocombustion of waste in local boilers influences the amount of unburned carbon in ash and its distribution into separate grain size classes. The paper is focused on determination of the ash melting point in dependence on the fuel quality.

1. Introduction

Boilers for local heating significantly contribute to air quality. The composition and quality of combusted fuel can be identified from the ash analysis. The fuels, the used equipment – the boiler, as well as the way of heating significantly contribute to the nature of the air pollution from the local heating. Approximately half a million households in the Czech Republic use Class 1 or 2 boilers according to ČSN (Czech National Standard) EN 303-5, which will become unsuitable in the future, and only about 150,000 households have boilers belonging to Class 3 or higher, which are suitable for the future use. In the Czech Republic, a number of measures have been implemented to reduce air pollution from small combustion sources up to the thermal output of 50 kW, so-called boilers for local heating combusting solid fuel and often waste. Reduction of air pollution will be achieved by preventing burning of combustible waste in the existing solid fuel boilers by installing automatic boilers for coal, biomass or both coal and biomass, where combustible waste cannot be batched. Unsuitable Emission Class 1 and 2 boilers can also be used for waste and non-compliant or forbidden fuels. Modern boilers with automatic stoking cannot be stocked with forbidden fuels, or it is very difficult.

The amendment to the Act on the Air Protection of 2012 introduced a mandatory inspection of boilers (Decree No. 194/2013 Coll. [1]) with effect from 1 January 2017. The newer Act No. 369/2016 Coll., which amends Act No. 201/2012 Coll., on the Air Protection, establishes the obligation of the operators of household combustion sources to allow persons authorized by the municipal authority with extended competences access to the source, its accessories and the used fuels [2].

The inspection of households from the point of view of combustion of fuels and wastes has been legislated by 13 of 22 European countries. The problem of the inspections is the missing methodology for identifying combusted fuel from ash from small combustion sources. The information on the bottom ash chemical composition is used to identify parameters that may be useful for determining the combusted fuel. The presented article deals with the possibility of identifying the combusted fuel based on assessments of ash energy parameters and chemical composition.

2. Chemical composition of fuels

The principal inorganic elements of the biomass i.e. simple inorganic compounds of Si, K, Ca, Mg, Na, P, S, and Cl are also present in coal, but there are some important differences in the way they are bonded to the minor inorganic component (ash) or the organic fuel component (table 1). In coal, some elements display affinity for the inorganic and organic component. The maceral trace element association could have effect on the enrichment of the trace elements in the bottom ash. Ni, As, and Hg have a strong affinity for vitrinite macerals. Inertinite macerals have a high affinity for Sn and Li and partial affinity for Ga and Se [3]. An important component of ash in coal consists of clay minerals that can bind: As, Al, Ba, Bi, Cd, Co, Cr, Cs, Cu, Ga, K, Li, Mg, Na, Ni, P, Pb, Rb, Sn, Sr, Ta, Th, Ti, U, V, iron sulphides: As, Ba, Cd, Co, Cu, Fe, Hg, Mo, Ni, Pb S, Sb, Se, Ti, W, Zn, and carbonates: Ca, Cd, Co, Mn, Zn [4].

The essential or nutritious ash forming elements for plants can be macronutrients, namely K, Ca, Mg, P and S, and micronutrients such as Fe, Mn and Cl. The data from a study related to the composition of 86 varieties of terrestrial biomass show that the major elements are commonly C > O > H > N > Ca > K, while the minor elements normally include Si > Mg > Al > S > Fe > P > Cl > Na, and occasionally Mn and Ti. The reference studies reveal that the contents of major and minor elements in plant ashes are Si > (Mg, P) > S > (Ca, K) > Na > Al > (Fe, Cl) > Mn > Ti [5].

Table 1. Chemical analysis of different kind of ash (normalized to 100%), wt%. [5].

	SiO_2	CaO	K_2O	MgO	P_2O_5	Al_2O_3	SO_3	Fe_2O_3	Na_2O	TiO_2	Mn	Cl^-
Paper plastic fluff	38.5	23.26	3.01	2.76	1.81	17.27	2.67	4.2	5.12	1.4		
Waste paper	28.62	7.63	0.16	2.4	0.2	53.53	1.73	0.82	0.54	4.37		
MSW	37.92	24.66	4.79	3.82	0.74	13.62	3.72	2.93	6.34	1.46	0.1	7.8
Furniture	52.17	13.78	3.74	3.25	0.5	17.14	0.99	5.59	2.34	0.5		
Demolition wood	33.55	20.96	5.17	11.8	3.15	9.66	4.91	5.8	2.26	2.74	0.16	1.72
Wood (conifer)	14.25	42.75	10.42	13.06	5.41	5.8	3.69	2.77	1.32	0.5	1.32	0.5
Pine sawdust	11.16	42.81	7.23	10.92	3.68	2.54	17.93	2.87	0.71	0.15	1.05	0.12
Pine wood	29.56	45.56	6.17	2.48	0.8	7.63	1.57	4.78	0.92	0.53	2.3	0.1
Pine bark	9.2	56.83	7.78	6.19	5.02	7.2	2.83	2.79	1.97	0.19	1.08	0.01
Pine branches	6.28	51.74	10.49	12.35	7.41	4.53	2.37	1.95	2.78	0.1		0.01
Biomass	39.67	19.57	2.6	4.03	2.18	16.15	3.16	6.27	2.62	3.75	0.16	
Coal - mean	54.06	6.57	1.6	1.83	0.5	23.18	3.54	6.85	0.82	1.05	0.049	0.14
Coal - minimum	32.04	0.43	0.29	0.31	0.1	11.32	0.27	0.79	0.09	0.62		
Coal maximum	68.35	27.78	4.15	3.98	1.7	35.23	14.42	16.44	2.9	1.61		
Lignite – mean	44.87	13.11	1.78	2.5	0.2	17.11	8.64	10.8	0.48	0.81	0.055	0.21

In bottom ash (BA) the oxide contents are normally SiO_2 > CaO > K_2O > MgO > P_2O_5 > Al_2O_3 > SO_3 > Fe_2O_3 > Cl_2O > Na_2O > MnO > TiO_2. The elements such as Ca, Cl, K, Mg, Mn, Na, P, and S in BA from natural biomass normally show higher contents than the respective values for coal ashes. On the other hand, elements such as Al, Fe, Si, and Ti in BA commonly reveal lower concentrations than the respective values for coal ashes [5].

Bottom ash from municipal waste incinerators is material rich in SiO_2 (34%) and CaO (22%). Iron as Fe_2O_3 (14%), aluminium as Al_2O_3 (10%), and Na as Na_2O (7%) are present in a smaller amount [6]. Significantly lower levels of Fe_2O_3 for ashes from municipal solid waste (MSW) incinerators were reported [5]. Iron content may be related to the level of municipal waste separation.

Table 2 lists the contents of the minor elements contained in bottom ash from combustion of biomass, coal and MSW incineration plants. The minimum and maximum contents of the elements mentioned in ashes from the Czech Republic derive from the combustion of brown coal (lignite) from the Most and Chomutov Coal Basins. Ashes from lignite combustion are characterized by higher contents of Co, Ni, and V. Higher contents of As, Cu, Cr, Mo, Sb, Sn, and Zn were found in ashes from MSW incinerators. When reaching a certain concentration, it is therefore possible to consider these elements as indicators for incineration of "undesirable admixtures in traditional fuels – coal and biomass". The sources of these elements in household wastes are:

As is used as a decolourizer and fining agent in the manufacturing of bottle glass and other types of glassware, wood preservative salts, in leather industry, as a paint pigment. Other utilization: doping agent in electro conductive polymers alloys, batteries, semiconductors, herbicides, insecticides [7].

Cu compounds are used as a wood preservative. Textile fibres are blended with copper to create antimicrobial protective fabrics. Copper oxide and carbonate are used to add colour in stain glass works. Integrated circuits and printed circuit boards increasingly feature copper in place of aluminium because of its superior electrical conductivity. Antimicrobial applications.

Cr is responsible for yield yellow, red and green pigments, but in traces it is utilized also for other pigments: from black to blue. Other utilization: wood protection, leather processing, steel.

Sb is utilized in production of paint, ceramic, glass and plastic production, white, yellow and orange pigments, rubber vulcanization.

Mo is used as an alloying element in steel and super alloys. Pigments in anticorrosive paints for steel and aluminium. Less toxic substitute for antimony oxide in flame retardant agents for PVC.

Sn is used in the manufacture of dyes, polymers, and textiles; as a food preservative; as an additive in perfumes used in soaps; and as an anti-gumming agent in lubricating oils. Tin oxide (SnO_2) is used in the manufacture of special kinds of glass, ceramic glazes and colours, perfumes and cosmetics, and textiles. Organotin is the most important compound for stabilising PVC. PVC compounds incorporating tin heat stabilisers are used in a diverse range of applications including sheet, bottles, profiles, injection moulded fittings, credit cards, blister packs, food containers – tin cans and display trays.

Zn in the waste fraction of textile/leather/rubber Zn content varies from around 30 to 18,000 mg/kg and in fraction of waste glass between 0 and 10,000 mg/kg. ZnO is also the main ingredient for the production of a flame retardant. Zinc borate is added to a number of materials and products, such as plastics (15 – 10,000 mg/kg), textiles and rubber (32 – 17.64 mg/kg). Zinc stearate, $Zn(C_{18}H_{35}O_2)_2$ is used as a heat stabilizer in PVC. Variety of PVC based plastics, such as potable water pipes, water bottles, cable covering and PVC toys are stabilized with the Ca/Zn. Cardboard/paper contain only 13 – 160 mg/kg Zn [8].

Table 3 lists the minimum and maximum concentrations for bottom ash and fly ash from biomass combustion and for fly ash from combustion of bituminous coal in the Czech Republic. Ashes from biomass combustion show significantly lower As and Cd contents in bottom ash, comparable content in bottom ash and fly ash is shown by Cr, Pb, and V. Other elements show a slight enrichment in fly ash. The highest contents of Co, Ni, and V were found in bituminous coal combustion products. In BA from biomass combustion, Zn occurs in a concentration about 5 times higher than in ash from coal combustion. In biomass fly ash, Zn concentration may be up to 10 times higher than in fly ash from coal. The highest metal contents were found in bottom ash from MSW incinerators: As, Cr, Cu, Mo, Pb, Sb, Sn, and Zn. These elements can be considered as indicators of combustion processes using waste.

Table 2. Chemical analysis of bottom ash (BA) and fly ash (FA) from biomass and coal combustion (mg/kg dry matter).

	Biomass									
	BA		BA-literature [9]		FA		FA-literature [9]		FA - lignite CR [4]	
	Min	Max	Min	Max	Min	Max	Min	Max	Min	Max
As	2.17	3.19	1.4		2.68	6.98	1.5	24	29.7	126
Ba	802	1400	1600	2200	797	2320	1200	4300	630	1537
Cd	0.15	0.46	<0.2	5.7	7.32	16.3	5.1	34	0.4	3.2
Co	4.23	7.3	6.7	11	5.79	9.69	11	13	37	77
Cu	64.6	111	69	200					87	120
Cr	24.9	69.5	64	220	26.5	62.7	32	290	152	235
Mo	1.06	1.84	1	5.87	1.46	4.29	6.6	9	1.38	11.5
Ni	27.4	38.6	22	200	22.4	52.5	19	54	89.5	190
Pb	4.74	79.8	4	40	10.7	73.8	25	470	25	56
Sb	0.49	3.48	0.86	2.3	0.72	5.83	1.7	3		
Sn	0.3	1	11	16	1.21	6.37	15	22	9.85	23.2
Sr	466	783	610	710	578	1240	750	2100		
Tl	0.12	0.47			0.978	1.98				
V	10.4	18.4	26	64	6.75	18.2	5.1	43	276	309
Zn	73.9	234	65	950	446	1120	370	1860	122	190

Table 3. Bottom ash and fly ash from electrostatic precipitators from MSW incineration plant.

	BA [10]	BA [11]	BA [12] Mean ± S.D.	BA [6] Mean	BA [13] Min - Max	FA[14] Min - Max	MSW ash Max
As			108 ± 42		0.1 - 190	6.4 - 27.9	0.1 - 190
Ba					400 - 3000	700 - 1100	400 - 3000
Cd			7 ± 1		0.3 - 70		0.3 - 70
Co			21 ± 2	47			47
Cu	700-1000	4600	315 ± 22	1801	190 - 8200	820 - 1550	190 - 8200
Cr			252 ± 42	537	23 - 3200	254 - 656	23 - 3200
Hg							
Mo			12 ± 2		2 - 280	10 - 24.7	2 - 280
Ni			48 ± 24	63	7 - 120	51.9 - 60.4	
Pb		550		234	100 - 13700	1370 - 2800	100 - 13700
Sb		9.2			10 - 430	128 - 773	128 - 773
Sn				128			128
V					20 - 120	10 - 35	20 - 120
Zn	1600-2500	3000	1922 ± 33	2357	610 - 7800	4210 - 14400	4210 - 14400
Zr				193			

MSW – Municipal Solid Waste, S.D. - standard deviation

If the identified concentration in ash from local furnaces is higher than: As (126 mg/kg), Cr (235 mg/kg), Cu (120 mg/kg), Mo (12 mg/kg), Sn (23 mg/kg) and Zn (1,900 mg/kg), it can be assumed that waste was combusted in the boiler in addition to the approved fuel.

3. Conclusion

As part of the bottom ash and fly ash analysis from combustion of biomass and coal, elements that clearly originate from coal combustion were defined: Co, Ni and V, or Zn from biomass (figure 1). As, Cr, Cu, Mo, Sn, and Zn will be used to identify waste combustion in local furnaces. The concentration that demonstrates the presence of waste in the combustion process is derived from the maximum occurrence of the given element in fly ash from bituminous coal combustion. These elements can be easily monitored by the X-Ray fluorescence method used by a number of accredited laboratories. In addition to the majority elements, chlorides and sodium (Na_2O) are of importance in demonstrating possible waste combustion. If the increased content of indicator elements is demonstrated in ash, the analysis can be refined by identifying the organic compounds by pyrolysis chromatography Py-GC/MS. This method will allow accurate identification of combusted waste. In the case of boilers that have a half emission limit for pollutants according to EN 303-5: 2012 – reduced from 150 or 125 mg/m^3 (Class 2 and 3 boilers) to 75 mg/m^3 (Class 4 boilers) or 60 mg/m^3 (Class 3 boilers), higher particulate capture will occur, which will cause an increase in the concentration of the majority and minority elements in ash. The proposed limit concentrations for the indicator elements will have to be increased by about one half. This assumption must be verified on an experimental device.

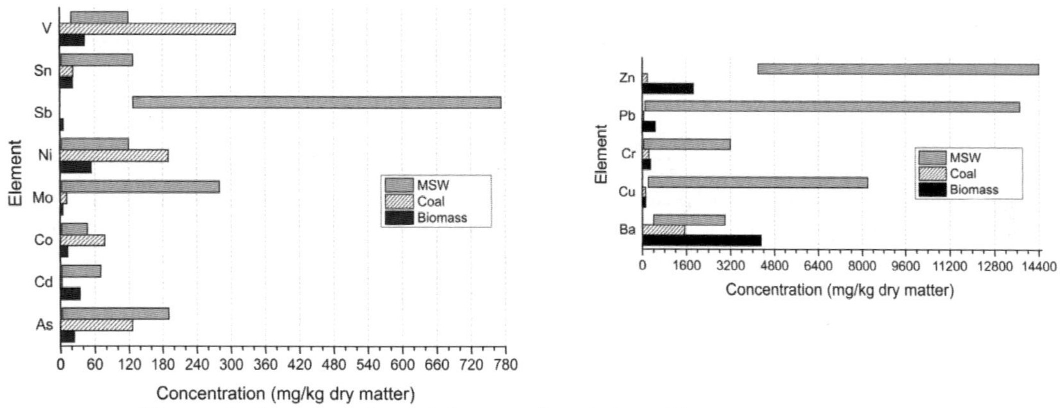

Figure 1. Concentration of elements in ash from biomass, coal and MSW.

Acknowledgement
This study was supported by the research projects of the Ministry of Education, Youth and Sport of the Czech Republic: The National Programme of Sustainability LO1404 – TUCENET, SGS 2017/35 Research in Selected Areas of "Smart Energy" of the 21st Century and the INTER-EXCELLENCE programme "European Anthroposphere as a Source of Raw Materials" LTC 17051

References

[1] Inspection of Boilers and Heat Distribution Decree No 194/2013 Coll *Ministry of Environment*
[2] Air Protection Act No 369/2016 Coll *Ministry of Environment*
[3] Oboirien B O, Thulari V and North B C 2016 *Appl. Energy* **177** 81–86
[4] Dolníčková D, Drozdová J, Raclavský K and Juchelková D 2012 *J. Pol. Miner. Eng. Soc.* **1** 59-68
[5] Vassilev S V, Vassileva Ch G, Song Y C, Li W Y and Feng J 2017 *Fuel* **208** 377–409
[6] Kowalski R F, Kasina M and Michalik M 2017 *Energy Procedia* **185** 56-62
[7] US EPA 1998 *Locating and estimating air emissions from sources of arsenic and arsenic compounds* (Research Triangle Park: EPA-454/R-98-013)
[8] Jones F, Bisaillon M, Lindberg D and Hupa M 2013 *Waste Manag.* **33** 2675–2679
[9] Maresca A, Hyks J and Astrup T F 2017 Recirculation of biomass ashes onto forest soils: ash composition, mineralogy and leaching properties *Waste Manag.* (*In press*)
[10] Xia Y, He P, Shao L and Zhang H 2017 *J. Environ. Sci.* **52** 178 – 189.
[11] Okada T and Tomikawa H 2013 *Waste Manag.* **33** 605–614
[12] Yao J, Qiu Z, Kong Q, Chen L, Zhu H and Long Y 2017 *Ecological Engineering* **102** 577-582
[13] Sabbas T, Polettini A, Pomi R, Astrup T, Hjelmar O, Mostbauer P, Cappai G, Magel G and Salhofer S 2003 *Waste Manag.* **23** 61-88.
[14] Raclavska H, Corsaro A, Hartmann-Koval S and Juchelkova D 2017 *JEMA* **205** 1-9

Long-term sustainability of the landscape in new climatic conditions

D Kubeckova[1] **and S Krocova**[2]

[1]VSB-Technical University of Ostrava, Faculty of Civil Engineering, L. Podeste, 1875, 708 33 Ostrava-Poruba, Czech Republic
[2]VSB-Technical University of Ostrava, Faculty of Safety Engineering, Lumirova 13, 1875, 703 00 Ostrava-Vyskovice, Czech Republic

Email: darja.kubeckova@vsb.cz

Abstract. The long-term sustainability of the landscape and its natural environment must be the decisive task of the public administration and, in the wider concept, of every citizen. In new climatic conditions, this need has intensified. The following article suggests in a basic scope whether the above-mentioned task can be accomplished, and what means of solution should be used.

1. Introduction

The natural environment is a very sensitive system. It has its own proper balance, which maintains the living conditions of flora, fauna and human populations at the same time. However, over time, the given conditions may change due to a number of natural or anthropogenic causes. Currently, one of the stages of a periodic change is under way. The world is undergoing a climate change that can affect not only the function of the natural environment itself, but also significantly change the living conditions of people on the planet as a whole. Its intensity is not known, but it can already be assumed from external manifestations, and in many cases it can be proven that it is most manifest in the environment of aquatic ecosystems [1, 2].

Depending on its intensity, a number of so-called negative phenomena will occur that the current human population will have to address urgently. The solution can take many forms and dimensions, depending on the type of solution in question and the maturity of the human society of a particular world region. The common denominator of all solutions, covering the overall concept, must be the activity of the public administration or the local authority. Its main task must always be to preserve the long-term sustainability of the landscape in a state in which the individual steps of partial solutions are supposed to take place continuously and the concerned concept is cemented into an organic whole. However, the solutions must not be segmented but must result from an overall risk analysis and the specific real threats to the country or its region. Current scientific knowledge in the field of performing risk analysis enables the state administration to select solutions that correspond to the real threat, cover its scope and intensity, while taking into account the technical and economic options of the solution. The following article suggests in a basic scope how to approach the issue dealing with the problems of long-term sustainability of the landscape in new climatic conditions.

2. Threats to the natural environment and aquatic ecosystems under new climatic conditions

The stability of natural conditions is one of the basic incentives for planning and addressing the long-term sustainability of territorial units. In the basic parameters, it determines the usefulness of a particular area for agricultural, forest-based or civil industrial activities. However, it is not a fixed factor, but often a highly variable quantity depending on the use of the territory for different human needs and activities. Basically, land used by the human population can be divided into the following categories and characteristics: agricultural and forestry management, civic needs of the housing stock, land for industrial and technical activities, and land with special protection of aquatic ecosystems and sources of raw materials.

Each of the above-mentioned and other activities inherently brings not only positive features, but also natural or anthropogenic threats and consequences [3]. Their characteristics are based on the intensity of the burden on the natural landscape with environmental and especially anthropogenic influences. Their extent differs depending on the type of the landscape, its resistance to ecological burdens and the possibilities for its elimination in real time.

2.1. Natural environment almost undisturbed by human activity

These are, as a rule, areas in countries and regions that have not been built up, or areas with minimum ecological burdens resulting from human activities. In the real environment, these burdens are represented by the following buildings and structures with minimum negative impacts on the original characteristics of the natural environment (as: rural and recreational buildings and regional sports facilities, agricultural and forestry buildings, construction of transport infrastructure, and structures and construction complexes of technical infrastructure).

However, these facilities and land developments are not entirely immune to the occurrence of extraordinary events and the long-term disturbance on the sustainability of the natural environment. It always depends on how it is conceived by the public and controlled by the public administration. However, the optimization of the relationships mentioned in the following chapter has the potential for satisfactory development

2.2. Industrial and intensively burdened landscape

Ensuring long-term sustainability and an acceptable natural environment in an industrial and intensively burdened landscape is significantly more complicated both from an ecological and a prognostic point of view. Since the second half of the last century, the area with the above-mentioned characteristics has increased. According to the forecasts of the UN and other scientific workplaces dealing with the demographic development of mankind, it is clear that in the 21st century, most of humanity will be concentrated in these localities.

Negative phenomena and threats to the natural environment will be based primarily on the following primary subjects and facilities: insufficiently treated urban and industrial water as far as substances are concerned, long-term reduction in the volume of flowing water in containers, increasing load of accumulated surface water with substances, the assumed lower infiltration of rainfall and snowfall into the water-bearing layers of the soil leading to a significant change in the quality and quantity of groundwater in most regions affected by climate change, significant enlargement of the area of large cities and the number of industrial areas and logistics areas, and significant expansion of transport infrastructure construction and increase in negative impacts with respect to maintaining the natural environment of the world's regions. These and other negative phenomena will be the major environmental threat of the 21st century. If the risk is not solved or if it is underestimated, potential negative threats will be deepened up to the critical point phase. Whether this threat occurs and to what extent will depend on a number of factors. One of the key ones, however, will be the activity of the public administration within the territory. It has irreplaceable means and authority to reduce or at least substantially eliminate most of the risks [4].

3. Public administration and its potential for reducing risks in the industrial landscape

In most countries of the world, the public administration is the state administration and self-government. Its role and function in the real life of citizens and legal entities is absolutely irreplaceable. One comes into contact with both systems throughout one's whole life. In some areas of territorialisation, basic needs and decision-making requirements are integrally increasing. These areas clearly include the industrial landscape and its technical infrastructure and protection of the aquatic ecosystems of the region in question. By its very nature, a number of operational and security threats and emergencies arise in the industrial landscape [5]. These events can have a different threat dimension and duration if they actually occur. Given that the activities of the state administration and decision-making, as opposed to the activity of natural persons, can be carried out only on the basis of the authorization by law, the public administration has at its disposal the tools resulting from the following figure 1 under the conditions of, for example, the Czech Republic.

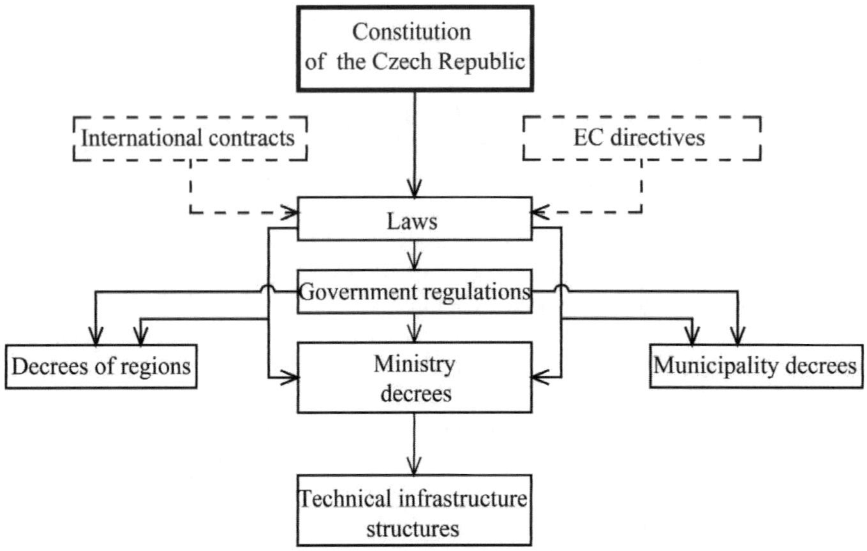

Figure 1. Sequence of legislative regulations in the Czech Republic.

The figure 1 shows the effect of the international treaties and EC directives on the drafting of laws, the Government Regulations and decrees in the drafting of legislative regulations of the Czech Republic. Their importance significantly increased after the Czech Republic joined the European Union, which included, among other things, the obligation to implement these supranational standards in the legislation of the individual countries. Under this configuration of law, the public administration must carry out its activities. For example, in the field of control and protection of aquatic ecosystems, "Water management authorities of regions and municipalities with extended powers" have the primary responsibility for the performance of the state administration. In the industrial landscape, irrespective of the specific local incorporation of the state administration in the legislative environment of the country concerned, water management authorities must focus their attention especially on the following areas of managing surface and groundwater and their long-term sustainability in the natural physical state: extraction of surface and groundwater by legal and natural persons, passive and active protection of aquatic ecosystems against contamination from industrial activities of individual subjects, keeping an overview of the old ecological burdens and their gradual elimination, reduction of water consumption for energy purposes in river basins endangered by hydrological drought, elaboration of studies and crisis scenarios for solving situations in increasing the consequences of hydrological drought, and control of public administration crisis plans and plans for crisis preparedness of entities in the water management sector and their possible updating in the light of new scientific knowledge of threats.

The optimal operation of water management authorities is the primary basis leading to the long-term sustainability of the natural environment in the industrial landscape. The outputs of this activity are not only discussed at the water management level, but they are a prerequisite for the development of the territorial units of the regions in all the countries of the world. It is necessary to realize that without water there is no life on Earth. This fact is generally known to everyone, from small children to specialists in various fields of human activity. Despite this common knowledge, however, it is not considered in practice. In forthcoming potential crisis situations caused by climate change and its impacts on aquatic ecosystems predominantly in the inhabited areas of the world's regions, there is no longer a timeframe for underestimating the global risk in question.

While scientific knowledge and states' technical capabilities do not have the full potential to solve this dangerous situation, they can, at least, mitigate it to a large extent by adopting preventive countermeasures [5]. The ways and means of responding to emerging environments are, to a fundamental extent, clear from the following subchapter of the article.

4. Means to increase the long-term sustainability of the natural environment in the countryside

Ensuring long-term sustainability of the natural landscape and the natural environment is a complex task. Often, in the attempt to achieve it, interventions were carried out in the past, which did not solve the situation but, on the contrary, often significantly worsened it. In particular, this concerns the construction of land amelioration structures with the aim of eliminating flooded land or amelioration of the various types of recipients in order to reduce the period of floods and their consequences. As an illustrative case, both interventions had their immediate effect, but they demonstrably failed in the complex of maintaining a natural balance between immediate effect and long-term benefits.

To reduce the above-mentioned and any additional risks of failure, the options and benefits should be based on the model shown in figure 2 and figure 3 below.

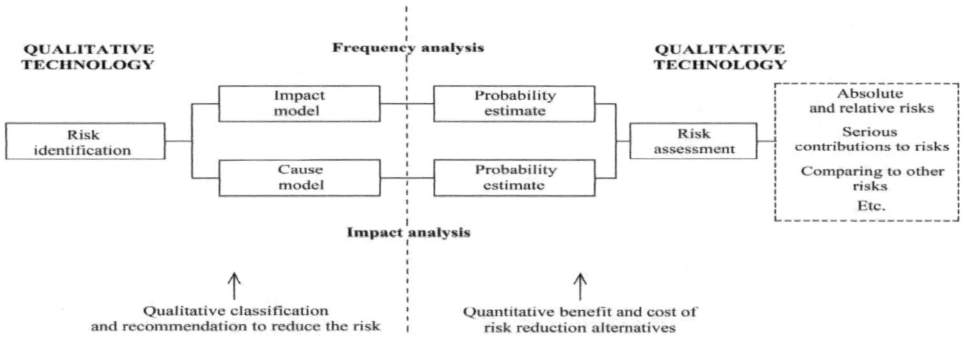

Figure 2. Sequence of legislative regulations in the Czech Republic [6].

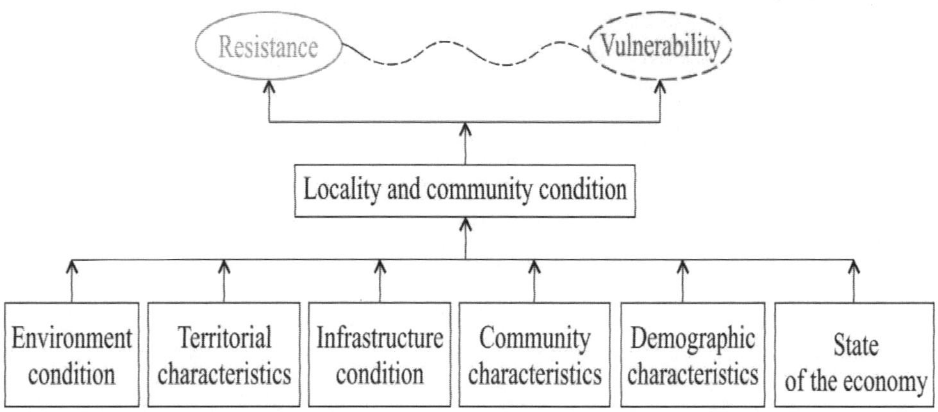

Figure 3. Sequence of legislative regulations in the Czech Republic [7].

Even with an optimal risk analysis process, the impossibility of an ideal solution and the resulting uncertainty rate must always be taken into account. It is quite probable that at least the number of sources of danger can be attained by the analysis. The diagram shows that when this risk management technique is used correctly by the public administration, it is possible to maximize the strengths of decision-making and, on the contrary, to overcome the weaknesses and the resulting risks. However, in order to make the decision as optimal as possible, it is always necessary to take into account the actual state of the landscape, its resistance and further vulnerability when underestimating the implementation of corrective measures.

With this technique of assessing operational safety risks and the adoption of alternative follow-up measures, the operators of the technical infrastructure of each country should also operate outside the public administration. Only their synergy has the prerequisites for achieving a real positive result in landscape protection. At the present time there is no longer room for theoretical considerations, but there must be a combination of theory and work. One of the ways to achieve this commitment is given in the article.

5. Conclusion

Both authors of the article deal with the issue on a long-term basis. Since the beginning of their professional careers, in their lectures for students at universities, they have been instilling knowledge about the unity of nature and the need to respect natural laws in human action. The incoming new generation of scientists and technicians will, among other things, have the task of rectifying a number of past mistakes in decision making. How to achieve this goal is indicated in the basic, significantly reduced scope in this article.

Acknowledgments

This work was supported by the research project VI20152019049 “RESILIENCE 2015: Dynamic Resilience Evaluation of Interrelated Critical Infrastructure Subsystems”, supported by the Ministry of the Interior of the Czech Republic in the years 2015-2019 and project “Development project 2017”, supported by the Ministry of the Interior of the Czech Republic in the years 2016-2018.

References

[1] Castany G 1978 Effects of drought on groundwater. Vulnerability of water dry layers. Bulletin du B R G M Section III 3

[2] *Drought Management Plan Report Including Agricultural Drought Indicators and Climate Changes Aspects. Water Scarcity and Drought Expert network* 2008 European Commission http://ec.europa.eu/environment/water/quality/pdf.dmp_report.pdf

[3] Krocova S 2017 Industrial Lanscape in the Period Drought. *Inz.Miner.* **39** 39-32

[4] Adamec V, Maléřová L and Adamec M 2016 How to assess teritory vulnerability *Sci. for Popul. Protect.* **1** 35-40

[5] Rehak D, Hrdina P, Senovsky M and Dvorak J 2011 Case Study on the Spatial Development Impact Assessment Tool Applied in the Spatial Development Environmental Security Assesment *Communications* **13** 32-36

[6] Prochazkova D and Riha J 2004 *Crisic management* (Czech Republic: Prague)

[7] Collective of Auhors 2007 *Perception of Security in a Territory* (Czech Republic:University of Pardubice)

The influence of the inverse character of the weather on concentrations of black carbon

M Kucbel[1, 2], B Svedova[2, 3], H Raclavska[2, 4], K Raclavsky[2], J Ruzickova[2] and D Juchelkova[3]

[1]Institute of Environmental Engineering, Faculty of Mining and Geology, VSB-Technical University of Ostrava, 17. listopadu 15, 708 33 Ostrava, Czech Republic
[2]ENET Centre-Energy Units for Utilization of Non-traditional Energy Sources, VSB-Technical University of Ostrava, 17. listopadu 15, 708 33 Ostrava, Czech Republic
[3]Department of Power Engineering, Faculty of Mechanical Engineering, VSB-Technical University of Ostrava, 17. listopadu 15, 708 33 Ostrava, Czech Republic
[4]Institute of Geological Engineering, Faculty of Mining and Geology, VSB-Technical University of Ostrava, 17. listopadu 15, 708 33 Ostrava, Czech Republic

Abstract. Black carbon (BC) is a product of incomplete combustion of fossil fuels, biomass and other fuels. The study is focused on the influence of an inverse character of the weather on the BC concentrations in the city of Ostrava during the period 2012–2016. The continuous measurement of daily concentrations of BC were performed by the portable Aethalometer, and PM_{10} by the DustTrak aerosol monitor. The average concentrations of BC and PM_{10} during the days with an inversion character of the weather, compared with the days without inversion, were increased up to 3.8 times and 3 times, respectively.

1. Introduction

Black carbon (BC) particles are an integral part of dust particles (PM) [1]. The BC concentration in Central Europe averages 5% in $PM_{2.5}$ for rural areas, 14% in $PM_{2.5}$ in urban areas, and 21% in $PM_{2.5}$ near roads [2]. Black carbon is generated by the incomplete combustion of fossil fuels, biomass, and other carbonaceous fuels [3]. BC adversely affects the environment, human health, and influences the climate [4]. PM and BC air pollution in urban environment varies, depending on the distance of industrial sources, traffic density, and other factors (population density, housing development, etc.). Meteorological factors also have a significant influence [5].

The highest PM_{10} concentrations occur during winter months due to frequent inversion situations [6].Temperature inversion is a natural phenomenon of a change in the normal tendency of air temperature to fall at higher altitude, as compared to normal in the troposphere with the temperature increasing with altitude [7]. During thermal inversions, atmospheric convection occurs; dispersion of pollutants is limited, local emission sources increase the concentration of pollutants [8]. Inversion represents a stable equilibrium of the atmosphere that prevents vertical movements in the atmospheric boundary layer [9]. In urban environment inversion there is a significant increase in BC concentrations over a few hours up to maximum levels. Explaining the causes of BC peak contrast occurrence can lead to better characterization of air pollution gradients, source identification, and assessment of meteorological impact on urban BC concentrations [8].

Measurement of the BC concentration took place in the Ostrava-Zábřeh urban district (Moravian-Silesian Region, Czech Republic). The influence of geomorphology of the area in combination with metallurgical production and numerous temperature inversions during the autumn and winter months leads to a high load of pollutants in Ostrava during the autumn and winter periods [10].

In Ostrava-Zábřeh, the permissible number of days exceeding the limit value for PM_{10} concentration is exceeded annually (daily value of 50 μg/m^3 may be exceeded 35 times per year according to the Act on Air Protection 201/2012 Coll. [11] the average annual air pollution PM_{10} limit is 40 μg/m^3). At the locality of Ostrava-Zábřeh, the daily PM_{10} air pollution limit value was exceeded in 2012 in 66 cases, 107 cases (2013), 92 cases (2014), 54 cases (2015), and in 2016, this air pollution limit value was exceeded in 48 cases [12]. In 2012, in the Ostrava-Karviná agglomeration, a smog situation was announced 7 times for a total duration of 985 hours, in 2013, a smog situation was announced 5 times (425 hours), in 2014, 5 times (240 hours), in 2015, in total 3 times (124 hours), and in 2016, twice (297 hours) [12].

The aim of the work is to determine the influence of the inverse character of the weather on the concentration of black carbon and PM_{10} in Ostrava-Zábřeh in 2012-2016.

2. Methods and samples

Ostrava-Zábřeh represents an urbanized part of the town, with typical panel buildings and houses on the outskirts. Measurement took place at Průkopnická Street, see figure 1. North of the measuring point, there is a major traffic artery - road no. 11 (Rudná) connecting the city of Ostrava with surrounding villages, and the D1 motorway in the western direction. In the east, Arcelor Mittal Steel metallurgical complex is located 7 km away, and Vítkovice Steel is located in the north-east. Black carbon concentrations were measured daily between 8-9 p.m. in the period from 9th January 2012 to 31st December 2016 (in total 1761 measurements) by optical method using the Magee Scientific Company AE-42 Aethalometer with a flow rate of 2 l/min. The concentration of BC was calculated from the optical attenuation measured at seven wavelengths caused by BC particles deposited at a filter tape, during 5-minute intervals. The concentrations of dust particles (PM_{10}) were determined by optical method, using the DustTrak^{TM} aerosol monitor model 8535 (TSI).

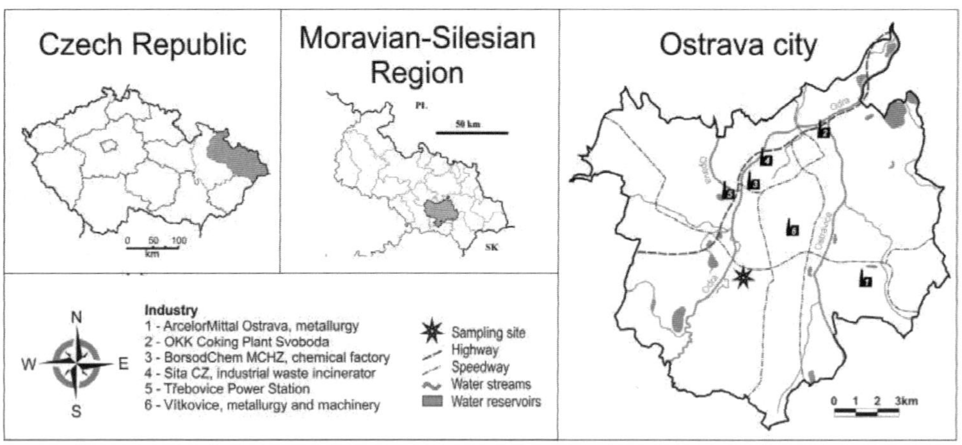

Figure 1. Situation of the Ostrava-Zábřeh sampling point.

3. Results

According to the Decree of the Ministry of Industry and Trade No. 194/2007 Coll. [13], the measurement period 2012-2016 was divided into heating season (September-May) and non-heating season (June-August). According to the Decree, the heating season begins on 1st September and ends on 31st May. The average BC concentration for the whole reporting period from 2012 to 2016 was 3.16 ± 3.79 μg/m^3, and PM_{10} concentration was 45.8 ± 36.5 μg/m^3. The PM_{10} concentration exceeded the average annual PM_{10} limit value (40 μg/m^3) permitted by legislation.

The observed BC concentration is common for the urban urbanized environment with industrial activity, and it is comparable, for example, to Poland. In Zabrze (Poland), the average BC concentration of 4.4 $\mu g/m^3$ was found in 2009-2011 [14].

The average BC concentration for the heating season (2012-2016) was 3.87 ± 4.14 $\mu g/m^3$ and the average PM_{10} concentration was 51.9 ± 39.9 $\mu g/m^3$. During the heating season, BC showed a normal distribution. 78% of all measured BC concentrations were less than 5 $\mu g/m^3$ and only 8% of BC concentrations were higher than 10 $\mu g/m^3$. BC in the non-heating season showed a different distribution, 99.8% of the measured BC concentrations were in the range of less than 5 $\mu g/m^3$, of which 57% of the BC concentration values were less than 1 $\mu g/m^3$ and only 0.2% occurred in the range of 5-10 $\mu g/m^3$). The average BC and PM_{10} concentrations for the non-heating season (2012-2016) reached the values of 1.05 ± 0.61 $\mu g/m^3$ and 27.8 ± 11.5 $\mu g/m^3$, respectively. Black carbon concentrations were found in 2011: in Racibórz (Poland), the value was 4.56 $\mu g/m^3$ during the heating season, and 0.99 $\mu g/m^3$ during the non-heating season [15].

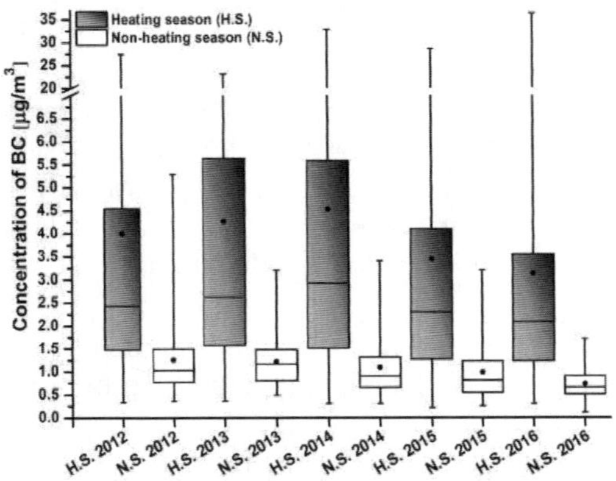

Figure 2. Variability of BC concentrations for the heating and non-heating season of the individual years.

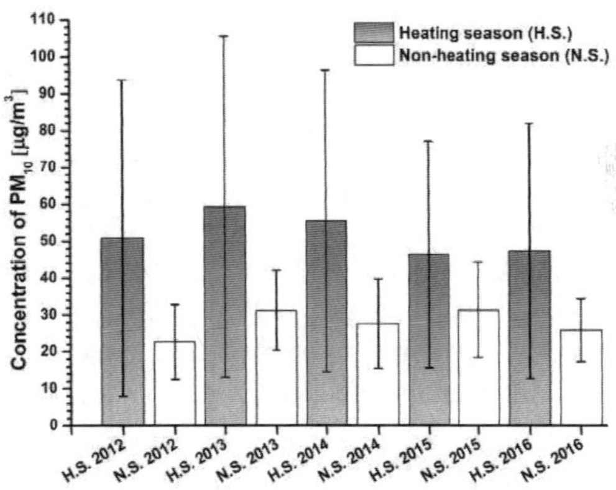

Figure 3. Average concentrations of PM_{10} including the standard deviation, for the heating and non-heating season.

The BC concentrations for the heating and non-heating season are shown in figure 2 and the average PM_{10} concentrations in figure 3. The ratio values of BC and PM_{10} concentrations (2012-2016) in the heating season and in the non-heating season were 3.69 and 1.87, which is due to the more frequent occurrence of impaired scattering conditions and increased combustion of fossil fuels and biomass in local furnaces. The ratios for the BC concentrations during heating and non-heating seasons ranged from 3.18 to 4.16, and the largest difference for the BC concentrations was recorded in 2014 (3.43 $\mu g/m^3$). The regression analysis between BC and PM_{10} showed a significant correlation dependence $r_s = 0.83^*$ during the heating season (2012-2016), in the non-heating season, it showed a mean dependence with $r_s = 0.48^*$. The contribution of combustion during the heating season from large energy, industrial, and transport sources, as well as from local heating plays a significant role in the BC and PM_{10} concentrations, and the origin of the particles is the same. During the warm season of the year, dust particles of PM_{10} grain size are largely made up of biogenic organic particles, resuspended particles, etc. The origin of PM_{10} and BC particles is therefore not entirely identical during the summer months, but the particles come from many different sources. The average percentage contributions of BC in PM_{10} for the period of 2012-2016 were $6.7 \pm 3.1\%$ in the heating season and $4.1 \pm 2.7\%$ in the non-heating season. The heating season of 2012-2016 was then divided into days with inversion ($PM_{10} > 50$ $\mu g/m^3$) and days without inversion ($PM_{10} < 50$ $\mu g/m^3$). It turned out that in 35% of cases, days with inverse character of the weather were recorded in the heating season (450 days with inversion character of the weather out of the total of 1302 days within the heating season). The distribution of the BC and PM_{10} concentrations for days with inversion and without inversion for the whole period of 2012-2016 is shown in figure 4; table 1 shows BC and PM_{10} concentrations for the individual years.

Table 1. Average BC and PM_{10} concentrations and standard deviation (S.D.) for the inverse character of the weather and no inversion in the heating season from 2012 to 2016.

	Days with inverse character of the weather				Days without inverse character of the weather			
	BC [µg/m³]	S. D.	PM_{10}	S. D.	BC [µg/m³]	S. D.	PM_{10}	S. D.
2012	8.39	± 5.11	100.3	± 46.1	2.05	± 1.21	28.0	± 10.9
2013	8.00	± 4.92	104.1	± 47.6	2.04	± 1.14	31.8	± 10.1
2014	7.95	± 5.75	88.7	± 44.5	2.07	± 1.33	31.5	± 10.3
2015	5.70	± 3.03	80.6	± 32.6	1.70	± 0.99	30.3	± 9.5
2016	6.27	± 5.01	83.6	± 45.0	1.84	± 1.33	32.0	± 9.6

The inverse character of the weather in the heating season 2012-2016 was recorded in 35% of cases. The average PM_{10} concentration in the heating season of 2012-2016 for days with the inverse weather was up to 3 times higher (92.0 ± 44.6 µg/m³) than for days without inversion (30.7 ± 10.17 µg/m³). The ratio the average BC concentrations for days with inversion (7.36 ± 5.02 µg/m³) and without inversion (1.94 ± 1.22 µg/m³) was even more significant (3.8). In Athens from December 2013 to February 2014, the BC concentrations for inversion days (4 µg/m³) were found to be almost three times higher (2.7 times) than for non-inversion days (1.5 µg/m³) [16].

The average percentage of BC in PM_{10} was slightly higher ($7.8 \pm 3\%$) for days with inverse weather patterns (range 1.4 - 19.2%) than for days without inversion ($6.2 \pm 3.1\%$) ranging from 1.2% to 29.9%. During the inverse character of the weather, a significant correlation relationship between BC and PM_{10} has been demonstrated with a Spearman correlation coefficient (r_s) of 0.77^* and for days without inversion ($r_s = 0.70^*$), which refers to the common origin of BC and PM_{10} emissions from combustion sources.

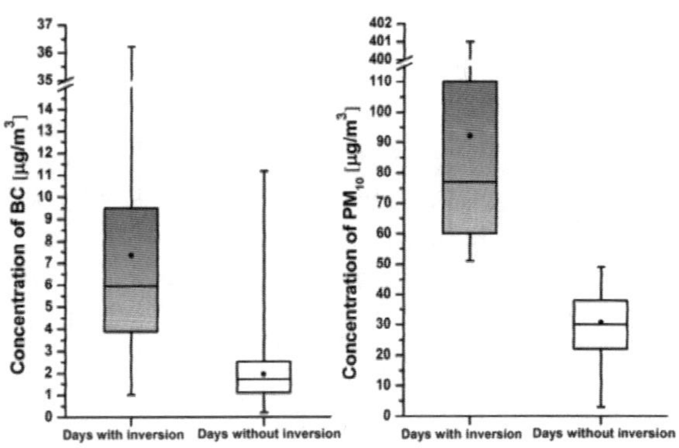

Figure 4. Distribution of PM_{10} and BC concentrations for the inverse weather and non-inverse days in the 2012-2016 heating season.

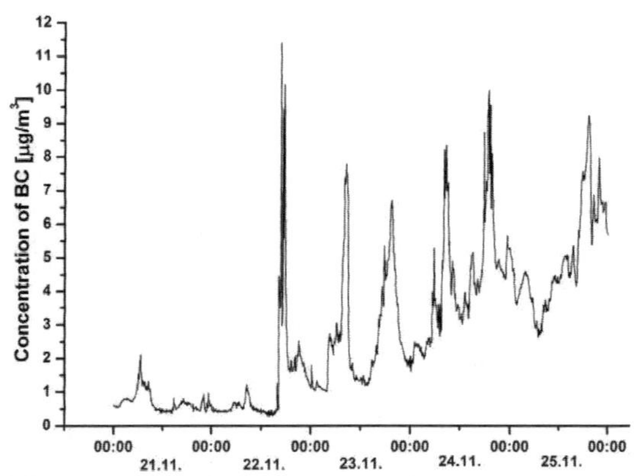

Figure 5. The course of 5 min BC concentrations during the period from 21 to 25 November 2016.

The study *Klejnowski and Błaszczyk, 2012* [14] in Zabrze (Poland) has shown that the daily pattern of BC levels is bimodal, with two peaks at equal level in the non-heating season, while in the heating season, the evening mode outreaches the morning peak approximately by about 2 µg/m³.

In the period from 21st to 25th November 2016, a continuous day-long measurement was carried out showing the beginning of the inverse character of the weather (from 22 November), see figure 5. The average BC concentration in the period from 21st to 25th November 2016 was 2.73 ± 2.20 µg/m³ with the minimum BC concentration of 0.261 µg/m³ up to the highest BC concentration of 11.40 µg/m³. On 21st November, BC concentrations were more than twice lower compared to the other days of inversion. In figure 5, two very distinct peaks may be observed during the morning hours (approximately from 4:00 a.m. to 10:00 a.m.) and during afternoon to evening hours (approximately 4:00 p.m. to 10:00 p.m.) and the total increase in BC concentrations throughout the day. Increased concentrations of BC are caused by morning and afternoon traffic peaks and the need for heating in local furnaces due to the drop in outdoor temperature and impaired scattering conditions.

The morning increase in BC concentrations is caused by emissions from road transport. The main source of high BC concentration in the evening hours observed in Warsaw and Racibórz are emissions from local heating [15]. During evening hours, the BC concentration increases faster in an inverse situation than in another part of the day, which is related to lower wind speeds [6].

4. Conclusions

The heating season of 2012-2016 in Ostrava-Zábřeh is characterized by an average 3.7 times higher contribution of the BC concentrations than in the non-heating season. The average PM_{10} concentrations are also up to three times higher in the heating season compared to the non-heating season. It turned out that in 35% of cases, days with inverse character of the weather were recorded in the heating season. During the inverse character of the weather, BC concentrations were almost 4 times higher than during days without inversion. The inverse character of the weather was found mainly in the autumn and winter months. Combustion sources, in particular, local furnaces, contribute significantly to the BC emissions; their effect is particularly evident at the time of inversion during the evening hours.

Acknowledgement

This study was supported by the research projects of the Ministry of Education, Youth and Sport of the Czech Republic: The National Programme of Sustainability LO1404 – TUCENET, and SGS 2017/35 Research in Selected Areas of "Smart Energy" of 21st Century, and SP2017/116 "Research of Various Types of Material, their Properties and Application Potential".

References

[1] Ning Z, Chan K L, Wong K C et al. 2013 *Atmos. Environ.* **80** 31-40
[2] Putaud J P, Van Dingenen R, Alastuey A et al. 2010 *Atmos. Environ.* **40** 1308-1320
[3] Wang Q, Huang R J, Zhao Z et al. 2016 *Atmos. Res.* **181** 163-171
[4] Wang R, Tao S, Wang W et al. 2012 *Environ. Sci. Technol.* **46** 7595-7603
[5] Clougherty J E, Wright R J, Baxter L K and Levy J I 2008 *Environ. Health* **7** 17
[6] Gramsch E, Cáceres D, Oyola P et al. 2014 *Atmos. Environ.* **98** 290-298
[7] Czarnecka M, Nidzgorska-Lencewicz J, Rawicki K et al. 2016 *J. Elem.*, **21** 1001-1015
[8] Tunno B J, Michanowicz D R, Shmool J L et al. 2016 *J. Exposure Anal. Environ. Epidemiol.* **26** 365-376
[9] Liu Q, Ma T, Olson M R, Liu Y, Zhang T, Wu Y and Schauer J 2016. *Scien. Rep.* **6** 33331
[10] Vossler T, Cernikovsky L, Novak J, Placha H et al. 2015 *Atmos. Poll. Res.* **6** 454-463
[11] Air Protection Act No. 201/2012 Coll. Ministry of Environment
[12] http://portal.chmi.cz/files/portal/docs/uoco/isko/grafroc/grafroc_CZ.html
[13] Decree no. 194/2007 Coll. Ministry of Industry and Trade
[14] Klejnowski K and Błaszczyk J 2012 *Air Prot. in Theory and Pract.* **2**
[15] Maciejewska K, Juda-Rezler K, Reizer M and Klejnowski K 2015 *Environ. Model. Softw.* **74** 212-226
[16] Fourtziou L, Liakakou E, Stavroulas I et al. 2017 *Atmos. Environ.* **148** 89-101

The influence of reclamation to malaco - assemblages of limestone quarry

J Kupka[1], H Svehlakova[1], J Beranek[1] and B Lokajickova[1]

[1]Institut of Environmental Engineering, Faculty of Mining and Geology, VSB – TU Ostrava, 17. listopadu 15, Ostrava Poruba, 710 00 Czech republic

E-mail: jiri.kupka@vsb.cz

Abstract. Quarries fundamentally change the character of the landscape, both in the mined parts and in its immediate surroundings. Inside the mined areas, there are many different types of microbiotopes, their character is very different from the surrounding ecosystem. Destruction of original biotopes is an opportunity for settlement by new plant and animal communities. To monitor these biodiversity changes in the landscape, appropriate groups of organisms are used. Only a bioindicator that is well processable from a taxonomic point of view and its demands on the environment and its distribution are well-known, gives reliable results. For these reasons, less moving molluscs are used in many works. The territory of the southern part of the Devonian limestone of the Moravian Karst is affected by the industrial mining of limestone. There is forest quarry in Lisen, which was ended in 1997 and a few years later reclamation work was started on parts of the mined area. The study of molluscs in this model territory was used to assess the impact of reclamation methods on the malaco - assemblage's diversity inside the quarry.

1. Introduction

The southern part of the massif of Devonian limestone of the Moravian Karst is affected by the industrial mining of limestone. Some valuable original biotopes, such as karst forest steppes, have been irreparably destroyed or preserved only on small areas (the natural monument Velka Klajdovka and Kavky). Local biodiversity is also disrupted by the Brno city district of Lisen, which is growing to foothills of Hady Hill. The increase of biodiversity through sensitive recultivation of local limestone quarries seems to be very appropriate and desirable.

Quarries, significantly affect the major ecological factors important for the occurrence of molluscs depending on the stage of succession. This is mainly the development of vegetation, the microclimate and properties of the substrate. Cover, structure, height stratification, and canopy of vegetation create a completely specific physical environment for molluscs. This environment is characterized by different types of vegetation shading, soil and air humidity, character of the plant litter, humus and soil. Molluscs are strongly bound to the habitat due to their limited mobility.

Abiotic conditions, ecological specifics and succession of vegetation in old quarries can be appropriately regulated and transformed by the reclamation techniques used. There are currently three main approaches to renewal. These are technical reclamation, spontaneous succession and controlled succession. By their proper combination and subsequent management, the mined areas can become very valuable sites as for natural value.

1.1. Characteristic of the study area

The study area is located north of Brno near the southern border of the Moravian Karst PLA. The Moravian Karst is the most extensive karst area of the Czech Republic with typical forms of surface and underground karst. From wider point of view study area is part of Macocha bioregion. The Macocha bioregion forms a narrow strip of limestone area in the centre of South Moravia and occupies the mesoregion Moravian Karst. It has an elongated shape in the north-south direction. Bioregion is created of limestone with rocky deeps. As for vegetation tiers there is oak communities of the 1st tier, but beech communities of the 4th tier and fir-beech communities od 5th tier prevails [1].

The Moravian Karst has a largely flattened surface, but it is divided by very sharp and deep valleys. Relief has the character of a plain highlands with a roughness largely of 150 to 200 m (up to 270 m only in Křtinsky creek). The lowest point is the Říčky valley at Muchova bouda (about 265 m), the highest point is area nearby Sosuvka (590m). Typical height is 300-530 m, outside of canyons 400-530 m [2]. The area is created by nappes of predominantly pure Devonian limestone, there are also the Granodiorite of the Brno massif and the loess in the southern part.

Soils in the area are heavier with frequent limestone gravel, shallow and dry, mineral rich. As a typical soil type, gray to dark gray rendzins are developed, which appear mainly on the slopes of deeps where they are enriched by carbonate material.

According to Quitt (1971) [3], the bioregion lies in climatic areas from the warmest (MT11) in the south, to the coldest (MT 3) in the north. The climate around Brno is warm and relatively dry (Brno 8.6℃; 547 mm); the northern areas are cooler and humid (Olomučany 7.7℃; 620 mm). The most northern part of the region has average temperatures of 6.6℃ and precipitation up to 660 mm. The local climate illustrates sharp differences, especially the contrast of the cold bottom of the canyon and warm and dry top edges. The southern edge of the bioregion has a very warm climate, enhanced by the southern orientation of the Hady Hill slopes [1].

Hydrographically and hydrologically, the Moravian Karst is different from the surrounding landscape by a whole range of peculiarities. Allochtonous waters which coming from the non-karst parts of the Drahanska vrchovina are almost immediately lost to the undergeound on the geological boundary of Devonian limestones, that´s why the hydrographic conditions and hydrological regime of these waters are very complex. The southern part of the Moravian Karst is drained by Ochozsky, Hadek and Hostenicky creeks. The catchment area is 76 km2 with an average annual flow of 0.16 m3.s-1. The largest cave system is the Ochoz cave, whose famous part was formed due to flood regime of Hostenicky potok. The active underground water flow is only on the lower part of the Netopyrky cave. Immersion flows rise like the Ricka Creek after go through very complex hydrographic network [2].

According to the Regional Phytogeographical Division of Czechoslovakia [4], the study area is part of the phytogeographical district of the 70th Moravian Karst and belongs to the Bohemian-Moravian Mesophytic District. There are three floral subelements in the territory. Hercyn and Carpathian forest flora are mixed here and the thermophilic and steppe flora from the Pontic-Pannonian region is acceding from the south. The natural vegetation is created oak-hornbeams with vegetation stands of the Hercyn association *Melampyro nemorosi-Carpinetum* and the Carpathian *Cariri pilosae-Carpinetum*. There are also rare tree-less habitats with species-rich communities of rock steppes and northern slopes on the carbonate soils (*Seslerio-Festucion pallentis*) here.

The fauna of the region has the character of Hercyn subprovince, but is heavily influenced by the Carpathian element, especially beechwoods on limestone. This effect is manifested especially on molluscs, ground beetles or flesh flies, we can observe occurrence of endemic Carpathian species (eg molluscs *Monachoides vicinus*, *Vitrea transsylvanica*). We can find also other important species as *Rhinolophus hipposideros*, *Myotis myotis* from mammals, *Salamandra salamandra* from amphibians, *Lacerta viridis* from reptiles, *Aegolius funereus*, *Cinclus cinclus*, *Ficedula parva*, *Ficedula albicollis* from birds [1].

1.2. History of landuse

The history of industrial mining of limestones on the northern outskirts of Brno dates from the end of 19th century. The first quarries at the foothill of the southernmost ridges of the Devonian limestone were the Dzungle Quarry and the Ruzenin Quarry. Both of them were interconnected by the large Haje Quarry around 1960. The ever-increasing demand for raw materials, especially for the cement industry in the first half of the 20th century, led to the opening of Lesni Quarry in Lisna, which is located in the north - eastern direction near Ruzenin quarry.

Lesni quarry in Lisna. Lesni quarry lies in the area of the southern part of the Hadecka peneplain. The area is located approximately 300 m from the edge of the southern part of the Moravian Karst PLA at an altitude of about 370 m. Part of the surrounding quarry is an agriculturally used landscape and part is forested. Deciduous forest stands of oak-hornbeam on limestone substrate prevail here [5]. Lesni quarry is divided into a total of 5 storeys. Quarrying was ended in 1997. Technical reclamation began in order to reforest part of the quarry in 2002-2003. It was approximately 20 % of the area.

2. Methods

We have chosen limestone quarry near Brno with finished mining. Technical reclamation was performed on several sites of quarry. The remaining areas were left without reclamation interventions to natural spontaneous succession. The list of sampling areas with coordinates and information about reclamation techniques is summarized in table 1. Malacological samples from 14 sampling areas were collected from May to September in 2015. We observed also the link among character of biotopes, vegetation cover and reclamation techniques.

Table 1. List of sampling areas.

Sampling area	Date	Latitude	Longitude	Reclamation
1.	1.5.2015	49°13'16,5"	16°41'43,7"	Technical recultivation
2.	1.5.2015	49°13'15,6"	16°41'41,4"	Technical recultivation
3.	1.5.2015	49°13'25,1"	16°41'42,5"	Spontaneous succession
4.	16.5.2015	49°13'13,1"	16°41'32,3"	Technical recultivation
5.	16.5.2015	49°13'13,4"	16°41'32,4"	Technical recultivation
6.	16.5.2015	49°13'13,9"	16°41'37,1"	Technical recultivation
7.	23.5.2015	49°13'20,2"	16°41'28,2"	Technical recultivation
8.	24.5.2015	49°13'23,1"	16°41'31,7"	Technical recultivation
9.	24.5.2015	49°13'13,3"	16°41'27,9"	Technical recultivation
10.	30.5.2015	49°13'22,6"	16°41'51,3"	Spontaneous succession
11.	5.7.2015	49°13'20,4"	16°41'49,0"	Spontaneous succession
12.	19.7.2015	49°13'17,9"	16°41'40,2"	Spontaneous succession
13.	18.9.2015	49°13'21,7"	16°41'42,1"	Spontaneous succession
14.	24.9.2015	49°13'17,2"	16°41'43,9"	Spontaneous succession

We collected zoological samples using a uniform methodology. Each of sampling area was 10 x 10 m or 20 x 5 m if was located next to the quarry wall. We collected molluscs for one hour manually by one person. Collecting of molluscs was performed on trees, on vegetation, under stones and other objects on the ground, in the plant litter, and from mosses. We thoroughly inspected the soil substrate in four squares with an edge of 30 x 30 cm to a depth of at least 5 cm on each of the sample area. We determined molluscs by Ložek [6] and Horsák et al. [7].

Ecological classification: SI 1 - silvicolae (strict forest species), SI 2 - silvicolae (predominantly forest species), SIh - highly humid forest species, ST - steppicolae (steppe species), SS - silvisteppicolae (species of xeroterm bushes and sparse groves) MS - mesicolae (euryvalent species), SIp - agricolae (euryvalent species living on rocks and in forests too), RP - ripicolae (species with high humidity demands).

Conservation status: LC - least concern, NT - nearly threatened

3. Results

We found 514 in total individuals of terrestrial gastropods belonging to 20 species of 6 ecological groups in Lisna Quarry. Table 2 shows a summary of all species including the number of living individuals found on individual sampling areas, ecological distribution, and conservation status.

Table 2. List of Moluscs.

Eco-elements		Name of species	Status	1.	2.	3.	4.	5.	6.	7.	8.	9.	10.	11.	12.	13.	14.
		Aegopinella nitens (Michaud, 1831)	LC	7	~	~	3	~	~	~	4	1	~	~	~	~	~
1	SI	*Cochlodina laminata* (Montagu, 1803)	LC	~	~	~	~	~	~	~	~	~	~	~		~	~
	SI(MS)	*Alinda biplicata* (Montagu, 1803)	LC	~	~	~	~	~	~	~	~	~	~	~	19	~	~
	SI(MS)	*Arion fuscus* (O.F. Müller, 1774)	LC	5	8	~	7	14	2	1	~	2	~	~	~	~	~
	SI(MS)	*Discus rotundatus* (O. F. Müller, 1774)	LC	1	~	~	~	~	~	~	~	1	~	~	13	~	4
2	SI(MS)	*Limax cinereoniger* (Wolf, 1803)	LC	~	~	2	1	~	1	1	~	~	4	~	~	~	~
	SI(MS)	*Monachoides incarnatus* (O. F. Müller, 1774)	LC	2	1	~	5	1	~	3	3	~	~	~	~	~	~
	SIth	*Helix pomatia (*Linnaeus, 1758)	LC	1	7	~	1	3	1	1	45	14	1	~	~	~	~
	SIth	*Merdigera obscura* (O. F. Müller, 1774)	LC	~	~	1	~	~	~	~	~	~	~	~	~	12	2
3	SIi	*Arion rufus* (Linnaeus, 1758)	LC	~	~	~	1	~	8	~	~	~	~	~	~	~	~
	ST	*Granaria frumentum* (Draparnaud, 1801)	NT	~	~	6	~	~	~	~	~	~	~	~	~	~	~
4	ST	*Xerolenta obvia* (Menke, 1828)	LC	~	~	2	~	1	~	3	~	1	21	18	18	19	28
	ST(SI)	*Cepaea vindobonensis* (Férussac, 1821)	NT	2	4	7	~	2	1	8	16	17	15	5	9	6	2
5	SS	*Euomphalia strigella* (Draparnaud, 1801)	LC	~	~	~	~	~	~	7	7	2	~	~	~	~	~
	MS	*Arion distinctus* (Mabille, 1868)	LC	5	7	~	~	~	~	~	~	~	4	~	~	~	~
	MS	*Arion vulgaris* (Moquin-Tandon, 1855)	LC	~	1	1	1	~	4	3	1	~	12	~	~	~	~
	MS	*Boettgerilla pallens* (Simroth, 1912)	LC	~	1	~	~	~	~	~	~	~	~	~	~	~	~
7	MS	*Oxychilus cellarius* (O. F. Müller, 1774)	LC	~	~	~	~	~	~	~	~	~	2	~	~	~	~
	MS	*Vitrina pellucida* (O. F. Müller, 1774)	LC	~	~	1	~	~	~	1	~	~	~	~	~	~	~
	SIp	*Laciniaria plicata* (Draparnaud, 1801)	NT	~	~	~	~	~	~	~	~	~	~	26	~	~	~
	SIp	*Limax maximus* (Linnaeus, 1758)	LC	~	~	~	~	~	~	1	2	~	~	~	~	~	~
		Σ of individuals														514	

4. Conclusion

The mining in the Lesni Quarry in Lisna was ended in 1997. After the mining some areas were left to spontaneous succession and part of the quarry was technically reclaimed. Technically reclaimed areas can be described as places with predominantly bushes and trees, with lots of fallen wood with a rich herbaceous ground cover. Areas of spontaneous succession we can characterize as places with minimal occurrence of trees and discontinuous grasslands.

We found 514 live mollusc individuals belonging to 20 species of 6 ecological groups in 14 sampling areas. A relatively large diversity of microbiotopes within the quarry is illustrated wide representation of ecological groups suggests. The largest group contains predominantly forest species, followed by equivalent species and steppe species. The presence of steppe species is not surprising. Approximately 15 years have elapsed since the mining, and most of areas are in early stages of the succession. There are habitats with rather "open" character that suits steppe species in these quarry. However, steppe species probably disappear due to progressive overgrown of quarry within few decades. We found most often species *Cepaea vindobonensis*, *Xerolenta obvia* and *Helix pomatia*. According to the Red List of Endangered Species of the Czech Republic by Farkač et al., 2005 [8], the species *Granaria frumentum*, *Cepaea vindobonensis* and *Laciniaria plicata* have conservation status "Near Threatened" (NT). Other species are included in the group "Least concern" (LC). Problematic can be occurrence of invasive species *Arion vulgaris* in half of the sampling areas.

Six from 26 species of gastropods we found only on technically reclaimed areas (*Aegopinella nitens*, *Arion fuscus*, *Monachoides incarnatus*, *Arion rufus*, *Euomphalia strigella*, *Limax maximus*). Predominant frequency of occurrence on reclaimed areas have also *Helix pomatia*. Five species (*Alinda biplicata*, *Merdigera obscura*, *Granaria frumentum*, *Oxychilus cellarius*, *Laciniaria plicata*) we found on areas of spontaneous succession. As for remaining species, there is the dominant frequency of *Xerolenta obvia* or *Discus rotundatus* on open spontaneous successive areas.

The diversity of the occurrence of mollusc's species on technically reclaimed areas and areas with spontaneous succession shows how important the heterogeneity of the environment is for regeneration and increasing the biodiversity of sites disturbed by mining.

5. References

[1] Culek M 1996 *Biogeographical division of the Czech Republic* (Praha: Enigma) p 347 s.

[2] Musil R et al 1993 *Moravian Karst - Labyrinths of Knowledge* (Adamov: J. Bližňák, GEO program) p 336

[3] Quitt E 1971 *Climatic regions of Czechoslovakia* (Brno: ČSAV) p 73

[4] Skalický V 1988 *Regional phytogeographical division* Flora of the Czech Socialist Republic (Praha: Academia)

[5] Tichý L 2003 *Botanical characteristics of the "Lesní Quarry"with regard to the occurrence of protected and endangered plant species* (Brno: Archív ZO ČSOP Pozemkový spolek Hády)

[6] Ložek V 1956 *The key to determining Czechoslovak molluscs* (Bratislava: SAV) p 437

[7] Horsák M, Juřičková L and Picka J 2013 *Molluscs of the Czech and Slovak Republics* (Zlín: Kabourek)

[8] Farkač J, Král D and Škorpík M 2005 *Red list of threatened species in the Czech Republic, Invertebrates* (Praha: Agentura ochrany přírody a krajiny ČR) p 760

IOP Conf. Series: Earth and Environmental Science **92** (2017) 012035 doi:10.1088/1755-1315/92/1/012035

Possibilities of application of artificial groundwater recharge in the catchment areas of the Odra and Bečva rivers

M Kyncl[1], R Heiduk[1] and S Heviánková[1]

[1]VŠB – Technical University of Ostrava, Faculty of Mining and Geology

E-mail: miroslav.kyncl@vsb.cz

Abstract. The paper deals with supplies of drinking water in times of drought. Facing the climatic changes, it is essential to look for alternatives sources of drinking water. Artificial infiltration ranks among such options. Two sites, Hranice na Moravě and Odry, have been monitored and evaluated. In those sites, it would be possible to use artificial infiltration and increase supplies of ground water.

1. Introduction

In times of drought, one should manage yield and quality of ground water that is used as a source of drinking water for people.

Badly-advised straightening of water courses caused by agricultural self-interest (mostly in the 1970s) and construction of industrial plants/other buildings has resulted in faster outflow of surface water from the Czech landscape, reducing, in turn, natural infiltration.

Infiltration is a part of a water cycle in nature. Water penetrates into soil and permeable rock. Infiltration is, in addition to condensation of water vapours in soil and condensation of water vapours in magma, the key source of ground water. One differentiates between natural infiltration and artificial infiltration. In case of infiltration, the penetrating water results from rainfall, surface water (embankment infiltration) or thawing snow. Artificial infiltration is when the infiltration is caused by artificial flooding of the landscape surface [1]. Water, while passing through rock, is being filtered and physical/chemical/biological contaminants are being removed [2].

Infiltration facilities that have been used most frequently include drainage tanks, infiltration bores, infiltration pipes (drainage), surface splashing and surface spraying.

The ground water will increase its supplies thanks to artificial infiltration only if the quality of original raw water is sufficient and only if suitable rock environment is available [3].

A coefficient of filtration is used to describe sufficient permeability of soil or the rate of water flow through rock in m/s (table 1).

Recently, the artificial infiltration has been becoming more and more popular. This technology is able to response promptly to such opposite water management problems as floods and drought [5]. It may solve many water management issues and environment problems. It may support management of economic and national problems. This process is really unique.

Table 1. Reference value of the coefficient of filtration for soil [4].

Soil	Permeability in m/s	Coefficient of filtration
Coarse grain gravel	0.1 to 0.005	10^{-1} to 5 x 10^{-3}
Fine through medium grain gravel	0.03 to 0.0005	3 x 10^{-2} to 5 x 10^{-4}
Sand gravel	0.01 to 0.0001	10^{-2} to 10^{-4}
Coarse grain sand	0.004 to 0.0001	4 x 10^{-3} to 10^{-4}
Medium grain sand	0.001 to 0.00006	10^{-3} to 6 x 10^{-5}
Fine grain sand	0.0004 to 0.000006	4 x 10^{-4} to 6 x 10^{-6}
Clay sand, sandy clay	0.000075 to 0.00000005	7,5 x 10^{-5} to 5 x 10^{-8}
Clay	0.000005 to 0.0000000001	5 x 10^{-6} to 10^{-10}
Argillaceous clay	0.000004 to 0.0000000001	4 x 10^{-6} to 10^{-10}
Clayey loam	0.00000001 to 0.0000000001	10^{-8} to 10^{-10}
(pursuant to DWA-A 138, April 2005)		

2. Selecting a site for artificial infiltration

This paper deals with selection of a site for potential installation of artificial infiltration for water purposes in river basins of the Odra River and Bečva River. In each location, there should be a shallow well or hydro-geological drill hole that could be used as a potential collection facility.

The first place where the artificial infiltration for water systems could be used is the Čs. legií Park in Hranice na Moravě (District of Přerov, Region of Olomouc) (figure 1). The site is located close to the Bečva River and was used in past for artificial infiltration. Another possibility is to use the land next to groundwater drill bores in Odry (District of Nový Jičín, Moravian-Silesian Region) (figure 1). Fluvial rock environment next to the Odra River can serve as a potential supply of infiltered water.

Figure 1. Blank map of water systems in the Czech Republic [8]. Odry on the Odra River and Hranice na Moravě on the Bečva River.

2.1. Hranice na Moravě on the Bečva River – the Čs. legií Park

Hranice na Moravě is located in the geomorphological system that is referred to as the Moravian Gate. Its territory is dewatered by the Bečva River which is the biggest left-hand tributary of the Morava River [6].

The average annual flow in the profile measured at Teplice nad Bečvou (at 41.4 river kilometer) is 15.5 m^3/s [7]. The Čs. legií Park where the artificial infiltration could be installed is located at 39.4 river kilometre.

Water management of the territory. The key source of drinking water for Hranice na Moravě is the water conduit from the Ostrava Area Water System. It supplies water from the water treatment plant in Podhradí. Hranice also receives underground water from the spring area in Lhotka. Valšovice, which is a part of Hranice, receives groundwater from the spring area of Ústí. Středolesí, another local part, is supplied water from the Radíkov water treatment plant that treats surface water. The group water system Hranice a Záhoří is owned by Vodovody a kanalizace Přerov a.s. [9].

Hydro-geological composition of the territory. The sedimentary fill of the Bečva River valley consists of bulk river gravel and sand. From the mechanical point of view, it can be regarded as sand gravel with round stones being as much as 30 cm big. The typical size is, however, 2-6 cm. Towards the rock cover, the gravel sand changes into humus medium-grain sand which forms the surface of the territory. The layer of sand/gravel is 7-8 mm at least. Approximately in that level under the surface, they sediment on sea deposits of Miocene Epoch. The gravel sand represent a water horizon which is relatively easy permeable. The shallow groundwater forms a single flow with a slightly tense level. The permeability corresponds approximately to 6.10^{-4} m/s. From the chemical point of view, this is alluvial water that contains iron and manganese ions [10].

Supplying the drinking water through artificial infiltration in past. The Hranice weir on the Bečva River with its storage level of water plays an important role in collection of shallow groundwater. The wells that are positively influenced by embankment infiltration have a relatively higher specific yield – about 10 l/s [11]. Water from alluvial sediments in the Čs. legií Park was collected in two collection wells (3 m and 4 m diameters) in ca. 8 mm depth. Then, the water was supplied to the "Nový odbyt" pumping station with a collection well that was used as both recharging and collection facility. Downstream the facility, the water was pumped into a ground reservoir on a hill.

Since 1960 consumption of drinking water in Hranice has been increasing gradually and significantly. From 1960 until 1965 consumption of drinking water increased almost by 50 % [12]. In order to improve the yield of water sources, an artificial infiltration system was designed for the Čs. legií Park. It was in operation from 1965 until 1985.

A submersible pump transported water from the Bečva River through a steel pipe to the infiltration area which was located in a 100 m distance. The water flowed onto a grass area. This created a wetland where water infiltrated into rock environment. A suction pipeline sucked water from existing wells into a pumping station. Using the artificial infiltration, it was possible to increase the capacity of a water source by 5 l/s. That process was used until new sources of water were discovered in Lhota. Now, the shallow water from the Čs. legií Park is not used anymore as storage of the drinking water to people.

Conclusion. Recommended solutions. First proposals to renew the artificial infiltration include the project focused on renewal of the original infiltration in the Čs. legií Park. The second project for artificial infiltration is infiltration of surface water from the Bečva River in the "Nový odbyt" pumping station. There is a subsidence in landscape and it is possible to flood the lend. Between the Čs. Legií Park and pumping station, there is a road. The infiltered water could be collected using the monitoring bore P3 (7 m deep) or the recharging/collecting well in the pumping station. The artificial infiltration in the pumping station can be improved, it a water reservoir (ca. 20 m^3) is used for storage of water to be infiltered.

Possible risks. Since 1980s, intensity of road transport has been increasing (more vehicles, more traffic on roads). This increases the risk of contamination of water for infiltration with oil and oil substances from nearby roads. Another threat is increased biological and micro-biological activation of water sources in times of drought.

2.2 Odry on the Odra River

Odry is a town located on both banks of the Odra River in the Odra Fold. Springs of the Odra River are located in Oderské vrchy, Odra hills, in a military zone (Libavá). The average annual flow of the Odra River in Odry is 3.58 m^3/s [13]. Close to the town, there are several tributaries: Vítovka and Zlatý potok on the left-hand bank and Mlýnský potok and Stodolní potok on the right-hand bank.

There is also a 6 km long millrace that supplies water to five ponds [14]. The Odra Fold that opens towards the Moravian Gate is approximately of a rectangular shape, 2 x 3 km long, where the longer axis is oriented from south-west to south-east [15, 16].

Water management of the Odry Fold. Drinking water for Odry is supplied from drill bores. There are four drill bores: OVHS 1 (d = 159.6 m), OVHS 2 (d = 250 m), NP 757 + HV2 (d = 260 – 270 m) and NP 769 (d = 125 m). The raw water from three drill bores OVHS 1, OVHS 2, NP 757 is pumped into the dumping station in Odry. There is an aeration tower and a 30 m^3 reservoir. Then, water is pumped into water tanks. Raw water from NP 769 is supplied for consumption in Mateiciuc (producer of plastic pipelines) [17].

Hydro-geological composition of the Odry Fold. Approximately 25 l/s of ground water is used now in the hydro-geological structure of the Odra Fold. Natural resources of ground water in the Neogene and Quarternary sediments were estimated by Hrouzková et al. (1984) to reach as many as 70 l/s.
The following water-bearing strata (a., c., d.) are located in the surroundings:

- Shallow strata of Quarternary clayey gravel with free level. Collector with 6.5 – 9 m thickness.
- Insulator: lime clay of the Lower Badenian layer.
- Depth level formed by filled-in sand in politic facies (clay rock) and basal clastics (detrite) of Lower Badenien layer located 130-290 m under the landscape with tense level.
- Deep fissure rangers in culm.

Considering the area under study, what is important for us is the shallow circulation of ground water (see a.).

The shallow Quarternary aquifer on the site is linked to fluvial clay gravel of the main platform in the original riverbed of the Odra River. The ground water has free surface there and does not seem to be linked directly to the level of the nearby Odra River. The embankment infiltration occurs only if there is increased level of water in the Odra River. The collector receives water mostly from rainfall. The general direction of ground water flow in the shallow circulation in fluvial sediments of the Odra River corresponds to the landscape slope and direction of flow of the Odra River, from north to south. Permeability of the collector that consists of the fluvial clay gravel is estimated to reach 3.10^{-4} m/s. Potential sources of water that are related to the shallow aquifer tend to change the yield considerably with each change in outflow in the shallow collector (any construction, excavation, burying of new or reconstruction of old pipelines/ducts in deeper layers…). [18].

Conclusion and recommended solutions. The drinking water for people living in Odry is supplied from bore holes drilled down to 125 - 270 m deep in the basal clastics collector in the Lower Badenian layer. In the Odra Fold, there are several monitoring bores (both deep and shallow ones).
When employing the artificial infiltration for supplies of drinking water in times of drought, it is possible to use the existing shallow bores for collection. That is why we are looking for shallow Quarternary sediments close to the Odra River. In the Odra Fold, there are either fluvial or Aeolian sediments. The fluvial sediments consist often from clayey gravel that is covered with a layer of flood clay. The total thickness of layers ranges between 4 and 7 m. The Aeolian sediments consists of floury soil, the maximum confirmed thickness of which is 12 m – to the east from Odry [15].

The best location for the artificial infiltration seems to be the surroundings of the drill hole NP 769 where a shallow hole, V36, is located (6.5 m). Another advantage is that there are not any obstacles such as roads or buildings between the drill hole and Odra River (which is not often the case with other drill holes). The distance between the Odra River and the infiltration area would be 150 m across the grass area.

Aknowledgements
This article was written within the project QJ1620148 “Research in possible use of artificial infiltration to increase the capacities of ground water resources in dry periods” financed by the Ministry of Agriculture of the Czech Republic

References
[1] Hrádek F and Kuřík P 2002. Hydrology – Infiltration (Czech agriculture University Prague) pp 81–85

[2] Czech geology service 2017 Ground water – working lists http://www.geology.cz/svet-geologie
[3] Vodárna Káraný a.s. 2017 Waterwork in Káraný www.vodarnakarany.cz
[4] Deutsche Vereinigung für Wasserwirtschaft, Abwasser und Abfall 2005: Arbeitsblatt DWA-A 138. Planung, Bau und Betrieb von Anlagen zur Versickerung von Niederschlagswasser.
[5] Hrkal Z 2016 Assessing the potential of river basins threatened by drought and water scarcity for the use of artificial recharge systems to strengthen water resources in the dry season. (VÚV T.G.M., v.v.i. Praha) p 17
[6] City Hranice na Moravě 2017 https://cs.wikipedia.org/wiki/Bečva
[7] Czech hydrometeorological Institute 2017 http://hydro.chmi.cz/hpps/hpps_prfbk_detail.php?seq=307352
[8] Geographical portal 2017 The blank map of the Czech Republic http://www.zemepis.com/reky4.php
[9] Water management of the city Hranice 2017 www.mesto-hranice.cz/cs/zivotni-prostredi/vodni-hospodarstvi-a-rybarstvi/vodni-hospodarstvi.html
[10] Engineering geological and hydrogeological survey se Žilina - Brno plant Hydrogeological survey Hranice na Moravě = Nový odbyt 1966
[11] GEOtest Hydrogeological survey 1968 Hranice na Moravě II Additional hydrogeological survey of the catchment area
[12] KSVK Ostrava Technical report 1967 Hranice – Water wells with siphon pipes in the park
[13] Czech hydrometeorological Institute 2017 http://hydro.chmi.cz/hpps/hpps_prfbk_detail php?seq=307097
[14] City Odry 2017 https://cs.wikipedia.org/wiki/Odry
[15] Hrouzková J 1985 Hydrogeological survey - Oderská kotlina Water resource se Prague
[16] The history of the city of Odry 2017 http://www.oderske-vrchy.cz/clanek/historie-oblasti/historie-mesta-oder/1/
[17] SmVaK Ostrava as 2013 Technical report p 402
[18] GEOoffice 2013 Loučky na Odrou - Strengthening the capacity of local groundwater resources Final report of the hydrogeological survey of deepwater - construction of the HV-2 well

IOP Conf. Series: Earth and Environmental Science **92** (2017) 012036 doi:10.1088/1755-1315/92/1/012036

Study of supply of drinking water in dry seasons in the Czech Republic

M Kyncl[1], S Heviankova[1] and T L Ch Nguien[2]

[1]VŠB – Technical University of Ostrava, Faculty of Mining and Geology
[2]Ton Duc Thang University, Vietnam

E-mail: miroslav.kyncl@vsb.cz

Abstract. Models of climate change assume a long-term growth in temperature. This poses a risk for the supply of drinking water to the population. This paper evaluates the possibilities of optimal use of surface and underground sources. It proposes the interconnection of surface resources and optimization of their use. In order to improve the yield of underground water sources, it recommends the use of artificial infiltration.

1. Introduction

Climatic changes associated with global warming have been and will be causing extensive deficiencies in the water cycle. Estimated climatic models predict a world-wide growth in temperatures and thus occurrence of droughts. No matter how the prognoses will turn out, it is vital to prepare measures in order to ensure water supplies in the dry periods.

Changes in the climate influence the quantity and quality of groundwater and surface water, and insufficient precipitation causes lower surface run-off and thus affects the quantity of groundwater. This leads to lower flows in watercourses, drops in the groundwater levels and lower yields in springs. Altogether, the phenomena eventuate hydrological droughts. In addition, higher temperatures cause higher evaporisation.

Drinking water supply belongs among basic infrastructure services as for population's needs. Climatic changes are a threat for which measures must be prepared. In the Czech Republic the water resource situation is very difficult as all water originates in precipitation. Problems associated with drought have manifested for several years and are assumed to aggravate. The strategy for dry periods should lie in such measures so that the water supply via public water mains is not interrupted. The measures may be classified into two areas. Measures on the part of resources mean that their protection must be increased, and additional sources of water must be searched for. On the part of water take-off, it is necessary to prepare measures leading to lower water take-off in order to ensure continuous water supplies in case of a water deficit.

The experience from the dry year 2015 shows that drought varies in its intensity in the different regions. In this we perceive a possible solution in the form of cooperation of the different surface water sources. Interconnecting the sources, the dependable yield of the sources may be increased. The interconnection is also advisable in larger scales. It means that whole water supply systems should be interconnected, which shall ensure water supplies also in regions with the current drought. The solution in groundwater resources may be artificial infiltration to store water for the dry periods.

IOP Conf. Series: Earth and Environmental Science **92** (2017) 012036 doi:10.1088/1755-1315/92/1/012036

2. Prognoses of climatic changes and their impact on water balance

The Czech Republic falls in the mild climate zone of the Northern Hemisphere. The climate is characterised by dominant westerly wind and intense cyclonic activity. The majority of the country receives annual 500 – 700 mm of precipitation; wetter regions from 700 – 1000 mm of precipitation a year. In the mountains, the maximum precipitation may reach 1700 mm. On the contrary, the driest sites have an annual precipitation of 410 mm only [1].

According to the estimates of the Czech Hydrometeorological Institute [2] the Czech Republic may expect an increase in the mean annual temperature of air of 1.4 – 1.8 °C between 2021 and 2050. Between 2071 – 2100 the temperature is assumed to grow by 3.3 – 3.7 °C. The data have been related to the reference period of 1961 – 1990. A future drop in precipitation is expected in summer. Extreme phenomena will occur on a more frequent basis. The expected threats are torrential rain and drought. The weather models estimate a future decrease in winter precipitation of as much as 20 %. In spring the precipitation is supposed to increase by 2 – 16 % depending on the different regions. The summer and autumn will experience a significant spatial variability in precipitation. Overall, there will be a drop of as much as 30 %; in places there may be an increase of as much as 10 %. The precipitation models show considerable fluctuations in precipitation in the different regions. As for reliability, the prognoses for the Czech Republic correspond to the prognoses processed by the German meteorological services [3]. For the period 2021 – 2050 they assume an increase in temperature of 0.5 – 1.5 °C and by 2100 a rise in temperature of 1.5 – 3.5 °C. Warming should be more apparent in winter and more frequent extreme phenomena are assumed, such as intense rain, sea storms or long droughts. The year 2015 is an example of a dry period. Within the territory of the Czech Republic it has been evaluated the hottest year for the past 134 years, i.e. the whole period under measurement records. There were 36 tropical days and the majority of the meteorological stations recorded new maximum temperature records. Lower precipitation was observed from the end of February and by the end of October the precipitation was lower by 165 mm, i.e. a drop of 30 %. Figure 1 shows the accumulated course of rainfall compared to long-term values [4].

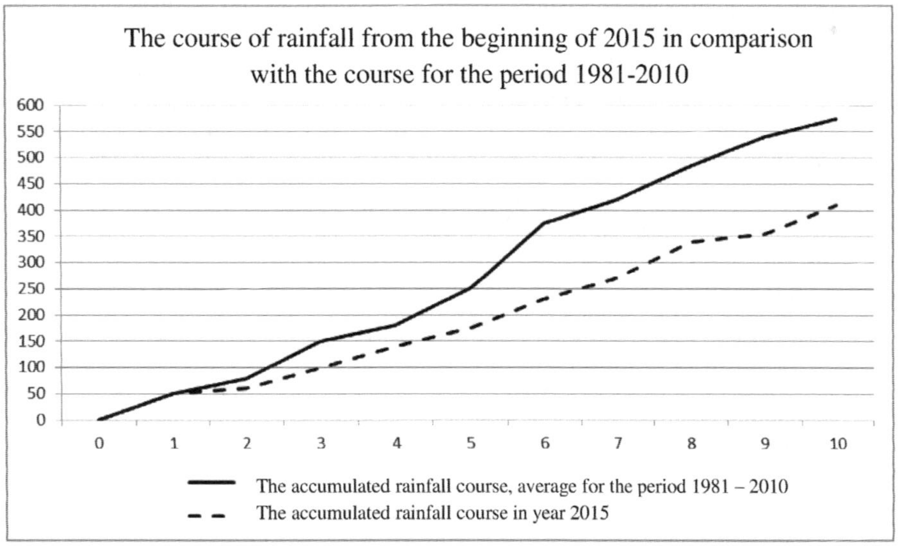

Figure 1. The course of rainfall from the beginning of 2015 in comparison with the course for the period 1981-2010.

The flow rates in rivers ranged from 10 to 50% of long-term mean annual discharge. The drought also manifested in the reserves of groundwater. 40% of all bores had seriously subnormal reserves. A more favourable situation was in water supply reservoirs on watercourses. The live capacities of the reservoirs were filled at 40 – 60%. This drought showed in the continuity of water supplies only partially, mainly due to large capacities of sources over the water demand.

3. An overview of drinking water supply

In the Czech Republic over 94 % of population take water from the public water mains. Annually, 624 mil. m^3 of drinking water are produced. One half of the produced volume originates in surface sources and one half comes from underground sources. The surface sources are mainly water management reservoirs on watercourses. The underground sources are both deep and shallow. The water discharge intensity for households ranges around 90 l per person and day. Good news is that nowadays the water source capacities have been used out of 60 %. The consumption of drinking water has dropped by half when compared to the period before 1990. After 1990 the Czech Republic has launched market economy. In terms of drinking water supplies, this meant that the previously state-subsidised water costs increased to cost price. Water take-off has decreased along with a rise in the price of water. For example, in 1990 the water discharge intensity for citizens was 170 l per person and day [5]. The experience of the year 2015 shows that the drought that lasted for the almost whole year did not significantly affect the continuity in drinking water supplies in water mains supplied from surface sources. However, there were problems with smaller groundwater sources. The experience of last century dry periods shows that there were 2-3-year droughts. Two-year droughts occurred approximately once in 10 years; three-year droughts occurred every 20 years. It is important to focus on longer periods when preparing measures for the dry periods [6].

4. Measures for water supplies from surface sources in longer dry periods

To assess the options for drinking water supplies in long dry periods, we used data and findings from the year 2015. To model how long certain sources of surface water may ensure continuous water supplies, we selected a locality in north-western Bohemia. The locality is situated in the rain shadow of the Krušné hory Mountains, where also the driest region of the Czech Republic, Žatec Basin, is situated. The mean annual precipitation ranges from 450 to 550 mm. As much as 84 % of population are supplied from the public water mains. The sources are water management reservoirs Přísečnice, Fláje and Žlutice. Water is treated for drinking water in water purification plants.

Přísečnice Reservoir has an overall volume of 54.7 mil.m^3, the live storage capacity is 46.7 mil.m^3, and the maximum allowable water system take-off is 960 l/s. The standard water system take-off is 490 l/s, which is 51 % of the maximum allowable take-off. The minimum long-term run-off of 50 l/s must always be ensured. At the minimum inflow into the reservoir in dry periods, the water supplies shall suffice for two years.

Fláje Reservoir has the overall volume of 23 mil.m^3, the live storage capacity is 19.5 mil.m^3, and the maximum allowable water system take-off is 510 l/s. The standard water system take-off is almost 500 l/s. The minimum long-term run-off was set as 75 l/s. In 2015 the storage capacity was used up from 48 % only. At the minimum inflow into the reservoir in dry periods, the water supplies would suffice for 286 days.

Žlutice Reservoir has the overall volume of only 15.6 mil.m^3, the live storage capacity is 10.4 mil.m^3, and the maximum allowable water system take-off is 190 l/s. In 2015 the standard take-off is 86 l/s. The minimum long-term run-off into the catchment area is 220 l/s, which is rather high with regard to the volume of the reservoir. This is to ensure water supplies into the catchment area below the reservoir. At the minimum inflow into the reservoir and maintaining the conditions, the water supplies would suffice for 235 days.

The situation in the reservoirs of Fláje and Žlutice is not very favourable as for ensuring continuous water supplies in dry periods. If the drought of 2015 had lasted longer, water supplies would have been ensured only for 8 months. The estimated droughts in the future may last from two to three years. This may be solved via interconnecting the water management systems of Přísečnice with those of Fláje and Žlutice. In such a case, the overall water reserves would last for 15 months. This way, it is a suitable measure for the future as new construction and capital repairs are needed. Prepared are also emergency measures, such as lower run-off from the reservoir into the catchment in order to prolong continuous water supplies to 2 years.

The model example shows that an interconnection of water management systems may ensure continuous drinking water supplies in times of drought. Considering longer dry periods, it is vital to prepare regulation measures to reduce drinking water take-off by customers. This is to ensure a continuity in drinking water supplies as any discontinuity leads to larger water losses.

In the Czech Republic, there are a number of extensive water management systems supplied mainly from surface reservoirs. This is another area, where solutions must be searched for, as the drought prognoses for the next 30 – 50 years are not favourable. However, this requires a solution in the form of investments that will be very costly. The implementation of such measures shall last longer than 10 years. To be specific, we may speak of interconnecting the Ostrava regional public water mains, where there is a permanent surplus in surface water sources, with regions in central Moravia. Central Moravia mostly has underground water sources and when compared with Northern Moravia (with its Ostrava regional public water mains) central Moravia has minimum rainfall. Another solution is an interconnection of a number of reservoirs, such as Přísečnice, Fláje and Žlutice and the North-Bohemian Lower Krušné hory public water mains, which is supplied from heavy rainfall from the Krušné hory Mountains. As an example, we may use the public water mains in Eastern Bohemia. It is a system predominantly supplied from underground sources. The system has been expanded into the Pardubice Region and Pěčín Reservoir was built. This shall ensure water supplies for the estimated period of 2030 – 2050. There are large reserves in the Vír regional public water mains to strengthen supplies for the wider Brno agglomeration. Another public water mains system is Southern Bohemia with its reserves. This system may be interconnected north-westwards with Příbram Region.

Apart from interconnecting the public water mains systems, the reliability of the systems is also important in case of emergency. If there is a failure in the local system, we cannot speak of cardinal problems. More serious problems may occur in case of failures of long-distance water management supply lines. Currently, the Czech Republic has more than 2700 km of long-distance supply lines (cross-section 400 – 1600 mm). One third of these is older than 50 years. Besides the prognoses of climatic changes, it is important to plan the revitalisation and modernisation of the existing water management systems for the future.

5. An increase in groundwater reserves for dry periods

In the Czech Republic, a half of drinking water reserves comes from the underground sources. However, in times of drought these groundwater sources are also under threat. This predominantly concerns more shallow underground sources, in which their yields fall quickly in dry periods. On the contrary, in deep underground water sources the yield falls less quickly, but filling in the sources make last many years. One option to improve the situation is artificial infiltration, which used to be applied more widely in the past. At present, artificial infiltration is applied in Kárany on the Jizera River.

Artificial infiltration may largely help water sources in dry periods. However, the selection of a suitable locality for artificial infiltration is not easy. Sources of surface water must be available for infiltration, which may be a problem in dry periods. Another important question is the quality of water to be infiltered underground. The choice of a suitable locality is thus crucial. Such suitable areas may be plots along the Bečva River. In the past, this method used to be applied also in Rožnov on the Rožnovská Bečva River and the Bečva River in Hranice and Lipník. Another suitable locality may be plots along the Odra River in the Odry Valley. This is supposed to be verified by research and technical solutions must be designed.

6. Conclusion

The development of the last decade has shown that the expected climatic changes are already taking place, namely there are longer dry periods and higher temperatures. The existing experience shows that in the Czech Republic this is a problem that may be solved. It is possible to interconnect the public water mains systems supplied from surface sources in order to optimise the different systems' capacities. Droughts do not always affect the country evenly and water supplies may be continuous also in drier regions. Artificial infiltration may be used in underground sources. However, it is vital to assess the possibilities of the different localities with respect to the water quantity and quality suitable for infiltration underground.

Acknowledgements

This paper was supported by the Ministry of Agriculture of the Czech Republic within Project QJ1620148 Research in possible use of artificial infiltration to increase the capacities of ground water resources in dry periods.

References

[1] Tolasz et al. 2007 Climate atlas of Czechia Český hydrometeorologický ústav Praha p 256
[2] Pretel J et al. 2001 Specification of existing estimates of the impacts of climate change in the sectors of water, agriculture and proposals for adaptations Final report on the project VaV SP/1a6/108/07 for the period 2007-2011
[3] Bundesregierung: Aktionsplan Anpassung der Deutschen Anpassungsstrategie an den Klimawandel vom Bundeskabinett am 31. Augustt 2011 beschlossen Berlin
[4] Czech Hydrometeorological Institute 2015 http://www.chmi.cz/files/portal/docs/meteo/ok/SUCHO/zpravy/2015/tyden37.pdf
[5] SOVAK ročenka 2017 SILVA s.r.o. Praha
[6] Ministry of the Environment 2014: Strategies to adapt to climate change in the Czech Republic

Contamination of potentially toxic elements in streams and water sediments in the area of abandoned Pb-Zn-Cu deposits (Hrubý Jesenník, Czech Republic)

J Lichnovský[1], J Kupka[1], V Štěrbová[1], P Andráš[2] and P Midula[3]

[1]Institute of Environmental Engineering, Faculty of Mining and Geology, VŠB – Technical University of Ostrava, 17. listopadu 15, 708 33, Ostrava 1
[2]Department of Environmental Management, Faculty of Natural Sciences, Matej Bel University, Tajovského 55, 974 01 Banská Bystrica 2
[3]Department of Biology and Ecology, Faculty of Natural Sciences, Matej Bel University, Tajovského 55, 974 01 Banská Bystrica 2

E-mail: jakub.lichnovsky@vsb.cz

Abstract. The deposits, located in Nová Ves and Zlaté Hory were well known and important sources of metal ore in Jesenniky region in the past. Especially the one in Nová Ves, which is recently the most important hydrothermal deposit of venous type in the whole area. The mining activity, aimed on lead and zinc minerals was practically permanent here from the middle-age to 1959. On the other hand, the site in Zlaté Hory is the most important ore deposit in Czech Silesia. The non-venous types of polymetallic, copper and gold deposits, evolved in the complex of metamorphic devon rocks are located on south and south-west directions of the area. Long and permanent mining industry caused remarkable changes in the local environment, creating mine heaps and depressions. The probability, that dump material contains potentially toxic substances that could be possibly leaked into surrounded environment is high.
This contribution presents the part of complex study results, aimed on evaluating of potential environmental impacts in above mentioned locations. It aims on contamination, caused by potentially toxic heavy metals (Pb, Zn, Cu, Ni, Fe, Mn, Co, Cd, Cr and As) at the sites, exposed to mining activity in the past. The study focus on the contamination of these sites and evaluate them as potential risk for surrounded environment.

1. Introduction

The Pb-Zn deposit in Nová Ves is the most important hydrothermal ore deposit of venous type in Jeseníky region from historical point. The mining activity was performed on eastern and south-eastern slopes of Soukenná (1026 m a. s. l.). The mining of polymetallic ores was well-known there from the Middle Age [1,2,4]. On the other hand, the mine region in Zlaté Hory is the most important ore deposit in Czech Silesia. The non-venous types of polymetallic, copper and gold deposits are situated here. Příčný hill, with altitude of 975 m a. s. l. is the highest peak of Zlatohorská highland. On this site, the rests of mining activity are still significantly present [3].

The historical dump-fields and mining enclaves in these areas could be on the present a potential risk for surrounded environment, which is a main topic of this work.

2. Materials and Methods

The main aim of the study includes the complex research of model areas (Soukenná and Příčný hill). The sampling sites were selected after the reconnaissance of terrain conditions, with the main focus on the description of interactions between the environment and mining enclaves. Figures 1 and 2 show individual localities of interest and sampling sites, marked as A (Soukenná) and B (Příčný hill).

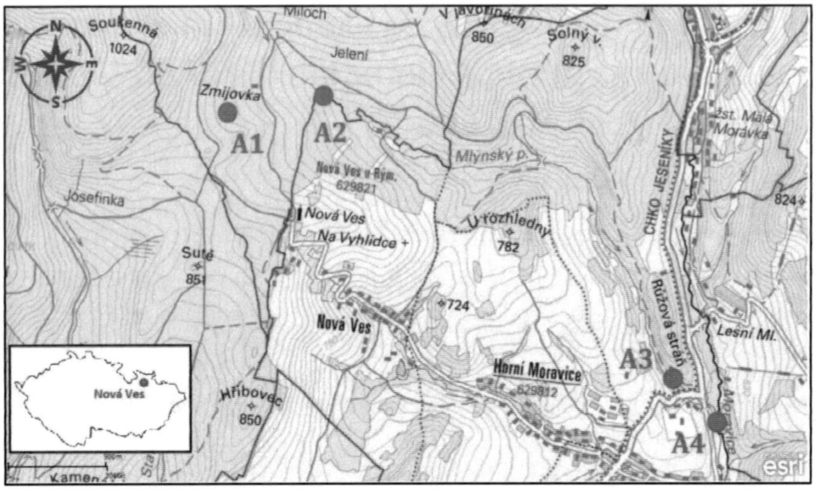

Figure 1. A1 – "Hereditary adit" (the name has just an informal character, since any project documentation, connected with this site was not available), A2 – Mlýnský brook, downstream from the junction with small creek (The creek flows from Dědičná shaft), A3 – Mlýnský brook, before reaching Moravice river, A4 – Moravice river, after the conjunction with Mlýnský brook.

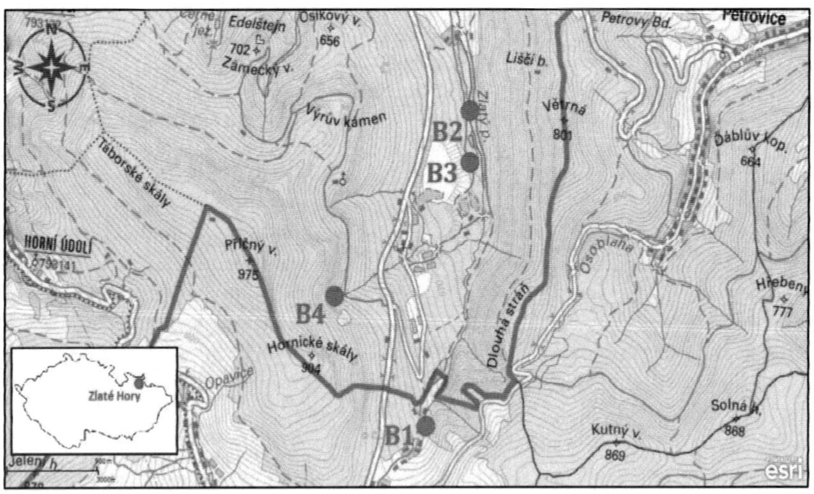

Figure 2. B1 – The spring site of Zlatý brook (in the valley under Příčný hill), B2 – Zlatý brook, B3 – small creek (confluent of Zlatý brook), B4 – Cascade of Modrý brook (this brook springs at Příčný hill).

Surface water and water sediments were regularly sampled every second month within the period of January – November 2016. The analyses of samples were performed in laboratories of VSB – Technical University in Ostrava by flame absorption, using atomic absorption spectrometer VARIAN AA 280 FS. Samples of soil sediment were dried at (105 °C) to gain equal weight and afterward grained to fine powder. After adding laboratory wax and pressing, the samples were analysed by roentgen fluorescent spectrometer WD-XRF. The final results of water were correlated with law

IOP Conf. Series: Earth and Environmental Science **92** (2017) 012037 doi:10.1088/1755-1315/92/1/012037

norm ČSN 75 7221, whereas the evaluation of sediments was realised according to Czech Ministry of Agriculture, law no. 9/2009 Coll., on fertilizers, auxiliary soil substances, auxiliary plant preparations and substrates and on agrochemical testing of agricultural soils (Fertilizers Act), as amended, and other related acts.

3. Results

All the values in this chapter are representing the highest measured contents for individual areas.

After the correlation of obtained results from surface water of Soukenná (table 1) with law norm ČSN 75 7221, the high concentrations were found for Zn and Cu. The highest concentrations of Zn (0.663 mg l^{-1}) were found at sites A1 and A2, what indicates a strong water pollution. At sampling site A2, the values decreased in the second period of research indicate decreased level of water pollution. The values, indicating the strong water pollution were also recorded for Cu (0.147 mg l^{-1}) at the same sites, however only in September. During the rest of research period, the concentration of copper decreased. There was not found any significance in the contents among the other investigated elements (Pb, Fe and Mn), indicating the water pollution. The contents of Zn and Cu at sampling sites A3 and A4 were lower, following the water flow to Moravice river.

Higher concentrations of investigated elements and consequently higher pollution was detected in Příčný hill (figure 2). The highest contents of potentially toxic elements (Zn, Cu, Fe and Mn) were found at sampling site B3. The contents of Zn and Mn reached here very high levels (0.447 mg l^{-1}; 5.93 mg kg^{-1}). Those elements show high values almost in each sample. The concentrations of Cu and Fe also reached in one sample values, indicating the pollution of water (0.292 mg l^{-1}; 9.02 mg l^{-1}). The sampling point B2 indicated in some measurements high water pollution, especially for Cu; Zn and Mn (0.124 mg l^{-1}; 0.123 mg l^{-1}; 0.638 mg l^{-1}).

In the case of one specific measurement, the water indicates even a very high level of water pollution. Accordingly to results, the water indicates water pollution also at sampling sites B1 and B4. Increased contents of Cu and Zn were found in a few samples (0.041 mg l^{-1}; 0.050 mg l^{-1}), as well as Fe in one sample (0.630 mg l^{-1}). Those values indicate the moderate water pollution.

Table 1. The pollution limits for surface water, according to law norm ČSN 75 7221, defined in water quality classes (I. – V.). • I. class: without pollution • II. class: light pollution • III. class: high pollution • IV. class: strong pollution • V. class: very strong pollution. The presented values are in mgl^{-1}.

Element	I. class	II. class	III. class	IV. class	V. class
Zn	< 0.015	< 0.050	< 0.100	< 0.200	≥ 0.200
Cu	< 0.005	< 0.020	< 0.050	< 0.100	≥ 0.100
Fe	< 0.500	< 1.000	< 2.000	< 3.000	≥ 3.000
Mn	< 0.100	< 0.300	< 0.500	< 0.800	≥ 0.800
Ni			no available		

In the comparison with the pollution limits in law no. 9/2009 Coll, the results of sampling and analyses from locality Soukenná (figure 1) show, that the highest contents are detected for Zn (5,667 mg kg^{-1}) at sites A1 and A2. Describing the other important elements, that overreached the permitted limit at these sites, there could be stated, that Co, Cu, Ni and As have following contents in sediments: 39 mg kg^{-1}; 238 mg kg^{-1}; 90 mg kg^{-1}; 39 mg kg^{-1}. Sampling site A2 was also specific with high concentrations of Pb and Cd (6,338 mg kg^{-1}; 27 mg kg^{-1}). The sediment at site A3 is characterized by light increased contents of Pb and Zn (558 mg kg^{-1}; 627 mg kg^{-1}). The sediments at the site A4 satisfy the limits of potentially toxic element content (except the softly increased level of Pb). The results of element contents in sediments from locality Příčný hill (figure 2) indicate the conditions without the pollution at site B1, since there was not found any overreached concentration of toxic elements. However, the highest contents of Zn; Cu; Co; Cd; Ni and As were detected at site B3 (4,953 mg kg^{-1}; 2 562 mg kg^{-1}; 542 mg kg^{-1}; 27 mg kg^{-1}; 447 mg kg^{-1}; 40 mg kg^{-1}). The site B2 was also specific, by overreaching the limits within elements Cd; Zn; Ni and Cu (18 mg kg^{-1}; 3 659 mg kg^{-1}; 199

mg kg^{-1}; 202 mg kg^{-1}). The concentrations of Pb and Cd in mentioned sediments were increased very lightly (B3). The overview of all measured elements and concentrations is presented.

Table 2. The limit concentrations of potentially toxic elements in dried matter of surface water sediments, according to law norm no. 9/2009 Coll.

Element	Limit value
Zn	600
Ni	80
Pb	100
As	30
Cu	100
Hg	0.8
Cd	2.5
V	180
Co	30
Cr	no available

4. Discussion

According to obtained results, there can be stated, that localities of interest (Soukenná, Příčný hill) are appropriate areas for this research. The Pb-Zn deposit is located in Soukenná and the deposit of Pb-Zn and mainly Cu is located in Příčný hill.

In 2016, numerous analyses were performed (912) to investigate 10 elements in surface water and 9 elements in sediments from this site. The high concentration of Zn and Cu was found in surface water of Soukenná, what could be caused by soil conditions in local dump-field, where Cu is present in oxidation zone in the form of secondary mineral. Besides Pb and Zn, in sediments were also proved the presence of Co; Ni; Cu; Cd and As. Those elements are mostly occurred as additions in galena (PbS), sphalerite (ZnS) and pyrite (FeS_2), that are present in local dump-fields [5]. In area of Příčný hill, the samples of water and sediment contain Zn; Pb; Cu; Mn; Fe; Ni; Co and As in high concentrations. The local deposits consist mainly of sphalerite (ZnS), galena (PbS) and chalcopyrite ($CuFeS_2$) [6]. Those minerals contains the investigated elements, thus they could be the sources of contamination in water and sediments.

The high concentration of Pb could be expected only in sediments, what is caused by high accumulation factor of this metal. In water it is hard-soluble and thus cumulative – galena is the exception among sulphides, it is not a subject of chemical and biochemical oxidation [7,8,9]. Zn is obviously leached into water by erosion processes of zinc minerals and accumulates on the floor of rivers and basins, but a little amount could be soluble in water [9]. Metallic form of Cu binds in soil to organic compounds, or to clay, what results in the prior copper occurrence in upper soil layers. This causes the leaching of Cu into surrounded environment. In water, Cu could be transported for big distance from the source, but more frequently is fixed to sediment [10]. The high concentration of Fe can be explained by the high mobility of this metal. It enters a water by Fe-mineral erosion [11]. Mn occurs in water with Fe, thus the presence of Fe-minerals indicates also higher content of this element. It leach to water from soil and sediments [12,13]. The source of Ni could be natural – it can enter the water by dissolution of a bottom sediments, or could be present in a rainfall. It binds to compounds, consisting of Fe and Mn [14]. Co could be absorbed by particles of soil and sediments. In compact state, this element is stabile in water [15]. As occurs as a secondary element, especially in Cu, Ag and Pb ore minerals. It leaches to the environment by mining industry those. It effectively accumulates in river sediments [16,17]. Cd is one of the most toxic elements in nature. It is mostly present in Pb, Zn and Cu ore minerals. It is not present in the elementary form [18].

5. Conclusions

Based on the obtained results, there is no any significant evidence that the investigated sites could be refer as a risk for the environment. The study of potentially toxic elements on these areas (Soukenná and Příčný hill) is still in the progress. The further research, including analyses of soil and biomass is required, including the stabilization of sorption capacity and the sorbents. The evaluating of impact, caused by potentially toxic elements could be improved by the performing of these steps.

Acknowledgments
This study was supported by projects SP2015/19 and SP2016/38. The authors are grateful to the crew of laboratory ICT for the assistance with sample preparation and analyses.

References

[1] Vencálek J 1998 District Bruntál (Bruntál: District Office Bruntál and School Office Bruntál)
[2] Vávra V and Štelcl J 2014 Significant geological locations of Moravia and Silesia (Brno: Masaryk University)
[3] Večeřa J and Večeřová V 2010 History of the Mines of Gold: History of minig in Zlaté Hory (Jeseník: Pinka)
[4] Pauliš P, Toegel V, Veselovský F and Hak J 2009 Corkit from Nova Ves near Rýmařov *Bull. mineral.-petrolog.* **17** 87-90
[5] Novotný P 2010 Significant mineralogical sites of the Protected Landscape Areas of the Jeseníky Mountains CAMPANULA: Proceedings of the Conference to the 40th Anniversary of the Protected Landscape Area of the Jesuit 54-59
[6] Kotris J 2015 Cleaning of mine waters in the Zlatohorský Ore Mountains *News of the Regional Museum in Olomouc* **309** 60-72
[7] Alloway B J 1995 Heavy metals in soils (Blackie and Son Ltd.: Glasgow) p 339
[8] Navrátil T and Rohovec J 2006 Lead: A difficult history of one of the heavy metals *Universe* **85** 518-521
[9] Pitter P 2009 Hydrochemistry (Prague: VŠCHT Prague)
[10] Makovníková J, Barančíková G, Dlapa P and Dercová K 2006 Anorganické kontaminanty v pôdnom ekosystéme *Chemické listy* **100**
[11] Horáková M et all 2003 Water analysis VŠCHT Prague p 335
[12] Sankar M S, Vega M A, Defoe P P, Kibria M G, Ford S, Telfeyan K, Neal A, Mohajerin T J, Hettiarachchi G M, Barua S, Hobson C, Johannesson K and Datta S 2014 Elevated arsenic and manganese in groundwaters of Murshidabad, West Bengal, India *Sci. of The Total Environ.* **488-489** 570-579
[13] Das A P, Sukla L B, Pradhan N and Nayak S 2011 Manganese biomining: A review. *Biores. Technol.* **102** 7381-7387
[14] Poonkothai M and Vijayavathi B S 2012 Nickel as an essential element and a toxicant *Inter. J. of Environ. Sci.* 1 285-288
[15] Kim J H, Gibb H J and Howe P D 2006 Cobalt and inorganic cobalt compounds *Concise International Chemical Assessment Document WHO* **69**
[16] Kafka Z and Punčochářová J 2002 Heavy metals in nature and their toxicity *Chemical sheets* **96** 611-617
[17] Smedley P and Kinniburgh D 2002 A review of the source, behaviour and distribution of arsenic in natural waters *Applied Geochemistry* **17** 517–568
[18] Neumann J, Lopuchovský J and Zapletal O 1990 Chemistry of Agriculture, Pharmacology and Toxicology (Prague: State Farming Publishing House) p 303

IOP Conf. Series: Earth and Environmental Science **92** (2017) 012038 doi:10.1088/1755-1315/92/1/012038

Comparing composts formed by different technological processing

B Lyckova[1], J Mudrunka[2], R Kucerova[1] and V Glogarova[1]

[1]VŠB-Technical University of Ostrava - Faculty of Mining and Geology, 17. listopadu 15/2172, 708 33 Ostrava-Poruba, Czech Republic
[2]VŠB-Technical University of Ostrava - Institute of combined study in Most, Budovatelů 2830/3, Most 434 01, Czech Republic

E-mail: barbora.lyckova@vsb.cz

Abstract. The presented article compares quality of composts which were formed by different technological processes. The subject to comparison was a compost which was created in a closed fermenter where ideal conditions for decomposition and organic substances conversion were ensured, with compost which was produced in an open box of community composting. The created composts were analysed to determine whether it is more important for the final compost to comply with the composting conditions or better sorting of raw materials needed for compost production. The results of the carried out experiments showed that quality of the resulting compost cannot be determined unequivocally.

1. Introduction

In accordance with legislation of waste management in EU, it is necessary to comply with so-called waste hierarchy which prefers use of waste to its disposal [1]. The biodegradable component forms a significant part of mixed municipal waste. Legislation established obligation for municipalities to ensure introduce and operate a system of separate collection of biowaste. The most wide-spread method of processing such waste is composting. Using this biowaste processing method is connected to suitable choice of composting technology.

2. Description of experiments

Within the experiments researching the quality of composts, the composts were compared from the point of view of two different technological processes. In the first case, composting was carried out in a so-called industrial way, in a closed system in an aerobic fermenter which had ideal conditions for decomposition and conversion of organic substances and vegetable products. In the second case, the compost was processed via community composting in a box in a community garden.

The created composts were analysed to determine whether it is more important for the final compost to comply with the composting conditions or better sorting of raw materials needed for compost production.

One of the experiments was a growing test when we examined growth of white mustard (*Sinapis alba*) and lettuce (*Lactuca sativa* L.). The plants were grown in cultivation containers in the mixture of compost and soil. After the growth period, the plants were harvested and analysed.

Another experiment was a germination test. This is a biological method of evaluating phytotoxicity of sample leachate using germination index of garden cress (*Lepidium sativum*) in the area of compost water leaching [2]. The test serves for evaluation of intensity of organic substances decomposition and the final compost maturity.

Also, results from laboratory analysis of both composts were compared. Heavy metals were determined as well as content of nutrients and elements. The results obtained from all analyses were assessed in accordance with valid Czech norms and Czech legislative requirements.

3. Growth test of mustard and lettuce

Two types of compost and arable soil were used for the growth test. The components were mixed in various rates into ten cultivation containers (pots). The composts were mixed with soil in the proportion 1:1 and 1/3:2:3. The test also included a control sample of the soil alone, without any compost. The varieties of white mustard and lettuce were used for the growth test; these were planted into the prepared substrates.

The test took place for six weeks. Germination and growth of plants was documented (see figure 1 – The plant growth in the third week of the experiment). In the sixth week, the growth trend remained constant. The grown plants were after the six-week period harvested, the underground aboveground parts were separated (roots). The plants were then dried at room temperature and weighed. The obtained values are recorded in table 1.

Table 1. The amount of harvested vegetable mass on individual mixtures.

Sample	Sample name	Mixture rate	Plant	Above-ground [g]	Roots [g]
A1	Industrial compost from the closed fermenter	1 : 1	White mustard	0.181	0.024
A2			Lettuce	0.076	0.069
B1		1/3 : 2/3	White mustard	0.117	0.019
B2			Lettuce	0.031	0.015
C1	Community compost from the community garden	1 : 1	White mustard	0.157	0.043
C2			Lettuce	0.228	0.055
D1		1/3 : 2/3	White mustard	0.475	0.059
D2			Lettuce	0.243	0.940
E1	Soil		White mustard	0.065	0.005
E2			Lettuce	0.226	0.139

Figure 1. Plant growth in the third week of the growth test.

The largest amount of harvested dried vegetable mass was weighed from the mixture of community compost and soil in the rate of 1/3 : 2/3. On the other hand, the smallest amount of dried vegetable mass was weighed in the above-ground part from the mixture of industrial compost and soil in the rate of 1/3:2/3 and in the underground part form the control sample in soil alone.

4.Germination test

The cress test of phytotoxicity allows qualitative evaluation of compost decomposition intensity. The phytotoxicity value directly reflects the content of toxic intermediate products formed during aerobic decomposition of organic substances [2]. In the Czech Republic, the cress test is carried out according to the manual of Pliva et al. [3]. The method is based on calculating germination index (IK) of garden cress in the area of water compost leachate. The absence of phytotoxins (IK is about100 %) indicates mature compost [4]. The values of germination index in our experiment are shown in table 2.

Table 2. The values of germination index.

Sample designation	IK (germination index)
Control	100.00 %
Industrial compost from a closed fermenter	119.27 %
Community compost from the community garden	97.77 %

The germination index shows that the industrial compost with IK of 119.27 % can be described as compost with growth stimulation properties, the community compost with IK of 97.77 % as well-matured compost.

Compost maturation also determines suitability of using the compost [3]. According to the presented sources, industrial compost is mature compost with growth stimulation properties and may be used as substrate for horticulture and floriculture. Community compost is well-matured compost and can be applied before sowing and planting.

5. Interpretation of results of chemical analysis of composts

The obtained composts were analysed in laboratories. Table 3 presents results of the analyses from the point of view of hazardous elements content, Table 4 presents results of agrochemical analyses from the point of view of basic nutrients and values of quality signs. The obtained values were compared to the limits of relevant norms – Decree of the Ministry of Agriculture No. 474/2000 Coll., on establishing requirements fertilisers, as amended and supplemented [5] and ČSN (Czech national standard) 46 5735 Industrial composts [6].

Table 3. Results of chemical analyses regarding content of hazardous elements.

Paramet- -er	Content unit in dry matter	Measured content		Limits according to the Decree No. 474/2000 Coll.	Limits according to the ČSN 46 5735
		Industrial compost	Communit . compost		
As	mg/kg	11.5	18.3	20	50
Be	mg/kg	0.784	1.09	–	–
Cd	mg/kg	0.535	0.432	2	13
Co	mg/kg	5.0	9.81	–	–
Cr	mg/kg	93.4	86.5	100	1000
Cu	mg/kg	43.6	33.4	150	1200
Hg	mg/kg	0.612	0.632	1.0	10
Mo	mg/kg	3.64	1.57	20	25
Ni	mg/kg	44.9	43.8	50	200
Pb	mg/kg	39.8	59.3	100	500

V	mg/kg	20.9	27.4	–	–
Zn	mg/kg	233	378	600	3000
Al	%	0.813	0.747	–	–
Fe	%	1.8	1.6	–	–
Mn	mg/kg	408	446	–	–
B	mg/kg	31.4	13.1	–	–
Ba	mg/kg	182	168	–	–

Table 4. Results of agrochemical analyses of composts regarding content of basic nutrients and values of quality signs.

Observed parameter	Units of measurement	Measured content		Standards value for composts according to ČSN 46 5735
		Industrial compost	Community compost	
Dry matter	%	49.4	69.1	min. 35 - max. 60
pH - H_2O		7.36	7.63	from 6.0 to 8.5
Combustible substances	% dry matter	39.4	11.8	min. 25
Cox total.	% dry matter	19.7	6.63	min. 12.5
N total.	% dry matter	1.64	0.638	min. 0.6
Rate C : N	% dry matter	12.0 : 1	10.4 : 1	max. 30 : 1
Non-degradable impurities	%	3.83	7.17	max. 2
Total content P	% dry matter	0.467	0.193	–
Total content K	% dry matter	2.03	0.537	–
Total content Ca	% dry matter	2.46	1.115	–
Total content Mg	% dry matter	0.422	0.295	–
Total content Na	% dry matter	0.063	0.026	–
Total content S	% dry matter	0.126	0.052	–

6. Evaluation of results

The results of the carried out experiments showed that quality of the resulting compost cannot be determined unequivocally. In the germination test, it was proved that the analysed industrial compost is mature compost with growth stimulation properties and is suitable for use in horticulture and floriculture. The observed community compost was evaluated as mature compost suitable for application before sowing and during planting.

Laboratory analyses of individual composts clearly show that industrial compost is convenient thanks to its content of hazardous elements and it has contents of acceptable nutrients, organic substances and values of other signs not only for growing plants, but also for its commercial use. Although the community compost is convenient considering the content of hazardous elements, but it is not convenient due to its content of organic substances, oxidable carbon, non-degradable impurities and dry matter.

The carried out experiments show better results in case of industrial compost, therefore it can be stated that complying with the composting conditions is the most important aspect, but the growth experiment showed that community compost had the best results.

7. Conclusion

Composting is natural aerobic decomposition of organic substance. Decomposition takes place with help of microorganisms which require optimum conditions to assist the whole process and make it as fast as possible. The main limiting factors include temperature, moisture content, airing, participle size, but also pH and carbon to nitrogen ratio. The product of the optimum process is high-quality fertilizer with a number of indispensable properties. The aim of the carried out experiments was to evaluate technological processes of composting from the point of view of quality of produced compost. Although the quality of analysed composts cannot be determined unequivocally, composting is a very important way of using biowaste as raw material for composting during which valuable substances can be returned back to soil.

Acknowledgements

This paper was compiled within the Project of Specific University Research (SGS) no. SP2017/31 Processing of animal excrements in vermicompost.

References

[1] Act. 185/2001 on waste and amendement of some other acts, in the wording of later regulations
[2] Bouda T, Formánková M 2014 Stanovení fytotoxicity kompostů p 35
[3] PLÍVA P et al 2006 *Příručka: Zakládání, průběh a řízení kompostovacího procesu Test fytotoxicity - Řeřichový test* (Praha: VUZT v Praze)
[4] Biom cz 2013 *Stanovení zralosti kompostu biologickou metodou* p 1
[5] Sbírka zákonů ČR 2000 *Vyhláška MZe č. 474/2000 Sb., o stanovení požadavků na hnojiva, v platném a účinném znění* pp 7494-7536
[6] ČSN 465735: 2000 Industrial compost

Using of mobile flood protection on the territory of the Moldova as possible protection of the community

L Malerova[1], J Pokorny[1], E Kristlova[1] and J Wojnarova[1]

[1]Department of Civil Protection, Faculty of Safety Engineering VŠB-Technical University of Ostrava, Ostrava-Výškovice, Lumírova 13/630, Czech Republic

Email: lenka.malerova@vsb.cz

Abstract. The Moldova is facing an increased incidence of floods. For these reasons, the involvement of state in flood protection is rising. Despite the fact that the Moldova is not a member of the European Union, its government follows legislation on prevention, risk assessment and flood protection which is closely connected with landscape. The article deals with the possibilities of using mobile flood barriers in this territory as possible protection of the community.

1. Introduction

Floods are one of the most common and common natural disasters. They can cause both life and animal losses, as well as losses on property and the environment. Countries across the globe are tackling different flood protection, which has become a very topical issue in recent years. This is due to the frequent occurrence of spring and lightning floods, global warming and anthropogenic interference in watercourses. Flood protection means measures to prevent and prevent damage to floods, carried out mainly by systematic prevention and implementation of preparatory measures within the municipality's preparedness for floods.

One of the countries facing the floods is Moldova. A part of the preventive measures in the flood protection area is the construction of mobile flood protection barriers ("FPB") in pre-selected locations. For the purposes of the article, the community is understood to mean the population living in the territory of Moldova and having some relations with each other.

2. The territory of Moldova

Moldova is located in an area with average annual rainfall of approx. 520 mm / year. The problem is the great unevenness of the distribution of these precipitations, as well as the insufficient retention capacity of the landscape (only 10% of Moldavia is covered with forests) [1]. The network of Moldavian rivers consists of two larger rivers - the Dnieper and the Prut, which flow into the lakes near the Danube and are connected with the Black Sea. There are 57 lakes and 3,500 ponds and reservoirs in Moldova [7][9].

Even though Moldova is not a member of the EU, it is subject to legislation on prevention, risk assessment and flood protection, similar to the EU member states. The issue of flood protection in Moldova is dealt with under the laws (eg, the Water Act No. 272/2011, provides a legal framework for the management, protection and efficient use of surface and groundwater, etc.) and European directives (eg the Floods Directive No 2007/60 / EC of 23 October 2007, etc.). The other legal regulation governing Moldova in the area of flood protection is "Commission Decision 2014/762 / EU of 16 October 2014 (hereinafter referred to as" the Decision ") laying down rules for Decision No

1313/2013 / EU of the European Parliament and of the Council Union Civil Protection Mechanism and repealing Commission Decision 2004/277 / EC, Euratom 2007/606 / EC, Euratom, in relation to the Civil Protection Module. An important module is Module 16 - Flood Control and Propagation. The task of this module is to consolidate existing (existing) facilities built up in recent years, or to build new barriers to prevent flooding from rivers, water tanks and rising waterways [5] [7] [9].

On the basis of the research (Study), the most vulnerable areas of the flood and identified for the FPB installation are: Vadul lui Vodă (suburb Chisinau) and Soroca (see Figure 1. Map showing vulnerabilities. Vadul lui Vodă is a suburb of Chişinău. It is located on the right side of the Dniester River with a population of 5,295 people. The village is an important place for tourism. Close to the village there is an airport that is popular for landing and tourist flights and a large number of tourist resorts with forest park and beaches. [7] The Soroca district is located in the northeast of Moldova with a population of 95,000 by 2016. The administrative district of Soroca is the city of the same name located on the right side of the Dniester. The town of Soroca is a very interesting and popular place for tourism and recreation not only for the locals but also for tourists [2][4] [6].

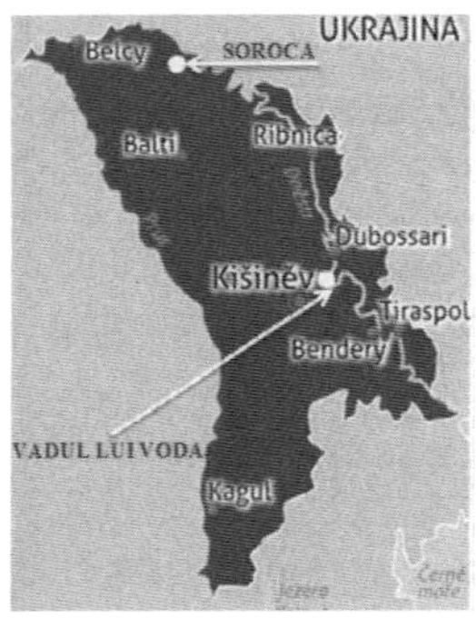

Figure 1. Map showing vulnerabilities [4][5].

3. Problems of bagging and mobile barriers

Part of anti-flood measures is the ability to cope with unwanted water leakage into areas that need to be protected from flooding. Types of mobile flood protection include sandbags or FBP. The most widespread mobile flood protection is classical sacks with an inert mix such as sand, gravel, soil, etc. Sandbags can be used to build dams, seal doors, windows, ventilation spaces, canals, etc. The effectiveness of the dike built from sacks is greatly increased by foil. There are currently many types and types of FPB on the market. The flood protection barriers are divided according to the type of construction (supporting frame, chamber concept, filling mixture) and used material (steel, aluminium, plastic, glass, cloth and wood). [2]

Table 1. Comparison of bag system with internal mixture and FPB shows a comparison of so-called bagging versus FPB. Based on the comparison made for the installation of flood protection facilities in Moldova, FPB were selected. [3]

Table 1. Comparison of bag system with internal mixture and FPB [3].

parameter	Sacks with internal mixture	Mobile flood barriers
Speed of installation	TIME CONSUMING The lengthy sand filling and the need to construct the pyramid layout in the open spaces	QUICK Immediate installation, a system of modules that connect to each other, will allow quick installation. Unlimited barrier length and shape.
Practice	READY TEAM The need for a whole team of people who are gradually filling, carrying and sacking.	SOME PERSONS simple laying, about 5 people
Filling	PICK AND FILLING EQUIPMENT	It is NOT NEEDED or WATER (flood) - it is

material	Sacks must be filled with mixtures, the need for filling equipment.	filled with water, which erects the barrier.
Disposal	HYGIENIC AND NUCLEAR FOR TIME AND WORKING FORCE Inappropriate hygienic conditions, need for more workforce.	CLEAN AND NON-ANNUAL Water cleaning, roller / board layering / construction and storage for further use.
Performance	WORKING - The need for a filling device and a team of people who are gradually filling, carrying and sacking.	NONE or WATER (flood)
Disassembly and disposal	NON-HYGIENIC, NUCLEAR FOR TIME AND WORKING FORCE Inappropriate hygienic conditions, the need for more workforce	PURE AND NON-ANNUAL Simple disassembly by handling technique.
Stability	LIMITED STABILITY - Especially for higher barriers, the bags need to be layered into a pyramid, which is material and space-intensive. Limited barrier height.	VERY STABLE Stable system of connected heavy bags / construction with boards or pallets and tarpaulin / copying terrain. Possibility of stacking up rows.
Permeability	PERVIOUS Bags pass through water throughput.	MINIMUM PROPORTIONALITY True terrain copying, and tight junction of the bags / design with the plates ensure the barrier is impenetrable.

4. Analysis of the appropriate mobile flood protection

The choice of the appropriateness of the FPB has been taken into account in accordance with the Decision. Based on this, the technical parameters, criteria and description were set for the selection of FPB, see Table 2 [1].

Table 2. Technical parameters for barrier selection [3].

Criterion	Description
Price	The total cost of an FPB does not exceed the total value of 380 000 Euro.
Protective height	Ability to cover water up to a minimum height of 1000 mm.
Length	Construction of FPB with a length of 1000 meters.
Intervention	Ability to simultaneously interfere with at least 2 locations in the areas (in the case of the Study it is a place of the village of Vadul lui Voda and the city of Soroca).
Transport	Easy transportability of FPBs, including all accessories (hand pallet trucks and forklifts) on lorries from the place of storage to an endangered flood site.
Installation	Rapid construction and installation variability without the use of stationary anchor points, base pads, foundation belts, easy disassembly, no specific grounding requirements (high load-bearing capacity). It must be possible to install on different terrain, including unpaved surfaces (gravel, dirt, grass, concrete, asphalt, concrete, etc.) on a flat weight basis. Possibility of parallel construction (building on multiple sections with subsequent interconnection). FPB must limit leaks at the point of contact of the bottom of the FPB and the terrain. Construction interval is set at 1000 meters / 15 persons / 1 hour
Shape variability	Shape variability of FPB - straight sections, arcs, possible solutions by combining individual types of FPB. The FPB parts must be connected to an unlimited length while retaining the ability to retain water. The FPB must be stable even when the sealing elements are punctured / broken.
Construction	In order to ensure the FPB maintenance, the structure is specified on the support structure, the type of support frame.
Lifetime	Minimum 10-year FPB lifetime (excluding sealing elements) for installation and repeatable use of FPB core components. FPB resistant to abrasion and mechanical

	damage (e.g. floating objects). FPB resistant to floating objects. FPB must perform its function at temperatures of -10 to + 50°C.
Control	Possibility to perform simple screening controls, without the need to involve other entities. Regular maintenance of FPB does not require the intervention of a supplier or manufacturer of FPB and can be performed by a trained user.

The analysis was performed using the multi-criteria pairing method of the Füller triangle, which was used for a number of criteria, namely 7. The method is used in the determination of weights with a larger number of criteria, where it is difficult to compare each other. Based on the use of this method, the coefficients were calculated, see Table 3.

Table 3. Selected coefficients for the criteria [3].

Criterion	Coefficient
1. Price	0.18
2. Protective height	0.25
3. Transport	0.03
4. Installation	0.11
5. Shape variability	0.07
6. Construction	0.18
7. Lifetime	0.18

The next step was the assignment of values, see Table 4, which was then multiplied by the coefficients.

Table 4. Assigning values to the criteria [3].

Numeric values	Symbol	Description
5	X	did not meet the criterion
10	∞	information not identified
15	√	met the criterion

The resulting values for individual barriers were calculated according to formula (1) [3].

$$\sum_{n=1}^{7}(k_n \times h_n) \quad (1)$$

Where: k_n - coefficient from Table 3
h_n - the numerical value of the criterion in Table 4.
The resulting values of each species are recorded, see Table 5.

Table 5. Resulting values of individual types of barriers [3].

Flood protection barriers	Description	Result
One-chamber	construction type chamber design, used material strength PVC, filling material water, cylindrical type	9.45
Tubus	type construction of a chamber concept, tubes covered with polyester, tubular type	9.65
Filling with filler material	type with steel construction, material - covered with PES woven PVC, water or loose material	8.2
Bags	bags made of rubber-textile material, fitted with flanges for water filling and deaeration, filling material water	8.2
Barrier filling with flood water	polyethylene barrier, filling material flood water	10

Filling without filling materials	type of supporting structure, type of supporting frame, sealing rubber rim	13.9
Barrier carrier	type of supporting structure, type of supporting frame, sealing foil, possibility of using wooden pallets	15

Based on the analysis that arose from the market survey offered by FPBs, the scale for their evaluation was set, see table 6.

Table 6. Set barrier rating scale [3].

Range of values	Rating
5 – 10	does not meet the conditions, cannot be used as a barrier at selected locations
over 11	meets conditions, can be used as a barrier at selected locations

It is clear from the evaluation that a suitable type for the installation of FPB in the endangered territory of Moldova is the supporting structure, the type of support frame of the sealing foil or the sealing rubber rim.

5. Conclusion

This article provides a preview of flood protection issues related to mobile flood barriers (FPB) in the territory of Moldova. The article is based primarily on the Study, which dealt with the selection of FPB in the endangered community, Vadul lui Vodă (suburb of Chisinau) and Soroca. Based on the analysis, it was found that a suitable type of FPB for installation in endangered places in Moldova is the FPB of the supporting structure, the type of support frame of the sealing foil or the sealing rubber rim. FPB will serve Service of civil protection and emergency situations of the Republic of Moldova for the following purposes: protection against floods and storm water, protection of industrial buildings and family houses, enhancement of existing dams or rivers, regulation and collection of flood water. The FPB can be ambidexterous use for floods, not only at Vadul lui Voda and Soroca. The disadvantage is the possibility of incorrect installation of FPB. This option is unlikely for reasons of training (practical exhibitions of MPZ) 10 persons for 2 x 7,5 hours in the Czech Republic and 10 persons 2 x 7,5 hours in Moldova.

Acknowledgments

The article was prepared as part of the project of the Student Grant Competition "The proposal (the creation of) the categorization of resources enabling the management of extraordinary events related to the community". Project number SP2017/140.

Reference

[1] EM-DAT 1988 The International Disaster Database
[2] Climatechange post 2017 Moldova - River floods Moldova
[3] Maléřová L, Pokorný J, Kristlová E and Wojnarová J 2017 Studie mobilních protipovodňových hrazení v Moldavsku
[4] Moldovenii Soroca 2017 Raionul Soroca
[5] Ramsbottom, D et al. 2016 Developing a national programme of flood risk management measures: Moldova E3S Web of Conferences **7**
[6] The Globe 2014 Safari Moldova (USA: Lulu Press Inc.)
[7] Vadullui 2017 Voda
[8] Water governance in transboundary Republic of Moldova 2017
[9] Worldof Moldova Locationof Moldova and geographical data about the country 2017

Measurement and evaluation of percolation drainage systems capacity in real conditions

G Markovic[1] and M Zelenakova[2]

[1]Technical University of Kosice, Faculty of Civil Engineering, Institute of Architectural Engineering, Kosice, Slovakia
[2]Technical University of Kosice, Faculty of Civil Engineering, Institute of Enviromental Engineering, Kosice, Slovakia

Email: gabriel.markovic@tuke.sk

Abstract. The drainage system must ensure a safe disposal of the surface water without endangering the buildings and safety of people. Despite the common use of rainwater infiltration facilities, there is still only limited data available evaluating the long-term capacity of such systems especially for underground infiltration facilities. This study presents experimental measurements and evaluation of long-term infiltration efficiency in real conditions and emphasizes the importance of hydrogeological survey. The measurements of infiltration efficiency were applied to an existing percolation drainage system – infiltration shafts. Infiltration shafts were made in year 2007 so that its drainage operation takes more than 8 years. This study was started in 2011 and still continues and presents 5 years measurements of infiltration efficiency for this infiltration facility.

1. Introduction

Urban drainage systems can be divided into two most commonly used; a combined sewer system and a separate sewer system. Combined sewer systems convey storm water and waste water away in a single pipe. Where the combined systems are used, the risk of combined sewer overflows appears (CSO) represented by transfers of untreated waste water into receiving waters [1]. On the other hand, separate sewer system carry storm water and waste water in the separate pipes, usually laid side-by-side [2].

A serious problem in recent years of waterworks praxis has appeared to be the one with growing necessity for surface water detention. Continuous growth of natural terrain coverage for building construction, industry, free time activities and transport reasons leads to widening of the area of build surface and lack of natural terrain with capability of natural rainfall water infiltration. This causes serious damage to natural water cycle.

Waste-water disposal system is very often overloaded with water coming from heavy rainfall which means that it cannot be used for the purpose of rainwater management. Higher disposition of manhole covers helps to keep sewerage water within safe level and minimize the risk of endangering sensitive ecosystems. Waste-water disposal systems, sewage treatment plants and recipients of water flows are overloaded and the risk of floods grows. This represents another reason for waterworks companies to have the possibility of forbidding the disposal of rainfall water from surface detention into the waste-water disposal system already during the creation and fulfilment of the development plan. This leads to the necessity of using rainwater infiltration systems or capturing rainwater for further reuse [3].

Infiltration capacity of soil mainly depends on two factors; the particle size and the moisture content of the soil [4].The permeability of infiltration zone is an essential qualitative and quantitative prerequisite for infiltration of rainwater. Permeability is represented by a filtration coefficient kf, which represents the effectiveness of infiltration facilities, respectively ability of subsoil infiltrate incoming rainwater. Therefore, the most important design parameter of the infiltration facilities is to determine the filtration coefficient kf on-site. Another important parameter for design of infiltration facilities is level of groundwater. The Mangangka´s research shows the decline of soil infiltration capacity due to high elevation groundwater [4]. For aboveground facilities, designer have to consider with influence of rainfall patterns on the instability of slopes [5].

Of course there is another very important factor for right operation of infiltration facilities – maintenance. According to Lindsey, Roberts and Page research, analysis of field inspections reveals that the condition of facilities declined significantly in four years [6]. Without maintenance of infiltration facilities, especially above-ground facilities, decrease of performance of infiltration facilities have been identified [7].

The primary threat to the long-term hydraulic performance of stormwater infiltration systems is their inclination to become clogged due to the accumulation and deposition of sediments on the system surface during its operation [8]. One main reason for clogging is a demonstrable lack of regular maintenance [6]. Balades et al. showed that coping effectively with this problem requires regular maintenance as a preventive measure to sustain the infiltration capacity of porous pavements.

2. Materials and Methods

2.1 Site description

The article describes experimental measurements and evaluation of the infiltration efficiency in the real conditions.

Our research and own measurements in the field of storm water quantity and quality parameters has been started in the campus of Technical University of Košice as a part of the management of storm water project. The objects of our research are two infiltration shafts in TU campus in Kosice, already existing prior to our research. These infiltration shafts were designed as a drainage solution for a real school building PK6. Two vertical shafts are located next to the PK6 building. All run-off rainwater collected on the roof flows into the underground pipes (figure 1) [3].

Figure 1. The location of drainage shafts near the PK6 building [3].

The initial measurement started in March 2011 and has continued in the infiltration shaft A so far, followed by the measurement of the inflow of rainwater runoff from the roof of the building PK6. The research was extended of measurements for second shaft (shaft B) in March 2012. It should be noted that the project of building drainage and project of design and realization of infiltration shafts is not available. These infiltration shafts were made before the start of the research and measurements in year 2007.

3. Results and discussion

The fills of the area of interest consist mostly of gravel clays, building waste and natural gravels. The exploratory bores were used to verify the thickness of these fills from 0.5 to 0.6 meters. Under the fills, the sediments of Hornád River were identified. Right under the backfill the continuous layer of clay with a thickness of 4.0 to 4.5 meters was identified. Under the flood sediments the fluvial gravel sediments of thickness 5.0 to 7.0 meters were identified and the gravels blended with fine-grained soil. The bottom layer consists of clay-gravel with a thickness of 0.7 to 2.7 m (The final report of geological survey for object Technicom, 2010).

The validation of the hydrogeological survey of the site, respectively verification of the infiltration coefficient kf of the soil in studied infiltration shafts near the PK6 building, was performed by taking the samples of soil from the bottom of the infiltration shafts. With the use of laboratory tests, the samples were evaluated as gravel blended with a fine-grained soil and infiltration coefficient set at 10^{-3} m/s which confirms the hydrogeological survey of the site conducted for the object of Technicom in the campus of TU Košice.

Figures 2-5 represent a typical process of flow rate and subsequent percolation of rainwater during the rainfall events in the percolation shafts [3]. Figures 2-5 show 2 selected rainfall events from the period of 2012 to 2013 with high rainfall intensity. All data from the research show that the total infiltration of runoff inflow into the infiltration shafts from the roof of PK6 building takes place at the same time within the duration of rainfall events, respectively very short-time after. This represents a high infiltration rate of these infiltration shafts determined by the coefficient of infiltration of soil at the bottom of the shaft $k_f = 1.10^{-3}$ m/s.

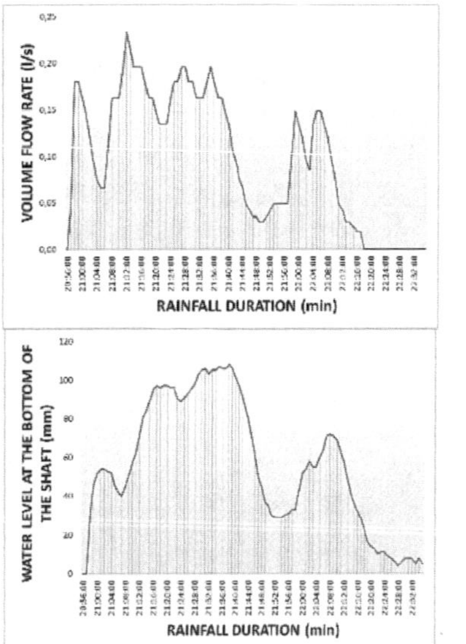

Figure 2-3. Volume flow rate and water level changes at the bottom of the shaft A on a rainfall of 24 April, 2012.

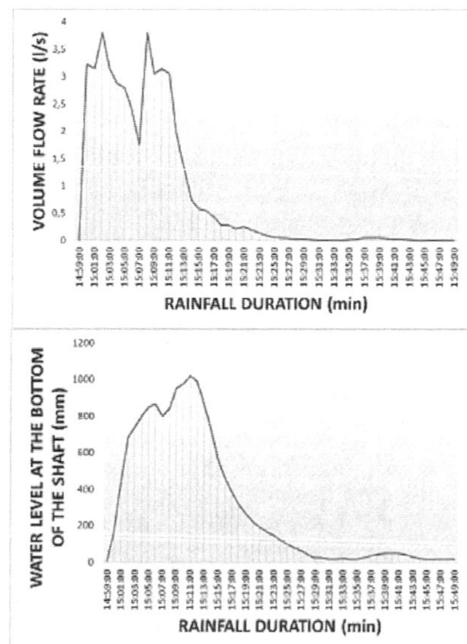

Figure 4-5. Volume flow rate and water level changes at the bottom of the shaft A on a rainfall of 30 July, 2013.

As resulting not only from figures 2-5, but also from the overall measured data during the research, the total infiltration of rainwater in the infiltration shaft, take place at the time of termination of rainfall events, respectively short-time after, which represent a high infiltration rate of this shaft, given by the coefficient of infiltration of soil at the bottom of shaft. Therefore, despite the smaller surface for infiltration of infiltration shafts with comparison to other types of infiltration facilities, the infiltration coefficient of surveyed infiltration shafts $k_f = 1.10^{-3}$ m/s ensures safe disposal of surface runoff. The maximum water level at the infiltration shafts are shown in figures 6 and 7 [3].

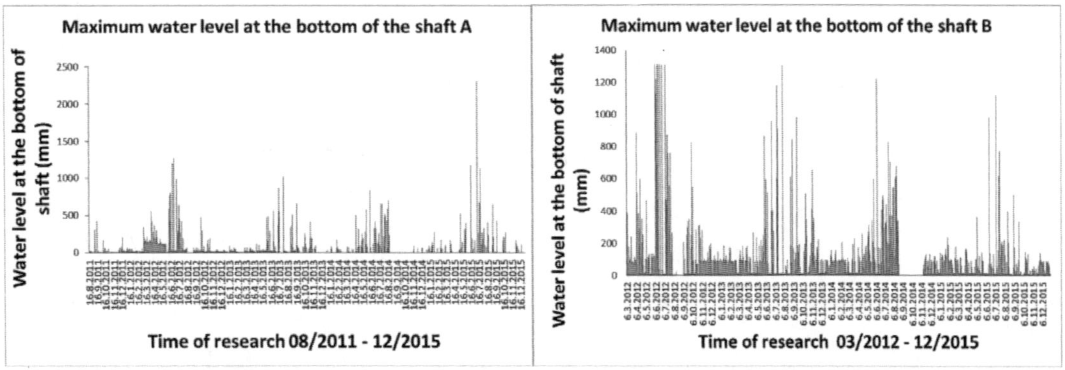

Figure 6-7. Values of maximum water level at the bottom of the shaft A from August 2011 to December 2015 and of the shaft B from March 2012 to December 2015.

The test of the infiltration capacity in both of infiltration shafts was done in 2014 - in shaft A where was used volume 1000 liters of water and second experiment in both of infiltration shafts (A,B) where was used volume 2000 liters of water.

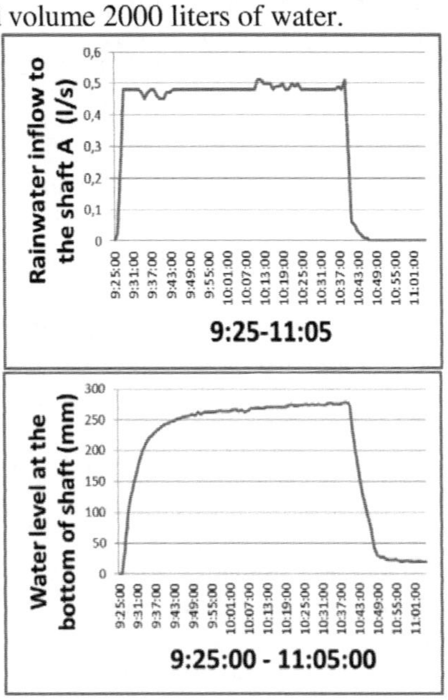

Figure 8-9. The experimental infiltration measurements of infiltration capacity in shaft A – volume of water used for the test - 2000 liters.

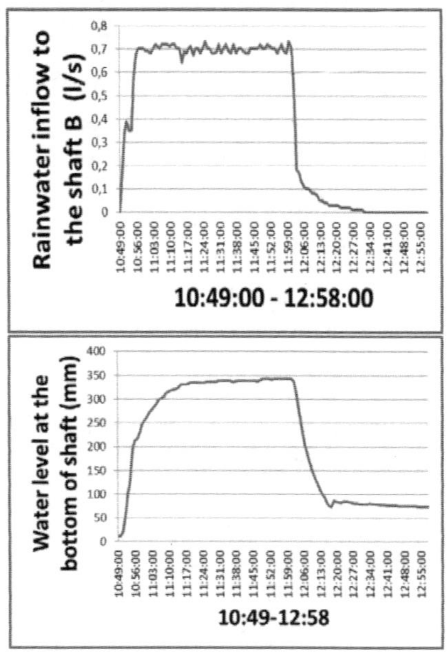

Figure 10-11. The experimental infiltration measurements of infiltration capacity in shaft B – volume of water used for the test - 2000 liters.

Table 1. Measured infiltration capacity from infiltration test.

Infiltration shaft	Volume of water used for the test (l)	Date	Year of infiltration facility operation	Infiltration capacity (l/min)
Shaft A	1000	2014	7	18.51
Shaft A	2000	2014	7	23.52
Shaft B	2000	2014	7	22.98

The course of infiltration capacity test is depicted on figures 8-11. Values of measured infiltration capacity are shown in table 1 and also were selected 5 rainfall events from every year of research and calculated the mean infiltration capacity for these infiltration shafts – table 2.

Table 2. Calculated infiltration capacity from selected rainfall events during the research.

Infiltration shaft	Date	Year of infiltration facility operation	Mean infiltration capacity (l/min)
Shaft A	2011	4	11.9
Shaft B	2011	4	-
Shaft A	2012	5	23.57
Shaft B	2012	5	35.26
Shaft A	2013	6	17.91
Shaft B	2013	6	32.64
Shaft A	2014	7	24.36
Shaft B	2014	7	18.99
Shaft A	2015	8	49.57
Shaft B	2015	8	24.93

4. Conclusion

Results from calculated data of selected rainfall events and from field test of infiltration capacity of our research show specific and various trend of infiltration capacity during operation of these infiltration shafts. The important fact is that there are no significant decreasing of infiltration capacity after 8 years of operation of these infiltration facilities and the operation of the facilities are fluent and free from complications. Of course, it should be noted that the infiltration shafts are underground infiltration facilities and inflow to these shaft was only from roof construction what represents less risk of contamination with high suspended solids in runoff.

Acknowledgements

This work was supported by the VEGA 1/0202/15 Sustainable and Safe Water Management in Buildings of the 3rd. Millennium

References

[1] Semadeni-Davies A and Bengtsson L 2000 Theoretical Background Chapter 1 in: *Urban Drainage in Specific Climates* (UNESCO: Paris)
[2] Butler D and Davies J W 2011 *Urban Drainage* (Spon Press: Taylor & Francis)
[3] Markovič G and Vranayová Z 2013 Infiltration as a means of surface water drainage (Košice: TU)
[4] Mangangka I R 2008 The Decline of Soil Infiltration Capacity Due To High Elevation Groundwater *Civil Eng. Dimen.* **10** 35-39
[5] Muntohar AS et al. 2013 Influence of Rainfall Patterns on the Instability of Slopes *Civil Eng. Dimen.* **15** 120-128
[6] Lindsey G, Roberts L and Page W 1992 Inspection and maintenance of infiltration facilities *J. Soil & Wat. Conser.* **47** 481-486
[7] Al-Rubaei et al. 2015 Long-term hydraulic performance of stormwater infiltration systems *Urban Wat. J.* **12** 660-671
[8] Dietz M E 2007 Low Impact Development Practices: A review of current research and recommendations for future directions *Wat. Air Soil Pollut.* **186** 351-363

The use of steel slag in concrete

P Martauz[1], V Vaclavik[2,3] and B Cvopa[1]

[1]Cement Plant - Považská Cementáreň, A.S., Ladce, Ladce, Slovakia,
[2]Institute of Environmental Engineering, Faculty of Mining and Geology, VSB-Technical University of Ostrava, Czech Republic,
[3]Institute of Clean Technologies for Mining and Utilization of Raw Materials for Energy Use Faculty of Mining and Geology, VSB-Technical University of Ostrava, Czech Republic

Email: vojtech.vaclavik@vsb.cz

Abstract. This paper presents the results of a research dealing with the use of unstable steel slag as a 100% substitute for natural aggregate in the production of concrete. Portland cement CEM I 42.5N and alkali activated hybrid cement H-CEMENT were used as the binder. The test results confirm the possibility to use steel slag as the filler in the production of concrete.

1. Introduction

Concrete is one of the most commonly used materials in the building industry. It is a universal building material (composite) that can be used in a variety of environments (dry, water, underground, in hot or cold climatic zones). Another advantage of this material is the possibility to change its properties depending on the requirements of the construction project and the price. Concrete consists of three main components: binder, filler and water. The most commonly used filler is natural aggregate of different fractions. The most frequently used aggregate fractions for structural concrete are 0/4 mm, 4/8 mm and 8/16 mm fractions. This article presents the results of an experimental research dealing with the preparation of concrete based on steel slag, which was used as a 100% substitute for natural aggregate using two types of binders. The first type of binder was Portland cement CEM I 42,5N and alkali activated hybrid cement H-CEMENT from Považská cementáreň, a.s. At present, we are familiar with the results of an experimental research on the use of steel slag in the production of concrete. It describes a partial replacement of natural aggregate by steel slag in the production of concrete with the strength of 30, 50 and 70 MPa [1].

A comparison of the physical and mechanical properties of concrete with a different content of steel slag as a substitute for natural aggregate with the maximum grain size of 12.5 mm and concrete based on natural filler – limestone are presented in [2]. The results of a research, where 60% of steel slag was used as part of the binder in combination with 20% of Portland cement and 20% of active magnesium in order to produce a concrete composite which was subsequently hardened in CO_2 environment and a pressure of 0.1 MPa are described in [3]. We are also familiar with research the results, where steel slag was used as a partial substitute for natural aggregate in the production of asphalt concrete [4]. The use of steel slag from Sisak Iron works Company in combination with natural aggregate road base layer and their strength parameters are described in [5]. The ecotoxic properties of concrete based on granulated slag from the production of gray iron are described in [6].

Steel slag is an additional component in steel production and has an irreplaceable role from many points of view. Nowadays, steel is mainly produced in oxygen converters and in electric arc furnaces. The presence of steel slag, depending on the used technology and the number of other parameters, such as the composition of the charge, the amount of added slag-forming additives, etc., varies within relatively wide limits, ranging from app. 20 to app. 250 kg per ton of crude steel. Depending on the nature of the process and its course, the chemical composition of the resulting slag types and their properties differ considerably as well. However, steel slag is not a mixture of oxides, but a complex system in which we can identify more than 100 components that cannot be mechanically separated.

When using steel slag as an artificial aggregate in the building industry, the following obstacles must be taken into account:

- high specific density of steel slag is beneficial in some areas of application, but has an adverse effect on the cost of slag transportation, especially at distances over 50 km;
- the content of free oxides of CaO and MgO is associated with lower volumetric stability of steel slag. These free oxides may hydrate with the access of moisture and gradually increase their volume, resulting in the disintegration of slag.

2. Material and methods

2.1. Steel slag

The experimental research of a 100% substitution of natural aggregate with steel slag took advantage of cast converter slag with the fr. of 0/8 mm, which was subjected to the tests of grain size, volume density and water absorption, as well as an elemental analysis. The results of the grain analysis, which was carried out according to CSN EN 933-1, are presented in figure 1.

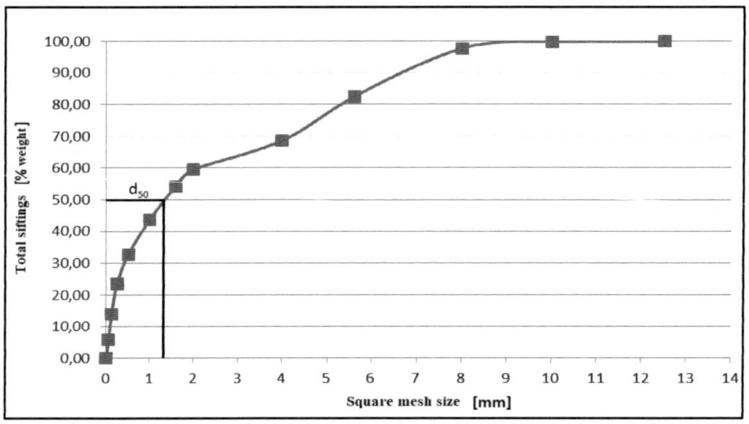

Figure 1. Cumulative grain size curve of 0/8 mm steel slag fraction.

Figure 1 shows the cumulative grain size curve of 0/8 mm steel slag fraction, which was obtained by sieving of steel slag through a set of screens with the square mesh size of: 0.063; 0.125; 0.25; 0.5; 1; 1.6; 2; 4; 5.6; 8; 10 and 12.5 mm. It should be noted that prior to the grain size test, the steel slag was not washed with water on 0.063 mm screen. The share of fine particles sifted through 0.063 mm screen is 5.7%. The mean grain parameter d50 as an important parameter characterizing grain size is 1,3 mm what means the value of the particle diameter at 50% of cumulative distribution.

The volume density and water absorption of steel slag with the fr. of 0/8 mm were determined according to CSN EN 1097-6 by pyknometric method. The results of the determination are presented in table 1.

Table 1. Values of volume density and absorption power of steel slag with the fr of 0/8 mm.

ρ_a bulk density of grains	3.742 $Mg{\cdot}m^{-3}$
ρ_{rd} bulk density of grains after drying in a drier	3.439 $Mg{\cdot}m^{-3}$
ρ_{ssd} bulk density of grains saturated with water and surface dried	3.520 $Mg{\cdot}m^{-3}$
WA_{24} water absorption	2.34 %

2.2. Cement, additive and water

Two types of cements were used to design the experimental concrete formula based on steel slag. They were Portland cement CEM I 42,5N and alkali activated hybrid cement H-CEMENT. The properties of H-CEMENT are described in [7]. These two types of cements were supplied by Považská cementáreň, a.s. Company. The additives were: superplasticizer Sika® ViscoCrete®-225 Powder and retardant additive RETARDAL 540, which delays the beginning of the hardening of concrete according to the cement dose. Tap water from the water supply system was used as the mixing water.

2.3. Mix design

Three mixes with the designation of RL1, RL2 and RL3 have been designed within the scope of the experimental research.

- **RL1 mix** - contained 0/8 mm steel slag fraction as filler, 380 kg of H-CEMENT, 241 kg of water, Sika superplasticizer in the amount of 0.5% of the cement weight, and retardant additive in the amount of 0.4% of the cement weight.
- **RL2 mix** - contained 0/8 mm steel slag fraction as filler, 380 kg of Portland cement CEM I 42,5N, 241 kg of water, Sika superplasticizer in the amount of 0.5% of the cement weight, and retardant additive in the amount of 0.4% of the cement weight.
- **RL3 mix** - contained 0/8 mm steel slag fraction as filler, 380 kg of Portland cement CEM I 42,5N, 241 kg of water, Sika superplasticizer in the amount of 0.5% of the cement weight. No retardant additive was used.

2.4. Testing the concrete properties

The following fresh concrete properties were tested: the consistency of fresh concrete - the slump test according to CSN EN 12350-2 immediately after mixing and after 30 minutes, the density of fresh concrete according to CSN EN 12350-6. The strength characteristics of hardened concrete were tested according to CSN EN 12390-3. Cube strength of concrete was tested after 3, 7, 14, 21 and 28 days, prism strength was tested after 28 days. The modulus of elasticity was tested according to CSN ISO 6784 from the deformability properties.

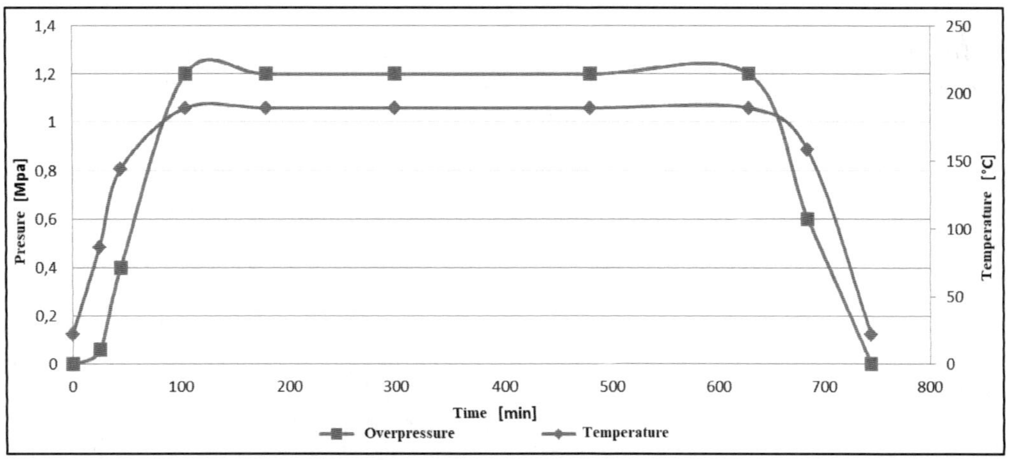

Figure 2. Autoclaving curves for testing the volume changes of concretes.

The durability of concrete based on steel slag was tested under conditions of higher temperature and pressure. The determination of the volume changes of concrete was carried out in a 540 litre laboratory autoclave at the maximum saturated steam pressure of 1.2 MPa and the maximum temperature of 189 °C. 150 mm cubes were used as the test specimens. The temperatures and pressures in the testing laboratory autoclave were set according to the autoclave curves presented in figure 2 in order to determine the volume changes of concrete based on steel slag.

3. Results and discussion

The results of the tested properties of fresh concrete mixture and the physical-mechanical and deformation properties of hardened concrete based on steel slag aggregate are presented in table. 2.

Table 2. Properties of fresh concrete mixture and hardened concrete based on steel slag.

Formula		RL1	RL2	RL3
Slump test immediately after mixing [mm]		258	255	249
Slump test after 30 min. [mm]		185	255	165
Density of fresh concrete [kg/m^3]		2728	2863	2717
Concrete cube strength[MPa]	after 3 days	9.1	21.2	18.0
	after 7 days	16.7	28.0	22.6
	after 14 days	22.3	31.8	27.0
	after 21 days	27.6	36.3	30.1
	after 28 days	29.1	37.4	34.6
	after 90 days	38.4	44.2	38.0
Concrete prism strength[MPa]	after 28 days	29.0	-	31.2
	after 90 days	37.5	-	34.3
Concrete modulus of elasticity[MPa]	after 28 days	32837	-	32509

The results of the slump test clearly show that the consistency of all concrete mixtures immediately after mixing (RL1, RL2, and RL3) corresponds to the slump degree of S5 and the slump degree of S4 after 30 minutes in the case of mixtures RL1 and RL3. There is an evident decrease in the slump value immediately after mixing and after 30 minutes in case of RL1 mixture, and it is 73 mm (28%). The value of RL3 is 84 mm (34%).

The density values of fresh concrete RL1 and RL3 are almost identical, with a visible increase in RL2 mixture. This increase was caused by the segregation of the steel slag grains during the production of the concrete mixture and the reflection of the mixing water during the production of the test specimens.

Table 2 clearly shows a gradual increase in the cube strength of the concretes based on steel slag and H-CEMENT (RL1) in comparison with the concrete based on steel slag and Portland cement CEM I 42.5 N (RL2, RL3). The strength of RL1 concrete after 3 days clearly shows a 2 to 2.3 times lower value of concrete strength compared to concrete RL2 and RL3. After 7 days, the strength of RL1 concrete is 1.4 and 1.7 times lower than that of RL2 and RL3 concretes. Even after 28 days, the RL1 concrete cube strength is lower compared to formula RL2 and RL3 based on Portland cement. After 90 days, the values of the cube strengths of samples RL1 and RL3 became equal.

Looking at the values of the modulus of elasticity of concrete based on steel slag (see table 2), it is obvious that the average values of the modulus of elasticity of the concrete based on steel slag and H-CEMENT (RL1 formula) and the concrete based on steel slag and Portland cement CEM I 42,5N (RL3 formula) are almost identical, although H-CEMENT belongs to the cement strength class of 32.5.

Figure3. View of the test specimen after autoclaving A) RL1 mix, B) RL3 mix.

The results of the durability tests of concrete based on steel slag have unambiguously shown the suitability of H-CEMENT as the binder for the production of concrete based on steel slag under conditions of higher temperature and pressure. The above presented fact is confirmed by figure 3, which compares the test specimens after the autoclave process. The picture clearly shows that the concrete based on steel slag and Portland cement disintegrated into concrete rubble. The concrete based on steel slag and H-CEMENT did not show any evidence of disruption of the test specimens after the autoclave process, on the contrary, the strength was increased from the initial value of 29.13 MPa to 43.03 MPa (i.e. by 32%).

4. Conclusion

The presented results of the experimental research have confirmed the suitability of the use of H-CEMENT binder in the production of concrete based on steel slag, which is represented in 100% volume replacement of natural aggregate. It has been proven that: the use of H-CEMENT in the production of concrete mixture based on steel slag results in an even increase in strength when compared to the concrete mixture based on steel slag with CEM I 42.5N. The cube and prism strengths of the concrete mixture with steel slag and H-CEMENT are the same. The modulus of elasticity of the concrete mixture using H-CEMENT and CEM I 42.5N is identical. After the autoclaving of concrete with steel slag and CEM I 42.5N, the test specimens were completely destroyed, confirming that CEM I 42,5N is unsuitable for the use as the binder for unstable steel slag in conditions of higher temperature and pressure.

Acknowledgement:

This article was written in connection with:

- Považská cementáreň, a.s., Ul. Janka Kráľa, 018 63 Ladce, Slovakia.
- project Institute of clean technologies for mining and utilization of raw materials for energy use - Sustainability program. Identification code: LO1406. Project is supported by the National Program for Sustainability I (2013-2020) financed by the state budget of the Czech Republic.

References:

[1] Abu-Eishah Samir I, Amr S EL-Dieb and Mostafa S Bedir 2012 Performance of concrete mixtures made with electric arc furnace (EAF) steel slag aggregate produced in the Arabian Gulf region *Cons. and Buil. Mat.* **34** 249-256

[2] Maslehuddin M, Alfarabi M Sharif, M Shameem, M Ibrahim and M S Barry 2003 Comparison of properties of steel slag and crushed limestone aggregate concretes *Cons. and Buil. Mat.* **17** 105-112

[3] Liwu M O, Feng Zhang, Min Deng, Fei Jin, Abir AL-Tabaa and Aiguo Wang 2017 Accelerated carbonation and performance of concrete made with steel slag as binding materials and aggregates *Cem. and Con. Com.* **83** 138-145

[4] Wu S, Xue Y, Ye Q and Chen Y 2007 Utilization of steel slag as aggregates for stone mas-tic asphalt (SMA) mixtures *Buil. and Env.* **42** 2580-2585

[5] Androjic I. and S. Dimter 2012 Compressive strenght of steel slag stabilized mixes. *Grade-vinar* **64** 15-21

[6] Hybská H, E Hroncová, J Ladomerský, K Balco and J Mitterpach 2017 Ecotoxicity of Concretes with Granulated Slag from Gray Iron Pilot Production as Filler *Mat.* **10** 505

[7] Strigáč J, P Martauz, A Eštoková, N Števulová and A Luptáková 2015 Bio-Corrosion Resistance of Concretes Containing Antimicrobial Ground Granulated Blastfurnace Slag *B. and N. H. H-CEM. Sol. St. Ph.* **244** 57-64

The proposal of recommendations for the operation of vacuum sewerage

J Mazák[1], T Dvorský[1], V Václavík[1] R Zajac[1] and P Hluštík[2]

[1]Institute of Environmental Engineering, Faculty of Mining and Geology, VSB-Technical University of Ostrava, Czech Republic,
[2]Institute of Municipal Water Management, Faculty of Civil Engineering, Brno University of Technology, Czech Republic

E-mail: tomas.dvorsky@vsb.cz

Abstract. This article deals with a comparison of vacuum sewerage system and gravity based sewerage system. It also includes the results of the comparison of both of these systems from various cities, and there are measures suggested on the basis of the findings focused on increasing the efficiency and reducing the operational costs of the selected vacuum sewerage system.

1. Introduction

Waste water discharge and its subsequent treatment has become an integral part of the technical society. However, gravity sewerage is still the most widespread waste water discharge system. This system has many technical, environmental and economic disadvantages [1,2,3]. Nevertheless, there are alternative systems, including vacuum sewerage system, which have recently been used more and more frequently. There are studies trying to compare the vacuum sewerage system with the most common conventional sewerage systems, namely gravity and pressure ones [4]. Another work tries to analyze vacuum sewers from the economic, technical, social, and environmental point of view and to suggest their best use in the areas with decentralized waste water discharge [5,6].

2. Description of vacuum sewerage system and its function

Nowadays, vacuum sewerage system is a commonly used representative of separate sewerage systems with an alternative way of waste water transportation through a pipeline. A diagram of the vacuum sewerage system is shown in figure 1.

Unlike the commonly used gravity sewerage system, waste water transported through a vacuum sewerage system takes advantage of the pressure difference between the atmospheric pressure and the mechanically maintained vacuum in the sewerage pipeline. Waste water from the individual buildings and properties is drained by means of a standard gravity sewerage connection into a collecting shaft, where it is accumulated in a collection sump. The collection sump is installed in the collecting shaft together with a suction valve, which forms the boundary-line between the atmospheric pressure and the vacuum in the sewerage pipeline. The vacuum in the sewerage system is maintained by means of vacuum pumps or ejector pumps located in a central vacuum station. Brief characteristics of the compared municipalities:

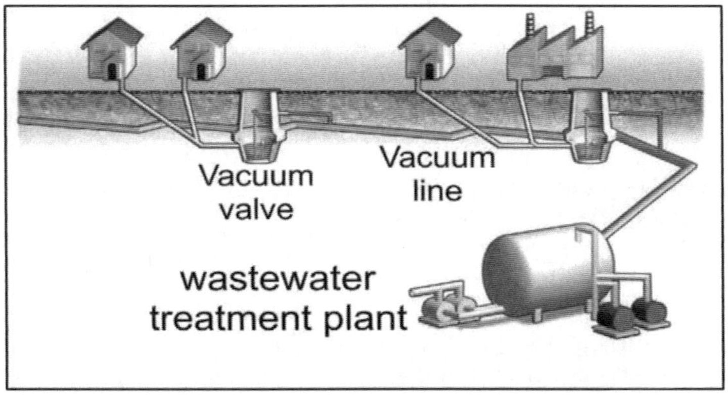

Figure 1. Diagram of vacuum sewerage system [7].

2.1. Zvole municipality – vacuum sewerage system
Zvole municipality is situated in the southern part of Šumperk district, between the cities of Zábřeh and Mohelnice. The cadastral area of the municipality has an area of 658 ha and there are 863 permanent residents registered in the municipality. The construction of a vacuum sewerage system ended in 2003. The total length of the vacuum sewerage system is 6.6 km. It consists of two main sewers A and B to which secondary sewers with the pipe profiles of DN 90, DN 110 and DN 160 are connected, and they are discharge into the local waste water treatment plant.

2.2. Loštice municipality – conventional gravity sewerage system
The municipality of Loštice is situated in the southern part of Šumperk district, the administrative area of the town has an area of 1 200 ha. Loštice has 3 026 permanent registered residents. The sewerage system was built in the second half of the last century, the WWTP was finished in 1999 and the separate sections were interconnected. Today, the uniform gravity sewerage system has the total length of 15.28 km and its arrangement is divided into the left-bank and right-bank parts. Concrete and reinforced concrete pipes with the most commonly used profile of DN 300 to DN 600 are used in most places.

3. Comparison of vacuum sewerage system with gravity one in terms of their preventive maintenance, failure rate, and direct costs

3.1. Preventive maintenance
A long-term and trouble-free operation of the sewerage system requires a suitable design and its execution, but also properly performed preventive maintenance and repairs of the property. These periodically repeated activities are crucial for the sustainability and adequate service life of the sewerage system. Preventive maintenance of a vacuum sewerage system can be divided into three areas: the maintenance of the vacuum station, the outdoor vacuum pipelines and the collection sumps.

The maintenance and control of the vacuum station, especially the vacuum source and pump technology, is the most time-consuming. This activity must be carried out in regular weekly cycles where, along with a routine inspection and maintenance, other operations with a longer, usually monthly, semi-annual and annual period are performed. The maintenance of vacuum sewerage sewers is limited to a visual inspection of the status of the sewerage pipeline protection zone, which is carried out once a year, and a semi-annual check of the functionality of the sectional vacuum pipeline shut-off valves. The preventive maintenance of collecting shafts and suction valves is not planned by the operator, but it is systematically carried out in such a way to check and clean all the collecting shafts, to check the function and status of the suction valves, and to set the control centre every year.

Preventive maintenance on the gravity sewerage system in Loštice is divided into the maintenance of the pumping stations, balancing structures, mountain gullies, sediment control and sewerage sewers cleaning, and the execution of a routine protective deratization. The inspection and maintenance of the potentially problematic objects, i.e. pumping stations and balancing chambers, is performed by the operator according to a maintenance schedule once a week. The mountain gully is checked and maintained quarterly and a routine protective deratization is performed twice a year. A revision of the relevant sections of the sewage system and a camera inspection of the pipeline conditions are carried out according to a plan as well.

3.2. Failure rate

The failure rate is an important part of sewerage system operation with a significant impact on its functionality. Emergency situations represent exactly the kind of the problem when failures in the provision of the waste water discharging service may occur. That is why the main task of the responsible operator is to solve the failure and emergency situations in the shortest possible time and to minimize the impact on the producers. You can see a list of the emergency situations of vacuum and gravity sewerage systems in table 1.

Table 1. Overview of emergency situations of vacuum and gravity sewerage systems.

Failure type		Unit	year 2015	year 2016
Vacuum sewerage system				
Clogged pump	vehicle operation	Km	1 456	1 140
	labour consumption	H	42	30
Open suction valve	vehicle operation	Km	2 444	1 928
	labour consumption	H	108	92
Total	vehicle operation	Km	3 900	3 068
	labour consumption	H	150	122
Gravity sewerage system				
Sewer shaft failure	vehicle operation	Km	67	25
	labour consumption	H	36	4
Sewerage system clogging	vehicle operation	Km	360	396
	machine performance	Eh.	10	9
Pump failure	vehicle operation	Km	0	15
	labour consumption	H	0	5
Total	vehicle operation	Km	427	436
	labour consumption	H	36	9
	machine performance	Eh.	10	9

*Eh. – engine hours

Gravity sewerage system showed a lower number of failures than the vacuum one during the monitored time period. The reason is that gravity sewerage system is less prone to negative impacts on its basic functionality due to the occurrence of failures or accidents of the sewerage sections. In case of vacuum sewerage system, the occurrence of a failure is usually detected immediately because of the loss of vacuum and limited waste water transport function. Objectively, it is necessary to note that common vacuum sewerage system failures are not demanding in terms of the consumed material and do not require special machinery or equipment.

3.3. Direct costs

In the case of the sewerage systems, the assessment of the level of operating costs is important for at least two reasons. The first one is the fact that the sewerage system provides a public service and the price for this service, according to the generally accepted rules, should not exceed the socially tolerable limits. The second reason is the relatively long service life of the whole system. A decision to select a particular drainage method is a commitment to continued maintenance and operation of the system for decades, and it is almost impossible to change the entire functional system.

Table 2 clearly shows that the level of direct operating costs of the assessed vacuum and gravity sewerage system is comparable in terms of their direct costs.

Table 2. Direct costs of the operation of vacuum and gravity sewerage system.

		Vacuum sewerage system		Gravity sewerage system	
Cost group	unit	year 2015	year 2016	year 2015	year 2016
Material for repairs	CZK	756.00	1 030.00	66 111.00	1 974.00
Electric energy	CZK	122 480.00	147 104.00	28 350.00	34 322.00
Sewerage system repairs	CZK	15 892.00	16 981.00	15 100.00	4 998.00
Personal costs	CZK	102 626.00	104 372.00	88 658.00	81 480.00
Transportation	CZK	65 926.00	65 342.00	71 443.00	74 721.00
Laboratory analyses	CZK	3 120.00	3 120.00	118 192.00	123 609.00
Total	CZK	312 815.00	339 965.00	389 869.00	323 120.00

4. Recommendation for the operation of a vacuum sewerage system

The operating recommendations can generally be divided into several areas, namely improving the awareness of the sewerage system users, building modifications of the vacuum sewerage system elements, suggestions concerning the maintenance changes and addition of a monitoring system.

4.1. Improving the awareness of sewerage system users

A significant share of the vacuum sewerage system failures and accidents is caused by the presence of undesirable objects (wet napkins, diapers, feminine hygiene products, textile pieces, etc.) in waste water. In order to sufficiently eliminate the problematic content of waste water and to improve the awareness of its producers, it is necessary to carry out a regular information campaign.

4.2. Building modifications of vacuum sewerage system elements

Freezing of the suction valves and control centres during the winter months is a major problem. The cause usually arises during the construction of a vacuum system, where the installation of the aeration pipelines on the gravity parts of the sewerage connections was not carried out, the aeration pipeline of the vacuum suction valve may be missing on the collection sump and the air is fed directly from the collection sump. It is advisable to add air intake pipelines to all the gravity parts of the sewerage system connections. This measure eliminates the need of the producers to provisionally suck air into the collecting shaft and the cause of the suction valves and control centres freezing during the winter period. The collecting shafts, which have not been equipped with an air supply pipeline for controlling and securing the cyclical opening and closing of the suction valve during the building of the sewerage system, must be completed by adding this construction.

4.3. Suggestions concerning maintenance changes

The maintenance of the selected problematic collecting shafts should be carried out once a year, in order to increase the efficiency and reliability, and in the case of other collecting shafts, it should be done continuously every four years. This intervention should, in addition to the inspection and cleaning of all parts of the collecting shaft, also include a removal of the suction valve and the control centre, cleaning and inspection of all the pipelines and hoses, the installation of another vacuum suction valve and control centre, which were dismantled, cleaned and lubricated in the workshop before.

4.4. Adding a monitoring system.

Achieving a consistently high level of vacuum sewerage system service is not possible without high-quality and useful information. The highest availability and informative value is associated with the data that can be obtained in real time and that include information about sufficiently large location and about the possible links.

5. Conclusion

The main problem related to the operation of a vacuum sewerage system is the frequency of failures and the high consumption of resources necessary for their identification. The analysis helped us to identify the most serious causes of the occurrence of emergency situations, such as the design deficiencies arising during the sewerage system construction, the presence of material in the transported waste water which cannot be discharged into the sewerage system according to the valid rules, and the incorrectly performed preventive maintenance of the suction valves. High consumption of resources necessary to solve the failure situations is also caused by the insufficient range of the collected operational records and their low informative value or by not using new technologies in the field of measurement and remote monitoring of the vacuum level and suction valve function.

Under suitable conditions, vacuum sewer system does not have higher requirements with respect to the inevitable operating costs and the necessary resources of the operator. Since this is a system of alternative way of waste water transportation to gravity sewerage system, one has to bear in mind that the meaning of the word alternative does not mean a general full-fledged substitution of the gravity sewerage system, but it offers another possibility of transporting waste water in locations where the conditions for the use of a gravity sewerage system are not favourable.

References

[1] Krocova S and Rezac M 2016 Infrastructure operation reliability in built-up areas *Communications* **18** 75-78

[2] Krocova S 2015 Threats of water contamination in an industrial lanscape *Inz. Miner.* **36** 125-130

[3] Elawwad A, Ragab M and Abdel-Halim H 2015 An economical, environmental, and social comparison between vacuum and gravity sewers in decentralized sanitation systems, with Egypt as a case study *Jour. Water Sanit. Hyg. Develop.* **5** 614 - 619

[4] Islam M 2017 Shafiqul. Comparative evaluation of vacuum sewer and gravity sewer systems. *Inter.Jour. Sys. Assur. Eng. Manag.* **8** 37-53

[5] Panfil C, Mirel I, Szigyarto I and Isacu M 2013 Technical, economical, social and ecological characteristics of vacuum sewage systém *Environ. Eng. Manag. J.* **12** 1017–1022

[6] Gikas, P, Ranieri E, Sougioultzis D et al. 2017 Alternative collection systems for decentralized wastewater management: an overview and case study of the vacuum collection system in Eretria town, Greece *Water Pract. Tech.* **12** 604-618

[7] *Vacuum Sewer System - Water Treatment Specialist – Coway EnTech* http://cowayet.com/eng/solutions/vacuum_sewer_collection.asp

Assessment of the suitability of trees for brownfields reuse in the post-mining landscape

J Mec[1], B Lokajickova[1], N Sotkova[1], H Svehlakova[1] and B Stalmachova[1]

[1]VŠB – Technical University of Ostrava, Faculty of Mining and Geology, Institute of Environmental Engineering, Czech Republic

Email: jan.mec.st@vsb.cz

Abstract. The post-mining landscape of Upper Silesian is deterioration of the original landscape caused by underground coal mining. There are huge ecosystems changes, which have been reclaimed by nature-friendly procedures. The aim of the work is to evaluate the suitability of selected trees for reuse of brownfields in this landscape and proposals for reclamation in the interest areas of Upper Silesian.

1. Introduction

The aim of this work is to evaluate the growth of trees in the areas affected by the mining of black coal. The areas on which the research is aimed are mainly mine dumps of Upper Silesia (figure 1). These post-mining landscapes themselves have been significantly affected by coal mining and subsequent piling of tailings (waste rock) from the pit to the delimited place of imposition. For the purpose of the research there were selected mine dumps of Ema, Heřmanice, Urx, together with the Barbora mine and František mine, as well as the post-mining landscape of Darkov Sea. Another area of interest are the stations of hygiene zone in the II. degree of protection of Bartovicky Forest waters. These habitats have been used as a reference area for fitness comparison of selected tree species. Reference area is located between intersections of mine dumps, which are selected for the purpose of research. In the surveyed areas are selected representatives of the most important tree species, including English oak *(Quercus robur)*, quaking aspen *(Populus tremula)*, birch *(Betula pendula)*, mazzard *(Cerasus avium)* and hawthorn *(Crataegus monogyna)*, black locust *(Robinia pseudoacacia)* [1].

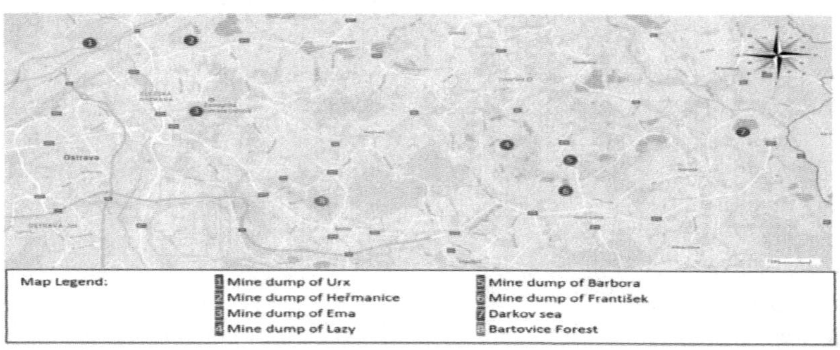

Figure 1. Map of selected areas with Map Legend (google.cz/maps - modified).

2. Methodology

Selection of habitats and species of juvenile stage trees represented take the form of stratified random sampling. Data collection at each selected site is being carried out by measuring the selected growth of a given tree, depending on the orientation towards the cardinal directions, the branch length, and the height at which the growth is located. Five representative growths are measured at each habitat/site. The collection of the data is mainly limited by weather conditions depending on the hazards associated with the working at heights. The selected habitat is marked and written first, before these data are collected. After that, five representative samples are selected at different tree heights and four measurements with different orientation on the cardinal directions. The fifth measurement is done on the tree terminal (growth peak). The length of each growth is measured in millimeters, followed by the total length of the branches in centimeters and followed by height from ground in meters at which the measurements were made. At the same time, the branches are marked with a marking spray, at the end of which branch the growth is measured, and the growth itself is marked by an arboretum mark with the designation of the cardinal direction and the order in which the growth is measured. After that, trunk perimeter is measured at a height of 130 centimeters from the ground.

The data are collected during the 2017 growing season. To accurate and safe measurements are used equipment for tree-lining and elevation work. The base of this equipment is an open or all-body seat with a protection and variable settings allowing the optimal distance from the tree trunk. In addition, static and dynamic ropes are used to protect and climb. The measurement itself is done by a tape measure, tape band and marker equipment. For successful and safe measurements are required three workers to measure, secure, and write data.

3. Selection of tree species

For research purposes, the most abundantly represented types of tree species that are in the maturity stage have been selected. Woods at this stage are characterized by generative reproduction. At this stage, the ideal growth of skeletal branches with other prolific overgrowth is typical.[2] In particular, the research areas include mainly these tree species: *Betula pendula, Robinia pseudoacacia, Cerasus Avium, Crataegus monogyna, Quercus robur* and *Populus tremula.*

Quaking aspen (*Populus tremula)*

In the Czech Republic it grows throughout the territory from the lowlands to the mountains, i tis the most widespread in the hilly areas. The aspen is a resistant pioneer tree, which first appears along with the birch on bare soils and clearings. It also grows in bright forests, pastures, abandoned pastures and forest ridges. It is resistant to frost and tolerant to soil reaction. It grows both on wet and dry soil, but mostly on sandy, rich in nutrients soil [3].

Birch (*Betula pendula)*

It appears in the Czech Republic plentifully up to altitude of about 1000m. Generally, it grows in Europe to the north to the Arctic Circle, to the south along the Pyrenees and the Apennines, to the east to the Lena River basin. Birches look for lighter forests and their edges, clearings, peat bogs, pastures, rocks, on the poorer, drier, even extremely acidic soils, bright habitats, while it is dying soon in the shade [4].

Black locust (*Robinia pseudoacacia)*

It comes from North America, eastern and central parts of the US, now extended all over the southern part of the North American continent. It was brought to Europe in the beginning of the 17th century, to Bohemia in 1710. It is cultivated in the mild belt of the whole world, it often runs wild, it spreads in many areas. It grows on the edge of forests, along roads, in alleys, around gardens and parks, near human settlements. It spreads to the natural communities of bright forests and shrubs, where it suppresses the original vegetation [5].

Mazzard (*Cerasus avium)*

It is spread across Europe, except in the southernmost regions. In our country it is rather abundantly widespread, throughout the territory of the republic. It is common in warmer areas and in lower mesophyll positions. Thick trunk is smooth when young, usually grayish brown, dark brown in old age bark that peels off in various widths and transverse bands [6].

Upright Hawthorn (*Crataegus monogyna)*

The area of the species occupies almost the whole of Europe. The Slovak Republic is the most widespread species of hawthorn from planar to sub-mountain level. It has a wide ecological plasticity, so it can grow from extremely xerothermic habitats to flooded alluvial soils. It is part of shrub communities, the edges of deciduous forests, often creating separate monocenoses, mostly in agricultural landscapes. It is most commonly found on limestone subsoil, very rarely on acidic or very acidic soils, such as peat [7].

English Oak (*Quercus robur)*

Oak belongs to light-requiring species. In water requirements, two ecotypes are distinguished. The widespread ecotype, found especially in the floodplain forests, has considerable demands on water, and can withstand spring floods. Second ecotype has the ability to grow in shallow, heavily drying soils in the summer and can be found in forest-steppe localities. Bottom water must be within the roots reach. It is a tree heavy on soils and it grows best on deep, clay soils found in floodplain forests or less (figure 2) [8].

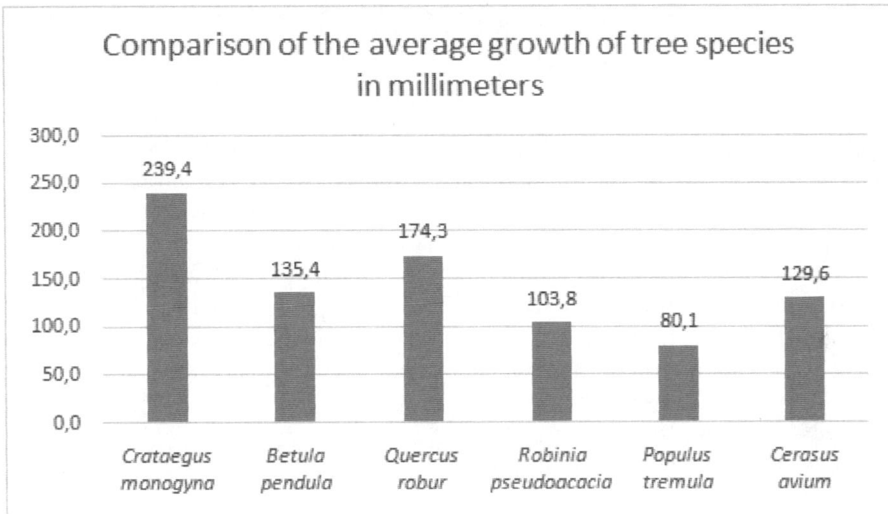

Figure 2. Comparison of the average growth of tree species in millimetres.

4. Results and discussion

The result of this research is the comparison of the fitness of juvenile stage trees occurring on the mine dumps of post-mining sites with the selected reference area of the Bartovice Forest. The available results show that dumps affect growth of individual trees. Selected tree additions grow into a shape of "fork", half of which withers in most cases. However, the average growth of post-mining sites is greater than the average growth of trees in the reference area. The largest average growth from habitats found on the mine dumps was measured during existing research at tree species *Crataegus monogyna*. Furthermore, according to the data gathered so far, the largest growth of wood species was in the mine dump Ema.

The lowest average growth was measured at tree species *Populus tremula*, which is considered as an optimal species for phytoremediation on industrial surfaces and on areas with bad air conditions (figure 3) [9]. Current research surprisingly indicates the inappropriate environment for the species *Populus tremula*. However, it's an original research which means there is no possibility to compare research data with others results. Research still continues and to complete the research is necessary collect more data and analyzes soil samples from sampling points on individual areas of interest.

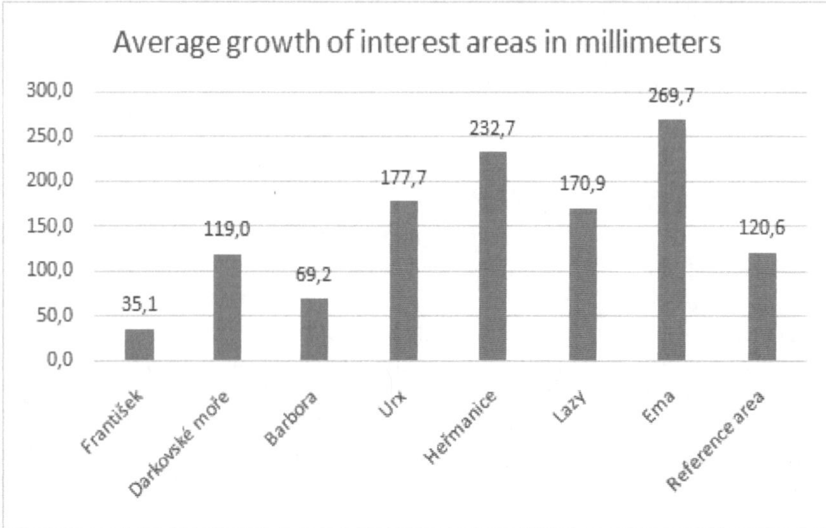

Figure 3. Average growth of interest areas in millimetres.

5. Conclusion

Research areas are characterized by the intensive deterioration of the original landscape elements and components caused by deep-scale coal mining. It is characteristic for significant changes in relief, changes in the water regime, changes in thermal and climatic conditions and extensive changes of ecosystems. However, the changes are not necessarily only negative. On the contrary, specific conditions which may suit certain plant species may be created at sites affected by mining. It is clear from the results that mine dump Ema is such an example. The largest growth of selected deciduous trees were found here, which means that some tree species complied more with habitat conditions at this site than on a reference area in the environment which wasn't directly affected by mining [1].

Acknowledgments

This article and poster were funded from sgs projects.

References

[1] Lokajíková B, Šotková N, Mec J, Švehláková H and Stalmachová B 2017 Fitness assessment by increment of the woody species in selected post-mining area of Upper Silesia SGEM 2017 (Albena Bulgaria)
[2] Larcher W 2006 Physiological plant ecology: Ecophysiology and stress physiology of functional groups (Berlin: Springer) p 513
[3] Populus tremula: topol osika L Dendrologie 2006 Databáze dřevin celého světa
[4] Divíšek J, Culek M and Jiroušek M 2010 Biogeografie: Multi.- Bříza bělokorá (Betula pendula Roth): Čeleď: Betulaceae – břízovité Geografický ústav Přírodovědecká fakulta University of Masaryk
[5] Divíšek J, Culek M and Jiroušek M 2010 Biogeografie Trnovník akát (Robinia pseudacacia L.): Čeleď: Fabaceae – bobovité Multi Geografický ústav Přírodovědecká fakulta Masarykova univerzita
[6] Prunus avium: slivoň třešeň L Dendrologie 2006 Databáze dřevin celého světa
[7] Crataegus monogyna: hloh jednosemenný 2006 Jacq. Databáze dřevin celého světa
[8] Divíšek J, Culek M and Jiroušek M 2010 Dub letní (Quercus robur L.): Čeleď: Fagaceae - bukovité Biogeografie: Multi Geografický ústav Přírodovědecká fakulta Masarykova univerzita

[9] Hillier J, Coombes A J 2007 The Hillier, Coombes Manual of Trees and Shrubs Hillier Coombes Coombes Nurseries (David & Charles: Winchester England)

Environmental sustainability assessment of family house alternatives and application of green technologies

A Moňoková[1], S Vilčeková[1], Ľ Mečiarová and E Krídlová Burdová[1]

[1]Technical University of Košice, Faculty of Civil Engineering, Vysokoškolská 4, Košice, Slovak Republic

E-mail: andrea.monokova@tuke.sk

Abstract. Transition to environmentally friendly technologies provides a comprehensive solution to problem of creating an economic value without destroying the nature. Buildings using green technologies lead to lower operating costs, healthier living and working environment and protect the environment more. The aim of this paper is to assess the environmental impact of two alternatives of family house designed as conventional building and building with green technologies. Evaluated family house are located in village Kokšov Bakša, which is situated 12 km south-east from city of Košice, a metropolis of eastern Slovakia. This analysis investigates the role of applied green technologies in single family houses for impact categories: global warming potential (GWP), acidification potential (AP) and eutrophication potential (EP) expressed as CO_{2eq}, SO_{2eq} and $PO_4^{3-}{}_{eq}$ within "Cradle to Grave" boundary by using the LCA assessment method. The main contribution of the study is a proof that green technologies have significant part in the reduction of environmental impacts. Results show that alternative of family house designed as green one contributes to CO_{2eq}, SO_{2eq} and $PO_4^{3-}{}_{eq}$ emissions by 81%, 73% and 35% less than alternative of conventional family house, respectively.

1. Introduction

LCA (Life Cycle Assessment) is a tool to evaluate the environmental impacts of building materials. The concept of environmental life cycle assessment (LCA) was developed from the idea of comprehensive environmental assessments of products, which was conceived in Europe and in the USA in the late 1960s and early 1970s. Originally, LCA was used as a tool by environmental consultants. The steps of most concern here are the life cycle inventory analysis (LCI) and initial stages of impact assessment. LCI involves detailed tracking of all the flows - in and out - of the system of interest, raw resources or materials, energy by type, water, and emissions to air, water and land by specific substance. LCI data can then be characterized in terms of impact potentials (e.g., global warming, ozone depletion, etc.). The aim of rating a building is to minimise its environmental impacts and to create more sustainable buildings. Thus, architects through their architectural design and selection of materials would create sustainable designs [1]. Study [2] states that construction activity introduces an essential role in socio-economic development of the country as it provides infrastructure set-out, on which all sectors of economy firmly depend. Therefore, it makes the building industry one of the most strategic sectors. In the European Union, this sector contributes to GDP roughly by 10% and it is also the largest industrial employer (employing 30% of the total number of employees in the industry), with an estimated 14.8 million employees and 3.1 million companies. Construction sector plays a significant role in the EU economy but on the other hand it is a major energy consumer and greenhouse gases emitter. For example, 42% of the total energy consumption in the EU, 35% of greenhouse gas emissions, about 50% of the mined crude, and 22% of production waste is related to the construction of all types of buildings [2]. The development of science and research allows building industries to move forward in the development of new materials on different basis. Building materials are thus gaining another dimension. While in the past natural materials were used exclusively, nowadays a great deal of them are produced artificially and often by technologies that have negative impacts on the environment. A part of the concept of sustainable development is the right choice of building materials for implementation of the selected object. By selection of environmentally friendly building materials a reduction in depletion of natural resources and factory emissions as well as creation of more suitable microclimate in building interior can be achieved [3,4].

The aim of this study is to analyse how different building envelope solutions as well as the usage of different building technologies may affect the impact on the environment. The LCA assessment of two alternatives of family house has been performed by One Click LCA software available online [5]. It's the highest rated LCA software for BREEAM, supporting up to 19 credits in various BREEAM schemes. For LEED, the software allows simple baseline comparison with ready-to-use templates to help achieve the Materials and Resources credits easily. Software One Click LCA allows by Life-Cycle Assessment to evaluate the effect on the environment of a product, service, or process over its entire life-cycle. This means that LCA takes into consideration all the steps that lead from raw material to manufactured product, including extraction of the materials, energy consumption, manufacture, transportation, use, recycling, and final disposal or end of life. It is a holistic methodology that quantifies how a product or process affects climate change, non-renewable resources, and the environment as a whole. Life-Cycle Assessment's strength lies in the fact that it takes into account what happens before and after the final product is used by customers, and can effectively measures effects over a long time of period.

2. Green technologies and green buildings

Green technology can be defined as the technology which is environmentally friendly, developed and used in such a way so that it doesn't disturb our environment and conserves natural resources. The construction industry has significant environmental, social and economic impacts on the society. As a result, the last decades have witnessed the rapid growth of the green building sector in order to mitigate the negative impacts associated with construction related activities. These include upfront cost vs. ongoing savings; and energy savings vs. building users' health and wellbeing. In China, it has been reported that some green buildings consume 26% less energy compared to conventional buildings. However, due to the incremental cost, it is not uncommon that enterprises and governments in China are unwilling to bear this kind of risk [6]. European climate strategy foresees measures to increase energy efficiency, competitiveness and the energy security of Europe by decreasing energy consumption. As buildings are responsible for 40% of the total energy consumption in the European Union, the Energy Performance of Buildings Directive sets energy consumption reduction targets for the member states [7].

3. Design of family house alternatives

3.1. Construction of building envelopes

The first alternative of family house uses solely conventional approaches and materials in construction, while the second one uses sustainable approaches with strong focuses on environmental and energy aspects. The alternatives are mentioned to be located in Kokšov Bakša, a municipality in Košice Region of Eastern Slovakia. The houses are placed in a flat terrain at an altitude of approximately 190 metres. They are designed as a single storey, detached family houses without basement. Interior layout consists of vestibule, hall, living room with kitchen and dining room, larder, three bedrooms, two bathrooms with WC, toilet, wardrobe, boiler room, terrace, garage (only in alternative 1) and a laundry room (only in alternative 2). Both houses are founded on strip foundation. The strip footing is 600 mm wide and at least 650 mm deep made of reinforced concrete of C16/20. Two rows shuttering form-work blocks are used on top of in-ground part. Aerated concrete blocks with the thickness of 300 mm and 250 mm are suggested for external and internal bearing walls in alternative 1. Double glazed PVC windows and doors with U=1.0 W/m^{2}K are proposed. Floors are designed as self-levelling poured screeds with ceramic and/or laminate finishing. The horizontal structures consist of reinforced concrete ceiling with thermal insulation of EPS liners above the ground floor and reinforced concrete ring beam wreaths and lintels made of C20/25 concrete class. The roof structure is proposed as flat with gravel ballast layering. External and internal bearing walls in alternative 2 are designed as CLT panels with thickness of 170 mm. Triple insulating glass windows and doors with U = 0.79 W/m^{2}K are designed as wooden - aluminium constructions. Floors are designed as self-levelling poured screeds with ceramic or wooden finishing. Horizontal structures consist of CLT panels with thickness of 170 mm above the ground floor. The roof structure is a flat green roof. The substrate thickness of 150 mm is proposed. Table 1 summarizes basic information about alternatives of family house.

Table 1. Information for designed alternatives of family house.

	Alternative 1	Alternative 2
Built-up area	250 m^2	224 m^2
Living area	98.06 m^2	117.11 m^2
Floor area	183.52 m^2	174.45 m^2
Built-up volume	1350 m3	986 m3

Figure 1 illustrates views for designed alternatives of family house.

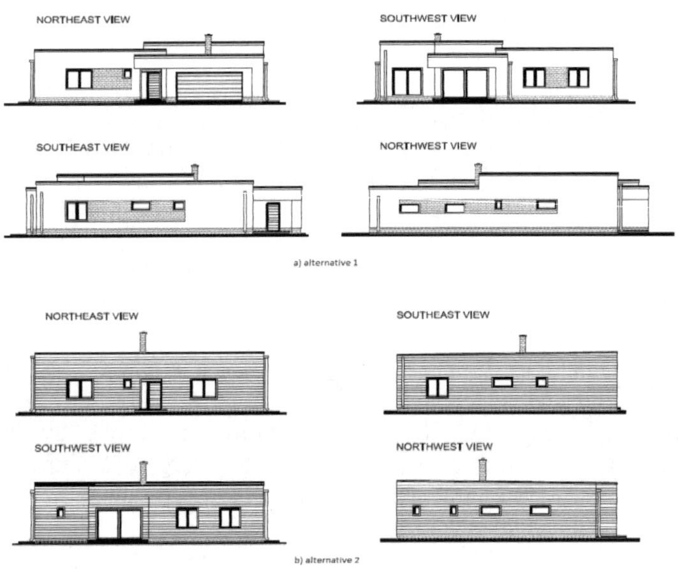

Figure 1. Views of designed family house.

3.2. Technologies of investigated houses

The first house is connected to public utilities except of sewage which shall be drained to a septic tank. A condensing gas boiler is used for domestic central heating and hot water preparing. Floor heating is installed in the whole house except of larder, boiler-room and garage. Supposed yearly gross energy consumption is around 22.0 MWh of which 14.8 MWh is consumed by heating the living part of the interior space during heating period (from October till April for a given latitude).

The second family house is connected to public utilities with sewage designed as a pressure sewage system. Heat pump is used as the source of heating and hot water preparation, which is stored in 300 l insulated hot water tank. The supposed total energy consumption per year is 7 MWh of which 2.5 MWh is used for heating.

4. Environmental assessment of designed alternatives

Table 2 presents results of the environmental impacts for individual phases of the life cycle of assessed family house alternatives.

Table 2. Indicators characterizing the environmental impacts.

Sector	GWP kg CO_{2eq}		AP kg SO_{2eq}		EP kg $PO_4^{3-}{}_{eq}$	
	Alt 1	Alt 2	Alt 1	Alt 2	Alt 1	Alt 2
Construction Materials	99,200	64,000	306.00	364.00	80.70	109.00
Transportation to site	7,560	4,150	20.30	12.40	4.59	2.66
Maintenance and material replacement	42,900	11,300	75.50	29.90	12.40	5.70
Energy use	343,000	150	1,200.00	0.96	108.00	0.14
Deconstruction	13,200	14,200	18.20	27.90	4.84	19.10
Total	505,860	93,800	1,620.00	435.16	210.50	136.60

Figures 2 and 3 depict the percentage ratio of designed materials for environmental impact expressed as GWP.

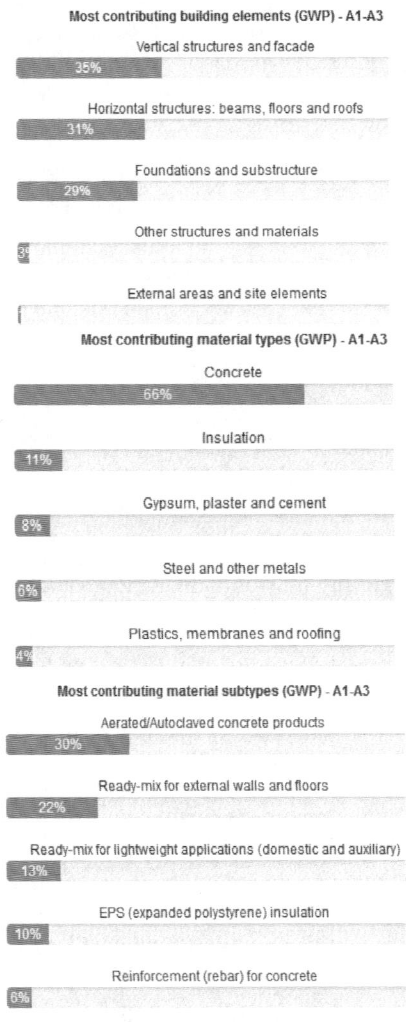

Figure 2. Alternative 1.

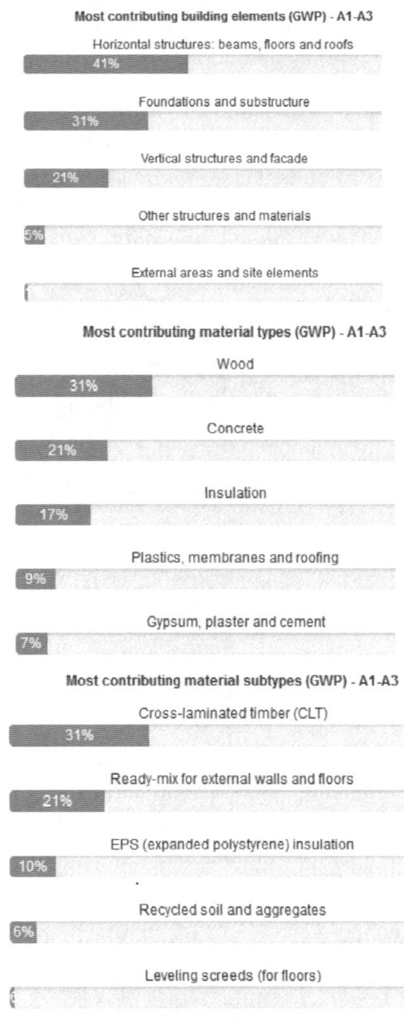

Figure 3. Alternative 2.

We can see that most contributing building elements are vertical structures and facade with value of 35% for alternative 1 and horizontal structures (beams, floors and roofs) with value of 41% for alternative 2. Most contributing material types are concrete with 66% for alternative 1 and foundation and substructure with value of 31% for alternative 2. And finally, most contributing material subtypes are aerated concrete products with 30% for alternative 1 and cross-laminated timber (CLT) with 31% for alternative 2. Figure 4 shows that alternative 2 contributes to CO2eq emissions by 81% less than alternative 1 and from SO2eq and PO43-eq by 73% and 35% less than alternative 1, respectively.

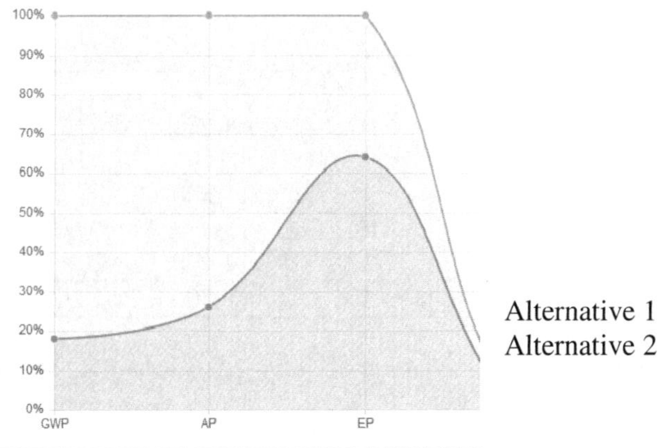

Figure 4. Comparing the impacts for both alternatives.

5. Conclusions

Based on calculations it can be seen that the house with the use of environmentally friendly materials and green technologies is more than a comparable alternative to the conventional design. This result is achieved by using the building materials, environmentally friendly technologies, as well as by modifications of roof to the green one. In Slovakia there are a number of buildings classified as green or high performance green buildings that are certificated from sustainability aspects. Design the high performance green buildings for the future of a sustainable life on Earth is indisputable. Therefore the certification of buildings from three dimensions of sustainability (environmental, social and economic) gives some assurance that the buildings do not burden the environment.

Acknowledgments

This paper was financially supported by the Slovak Grant No. 1/0307/16.

References

[1] Dekkiche H and Taileb A 2016 The Importance of Integrating LCA into the LEED Rating System *Procedia Engineering* **145** pp 844 – 851

[2] European Commission 2012 Reference Document on Best Environmental Management Practice in the Building and Construction Sector

[3] Green Technology 2015 Green Technology – What is it? *http://www.green-technology.org/what.htm(2015)*

[4] American Elements 2016 About Green Technology *https://www.americanelements.com/green-technology-alternative-energy.html*

[5] One Click LCA 2017 *http://www.oneclicklca.com/green-building-software/*

[6] Shi Q Yan Y Zuo J and Yu T 2016 Objective conflicts in green buildings projects: A critical analysis *Build. and Environ.* **96** p 107-117

[7] Sulakatko V Lilla I and Witta E 2016 Methodological framework to assess the significance of External Thermal Insulation Composite System (ETICS) on-site activities *Energy Procedia* **96** p 446 – 454

IOP Conf. Series: Earth and Environmental Science **92** (2017) 012045 doi:10.1088/1755-1315/92/1/012045

The presence of insect at composting

J Mudruňka[1], B Lyčková[2], R Kučerová[2], V Glogarová[2], J Závada[2], B Gibesová[1] and D Takač [2]

[1]VŠB-Technical University of Ostrava - Institute of combined study in Most, Budovatelů 2830/3, Most 434 01, Czech Republic
[2]VŠB-Technical University of Ostrava - Faculty of Mining and Geology, 17. listopadu 15/2172, 708 33 Ostrava-Poruba, Czech Republic

Email: jaroslav.mudrunka@vsb.cz

Abstract. During composting biodegradable waste, microbic organisms reproduce massively, most of which belong to serious biopathogens which are able to penetrate various environmental layers. Their vector species include dipterous insect (*Diptera*) which reaches considerable amounts in composting plant premises as well as home composting units, mainly during summer months. Therefore measures must be taken to eliminate or reduce this unwanted phenomenon (sanitisation, disinfection). For evaluating obtained results, relative abundance calculation was chosen.

1. Introduction

When handling waste, there is always real probability of environmental hazards because during decomposition of biodegradable materials, a lot of microorganisms of autochthonous type [1]. Pathogenic bacteria, fungi and their spores can contaminate not only at processing plants, but it also can, by wind or vectors, penetrate their surroundings and pose a risk for people, animals and plants. Mainly some species of dipterous insect (*Diptera*) the imagos of which lay their eggs into waste and also into maturing substrates and larvae of which develop and cocoon, can transmit a plenty of infectious diseases on their bodies or via their digestive tract [2]. Within a research which was carried out during 2016, selected species of this insect were detected and taxonomically determined at home composting unit and at composting plant.

2. Environmental hazards of composting

Composting is the optimum way of biowaste processing because it influences environmental layers only marginally and the final product can be used as quality natural organic fertilizer. However, during composting, the surrounding is polluted with microorganisms, volatile organic compounds of microbic origin and smell [3]. The unignorable risk factor is dipterous insect (*Diptera*) penetrating into composting plants premises and into home composting unit because biodegradable waste, as well as maturing substrate, is suitable environment for its eggs, larvae and nymphs. With increased presence of insect, humans or animals can be attached upon directly, while serious level of infestation with insect larvae (myiasis, pseudomyiasis) cannot be avoided. However, such risks are negligible, and they can be totally eliminated due to maintaining principles of proper hygiene [4].

On the other hand, there exists highly serious risk of subsequent transmission of microbial pathogens on a body and or in insect mouthparts. Biopathogens transmitted by them can be a potential source of infections and cause diseases not only by its infectiousness, but also by toxins produced. An immensely dangerous biopathogen is bacteria *Escherichia coli* phyla of which are commonly present in livestock guts and together with excrements can be taken into compostable biowaste. During infection, significant health complications may occur. Another significant thing is a danger of contaminating plant surfaces with this bacteria [5]. There can also appear pathogenic species *B. anthracis* from *Bacillus* phylum which attacks animals as well as humans while its impact is fatal. In composts, there can also be bacteria of *Streptococcus* phylum out of which several species are significantly pathogenic [6]. In such cases, the risk of transmission in insect mouthparts is particularly serious. Also the risk of contamination with bacteria *Clostridium botulinum* is relatively high as it produces botulinum neurotoxins with fatally toxic effects [7]. The presence of larvae of *Calliphoridae* family, in GIT of which there are colonies of the above mentioned bacteria in high concentrations. The premises and surroundings of composting plants are also contaminated with spores of various fungi (e.g. *Aspergillus*) which implies that their secondary transmission on insect bodies is highly probable. However, the risk of fungi and microscopic fungi and their transport via insects exists also during home composting or in connection of insufficiently hygienic collection and assembling of biowaste (including so-called kitchen) [8].

3. Characteristics of dipterous insects (*Diptera*)

Out of approximately over a million described insect species, about eighty six thousand belong to *Diptera* order (*Diptera)* all species of which have large compound eyes. Their mouthparts can differ – a rostrum might be missing or it might be reduced. The mesothorax is mightly vaulted and a single pair of fore wings grows out of it, while the hind pair is reduced to small halters. Each part is each body segment has a pair of differently shaped legs [9]. The twelve-part abdomen can have the last parts transformed due to easier reproduction. Dipterous insect are sexually dimorphic. Females lay glossy and mostly white eggs (length of about 1 mm and width of about 0.4 mm) at places with sufficient humidity and food source. After larvae hatch, three stages of their development take place and at the end a larva pupates. The perfect metamorphosis of *Diptera* in a pupa is finished with an adult emerging from it [10] [11]. Occurrence of this insect is tied to suitable life conditions which can mainly be found near human habitations because larvae and adults feed on plant or animal tissues, kitchen waste, feces and so on [12]. From the anthropogenic point of view, it is possible to divide this order into insignificant, useful and harmful species; the last to be named can damage plants, parasitize on animals and carry pathogens [13] [14]. Taxonomically, the order divides into several families out of which the most important ones, due to the issue presented her, are *Muscidae*, *Calliphoridae*, *Cordyluridae*, *Scatophagidae*, *Sarcophagidae*, as well as *Drosophilidae* [15].

4. The method of detection and insect determination

The research itself was implemented from April till October 2016 at the premises of a composting plant in Ustecky district and simultaneously in the garden of a family house in a garden suburb approximately 20 km far from the composting plant mentioned above. The composting plants processes 80,000 tons of biodegradable waste per year, the waste is laid in dumps in the building of operating halls. Food waste is also composted here, which is, after sanitising in a fermentation box together with cut grass and wood chippings mixed into common biowaste. Suitable trap devices were located in the area of the composting plant, namely "flychatchers" (glued paper bands with chemical attractant, and plastic bowls of white and yellow colour which were filled with water with formaldehyde (4%) and detergent. The home compost is a usual one: it is a fill of 2 x 1 m, laid directly on the ground with board fender lining where usual biowaste from household and garden is put. Trap devices were also installed near the compost site ("flycatchers" and bowls). The trapped insects were killed in a killing jar and they were taxonomically determined.

One of the most substantial factors which has crucial influence on the process of entomologic sampling is weather at the given site and at the given time. *Diptera* mostly belong to sun-loving insect species therefore their occurrence at the sites is at its fullest during sunshine. At the time of low temperatures, insect is rather inactive but this also applies at the time of extremely high temperatures.

Also during rainfalls, most insects are hidden. Also wind is very important for detection of insect individuals itself as in case it is stronger or in case of turbulence, insect is carried by airstream. All the weather data were obtained via our own measurements, personal observation (rainfall, cloud amount) or from Czech Hydro-meteorological Institute database in Prague (wind force, wind direction).

Due to the fact that large amounts of various insects were found in bowls and flycatchers, we selectively chose only such species which taxonomically belong to Diptera order, and also number of which exceeded numbers of other insect species caught and at the same time, indeterminable specimens were excluded (destroyed body, etc.).

5. Evaluation of the obtained results

For evaluating obtained results, relative abundance calculation was chosen. This term of ecological statistics expresses amount of insect individuals of the given species related to certain predetermined type of quantitative sampling method. In this particular case it was the total amount of caught specimens to the total amount of obtained samples. Calculations were implemented via below mentioned Equation (1) while the total amount of individuals of the observed species was determined as the total sum of all insect individuals of the given species caught in calendar month and the total amount of obtained samples was the total sum of individual samples in one month (one individual sample per one day):

$$RA = \frac{S_n}{S_{vz}} \quad (1)$$

where: RA = relative abundance; S_n = number of individuals of the observed species; S_{vz} = total amount of obtained samples.

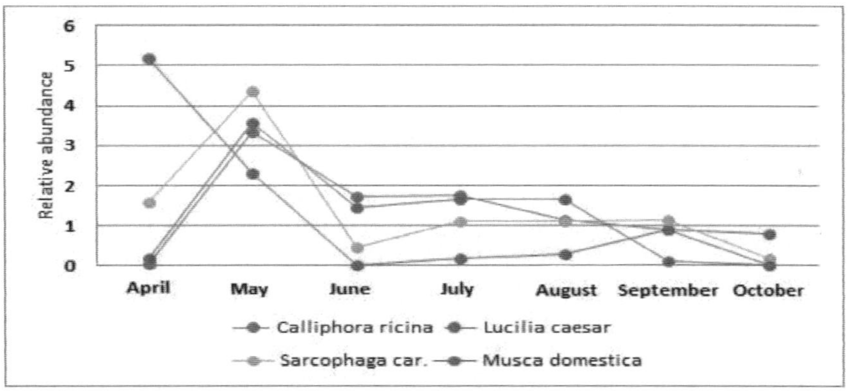

Figure 1. Values of relative abundance – composting plant.

As evident from relative abundance calculations, (see figure 1 and figure 2), numbers of all species of *Diptera* were during the observed period of time considerably higher in the premises of the composting plant. Mainly in case of *Sarcophaga carnaria* and *Musca domestica* (housefly), the difference is particularly obvious. Increased amounts of insect in the composting plant are certainly caused by much bigger volume of composted waste material than it is in case of home composting unit. Also, muck is not included into a fill in a private garden which explains increased amount of insect in the composting plant because such waste is normally processed there. Also very small amount of food waste which gets into a home composting unit is the reason why *Sarcophaga carnaria* only rarely occurs here. In operation halls where constant air circulation (draught) is apparent, amount of caught insects is substantially lower that it is at outside dumps of still unprocessed biowaste. If we compare data on climatic conditions in the given area with records on amount of detected insects, we can confirm the known information about life cycle of the given insect species.

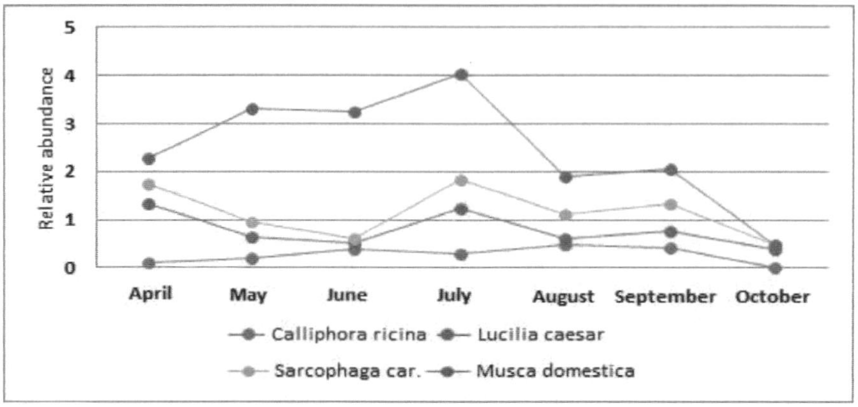

Figure 2. Values of relative abundance – home composting unit.

For instance, *Calliphora vicina* R. D. is mostly active at temperatures of about 27 °C while cryophilic *Lucilia caesar* L. can be observed already at temperatures under 10 °C. Thus it may be claimed that data obtained via the performed entomologic sampling confirm eventual assumption about abundant occurrence of *Diptera* during composting. It is evident that amount of *Diptera* can, mainly during summer months, reach considerable values, namely at premises of industrial composting plants during processing so-called food waste or muck.

6. Sanitation options

Based on findings of increased and undesirable occurrence of *Diptera* at places where biowaste is being processed via composting, it is necessary to take such measurements which would eliminate or substantially reduce this feature. Mainly it is essential to avoid penetrating of adult individuals to the collected biowaste. This will avert the risk of transmission of microbial pathogens on insect bodies into surrounding parts of the environment, simultaneously it limits the possibility of laying insect eggs.

All containers in which biowaste is transported into a composting plant have to air-proof lockable and sealed and they have to be sanitized properly in exactly given intervals. This requires suitable logistic plans and operating instructions and attend to their following, together with operational alternations (e.g. according to the development of meteorological conditions, etc.). It is obvious that transmission of biopathogens via insect can normally happen during further operational processes too (biowaste grinding, basing dumps, etc.). Sanitization of all technological devices (machinery, tools, and vehicles) therefore appears to be necessary. However, sanitization itself is focused only on elimination of microbial pathogens concentration in biowaste and substrates and does not concern presence of insect in the premises of a composting plant or in the surroundings of a home composting unit. Therefore it is required to ensure a suitable disinsection method too. Disinsection methods are generally divided into precautionary (prophylactic), repressive (exterminating) and integrated. The above mentioned technical precautions and sanitization can be considered a part of group 1 because they help to create such environment state where insect reproduction is limited. Repressive methods can be mechanical (flycatchers, traps, collectors and so on). This procedure is long and ineffective. Even the application of the most modern methods (collectors with pheromones and other attractants) did not make this method more effective, therefore they are used only for obtaining entomological samples. An interesting and usable in smaller composting plants repressive method can be treatment of biowaste or substrates with hot steam which kills insect eggs, larvae and imagoes. In the conditions of home composting unit, shedding the base with boiling water can be recommended as realistic because the boiling temperature is lethal for all development stages of insect.

Due to undesirable secondary effects on environmental layers and on human health and also to negative impacts on life cycle of necessary microorganisms in substrate, direct application of chemical insecticidal substances into compost cannot be recommended. Also coating of compost site walls or fencing the composting plant or spraying collecting vessels with repellents are not suitable due to their toxicity. On the contrary, disinsection performed using biological methods during which natural enemies of insect species are used is optimal. In case of *Diptera* these are parasitic wasps of *Spalangia* L. Also application of simple mechanical means, e.g. covering the base and substrates with permeable technical cloths or fine netting is able to prevent penetrating insect. This protection method appears to be suitable in conditions of home composting unit. Also adding fresh-cut grass to biowaste is useful because fresh-cut grass decreases prosperous life conditions for pathogenic microorganisms and neither *Diptera* which seeks decaying organic substances is attracted to a base treated this way. This simple method can effectively reduce the presented environmental risks.

7. Conclusion

Emissions of bio-aerosols and transmission of infectious diseases agents via *Diptera* can be considered significant environmental risk threatening not only workers at the processing plants in question, but also residents, animals and plants in the surroundings. This fact gives rise the need to prevent insect penetrating into compost and necessity to implement disinsection and protection interventions. For the present, neither precepts of law, nor technical and methodological regulations do not make provisions for such risk and only specify limits of concentrations of chosen microbial pathogens in composting substrate. They do not concern with expositions of bio –pathogens into atmosphere of composting plants, neither have they determined mandatory regulations for eliminating their spread via insect.

Even though entire elimination of the environmental risk in question is due to the character of the risk factor – i.e. *Diptera* – fundamentally impossible, performing protection and elimination precautions during composting biowaste is absolutely necessary because transmission of microbial pathogens via insect can be considered undoubtedly undesirable feature.

Acknowledgements

This paper was compiled within the Project of Specific University Research (SGS) no. SP2017/31 Processing of animal excrements in vermicompost.

References

[1] Awasthi M K, Pandey A K, Khan J, Bundela P S, Wong J W C and Selvam A 2014 Evaluation of thermophilic fungal consortium for organic municipal solid waste composting *Bioresour. Technol.* **168** pp 214–221
[2] Albrecht A, Witzenberger R, Bernzen U and Jäckel U 2007 Detection of airborne microbes in a composting facility by cultivation based and cultivation-independent methods *Ann. Agric. Environ. Med.* **14** pp 81-85
[3] Fischer G, Albrecht A, Jackel U and Kampfer P 2008 Analysis of airborne microorganisms, MVOC and odour in the surrounding of composting facilities and implications for future investigations *Int. J. Hyg. Environ. Health* **211** pp 132–142
[4] Zimová M and Matějů L 2014 *Kompostování odpadů a potencionální riziko mikrobiální kontaminace* (Praha: Státní zdravotní ústav v Praze) p 5
[5] Kůdela V 2012 Úloha rostlin v epidemiích způsobovaných toxigenními kmeny střevní bakterie Escherichia coli *Živa* **3** pp 110-112
[6] Schindler J 2010 *Mikrobiologie pro studenty zdravotnických škol* (Praha: Grada Publishing, a.s.) p 224
[7] Ltibke U 2016 *Microbial Inoculants for Controlled Composting of Organics Materials* (Peuerbach: Controlled Microbial Composting Group, Austria) p 4
[8] Herr C E, Zun Nieden A, Stilianakis N I and Eikman T F 2004 Health Effects Assiciated with Exposure to residential Organic dust *Am. J. Ind. Med.* **46** pp 381-385
[9] Hudec K, Kolibáč J, Laštůvka Z and Peňás M 2007 *Příroda České republiky-průvodce faunou* (Praha: Academia) p 440
[10] Mai M and Amendt J 2012 Effect of different post-feeding intervals on the total time of development of the blafly Lucilia sericata (Diptera: Calliphorodae) *Forensic Science International* pp 65-9

[11] Genard Zuha R, Razak M, Ahmad T A and Omar B 2012 Interaction effects of temperature and food on the development of forensically important fly, Megaselia scalaris (Loew) (Diptera: Phoridae) *Parasitology Research* p 2187.

[12] Byrd J H and Castner J L 2010 *Forensis Entomology: The Utility of Arthropods in Legal Investigations* (BOCA Raton: CRC Press) p 681.

[13] Laštůvka Z et al. 2001 *Zoologie pro zemědělce a lesníky* (Brno: Konvoj) p 267

[14] Showler A T and Osbrink L A 2015 Stable Fly, Somoxys calcitrans (L.). Dispersal and Governing factors *International Journal of Insect Science* **7** pp 19-25

[15] Javorek V 1967 *Kapesní atlas dvoukřídlého hmyzu* (Praha: Státní pedagogické nakladatelství) p 269.

Hydro morphological assessment of the Porubka river

J Novakova[1], I Melcakova[1], H Svehlakova[1], L Marcakova[1], T Matejova[1] and L Klimsa[1]

[1]VSB – Technical University Ostrava, 17. Listopadu 15/2172, 708 33 Ostrava-Poruba, Czech republic

E-mail: jana.novakova@vsb.cz

Abstract: Results of monitoring hydromorphological assessment of the Porubka watercourse, use the method HEM - Hydroecological monitoring, are presented in this article. HEM is based of the evaluation not only of main stream, as morphology and hydrology parameters, but of the same evaluation flood plain. The ecological status of river ecosystems is usually evaluated following hydromorphological features of riverbed, flow regime and river continuity, and was applied for the effect of proposed restoration measures. Ecohydrological quality by the method HEM classes was established from 1 (high) to 5 (bad), characterizing the state of the water course.

1. Introduction

The issue of hydro morphological assessment of watercourses is based on scientific disciplines such as fluvial geomorphology, whose foundations are in the monograph Demek [1]. Today, the authors follow him as Králová [2] ; Šindlar [3,4] ; Hradecký [5] ; Máčka [6] and foreign authors as Rosgen [7]; Goudie [8]; Naiman [9]. The Czech Ministry of the Environment approved individual binding methodologies to monitor and evaluate the components of surface water ecological status: Kokeš, Němejcová [10]; and Langhammer [11,12]. The European Water Framework Directive (WFD) became the fundamental basis for any water policy-related action of the European Community[13].

The method Langhammer [11,12] is complex assessment method integrating morphological, hydrological elements, of river basin and of flood plain for evaluating eco hydrological quality river and its basin. The analysis of hydrological alterations that occur at shorter time scales, that have very important effects on ecological communities.

2. Materials and methods

2.1. Locality selection

Hydromorphological monitoring data are collected in case study areas river basin Porubka. The Porubka River is one of the smaller tributaries of the left-middle portion of the Odra. It ends into the Odra in the city of Ostrava-Svinov at an altitude of 210m above sea level, and rises in 390m below the village Pusta Polom. The total length of the river Porubka, from it´s source to the mouth, is 20.1km. The length of mapping main stream, in case study areas, is 11.00 km, from 8.46 flow km to 19.46 flow km. It was mapping 29 river sections, were selected according the method Langhammer [11,12]. There are the key criteria for mapping homogeneity of river sections, as typology of watercourse and channel route and the character of the riparian zone and the floodplain. The gradient changes were selected as a new river section.

2.1.1. The Porubka watercourse in select section 016, HOD_0170_POR_016 (cadastral area Dolni Lhota – Ostrava)

The section 016 (figure 1) (the 16th sections from the source Porubka, upper section in river basin) has the length section 550m, because the change homogeneity is after 550 m. Channel flows parallel to the tram line, in this section watercourse branches two parallel channels (016a and 016b) and its left branch supplies the ponds system. The land use in this section are meadows, line vegetation, forests.

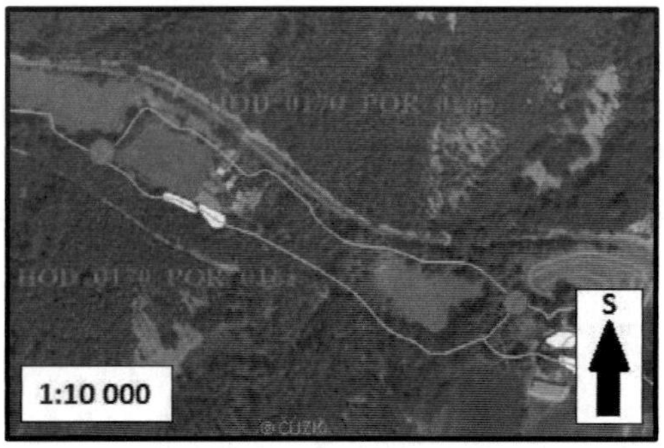

Figure 1. Locality select section 016 - 016a and 016b, (geoportal.gov.cz, 2015).

2.2. Methodology HEM

HEM methodology considers 17 parameters in 3 mapped zones (channel, bank zone, inundated land). An overview of parameters assessed according to HEM, weight of individual parameters and zonation, see Table 2. The different weight is set to represent the impact of parameters that are of major importance to hydromorphology of rivers. Then the overall hydromorphological quality value for each reach is calculated as the arithmetic mean of 3 zones.

2.2.1. Laboratory practice

These assignments were suggested for laboratory practice (inside). Preference a suitable watercourse, evaluating significance, diversity, availability, overall status of suitable river. To assign of the watercourse type according to the Langhammer methodology [14], which is combination of 4 parameters (sea level, altitude, geological subsoil, Strahler stream order - numerical measure of its branching complexity). Its necessary theoretical determine of the river sections, according the river length in topography maps (1:10000) of the case study areas: minimum size of the river length is 100m, for a small watercourses, the channel width up to 10m. Theoretical preparation for stream evaluation (e.g. river curvature or sinuosity or braiding), use of distance data (e.g. historical maps of I and III Military mapping, orthophotomaps, geoportals) [11,12].

2.2.2. Terrain practice

These assignments were suggested for terrain practice (outside). Terrain mapping is for verification of map data, measuring with GPS (eTrex Vista HCx, Garmin, and laser rangefinder (PMDL 5 A1 Multifunction Detector), camera (Olympus FE-46). Mapping to prescribed forms, these forms are in the method HEM [11,12]. Prescribed forms design morphological parameters, e.g. channel route (TRA), depth variations (VSK), type of profile (VHL) , depth of profile (VHP), width variation (BMK), substrate type (DNS), bed-fixing (UDN), dead wood (MDK), special bottom structures (STD),bank impairments (UBR), bank vegetation (BVG), land use in flood plain (VNI).Prescribed forms design hydrological parameters, e.g. flow variations (PRO), hydrological regime discharge (OHR). Necessary to design river continuity, e.g. migration barriers (PPK) and flood plain barriers (PIN) [11,12].

Hydro morphological characteristics, by method HEM, All 29 river sections were mapping from May 2015 to November 2016.

Table 1 (complete 17 parameters, e.g. morphological conditions,e.g. hydrological regime, e.g.continuity equation) were mapped in three zones of the Porubka watercourse, as channel, riparian zone - outside banks and widens its valley, respectively flood plain. The selected parameters are evaluated for the right bank and the same for the left bank. The mapping results must be written to prescribed forms [11,12]. All 29 river sections were mapping from May 2015 to November 2016.

Table 1. Mapping parameters of the method HEM, modified [11,12].

Zone	Parameter	RSV Quality Components
Channel	Channel way adaptation (TRA)	morphological conditions
	Variability channel width (VSK)	morphological conditions
	Variability recess longitudinal profile (VHL)	morphological conditions
	Variability depths cross profile (VHP)	morphological conditions
	Bottom sediment (DNS)	morphological conditions
	Bottom upgrade (UDN)	morphological conditions
	Dead wood in channel (MDK)	morphological conditions
	Bottom stuctures (STD)	morphological conditions
	Flow stream (PRO)	hydrological regime
	Hydrological regime influence (OHR)	hydrological regime
	Longitudinal permeability (PPK)	continuity equation
Bank zone	Bank management (UBR)	morphological conditions
	Bank vegetation (BVG)	morphological conditions
	Land use outside bank (VPZ)	morphological conditions
Inundated land	Land use in flood plain (VNI)	morphological conditions
	Permeability flood plain (PIN)	continuity equation
	Bank stability and lateral migration of the channel (BMK)	morphological conditions

2.2.3. Evaluation methodology

The method HEM calculated as the weighted mean of all parameters, as morphological conditions, as hydrological parameters, as continuity equation. The different weight is set to represent the impact of parameters that are of major importance to hydromorphology of rivers. The quality of the river section is calculated as HMS (hydromorphological score), see equation (1), weighted average of individual indicators. The methodology accurately determines either universal or specific evaluation matrices (universal for all types of watercourses, specific to the typology of the river stream, e.g. kdns_typ, matrix value based on the river stream type of the bottom substrate. For parameters where left bank and right bank are evaluated in particular, the worst score value is used higher number. The weights of individual pointers for different watercourse typology groups are set based on expert estimation and subsequent calibration, based on sample flow mapping of each type [11,12]. The weight variability for weight scoring shows the importance of each parameter.

Calculation of hydromorphological score **HMS**

$$HMS = \frac{\begin{array}{c} TRA*ktra_{typ}+VSK*kvsk_{typ}+VHL*kvhl_{typ}+VHP*kvhp_{typ}+ \\ DNS*kdns_{typ}+UDN*kudn_{typ}+MDK*kmdk_{typ}+STD*kstd\ \ typ+ \\ PRO*kpro_{typ}+OHR*kohr_{typ}+PPK*kppk_{typ}+UBR*kubr_{typ}+ \\ BVG*kbvg\_typ+VPZ*kvpz\_typ+VNI*kvni\_typ+PIN*kpin\_typ+BMK*kbmk\_typ \end{array}}{4} \quad (1)$$

$$HMK_{VU} = \frac{\sum_{i=1}^{n} HMSi*Li}{\sum_{i=1}^{n} Li} \quad (2)$$

HMKVU is the hydro morphological quality of the watercourse, see equation (2), which is calculated as the weighted mean of the HMS value but the weight is channel width (L) and (n) is the number of the assessed sections within the water[14]. The overall state of the watercourse, water body is ranked in the grade of 1-5, where the score≥1 <1.5: close to nature (grade 1), ≥1,5 <2,5: poorly modified (grade 2), ≥2.5 <3.5: moderately modified (grade 3), ≥3.5 <4.5: highly modified (grade 4), ≥4.5 <5: strongly modified (grade 5) [11,12].

3. Results

The testing of this methodology HEM, took place in the Porubka watercourse, at 8.46 flow km to 19.46 flow km, overall 29 sections with a gradient of change of the above mentioned parameters were mapped. The type of watercourse is 222, rivers of medium heights (200-500m.n.m), the Baltic, the sedimentary rocks (sandstone, claystone, quaternary).A sample of mapping and scoring of the Porubka watercourse is shown in the example of the section 016, see Table 2. It isn't possible to describe all results, because there's limit length of the article.

Table 2. Example of the scoring section 016 (016a and 016b), of the Porubka watercourse.

Section number			016a	016b
Lenght [m]			548	550
Zone	Parameter	Weight		
	TRA	1	1	1
	VSK	0.10	1	1
	VHL	0.10	1	2
	VHP	0.10	2	2
	DNS	0.10	2	2
Channel way	UDN	0.25	1	1
	MDK	0.10	2	3
	STD	0,15	3	3
	PRO	0.10	1	1
	OHR	0.10	1	1
	PPK	0.50	1	2
Riparian zone	UBR	0,25	1	3
	BVG	0.15	2	2
	VPZ	0,40	2	2
	VNI	0.30	2	2
Inundated land	PIN	0.15	1	5
	BMK	0.15	2	3
HMS			1.400	1.888

HMSPOR_016a, section 016a, has a value of 1,400, see Table 2, which puts in the first class, its status is close to nature. HMSPOR_016b, section 016b, has a value of 1,888, see Table 2**Chyba! Nenašiel sa žiaden zdroj odkazov.**, which puts in the second class, its state poorly modified. The worse quality is given by the parallel tramway, respectively limitation of the floodplain and water management of the right bank. The best results were given in individual zones, respectively channel and the Porubka planform, classified in quality class 1 and quality class 2. The worst quality status is on the banks, where all the quality classes are represented as well as class 5, because the banks impairments. If it was scored by scoring of individual sections, the worst quality class is 3, in the intravilan. 13 sections are assessed by quality class 1. The Porubka watercourse is generally evaluated by the second class of hydro morphological quality (HMS), with a total length of 11.3 r.km [15].

4. Conclusions

The problem of the methodology [11], is with the scoring assessment, where the quality class is up to 1 point higher, it is necessary to perform field practice. That are time consuming, because the seemingly the same sections, e.g. extravillan, may have different weight parameters that affect the quality of these parameters. Due to the size of the article, it is not possible to describe all in detail. It is possible to simplify such a comprehensive methodology by summarizing the positive and negative aspects of this methodology.

Positive aspects are: Implementation of legislation and standards, already mentioned above. Mapping form is brief, clear. Well-handed manual for prescribes forms. Calculation of the score uses weighted average method. Negative aspects are: Name of the methodology, it isn't hydro-ecological monitoring but its suitable title as hydro morphological monitoring, because this method maps mainly hydro morphological quality. Mapping period recommended according to methodology (September-October), but to map hydrological regime inappropriate due to minimum depths. Mapping parameter (only supplemental) as invasive species poorly identifiable, some species are without inflorescence. Estimates of some parameters – e.g. depth variability can´t be estimated, but must be measured. Determination of bottom substrate fractions is required by grain analysis. Scoring principle - according to the methodology the higher the number of categories, the higher the quality, which may not be relevant. For example, natural flows may not be such variability and the quality of morphological and hydrological parameters is high. HMS has RSV significance in the environmental assessment of watercourse quality. This methodology, mainly in the mapping form, doesn't distinguish small river as torrent. The main objective of this methodology is the creation of documents for revitalization studies, as part of flood control measures. The method seems to be applied to all water bodies at least in CR. The Czech HEM method that additionally comply with the requirements of the European Standard EN 14614 [16].

References

[1] Demek J 1988 *Obecná geomorfologie (*Praha: Academia) p 480
[2] Králová H et al. 2001 *Řeky pro život. Revitalizace řek a péče o nivní biotopy* (Brno: ZO ČSOP Veronica) p 440
[3] Šindlar M and Vlček L 2002 Geomorfologické typy vodních toků a jejich využití pro revitalizace *Vodní hospodářství* **52** pp 172-176
[4] Šindlar M 2008 Přírodě blízká protipovodňová opatření na tocích a v nivách *Stavby vodního hospodářství a krajinné inženýrství* p 80
[5] Hradecký J 2016 *Landscapes and Landforms of the Czech Republic* World Geomorphological Landscapes (Switzerland: Springer) pp 291-304
[6] Máčka Z 2016 *Litovelské Pomoraví - Landscape around anastomosing river* World Geomorphological Landscapes (Switzerland: Springer) pp 291-304
[7] Rosgen D L 1994 *A classification of natural rivers* (USA: Catena) pp 169-199
[8] Goudie A 2004 *Encyclopedia of geomorphology* (London: Routledge) p 578
[9] Naiman R 2006 *Riparia. Ecology, Conservation, and Management of Streamside* Communities pp 430
[10] Kokeš J and Němejcová D 2006 *Metodika odběru a zpracování vzorku makrozoobentosu tekoucích vod metodou PERLA* (Praha) p 10
[11] Langhammer J 2014 *Metodika monitoringu hydromorfologických ukazatelů ekologické kvality vodních toků* (Praha: MŽP ČR) p 72 http://www.mzp.cz/cz/prehled_akceptovanych _metodik_tekoucich_vod
[12] Langhammer J et al. 2014 *Metodika typově specifického hodnocení hydromorfologických ukazatelů ekologické kvality vodních toků.* (Praha: MŽP ČR) http://www.mzp.cz/cz/prehled_akceptovanych_metodik_tekoucich_vod
[13] WFD 2000 Water Framework Directive 2000, 2000/60/EC. 327 p 1–72
[14] Langhammer J et al 2009 *Vymezení typů vodních toků* (Praha: Univerzita Karlova v Praze) p 29
[15] Marčáková L 2016 *Testovanie metodiky hem na vodnom toku porubka.* (Ostrava: VŠB-TUO) p 93
[16] ČSN EN 14 614: 2005*Jakost vod – návod pro hodnocení hydromorfologických charakteristik řek*

Development of composting plant rock surrounding pollution. Case study – Točna in Příbor, The Czech Republic

I Ondrasikova[1] **and L Stancl**[1]

[1] AZ GEO s.r.o., člen skupiny Valbek, Kořenského 1262/40, 703 00 Ostrava, The Czech Republic

Email: ivana.ondrasikova@azgeo.cz

Abstract. Composting plant Točna utilize and modifies biologically biodegradable waste by aerobic fermentation method. After emergency leakage of leachate water in 2010 the quality of rock surrounding is being regularly monitored in the range of indicators: pH, conductivity, COD (chemical oxygen demand, Cr), HCO_3^-, Cl^-, $N\text{-}NH_4^+$, $N\text{-}NO_3^-$, $N\text{-}NO_2^-$ and humic substances. Technical adjustments of the interest area has been also made to prevent leakage of contaminated water. Locality monitoring system is ensured separately for shallow backfill aquifer and deeper quaternary aquifer. Protective hydraulic barrier is running at the same time to prevent effluent of shallow water to groundwater. Nitrogenous substances (above all ammonium ions) and humic substances belongs among main groundwater contaminants of the interest area. Peak concentration level of this substances is connected with shallow backfill aquifer, near south-east and east edge of the locality, not in general quaternary water direction. From long-term monitoring results follows that concentration of monitored substances is gradually decreasing, especially in water connected with backfill layer. Drier weather of last year helps to this development. Quality of quaternary aquifer is not influenced significantly, only locally in spots, in the places of both aquifers connection where increase of ammonium ions occurs as result of humic substance decomposition. In effect, monitored substances are not disseminated in quaternary aquifer.

1. Introduction

Composting plant Točna utilize and modifies biologically biodegradable waste by aerobic fermentation method. Fermentation area of composting plant is made of continuous paved asphalt concrete surface of c. 3 000 m^2. The area is declined to the gathering tanks where liquid products of fermentation (leakages) and rainfall water coming from area are collected.

Frequent leakages of dark brown water coming from shallow rock surrounding onto the surface used to occur in the past (before 2009). Brown water appeared in the trough which was designed to collect rainwater and is located along the entrance road leading to the composting plant. Brown water outflowed from the through to the road ditch and to natural surroundings.

Remedial measures included building of the monitoring system, introduction of the regular monitoring of groundwater quality in the range of pollution indicators: pH, conductivity, COD (chemical oxygen demand, Cr), HCO_3^-, Cl^-, $N\text{-}NH_4^+$, $N\text{-}NO_3^-$, $N\text{-}NO_2^-$ and humic substances, the operation of hydraulic barrier and partial reconstructive works.

The monitoring system of the locality was built in 2009 (Zakova, 2009) and consists of PN boreholes (no. 1 to 7) for shallow backfill aquifer and from PK boreholes (no. 1 to 4) for deeper quaternary aquifer. Protective hydraulic barrier is being operated in PN-1 borehole. Due to the shallow groundwater pumping, the absence of shallow water leakage into the ditch and spreading of the water onto the terrain surface is ensured. Reconstructive technical works of the area and of the tanks have been made during the year of 2010 to 2015 which ensured tightness of the surface of the composting area and of the leachate water retaining tanks.

The interest area is located in the Czech Republic, in Moravian-Silesian region, at the southwestern edge of the town Příbor, in the area of the former agricultural company. The terrain surface is plane. Altitude varies from 296 to 299 m above the sea level. The nearest recipient of the surface water is a nameless left-side inflow of Lubina that flows c. 300 m eastwards the locality. The interest area layout with its monitoring system is stated in the following picture.

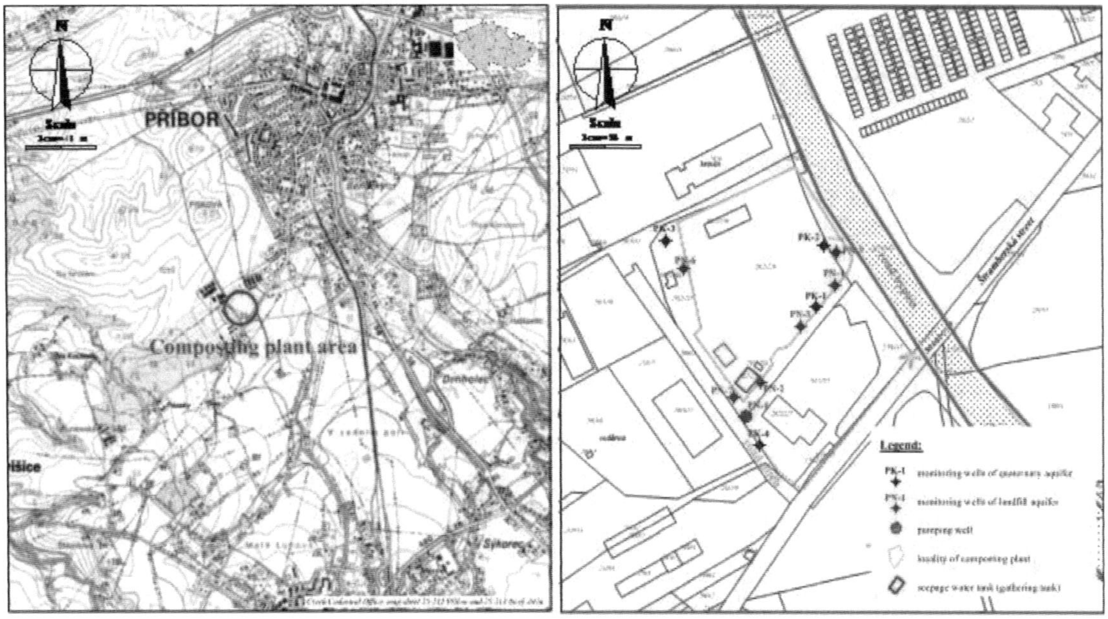

Figure 1. Locality of composting plant with monitoring wells.

2. Geological and hydrogeological locality conditions

The following geological and hydrogeological facts have been verified by investigation works implemented to design remedial measures (Ondrasikova, 2008, 2010; Zakova, 2009).

- The area under the composting plant was excavated (brick clay mining) and filled by the material of unspecified type as well as the near old landfill 'Na Točně'. Original locality geological sequence is: humous loam of 0,4m thickness in the top, clayey sandy loam layer of c. 3.0 m thickness, clayey sandy gravel layer of c. 3,0 to 3,5m thickness and finally surface of the basement rock at about 7m under the terrain (claystone of frýdecká layer of the late Cretaceous).
- Heterogeneous backfill layer of stony character with admixture of the slag, bricks, concrete and gravel has been verified in the area of the composting plant locality. The backfill layer is locally covered by discontinuous clayey backfill layer which is very often water - bearing by highly polluted water corresponding to the seepage water of the composted material. Original rusty brown loess loam was verified below backfill layer. Original clayey loam was removed in the Eastern part of the composting plant (around PK-1 and PN-3 boreholes). Backfill is located just on the permeable sandy gravel layer of Saale glaciation. The clayey layer separates backfill and quaternary aquifer.
- Hydraulic conductivity of backfill material was determined in the range of 2.10^{-4} to 4.10^{-7} $m.s^{-1}$ that shows to rather high to weak permeability of the material. According to its granularity,

bottom loess loam layer corresponds to clay of 1.10^{-9} $m.s^{-1}$ hydraulic conductivity of slightly permeable surroundings. On the basis of the express pumping tests results, PN-1 borehole yield was determined at the level of c. 0,05-0,1 l/s, that is c. 4,3 to 8,6 m^3 of contaminated water per day.

- General groundwater flow direction is from southwest to northeast, towards the Lubina river valley that creates local erosion base. Maximum groundwater level in shallow aquifer is connected with the borehole PN-3 and its surroundings (about 297,5 m above the sea level). Groundwater flow direction of the backfill layer goes rather from southwest to northwest. According to the geological profile of the interest area, both aquifers are connected. Overflow from backfill to quaternary aquifer occurs in the area of PK-1 and PN-3 boreholes. Due to the construction of the bypass road in 2011, PN-4 and PN-5 boreholes and partially earth cut along the original road were dewatered. Groundwater level is closely connected with the amount of precipitation in the interest area.

3. Methodology of work – field measurements, analytical method and evaluation

Groundwater sampling was carried out 3 times a year – in the spring, before the beginning of composting, during intensive operation in summer and after composting conclusion in autumn. During evaluation the data from 2010 to 2016 were implemented (Ondrasikova 2010-2016). Long-term evaluation of rock surrounding quality is focused on humic substances and ammonium ions.

Sampling was implemented in accordance with the relevant standards and regulations. Groundwater samples were taken dynamically. Groundwater level was measured (an acoustic level meter G-20) before pumping in every well. Physio – chemical parameters of groundwater were observed during pumping. Measurement was made by field measuring device CyberScan in the range of pH, conductivity, temperature and total dissolved solids. Samples were taken to standard samplers after physio – chemical parameters stabilization or after drawing off of 3 volumes of borehole water. The analyses of groundwater and surface water samples were made by ALS Laboratory Group, ALS Czech Republic s.r.o. accredited by Czech Accreditation Institute. The methodology of indicators determination was:

- dissolved substances 105°C by gravimetry
- chlorides, bicarbonates by titration
- COD (chemical oxygen demand, Cr), Biochemical oxygen demand by titration
- Fe atomic absorption spectrometry
- $N\text{-}NH_4^+$, $N\text{-}NO_3^-$, $N\text{-}NO_2^-$ photometrically
- N-org by spectrophotometry
- humic substances by spectrophotometry

4. Results and Discussion

The primary contaminants of the investigated area were identified on the basis of seepage water quality that originates from the composting plant (Zakova, 2009). Values of chosen seepage water indicators are stated in following table. Seepage water has high content of humic substances, bicarbonates, ammonium ions, chlorides and potassium. Reaction of water is neutral to slightly alkaline.

Specific values of some substances occurring in seepage water from composting plant is the result of the used composting technology. Increased content of potassium mostly indicates existence of animal waste that can be added to improve final ratio of carbon / nitrogen in the compost. The value of pH should vary from 6 to 8 to ensure the activity of microorganisms. Adjustment of acid pH that is typical for organic matter mineralization is commonly implemented by burned lime addition.

Table 1. Selected indicators of leachate water.

Indicator	Concentration
humic substances	11 100 mg/l
HCO_3^-	3 965 mg/l
NH_4^+	170 mg/l
Cl^-	1 277 mg/l
SO_4^{2-}	316 mg/l
K	3 140 mg/l
Ca	300,6 mg/l
Fe	21.5 mg/l
Mg	109.44 mg/l
pH	7.6
conductivity	1073 mS/m

The highest concentration of nitrogen and humic substances that represent main locality contaminants is connected with PN-1, PN-2 and PN-3 boreholes area, i.e. surrounding of shallow backfill aquifer. Concentration of humic substances exceeded 1000 mg/l and of ammonium ions exceeded 10 mg/l in the past. At the present, concentration of humic substances varies from units to first tens of mg/l, occasionally exceeds 100 mg/l. Concentration of ammonium ions decreased to units of mg/l.

Long-term development of humic substances (PN-2) is shown in the following picture, concentrations of selected indicators of ground water (PN-2) are stated in following table.

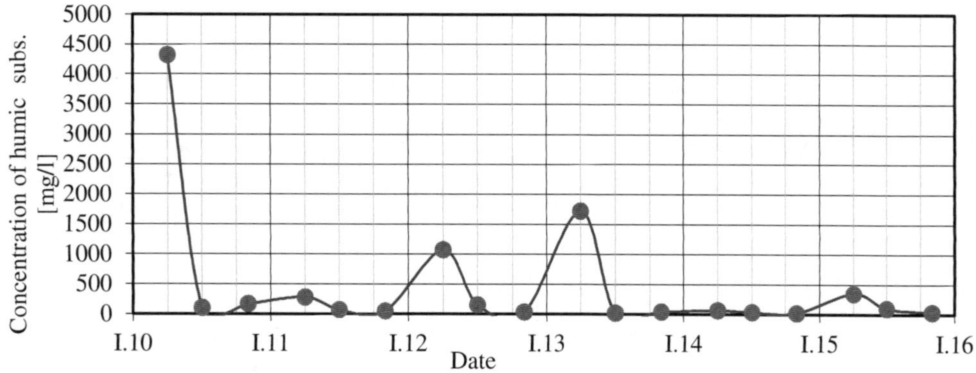

Figure 2. Long-term development of humic substances in groundwater (PN-2).

Character of backfill shallow water is influenced by increased surface rainfall recharge in the area of PN-1, PN-2 and PN-3 boreholes. It was observed via water temperature changes. The temperature varies from 6 to 25 degrees of Celsius with maximum level in July and minimum in April. Temperature of quaternary aquifer is more balanced and ranges from 9 to 14 degrees centigrade. From the long-term point of view we can observe declining concentration of almost all indicators in shallow water except for HCO_3^- and pH. Occasionally increased values of primary contaminants can be observed in connection with higher precipitation.

Humic substances almost do not occur in groundwater of quaternary aquifer. Only occasionally its concentration exceeds 10 mg/l. In the contrary, high content of ammonium ions occurs in quaternary aquifer. Its level is commonly around units of mg/l, locally about 10 mg/l (PK-2 borehole). Chemism change of the quaternary aquifer can signal degradation of humic substances whose partial degradation of side chains can cause the appearance of ammonium ions. Ammonia nitrogen is final product of mineralization of the plant matter. Further in the direction of the ground water flow, ammonia ions are subject of nitrification in oxidizing environment.

Table 2. Selected indicators of ground water (PN-2).

	Concentration [mg/l]				
Date	Cl^-	HCO_3^-	$N\text{-}NH_4$	N-org	Humic s.
27.4.2010	340	945	60.2	65.5	4320
5.4.2011	70	378	8.7	9.53	282
11.4.2012	132	752	13.7	33.6	1070
15.4.2013	189	922	10.3	25.4	1720
2.4.2014	45.2	595	0.76	7.1	63.9
13.4.2015	87.8	846	14.2	30.6	347
21.4.2016	136	444	19.5	19.7	<1.0

5. Conclusions

There are two hydraulic systems, backfill and quaternary aquifer in the composting plant Točna. General groundwater flow direction of the quaternary aquifer is towards northeast. Within backfill aquifer, the highest groundwater level is situated around the PN-3 borehole where both, backfill and quaternary aquifer, are in contact and there is the occurrence of overflow to rock surrounding due to the increased infiltration.

Nitrogen substances (especially ammonium ions) and humic substances belong among the main groundwater contaminants in the interest area. Maximum concentration level of these substances is located around the PN-1, PN-2, PN-3, PK-1 and PK-2 boreholes, i.e. at the southeast and east edge of the composting plant locality. According to long-term results of groundwater monitoring, there is gradual reduction of observed concentrations, especially in the water of backfill layer. The influence on the quality of quaternary aquifer is not distinct, just in particular spots situated in the contact of both aquifers (PK-1 and PK-2) where ammonium ions are increased (decomposition of the humic substances). The observed substances almost do not disseminate in the quaternary aquifer. The amount of water leaking onto to terrain surface is distinctly limited by the hydraulic barrier (groundwater pumping in PN-1 borehole).

It is obvious that the situation in situ has gradually been stabilizing which is considerably supported by drier weather conditions in recent years. Thanks to implementation of the technical measures, there is no further groundwater pollution. We can observe gradual improvement of the groundwater quality in the rock surroundings.

6. Acknowledments

We would like to thank the city of Kopřivnice and Příbor for lending us the measured data it owns for the purposes of the study.

7. References

[1] Ondrasikova I 2008 Příbor-Composting plant–proposal of remedial measures (Ostrava: AZ GEO s.r.o.)
[2] Ondrasikova I 2010-2016 Příbor-Composting plant-monitoring (Ostrava, AZ GEO s.r.o.)
[3] Zakova L 2009 Příbor-Composting plant-system of monitoring (Ostrava: Unigeo a.s.)

IOP Conf. Series: Earth and Environmental Science **92** (2017) 012048 doi:10.1088/1755-1315/92/1/012048

Leaching of Ca, Si, Fe and Al from concretes, based on sulphate resistant cement, due to bacterial attack - a correlation study

V Ondrejka Harbulakova[1], A Estokova[1], A Luptakova[2] and M Kovalcikova[1]

[1]Technical University of Kosice, Faculty of Civil Engineering, Institute of Environmental Engineering, Kosice, Slovakia
[2]Slovak Academy of Science, Institute of Geotechnics, Kosice, Slovakia

Email: vlasta.harbulakova@gmail.com

Abstract. Investigation of biological corrosion of concrete requires that composition of material, surface and aggressive environment have to be taken into account. Sulphate resisting cement, as one of the resistivity improving factors, was used for preparation of concrete samples studied in this experiment. Dependences of leached-out quantities of selected ions regarding bacterial versus non-bacterial environments were investigated and evaluated using correlation analysis. Leaching trends of calcium, silicon and aluminium strongly depends on the aggressiveness of medium which are the samples exposed to. A weak correlation was found for the leaching trends of aluminium and iron in both bacterial and non-bacterial media.

1. Introduction

Presence of microorganisms in sewer systems is associated with several problems, including microbial corrosion of concrete, release of obnoxious odours to the urban atmosphere and emissions of sulphide gas to sewer workers [1,2,3]. Formation of biological induced corrosion of concretes is linked with bacteria *Acidithiobacillus thiooxidans*, identified in the sewers, which are responsible for sulphate attack and consequently, for the deterioration of concrete matrix [4]. The sulphate resistant cement is applied in concrete structures in order to increase the resistivity against the sulphate attack and durability of concretes.

Deterioration of concretes due to bacterial sulphate attack is often manifested by leaching of the main concrete components representing by calcium, silicon, iron, and aluminum compounds. Because sulphate attack on concrete is not characterized by a single chemical process, when considering the best preventive measures one has to take into consideration the processes involved in concrete production and the environmental conditions to which the concrete will be exposed throughout its service life. However, some basic rules apply in most cases. The most important among them are using of proper structural design, drainage, protective barriers/coatings, using of low w/c mixture proportions, adequate cement content, proper consolidation or using of low-calcium cements, including sulphate-resisting cements and cements containing pozzolans or slag [5]. The application of a multi proxy approach, including chemical, mineralogical, biological and isotopic analyses as a very powerful tool to investigate complex reactions induced by biogenic sulphuric acid attack was presented in paper by Grengg et al. [4,6]. According to Grengg et al. [7] *A. thiooxidans* are likely to dominate the oxygen rich zones close to the surface, while *A. ferrooxidans* are proposed to adopt the pore spaces within deeper, anaerobic, corrosion layers.

The results of calcium, silicon, iron and aluminium leaching under the influence of *Acidithiobacillus thiooxidans* bacteria from concrete composites during 90 days are presented in the paper. Dependences of leached-out quantities of selected ions regarding bacterial versus non-bacterial environments were investigated and evaluated using correlation analysis.

2. Materials and methods

2.1. Preparation of samples

Concrete samples were made of sulphate resisting cement CEM I 42.5 N-SR 0 produced with clinker with a very low tricalcium aluminate (C_3A) content. Concrete mixture contested of sulphate resisting cement (360kg), water (170L), three fractions of aggregates: 0/4 mm (825kg), 4/8 mm (235kg), and 8/16 mm (740kg) and plasticizer (3.1L) per 1 m^3 of fresh concrete. The proposed composition of mixtures resulted in water to cement ratio w/c = 0.47. Standardized concrete prisms with dimensions of 100 × 100 × 400 mm^3 were cured for 28 days in an aqueous environment prior to testing the required mechanical parameters. Consequently, the concrete samples were cut to prepare the test specimens of smaller dimensions. After removal of fine surface impurities by brush, the small specimens of dimensions approx. 50 × 50 × 20 mm^3 were immersed into distilled water for 2 hours, then dried on the air and subsequently immersed in ethanol for 24 hours to ensure sterile conditions. The samples were then placed on a sterile filter paper in an aseptic box. After drying for 2 hours at room temperature, the specimens were dried to a constant weight at 105 ° C using laboratory oven. Subsequently, samples of cement composites were immersed in the bacterial and reference media over a period of 90 days.

2.2. Experimental testing

Experimental simulation was focused on leaching the main components under biocorrosion caused by sulphur oxidizing bacteria *Acidithiobacillus thiooxidans* (samples marked as BM). Bacterial medium consisted of *A. thiooxidans* inoculum and cultivation medium in 1:4 ratio. Sulphur oxidizing bacteria *A. thiooxidans* were isolated from the mixed culture obtained from the mine water (the shaft Pech, the locality Smolník, Eastern Slovakia). The selective nutrient medium according to Waksman was used for the isolation and cultivation of bacteria at the temperature of 28 - 30°C and pH of 2.0 – 3.5 and as reference medium without bacteria as well (samples marked as CM).

Leaching trends of calcium, silicon, iron and aluminium ions measured in each liquid phase were studied by X-ray fluorescence analysis (XRF) using SPECTRO iQ II equipment (Ametek, Kleve, Germany) with a silicon drift detector (SDD) with a resolution of 145 eV at 10,000 pulses. The primary beam was polarized by Bragg crystal and Highly Ordered Pyrolytic Graphite—HOPG target. The sample measurement was performed at 180 s at a voltage of 25 kV with an applied current of 0.5 mA and 50 kV at 1.0 mA. The sample was flushed with helium. The intensities for all elements were corrected automatically for line interference and absorption effects due to all the other elements using the fundamental parameter method.

2.3. Correlation analysis

The leached-out amounts of Ca, Si, Fe and Al in milligrams per litre represented deterioration parameters have been used in the correlation analysis of the concrete's deterioration. Statistical method was used for evaluation of the trend of chemical element leaching, as well as for the description of a relation among the selected parameters. Information about the two-dimensional statistical data set gives a correlation coefficient R_{xy} [8], as shown in equation (1).

$$R_{xy} = \frac{n\sum_{i=1}^{n} x_i y_i - \left(\sum_{i=1}^{n} x_i\right)\left(\sum_{i=1}^{n} y_i\right)}{\sqrt{\left[n\sum_{i=1}^{n} x_i^2 - \left(\sum_{i=1}^{n} x_i\right)^2\right]\left[n\sum_{i=1}^{n} y_i^2 - \left(\sum_{i=1}^{n} y_i\right)^2\right]}} \quad (1)$$

The calculated R_{xy} values are in the interval <−1,1>. If $R_{xy} = 1$, the correlation is full linear; if $R_{xy} = -1$, then the correlation is inversely linear; and if $R_{xy} = 0$, the pairs of values are fully independent. Then, degree of the correlative closeness is defined as: medium, if $0.3 \leq |R_{xy}| < 0.5$; significant, if $0.5 \leq |R_{xy}| < 0.7$; high, if $0.7 \leq |R_{xy}| < 0.9$; and very high, if $0.9 \leq |R_{xy}|$. A high correlation means that two or more variables have a strong relationship with each other, while a low correlation means that the variables are hardly related. For the purposes of our assessment, the correlation coefficient was obtained using the function "Pearson" in Microsoft Excel.

3. Results and discussion

The concentrations of the main concrete′ components, measured by XRF analysis is given in table 1.

Table 1. Concentration of the main components of concrete.

	CaO	SiO_2	Fe_2O_3	Al_2O_3	MgO	P_2O_5	SO3	Cl	K_2O	TiO_2	MnO	other
[%]	14.49	62.81	5.51	6.94	1.61	0.06	0.16	0.01	1.1	0.33	0.08	6.90

The chemical composition of the concrete based on the sulphate resistant cement differed slightly from the concrete based on the ordinary Portland cement only [9]. The higher percentage of iron and silicon oxides has been probably connected to the differences in silicon and iron content in ordinary Portland cement (OPC) and sulphate resistant cement, respectively. In spite of the significant difference in aluminium content in mentioned binders, the almost same concentration at about 7 wt. % was observed in studied concrete samples regarding aluminium oxide than in OPC-based concretes.

The comparisons of the leached-out quantities of Ca, Si, Al, and Fe in bacterial and reference media each other are presented in the figures 1-4.

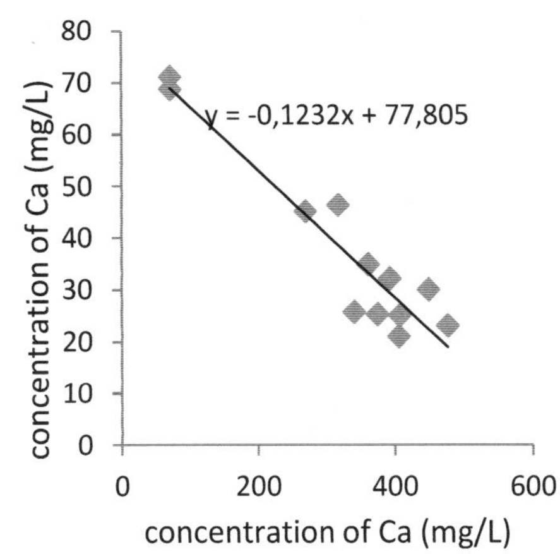

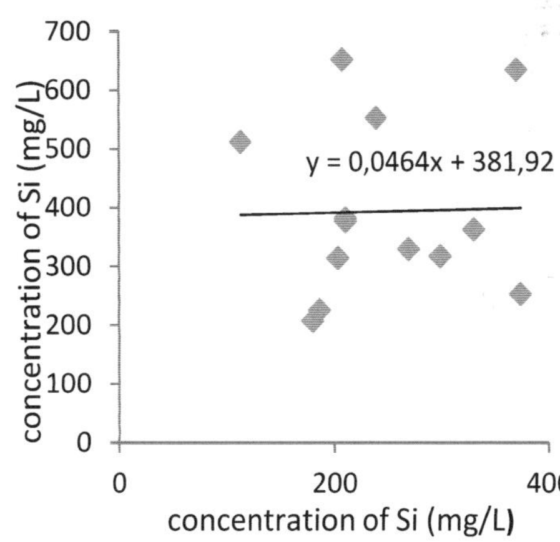

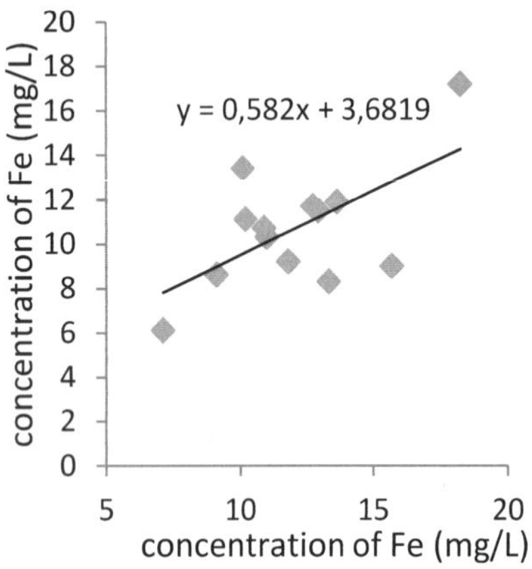

Figure 1-2. Functionality CaBM = f(CaCM) and Functionality $Fe_{BM} = f(Fe_{CM})$ (from up to down).

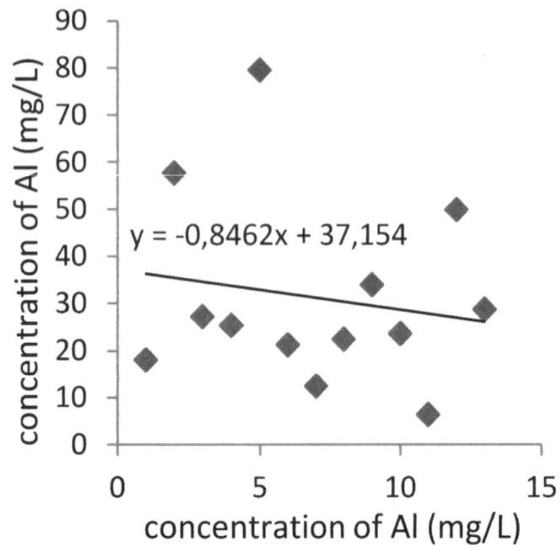

Figure 3-4. Functionality SiBM = f(SiCM) and Functionality $Al_{BM} = f(Al_{CM})$. (from up to down).

The X-axis presents the leached-out amounts of ions due to bacterial influence, in medium of *A. thiooxidans,* while the Y-axis gives the values of the leached-out quantities of ions in reference non-bacterial medium.

As it is seen in figures 1-4, no dependence regarding bacterial versus non-bacterial environments was found for the leaching trends of silicon and aluminium while significant and very high correlation was found for calcium and iron. The correlation coefficient regarding the relations mentioned, are given in table 2. The results of the correlation analysis point to the fact that the more calcium was dissolved due to bacterial attack the less calcium was dissolved in the reference medium. Our previous results confirmed that the bacterial exposure causes much more extensive leaching of the main concrete components. Based on the correlation analysis, according to our previous experimental findings [10,11] and other authors results [12], it can be concluded, that the leaching trends of calcium, silicon and aluminium strongly depends on the aggressiveness of medium which are the samples exposed to and the leaching mechanisms and behaviour of the ions in solutions differ when compared the bacterial and non-bacterial environments.

Table 2. Calculated correlation coefficients for the individual ions leaching.

	Ca_{BM}/Ca_{CM}	Si_{BM}/Si_{CM}	Fe_{BM}/Fe_{CM}	Al_{BM}/Al_{CM}
R_{xy}	-0.95	0.02	0.62	0.16

The leaching trend of iron seems to be independent on the medium, the leached-out quantities have been measured very low and oscillated in a very close concentration region 6 – 10 mg/L during whole experimental period.

The calculated Pearson correlation coefficients for the leaching trends Si/Ca and Al/Fe are reported in table 3.

Table 3. Correlation coefficients for the mutual relations of ions.

	correlation coefficient R_{xy}	
Ions	BM	CM
Si/Ca	0.17	0.05
Al/Fe	0.75	0.77

A weak correlation was found for the leaching trends of aluminium and iron in both bacterial and non-bacterial media. On the other hand, no linear correlation was found for the leaching behaviour of calcium and silicon, as assumed.

4. Conclusion

The paper aimed at the investigation of the concrete composites, based on the sulphate resistant cement, in terms of leaching the calcium, silicon, iron and aluminium compounds due to *Acidithiobacillus thiooxidans* exposure. The correlation analysis was used to evaluate the leaching trends of selected ions during the experimental period. The differences in leaching rates of calcium have been found when compared the bacterial and non-bacterial exposures. The same findings have been observed for silicon and aluminium.

Acknowledgements

This paper has been prepared within the Project of the Scientific Grant Agency of the Ministry of Education, science, research and sport of the Slovak Republic and the Slovak Academy of Sciences (VEGA) No. 2/0145/15.

References

[1] Wells T, Melchers RE 2015 Modelling concrete deterioration in sewers using theory and field observations *Cem Concr Res* **77** 82–96
[2] Senhadji Y et al. 2014 Influence of natural pozzolan, silica fume and limestone fine on strength, acid resistance and microstructure of mortar *Powder Technology* **254** 314–323
[3] Girardi F, Di Maggio R 2011 Resistance of concrete mixtures to cyclic sulfuric acid exposure and mixed sulfates: Effect of the type of aggregate *Cem Concr Compos* **33** 276–285
[4] Grengg C et al 2015 Microbiologically induced concrete corrosion: A case study from a combined sewer network *Cem Concr Res* **77** 16-25
[5] Boyd AJ, Skalny J 2007 Environmental deterioration of concrete chapter 5 WIT Transactions on State of the Art in Science and Engineering **28** 143-184
[6] Pietrucha-Urbanik K 2013 Multidimensional comparative analysis of water infrastructures differentiation Conference on Environmental Engineering IV Location: Lublin Poland 29-34
[7] Grengg C et al 2017 The decisive role of acidophilic bacteria in concrete sewer networks: A new model for fast progressing microbial concrete corrosion *Cem Concr Res* **101** 93-101
[8] Kreyszig E 2011 *Advanced Engineering Mathematics* (John Wiley and sons: United States of America)
[9] Estokova et al 2016 Testing Silica Fume-Based Concrete Composites under Chemical and Microbiological Sulfate Attacks *Materials* **9** (5) 1996-1944
[10] Ondrejka Harbulakova et al 2015 Using a Statistical Method for the Concrete Deterioration Assessment in Sulphate Environment *Chem Eng Trans* **43** 2221-2226
[11] Estokova et al 2011 Sulphur oxidizing bacteria as the causative factor of biocorrosion of concrete *Chem Eng Trans* **24** 1-6
[12] Chahal N, Siddique R, Rajor A 2012 Influence of bacteria on the compressive strength, water absorption and rapid chloride permeability of concrete incorporating silica fume *Construct Build Mater* **37** 645–651

IOP Conf. Series: Earth and Environmental Science **92** (2017) 012049 doi:10.1088/1755-1315/92/1/012049

Reagents and fractions impact on sulphide ore heap bioleaching at Smolnik mine

L M Oros[1] and J Zavada[1]

[1]VŠB-Technical University of Ostrava, Institute of Environmental Engineering, Czech Republic

Email: jaroslav.zavada@vsb.cz

Abstract. Mine Smolnik is one of the oldest sulphide ore mines in Europe and it is also an important part of bioleaching development. This paper follows previous attempts to extract residual metals from nearby heaps via variations in bioleaching reagents with regard to recent findings and needs in the related industry. Furthermore, economic and process relations between reagents and chosen heap fractions were also investigated in this case study.

1. Introduction

The hydrometallurgy is a well-known technique for metal extraction and recovery. There is a rich evidence of exploiting ore bodies via hydrometallurgy since middle ages all around the world. Various techniques were developed to recover metals from enriched mine waters including copper cementation on scrap iron considered as the oldest [1,2]. Hydrometallurgy has become emerging topic in past decades due to its low energy consumption as the increasing demand for metals caused by global human population growth necessarily led to marginal ore bodies exploitation [2,3]. Nowadays it is clear that phenomenon causing metal dissolution is caused either by an inorganic chemical reaction and biochemical activity of autochthonic microorganisms with varying influence depending on the conditions.

Sulphide mine Smolník located in Slovakia has been exploited for over six centuries and the ore mining continued until 1990 and the mine was flooded in 1994. During its existence, hydrometallurgical principals were applied as the meteoric (rain) water was directed into mine tailings and collected for further processing. To enhance this process acid mine waters (AMD) have been pumped onto tailings, however, no significant difference has been achieved [1,4,5,6].

This paper follows previous attempts to exploit residual metals (Cu) dispersed in mine tailings. The main aim of this study is to evaluate the influence of tailings fractions and different solvents to establish the most advantageous correlation between fraction and solvent under condition present in mine Smolnik. In general, leaching kinetics is affected by the particle size of the material that is leached and generally, a smaller particle size gives faster leaching kinetics since finer particles have a larger surface area. That means that a smaller particle size has a greater influence on a leaching process that is diffusion controlled [6]. However, disintegration is a highly energy-intensive process, thus the economic correlation between time and metal recovery is an important aspect of low-grade resources processing.

2. Materials and methods

Leaching took place in aerated 5L PET containers during 49 days. Samples for AAS spectrometry were taken every 7 days and pH was also measured.

2.1. Samples

Tailing samples have been taken randomly from heaps cumulated nearby mine Smolnik. Samples were grounded in a vibratory mill and homogenized via quartation. Homogenized samples were sieved into 3 fractions: <1 mm, 1-6 mm, 6-20 mm.

In order to compare different leaching solvents containing various microbial consortia contributing to bioleaching yield four solvents for each fraction were prepared:

- Pure bacterial culture (Solution: AT+AF). Consortium of *Acidithiobacillus thiooxidans* + *Acidithiobacillus ferrooxidans* was incubated for 8 days in 9K solution.
- Autochthonic microbial consortium AMD water from mine Smolnik (Solution: DK1).
- Autochthonic consortium (Solution: DK2). Consortium was incubated from precipitated sulfate salts found on the heap via dissolving of 1g of this material in 50 ml of water.
- Water acidified by H_2SO_4 (Solution: K).

In total 12 (3 fractions x 4 solvents) aerated reactors were simultaneously used for this experiment. Prior to pH adjustment by 2M H_2SO_4, Reactors were filled with water so that total volume in all 12 reactors was 3L as presented in table 1.

Table 1. Experimental conditions for exploited heaps bioleaching.

Fraction	< 1 mm	1-6 mm	6–20 mm
Tailings sample	25g	50g	100g
Inoculum AT+AF (50:50)	50 ml	50 ml	50 ml
Inoculum DK2	50 ml	50 ml	50 ml
Inoculum DK1	1.5L	1.5L	1.5L
pH	2	2	2
Total volume	3L		

Temperature 20°C was maintained during the experiment. Tap water was used in this experiment in order to avoid hypotonic environment caused by demineralized water and also to simulate in situ process conditions.

3. Results and discussion

Results of Cu recovery from different size fractions of exploited heap under different solvent conditions are presented in figure1-3. During the whole experiment, pH in all 12 reactors did not exceed 2.4 and remained stable.

As presented in figure 1 metal recovery from tailings sample fraction 0-1mm took place according to preconditions that presence of adapted bacterial cultures present in solutions contributed to the higher metal recovery. However, the impact on the metal recovery of autochthonic bacterial consortia wasn't significant and the extraction process was similar to inorganic leaching in an acidified water sample (K).

Results of the metal recovery from tailings sample fraction within the range 1-6 mm presented in figure 2 are similar to results mentioned above. Bigger fractions slowed the diffusion controlled process of leaching as expected but after 30 days of leaching achieved similar recovery results as in figure 1. Comparing figure 1, figure 2 and figure 3 representing tailings samples within the range 0-20 mm the significant contribution of biochemical leaching to overall recovery process arises. Acidified water (K) and pure culture solvents (AT+AF) had a very similar process, however, autochthonic cultures originating from AMD (DK1) and precipitated sulphates from Smolník achieved higher recovery rates. In total DK1 culture achieved the highest metal recovery rates in all examined fractions [6,7].

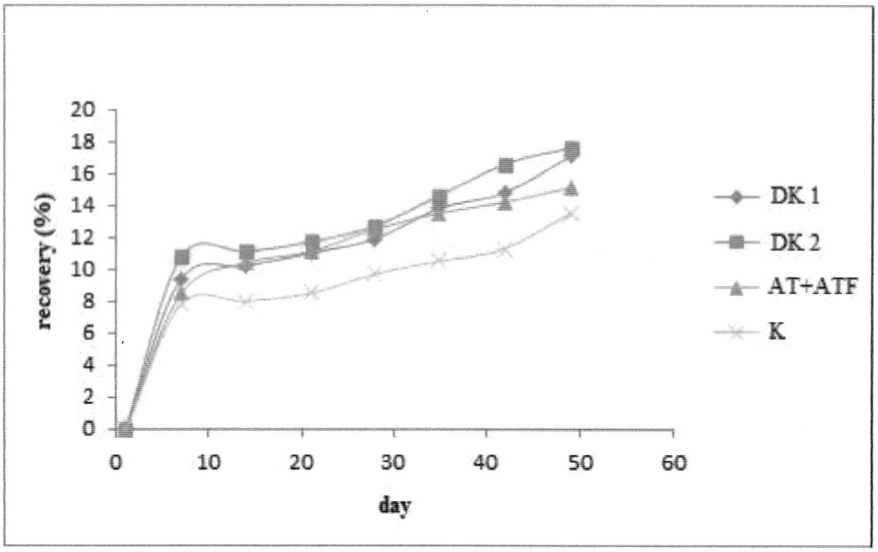

Figure 1. Recovery of Cu into solution from particles 0-1 mm.

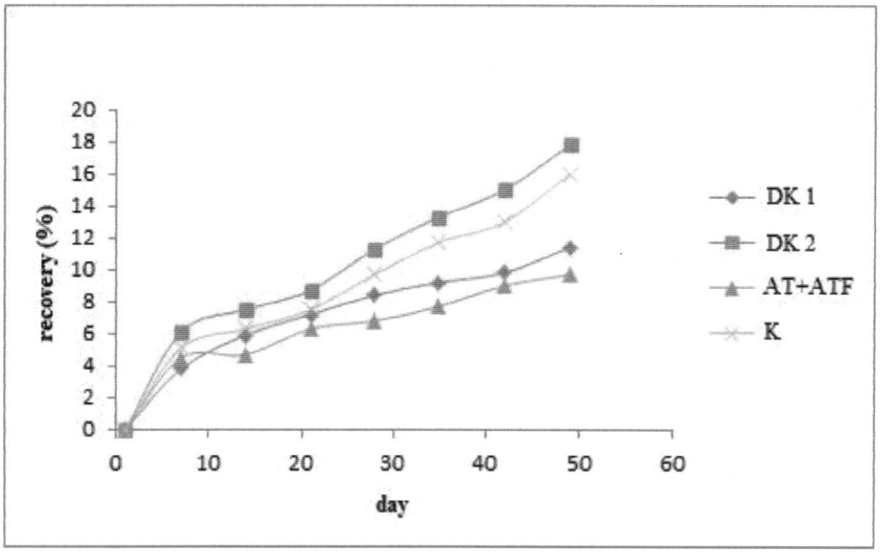

Figure 2. Recovery of Cu into solution from particles 1-6 mm.

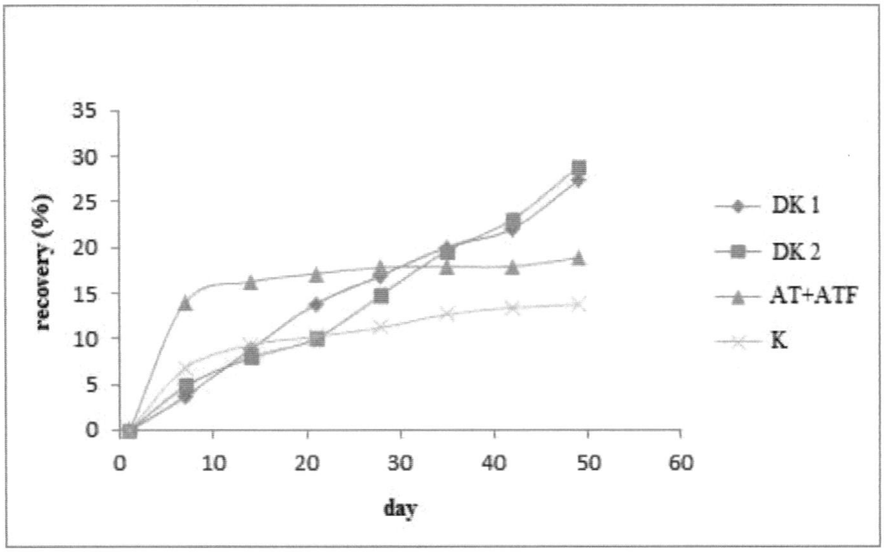

Figure 3. Recovery of Cu into solution from particles 6-20 mm.

The positive influence of adapted bacterial cultures on metal recovery is observed. The most interesting is the correlation between fraction size and biochemical leaching. In other words, the contribution of biochemical leaching to overall leaching process is rising with the higher granularity of the sample within the examined range. It appears that bioleaching of the exploited tailings in mine Smolnik requires determination of specific surface area that needs to be considered in potential further exploitation and the principals of the diffusion controlled process doesn't fully describe contributions of biochemical activity of autochthonic microorganisms [8-11].

4. Conclusion

In conclusion, it appears that contribution rate between inorganic leaching and biochemical leaching to overall metal recovery process depends on the specific surface area. Within the examinated range of exploited tailings samples, contribution of biochemical leaching is rising with the increasing particle size.

While leaching the finest sample fraction the kinetics of metal mobilization into solution via chemical and via bio-chemical leaching had a similar process. Difference between biochemical (DK1, DK2) and chemical (K) metal recovery values did not exceed 4%. Interestingly, as the particle size was rising the leaching process differences arisen in-between all applied leaching conditions. The results are clear that the highest recovery rates were achieved by autochthonic microbial cultures originating from Smolnik mine area. In particular medium DK2 which achieved highest recovery rate in all fraction sizes.

References

[1] Lintnerová O, Šottník P and Šoltés S 2008 *Est. J. of Ear. Sci.* **57** pp 104–110
[2] Johnson D B 2013 *Env. Sci. and Pol. Res. Int.* **20** pp 7768–7776
[3] Pradhan N, Nathsarma K C, Rao Srinivasa K, Sukla L B. and Mishra B K 2008 *Min. Eng.* **21** pp 355–365
[4] Grecula P and Slavkay M (ed) 1995 *Min. Dep. of the Slo. Ore Mount.* (Bratislava: Geocomplex)
[5] Lintnerová O, Šottník P and Šoltés S 2006 *Geo. Carp.* **57** pp 311–324
[6] Abdel-Aal E A 2000 *Hydrometallurgy* **55** pp 247–254
[7] Blažek V, Závada J, Bouchal T, Lebr J and Fečko P 2012 *Inzynieria Mineralna* **13** pp 1–7

[8] Šimonovičová A, Peťková K, Jurkovič Ľ, Ferianc P, Vojtková H, Remenár M, Kraková L, Pangallo D, Hiller E and Čerňanský S 2016 *Wat., Air, and Soil Pol.* **227** p 336

[9] Šimonovičová A, Ferianc P, Vojtková H, Pangallo D, Hanajík P, Kraková L, Feketeová Z, Čerňanský S, Okenicová L, Žemberyová M, Bujdoš M and Pauditšová E 2017 *Chemosphere* **171** pp 89–96

[10] Kolenčík M, Vojtková H, Urík M, Čaplovičová M, Pištora J, Cada M, Babičová A, Feng H, Qian Y and Ramakanth I 2017 *Wat., Air, and Soil Pol.* **228** p 224

[11] Vojtková H, Hlavňová R and Babičová A 2015 *15th Int. Mul. Sci. Geo. SGEM 2015 (Albena)* **1** (Sofia: STEF92 Technology Ltd.) pp 305–310

IOP Conf. Series: Earth and Environmental Science **92** (2017) 012050 doi:10.1088/1755-1315/92/1/012050

Research of cement mixtures with additions of industrial by-products

R Papesch[1, 2], L Klus[2], J Svoboda[2] and R Zajac[2]

[1]Technical and Testing Institute of Construction Praha, s.p. – branch Ostrava, U Studia 14, 700 30 Ostrava
[2]Institute of Environmental Engineering, Faculty of Mining and Geology, VSB-Technical University of Ostrava, Czech Republic

Email: papesch@tzus.cz

Abstract. The main goal of the article is the comparison of the possible use of secondary energy products. Used fly ashes, respectively steel dusts in cement mixes derive from production in Moravian-Silesian Region. The research focused on their influence on the chemical and physico-mechanical characteristics of the fresh and solid mixture. The aim was to find suitable formulations for grouting works, highway construction possibly rehabilitation of underground cavities created by mining activities. The introduction is mentioned the history of waste utilization up to current use as a product and the overall state of the problem. The conclusion is an evaluation of possible use in practice, including recommendations to carry out further tests.

1. Introduction

In the Czech Republic, the production of fossil fuels (especially brown coal) has the highest part of total production. In the heating plants and power plants is burned mainly brown coal extracted in underground or open pit mines in northwest and northeast parts of Czech Republic. This involves with high production of solid residues after combustion or accompanying products from desulphurization. The forming of dumps, where is stock-pilled industrial waste arising from the diversified industrial processes, is a common fact. There are several dumps in the Czech Republic. Nowadays, these products, not dangerous waste as in past, can further be used because of their positive properties. The fly ash has more or less distinctive puzzolanic properties in depending on his contents. These properties make fly ash suitably material for construction purposes. Use of this material would become a solution to the required reduction of natural resources extractions such as aggregate and limestone.

For their use, for example as solid backfill material is important to verify their properties to meet all standard requirements. A large scale of secondary materials is being tested today to improve the resulting product properties. And one such example is the use of organic waste as filler in the production of mortar [1, 2, 3, 4] Another such example is a similar issue, the use of waste recycled complementary product on a natural basis into building materials, was dealt with in an article entitled "Implementation of waste cellulosic fibers into building materials [5, 6]. Research on this topic is deal with research of Concrete Containing Recycled Concrete Aggregate with Modified Surface in combination with the efficient use of resources and the achievement of the material properties of the resulting of these composite´s [7, 8].

2. Description of mortar components and their properties

2.1. Cement fine
It is a mixture of mineral filler with a grain size 0 - 0.7 mm, portland cement and additives that improve processing and use properties of the product. Cement is most common used hydraulic binder.

2.2. Fly ash
Fly ash is material incipient by the combustion of extracted coal for the purpose of electric production in power plants. The fly ash has more or less distinctive puzzolanic properties in depending on his contents. These properties make fly ash suitably material for construction purposes. Ash properties depends primarily on the type of combustion coal, grain size and combustion conditions, as well as on the chemical and mineralogical composition of used fuel. Physically, fly ash is a light brown to dark gray, very fine dust with a particle size 1 to 400 microns, with glassy structure and in most cases with rounded grain shapes resulting from steep cooling during the combustion process. The chemical composition of the fly ash consists primarily of silicon, aluminum, calcium, iron, magnesium, sulfur and alkaline oxides, as well as residues of combustible materials. In more detailed description of fly ashes, emphasis is placed on two of his properties that greatly influence its usability: the content of active puzzolanic SiO2 and CaO and the grain size. It has been found, that a at smaller grain size of fly ash (25 microns) is positively affected its puzzolanic characteristic, therefore these fine fractions are preferred over the coarse fraction. According to puzzolan activity of fly ash is determined ability to produce a solid mixture in the presence of CaO. Puzzolanic activity also could be directly affected by aluminium content, grain size, loss by ignition or alkali content (Na_2O). Granular fly ash: These kind of fly ash is made by "classic" fuel combustion process at a temperature about 1400 to 1600 °C. They are distinguished mainly by β-quartz and mullite ($3Al_2O_3 \cdot 2SiO_2$) together with a glassy phase in an amount usually greater than 50 %, which in essential way under normal or higher temperatures contributes to the reactivity of fly ash with CaO or cement. Under hydrothermal conditions mullite in low extents also enter into reaction. Fluid fly ash: Some types of power plants or heating plants use fluidized combustion technology at atmospheric pressure. Ground fuel with the addition of limestone or dolomite is burned in the circulating layer at 860 °C. During the disintegration process, from fuel is released SO2 and is bound to CaSO4. This is particularly important from an ecological point of view, because otherwise sulphur dioxide leak into the atmosphere, resulting in very harmful acid rains.

These improved processes, thanks to transported burnt gas from furnace area, lead to separation of individual fractions, fine parts in form of flue dust are carried away and coarse parts remain in the combustion chamber. Flue dusts of solid substances are eliminated from burnt gas by classical technological processes, by cyclone clarifiers or by filters.

2.3. Steel plant dust
Steel plant dust is an industrial material produced in furnaces during the production of castings from grey iron or ductile cast iron. The gases from furnace contain solid particles that are trapped, condensed and trapped in electrostatic precipitators. It is a dry and loose material whose grains have a spherical shape and a maximum size of about 0.1 mm.

3. Experimental part
The objective was to verify the properties of fly ash(steel plant dust)-cement mixtures. Above all, their strength characteristics and frost resistance, depending on the amount of added fly ash (steel plant dust), which could be subsequently used for proposal of various kinds of composite materials for specific works. As a major component, to which 5%, 10% and 20% part of fly ash or steel plant dust was added, was chosen a fine cement, class C25.

The following types of secondary energy products were chosen: granular fly ash from power plant – granulating furnace, desulphurization by wet limestone method, fly ashes; fluid fly ash from power plant – fluid furnace, fly ashes; steel plant dust – electrical precipitator. Constitution of individual composition:

Sample No. 1 – reference material, 100 % of fine cement;
Sample No. 2 – 5 % of granular fly ash + 95 % of fine cement;
Sample No. 3 – 10 % of granular fly ash + 90 % of fine cement;
Sample No. 4 – 20 % of granular fly ash + 80 % of fine cement;
Sample No. 5 – 5 % of fluid fly ash + 95 % of fine cement;
Sample No. 6 – 10 % of fluid fly ash + 90 % of fine cement;
Sample No. 7 – 20 % of fluid fly ash + 80 % of fine cement;
Sample No. 8 – 5 % of steel plant dust + 95 % of fine cement;
Sample No. 9 – 10 % of steel plant dust + 90 % of fine cement;
Sample No. 10 – 20 % of steel plant dust + 80 % of fine cement.

3.1. Produce of fresh and solid mortar mixtures and subsequent tests
Mixtures were produced in laboratory of building materials. Mixing of the mixture was carried out under laboratory conditions, temperature (20±2) °C and relative humidity (60±5) %. Sample of fresh mortar was prepared on presribed value of spilling. The amount of water required to achieve the right consistency was determined on the basis of experimental tests. The water content of the mixture, suitable for grouting has always been determined so that the spill of the mixture performed by shaking table BetonSystem RS-15 in accordance with standard EN 1015-3 was 230 mm. The mixture was prepared by the MI-CM5A Concrete Mixer. The mixing time and speed were in accordance with EN 1015-2, it means mixing for 90 seconds at low speed.
Optimal mixtures were placed into moulds with dimensions 40x40x160 mm. The content was compacted in the middle of the mould by 15 strikes by rammer made from non-absorbent material square cross-section. The filled moulds for strength and resistance tests were aligned by spatula, covered by oiled glass and placed to humidity chamber with a prescribed humidity 95 % and temperature 20 °C. Two days later after mixing samples were took out from moulds and stored into environment with prescribed humidity. The strength characteristics were calculated from measured value of acting force at which the samples was broken. For testing was used press machine BetonSystem with a maximum load 10 kN for bending tensile strength and 300 kN for compressive strength with calibration till 05.03.2019. On samples were performed the following physico-mechanical properties: Determination of consistence of fresh mortar (by flow table) – EN 1015-3; Determination of air content of fresh mortar – EN 1015-7; Determination of bulk density of fresh mortar – EN 1015-6; Determination of flexural and compressive strength of hardened mortar – EN 1015-11; Testing of frost resistance of mortar – ČSN 72 2452.

4. Evaluation

4.1. Evaluation of fresh mortar mixture
The bulk density of each mixture is almost the same, but the amount of water required to achieve required consistency considerably vary. The highest amount of water is needed for optimal de flocculation of the mixture with the addition of steel plant dust and fluid fly ash. The air content decreases depending on the amount of added additive, which has a much smaller grain size for all three types of additives that results in better cement grain coating and reduced air volume in the mixture (table 1).

Table 1. Composition of each fly ash (steel plant dust)-cement mixtures and their essential characteristics.

Sample No.	Cement fine C25 [g]	Secondary energy product [g]	Water [g]	Water-cement ratio	Bulk density [kg·m^{-3}]	Air content [%]
1	3000	0	570	19	2041	5.4
2	2850	150	540	18	2041	4.9
3	2700	300	540	18	2045	3.9
4	2400	600	540	18	2050	2.4
5	2850	150	600	20	2075	4.4
6	2700	300	660	22	2076	3.0
7	2400	600	780	26	2079	2.0
8	2850	150	600	20	2042	4.2
9	2700	300	705	23.5	2048	3.8
10	2400	600	900	30	2016	1.8

4.2. Evaluation of solid mortar mixture

Verification tests of tensile strength in bending and compressive strength were performed on solid test specimens after 2, 7 and 28 days after mixing. Within the frost resistance test, the strengths tests were performed until the sample decay, maximum number was 100 test cycles. Before the final results were put into graphs, were corrected by calculating of statistical characteristic of random choice (arithmetic mean, standard deviation, variation coefficient), so that all of the strengths comply with interval X ± 3s.

Mixtures with additions of granular fly ash prove higher initial strength than the reference sample. This property may be caused by higher content of glassy phase that significantly affects the reactivity of fly ash with CaO. However, after 28 days these samples show lower strength than for reference sample. Mixtures with addition of fluid fly ash prove lower initial strength, but the strength after 28 days are higher than for reference sample that may be caused due to higher content of free CaO. For steel dusts is possible to notice with their increased value lower initial strength and also compressive and bending strength after 28 days. Is known improvement of strength characteristics for long term samples (3 months and more). Currently are in laboratory of building materials stored samples for verification of strength after 3 and 6 months.

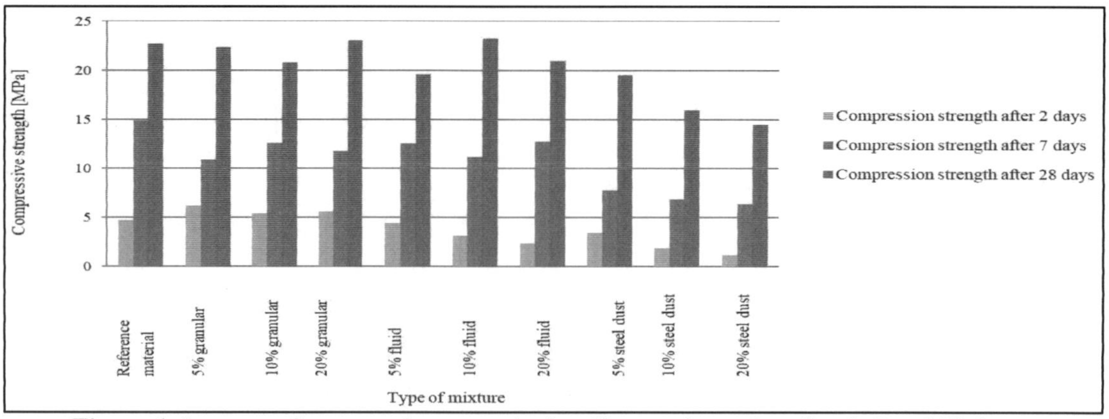

Figure 1. Increase of compressive strengths for all type of samples after 2, 7 and 28 days.

For steel dusts is possible to notice with their increased value lower initial strength (figure 2) and also compressive and bending strength after 28 days (figure 1). Is known improvement of strength characteristics for long term samples (3 months and more). Currently are in laboratory of building materials stored samples for verification of strength after 3 and 6 months.

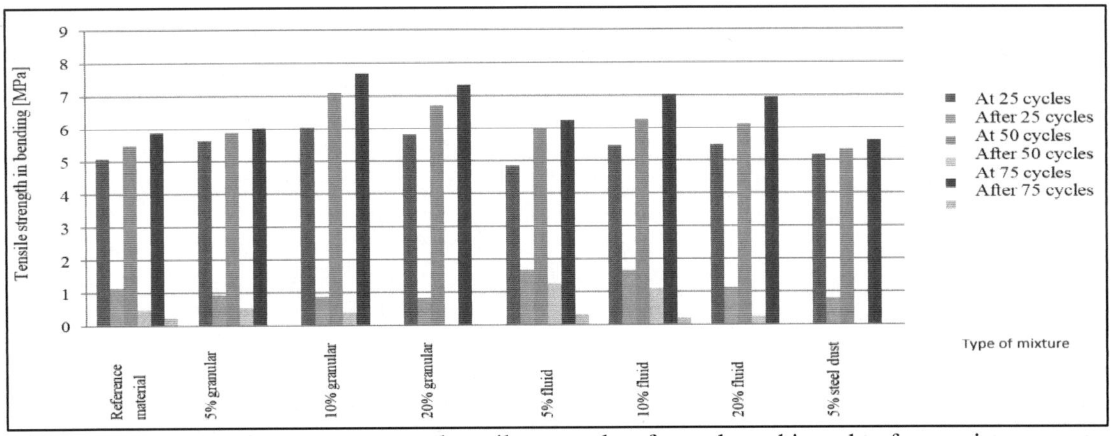

Figure 2. Difference between measured tensile strengths of samples subjected to frost resistance test and reference samples removed from climatic chamber.

Table 2. Coefficients of frost resistance to reference samples

Number of freeze-thaw cycles Type of sample	25	50	75
Reference sample	23 %	9 %	4 %
5% gr.fly ash	17 %	9 %	/
10% gr.fly ash	14 %	5 %	/
20% gr.fly ash	13 %	/	/
5% fluid fly ash	35 %	20 %	4 %
10% fluid fly ash	33 %	17 %	2 %
20% fluid fly ash	21 %	4 %	/
5% steel plant dust	16 %	/	/

For frost resistance (table 2) was intended to expose samples to 100 freeze-thaw cycles divided into 4 periods. Final period was not achieved due to considerable decomposition of all samples. After 25 cycles, the resulting strength of all mixtures did not particularly distinct from the reference sample. Samples with 5% and 10% fluid fly ash prove about 25 % higher values than strength of reference sample. For mixtures with 20% granular and 20% fluid fly ashes and 5% steel dust fine cracks have been occurred on surface in length up to 5 mm and deep maximally 1 mm. For samples with 20% of fluid fly ash occur the crumbling of edges of the sample.

After 50 freeze-thaw cycles the 5% and 10% fluid and 5% granular fly ash prove higher strength than reference sample. Strengths for mixtures with 20% of fluid fly ashes decreased in comparison with reference sample about more than 50 %, strengths of mixtures with 20% granular fly ash and 5% steel plant dust was not possible to measure, for these samples instead of fine cracks occurred the distortion of whole sides.

Performed 75 cycles was reached only with 5% and 10% fluid fly ashes, with very low but comparable strengths in comparison with reference sample.

5.Evaluation of experimental part

The main aim of the thesis was to verify the influence of secondary energy products and their relative proportions in the mixture (granular and fluid fly ashes produced by power plants, steel plant dust) to the resulting rheological properties of the mixture. At the same time was observed their effect on the development of tensile strength and compressive strength in different times and after freeze-thaw cycles. On the basis of obtained results, is possible to state, that if are not required rapid increases of initial strengths is possible to make cement mixes with up to 20 % compensation of fine cement by a suitable type of secondary energy product (mainly granular or fluid fly ashes, but also steel plant dust). The tensile strengths in bending and also compressive strengths are not significantly affected. Using of this substitute is possible to assume cost reduction in comparison with reference mixture. **Fly ashes:** If is the main aim to achieve higher initial strengths of ash-cement mixtures to mixtures without fly ash, is necessary to reduce the value of water coefficient, which subsequently affects the workability. The fine fraction of used fly ash with higher specific surface area contributes significantly to get higher values of strengths. It is generally known that the negative influence of the aggressive environment on the life cycle of the grout is possible to restrain by the addition of fly ash. The positive effect is mainly observed in fly ash mixtures exposed to sulphate solutions. **Steel plant dust:** For test mixtures with steel plant dust, the initial strengths increased slowly. Although the strengths after 2, 7 and 28 days showed low values, according to available information's, the resulting 90 days strengths, respectively 150 days strengths of the mixture were comparable with strengths of reference mixture after 28 days.

Due to presence of steel plant dusts in tested mixtures, retardation was observed during setting, especially in first 24 hours, during which the setting of fresh mortar was very slow. The effect of retardation begins to retreat after 48 hours of aging. The presence of dusts prove after approximately 7 days when the strength of mixtures with 10% and 20% part of dust already show lower differences to reference mixture.

6. Conclusion

Use of mixtures with high mentioned additions should always be considered, especially according to place of use, conditions which affect location and with properties we want to achieve. For general recommendation of possible use as a substitution of cement in mixtures is certainly advisable to perform several other tests, e.g. determination of permeability, determination of adhesion and heat expansion, determination of initial and final setting times under higher temperatures or resistance to aggressive environments. It would also be worthwhile to research the ecological burden of secondary energy products on environment, especially after the end of its life cycle. Known is facile leachability of toxic metals such as arsenic or chromium or extremely high pH values for water waste.

References

[1] Václavík V et al. 2014 The Use of Industrial Waste as a Secondary Raw Material in Restoration Plaster with Thermal Insulating Effect *Advanced Materials Research* **897** 204-214

[2] Václavík V, Valíček, J, Dvorský T et al. 2012 A method of utilization of polyurethane after the end of its life cycle Rocznik Ochrona Srodowiska 96-106

[3] Hybská H, E Hroncová, J Ladomerský, K Balco and J Mitterpach 2017 Ecotoxicity of Concretes with Granulated Slag from Gray Iron Pilot Production as Filler *Materials* **10** (5) 505

[4] Ladomerský J et al. 2016 One-year properties of concrete with partial substitution of natural aggregate by cupola foundry slag In Journal of Cleaner Production **131** 739-746

[5] Dvorský T et al. 2015 Research of the use of waste rigid polyurethane foam in the segment of lightweight concretes 51-56

[6] Stevulova N, I Schwarzova, Hospodarova V and Junak J 2016 Implementation of waste cellulosic fibres into building materials **50** 367-372

[7] Junak J, Sicakova A 2017 Concrete Containing Recycled Concrete Aggregate with Modified Surface Procedia Engineering **180** 1284-1291

[8] Stevulova N, Junak, J, Grul, R, Bacikova M 2008 Utilization possibilities of selected waste kinds in building materials preparing 193-200

IOP Conf. Series: Earth and Environmental Science **92** (2017) 012051 doi:10.1088/1755-1315/92/1/012051

Multi-criteria Analysis of Factors for Application of Concrete Composites Considering Their Environmental Harmfulness

A Paulikova[1], A Estokova[2] and J Mitterpach[3]

[1]Slovak University of Technology in Bratislava, Faculty of Materials Science and Technology in Trnava, Jana Bottu 25, 917 24 Trnava, Slovakia
[2] Technical University of Kosice, Faculty of Civil Engineering, Institute of Environmental Engineering, Vysokoskolska 4, 042 00 Kosice, Slovakia
[3]Technical University in Zvolen, Faculty of Ecology and Environmental Sciences, T.G.Masaryka 24, 960-53, Zvolen, Slovakia

Email: paulikova.alena@gmail.com

Abstract. The analysis of factors is important in insight of the selection of proper building material with environmental added value. A comprehensive solution is possible if at the beginning there are all the relevant factors in detail characterized predominately that have got a major impact on the area in terms of environmentalharmfulness prevention. There are many groups of environmental factors. In this article only four factors are considered, i.e. contain of Cr^{VI} (mg/kg) and index of mass activity for radionuclides (Ra, Th, K) which are the most harmful. These factors can be evaluated by means of a supplementary tool, e.g. multi-criteriaanalysis, which improves and supports decision processes in the framework of construction bybuilding management,etc.

1. Introduction

In general, it is difficult to select criteria for evaluation of building material harmfulness because of a wide range ofhuman health-linked indicators and methods. There are many various approaches to evaluate the negative or dangerous even toxic impacts of building materials from the simple lists or calculation to the sophisticated software [1,2,3,4]. However, the evaluating methods are usually based on the inputs data from the commonly used databases which give often only the average values.

Therefore, the calculations based on the real experimental data are still of great importance.

The current trend in the use of secondary fuels in cement production and the use of secondary raw materials in the production of concrete and cement composite materials has drawn attention to the environmental safety of building materials. Special attention is paid to leaching of environmentally hazardous substances, in particular toxic heavy metals, from water-borne structures [5,6] and the radioactivity of building materials.The aforementioned factors are involved in the negative effects of building materials on human health.

This paper presents a part of a proposal of methodology enabling to obtain data for the relevant selection of application material, where environmental safety and protection measures should be applied at first. The harmfulness evaluation, in the presented paper, aims at comparison of the radioactivity and leachability of hexavalent chromium from selected cement composites using multi-criterial approach.

2. Methodology of complex assessment

The paper aims at the calculation of order for priorities concerning environmental harmfulness of concrete composites based on ordinary Portland cement as well as composites containing secondary materials such as silica fume and zeolite. The evaluation of environmental harmfulness was performed onfour concretes types whereas all mixtures contained the same amount of CEM I 42.5 N cement (360 kg). The K1 and K4 samples were of almost the same compositions, based on ordinary Portland cement, differing only in water and plasticizer amounts. K4a and K4b samples differ from the point of treatment after the hardening. The K4a sample was cured in water while K4b sample was cured in the laboratory atmosphere. K2 concrete sample has been enriched by silica fume (20 kg) while K3 sample by mix of silica fume and zeolite (20 + 20 kg). The recipes of the concretes investigated were designed in our previous research regarding the durability of concretes testing [7,8].

The proposed methodology of complex environmental harmfulness for human healthwas based on the quantification of qualitative parameters using multi-dimensional harmfulness assessment indicators in recommended composition according to the evaluation purpose.

The recommended number of quantifiable indicators defined and monitored by experts is 20÷25. As an input to the problem solution serves a matrix of data values for individual j-indicators with determination of the permissible standard for their values u_j, evaluated from i-th number of sites. The calculation includes a triangular matrix of correlation indexes r_{ij}, as well as the values of the arithmetic mean y_j, a standard deviation s_j from a set of individual indicators and the Student's t-test values for these indicator values.

The indicators are further processed using an original mathematical and statistical method, based on vector, factor and discriminant analysis, correlation calculus and tests of significance so that a complex of environmental harmfulness value for human health or building material (concrete) application is determined by a dimensionless criterion Q_i.

The value behaviour of Q_i can be well-arranged demonstrated graphically on a map of the evaluated building application or it follows its course in the dynamic of environmental protection.

This methodology can be applied as a supplementary tool for the building developers' management, state government offices, the Ministry of Environment, Ministry of Transport and Construction and Ministry of Health. It allows a complex view of multi-criteria environmental issues. Significance tests also allow a partial or detailed detection of critical environmental harmfulnesswith regard to the human health and building application.

The selection of descriptors, i.e. selected evaluating factors as well as their arrangement in a matrix (an order of columns in a data sheet) depends on evaluator's professional orientation. This choice influences the final level of the actual environmental harmfulness. In our evaluation, the radioactivity and the heavy metals toxicity have been considered as the most dangerous to human health. Isotopes of potassium (^{40}K), uranium (^{238}U) and thorium (^{232}Th) are the most important radionuclides when assessing the radioactivity of building materials due to their primary origin in raw materials [9]. Heavy metals present in the cement composites can be released to the surrounding aqueous medium which acts on the material by direct leaching or dissolution and diffusion in the porous liquid. These processes are influenced by the structure of the pores and the porosity of the building material. The solubility of heavy metals depends on their binding to the hydration products, the oxidation number in which they occur as well as their ability to form complex compounds and precipitates. The negative effects of heavy metals on human health are widely known. Chromium, commonly present in cement raw materials, belongs to the most dangerous metals. A very special toxicity is connected with the hexavalent form of chromium. Therefore, the following individual factors have been selected for the multi-criterial evaluation of concretes:

- mass activities of radionuclide ^{226}Ra [Bq/kg],
- mass activities of radionuclide ^{232}Th [Bq/kg],
- mass activities of radionuclide ^{40}K [Bq/kg],
- leachability of Cr(VI) [mg/kg].

3. Results

Table 1 presents input data for the concrete composites: K1, K2, K3, K4a and K4b (for their characteristics see Section 2 "Methodology of complex assessment") with the selected monitored and measured characteristic factors which have got various level of environmental harmfulness. The values of the selected indicators, chosen for the calculation, were achieved in our previous research focused on the durability of cement-based materials [10,11]. The order of columns with selected factors is arranged according to environmental significance, which was determined by an environmental specialists (three independent evaluators with various universities). These evaluators are strongly environmental oriented to support of environmental protection for the new application of concrete composites. The individual factors A, B, C and D have got an amplifying effect on environmental harmfulness and they have got a positive polarization of criteria.

Table 1. Input data - individual factors for a multi-criteria analysis.

individual criteria of environmental harmfulness for concrete composite "j"		A	B	C	D
		mass activities of radionuclides			leachability
		^{226}Ra	^{232}Th	^{40}K	Cr(VI)
units		[Bq/kg]	[Bq/kg]	[Bq/kg]	[mg/kg]
		maximum value of criteria "a_{ij}" for individual concrete composites			
labelling of concrete composite "i"	K1	13.98	83.00	32.50	0.84
	K2	9.55	508.70	14.10	1.44
	K3	7.84	388.60	21.80	0.39
	K4a	14.70	347.20	27.60	1.24
	K4b	10.74	759.40	23.10	0.98

Legend:
i – labelling of concrete composite,j – factors (criterions),A – mass activities of radionuclide ^{226}Ra,B – mass activities of radionuclide ^{232}Th, C – mass activities of radionuclide ^{40}K, D – amount of Cr(VI).

The results of multi-criteria prioritization for level of environmental harmfulness of the selected concrete composites are presented in figure 1 and table 2. The figure 1 shows a comparison of the calculated level of environmental harmfulness (Qi) of evaluated concrete composites according to the above-mentioned methodology and using the calculation performed by a computer algorithm.

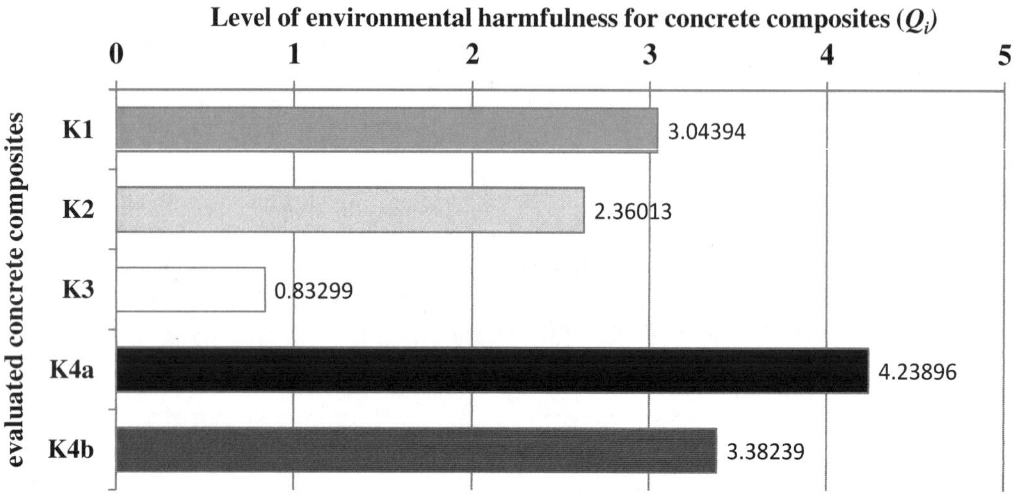

Figure 1. Environmental harmfulness of concrete composites.

Based on the calculated level of environmental harmfulness (Q_i), i.e. dimensionless variable which considers with harmful influences of concrete composites in their application sphere, table 2 determines the order of the concrete composites regarding their harmfulness.The multi-criteria calculation taking into consideration environmental harmfulness of concrete compositecan help to determine the priority for their application in new buildings.

Table 2. Multi-criterial analysis of the concretes harmfulness.

Individual concrete composite	Level of environmental harmfulness Q_i [dmnl]	Application priority for construction/buildings
K1	3.04394	III.
K2	2.63013	II.
K3	0.83299	I.
K4a	4.23896	V.
K4b	3.38239	IV.

The concrete composite K3 with mix of silica fume and zeolite has been identified to have the lowest human health impact in terms of evaluated parameters. The calculated level of environmental harmfulness ($Q_3 = 0.83$) was significantly lower than those of calculated for the other composites. As seen in table 2, the most negative human health impact ($Q_{4a} = 4.24$) has been calculated for the K4a composite based on ordinary Portland cement (OPC). When comparing the Q_3 to Q_{4a}, the found harmfulness parameter of K3 was even 5 times lower. The other composites based on ordinary Portland cement reached the close values of 3.04 and 3.38 for K1 and K4b, respectively. The cement composite with only silica fume addition was found to be more dangerous than K3 but less dangerous than OPCs.

As presented in table 2, the order of cement composites regarding their harmfulness, according to the used methodology, can be as follows:K3 < K2 < K1 < K4b < K4a.

4. Conclusions

The paper presents the results of the multi-criterial analysis of the harmfulness of the selected cement composites. To evaluate the suitability of cement composites to practical application, a more complex approach is needed considering another mechanical, chemical or technological indicators. The extended multi-criterial evaluation of the cement composites is of our interest at present. These complex evaluation could be used in real decision-making processes within a building developers' management.

Acknowledgment

This paper was elaborated in the framework of the Project of the Scientific Grant Agency of the Ministry of Education, science, research and sport of the Slovak Republic and the Slovak Academy of Sciences (VEGA) No. 1/0648/17 Study of significant environmental parameters of building materials regarding their environmental safety.

References

[1] Pauliková A and Železník O 2014 Multicriterial analysis of factors considering intensity and extent of floods ICCC 2014 - 15th International Carpathian Control Conference 418-42
[2] Ocelíková E 2004 Multikriteriálne rozhodovanie (Košice: Elfa sro) p 87
[3] Nováková R and Ovsenák V 2016 The importance of marketing for effective management of quality indicators *Inter. Conf. on Eng. Sci. and Prod. Manag.* p 229-234
[4] Linkov I and Moberg E 2011 Multi-Criteria Decision Analysis: Environmental Applications and Case Studies (Environmental Assessment and Management) CRC Press p 204
[5] Hohberg I 2003 Characterization, modelling and assessment of the leaching behaviour of environmentally relevant inorganic compounds from cement based building materials *Deut. Auss. für Stah. Wilh. Ernst & Sohn* **542**
[6] Mullauer W, Beddoe R E and Heinz D 2015 Leaching behaviour of major and trace elements from concrete: effect of fly ash and GGBS *Cem. and Conc. Comp.* **58** p 129-139
[7] Eštoková A et all 2016 Testing Silica Fume-Based Concrete Composites under Chemical and Microbiological Sulphate Attacks *Materials* **9** 324
[8] Kovalčíková M and Eštoková A 2014 Leaching of calcium and silicon from cement composites in the aggressive environment *Poll. Per.* **9** p 123-130
[9] Cabanekova H 1996 Concentration of natural radionuclides in various types of building materials in Slovakia Bratislava *J. of Radio. and Nucl. Chem.* **209** p 301-306
[10] Eštoková A and Palaščáková L 2013 Assessment of natural radioactivity levels of cements and cement composites in the Slovak Republic *Inter. J. Environ. Res. and Public Health* **10** p 7165-7179
[11] Eštoková A, Oravec J and Kovalčíková M 2015 Environmental impact assessment of the concrete composites in terms of the selected toxic metals leaching *Chem. Eng. Trans.* **43** p 1915-1920

Combined deterministic – stochastic forecasting of monthly river flows for water management

E Peksova Szolgayova[1,2], R Vyleta[3], J Szolgay[3] and Z Lukac[4]

[1]Na Veselí 22, 14000 Praha, Czech Republic
[2]formerly at The Centre for Water Resource Systems, Vienna University of Technology, Karlsplatz 13, 1040 Vienna, Austria
[3]Dept. of Land and Water Resources Management, Faculty of Civil Engineering, Slovak University of Technology, Radlinského 11, 810 05 Bratislava, Slovakia
[4]Bratislava Water Company, Prešovská 48, 826 46 Bratislava, Slovakia

E-mail: jan.szolgay@stuba.sk

Abstract. Monthly discharges can be modelled and predicted by the decomposition of the runoff process model into two components – deterministic and stochastic. For such approach the term hybrid was often adopted. In this study a hybrid (deterministic-stochastic) modelling approach for one step ahead forecasting of mean monthly discharges at the gauging stations Banská Bystrica and Brehy on the Hron River in Slovakia was developed. The aim was to join the conceptual monthly rainfall runoff model KVHK and several time series models of the forecasting error time series of the conceptual model into a hybrid framework. Since these rainfall-runoff model error series may exhibit nonstationarity and heteroscedasticity, beside traditional ARMA models, GARCH type nonlinear time series models were also considered.

1. Introduction

Reliable forecasts of monthly inflows to reservoirs are of particular interest for scheduling of releases for many water resources management tasks, e.g. for hydropower production or low flow augmentation. Approaches used for monthly river flow forecasting include a wide range of methods from deterministic (usually conceptual water balance models) through stochastic approaches (mainly in the time series domain) to methods of artificial intelligence (e.g. neural networks). Especially when quantitative information on the uncertainty of predictions is required, reliable predictions remain a difficult task for physically based monthly watershed models owing to the complexity of the involved meteorological and hydrological processes. In the recent past therefore the so-called hybrid (combined deterministic-stochastic) modelling approaches gained increased interest, which try to combine the advantages offered by both deterministic and stochastic methods [1,2].

In this study a hybrid (deterministic-stochastic) modelling approach for one step ahead forecasting of mean monthly discharges at the stations Banská Bystrica and Brehy on the Hron River in Slovakia is presented. First the two modelling approaches, the conceptual monthly rainfall runoff model KVHK and several time series models of the KVHK model forecasting error time series are described. Since these error processes are in general nonlinear and in addition to that such forecast error series may exhibit nonstationarity and heteroscedasticity [3, 4, 5], beside traditional ARMA models, GARCH type nonlinear time series models are also considered here in the hybrid framework. We produce one - step - ahead forecasts from the fitted models and compare the forecasting performance of nonlinear time series models with that of linear error series models, since there is limited knowledge of their predictive capabilities concerning the particular problem of forecasting modelling error series.

2. Data

The forecasting performance of the hybrid methodology was tested for one month ahead flow forecasting (without considering the uncertainty of the forecast) on the Hron River in Slovakia. From the practical point of view the flow forecasting problem in the upper part of the Hron basin is complex [6]. In the alpine mountain regions rainfall runoff processes are dominant in flow generation. Rainfall of cyclonic origin represent main danger to major cities and industrial areas along the river. The hydrological modelling therefore concentrates on rainfall runoff processes. While daily time step is sufficient for hydrological forecasting for flood protection, the monthly time step is needed for water resources management tasks.

The Hron River springs in the Low Tatry Mountains at an elevation of 980 m asl. With its length of 298 km it is the second largest river in the country, its basin has size of 5465 km2. The map of the Hron River basin is shown in figure 1.

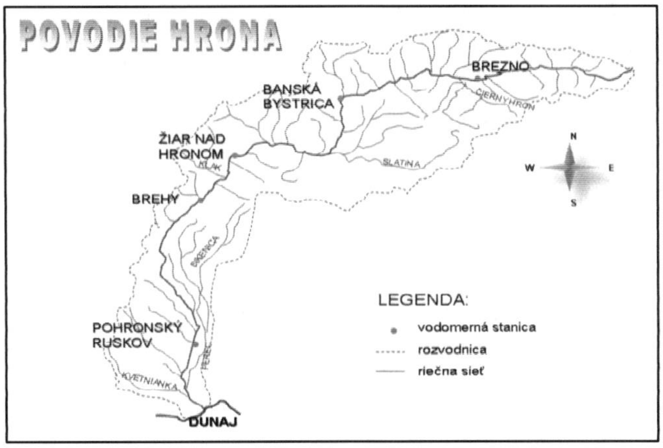

Figure 1. The Hron River (red dot – gauging station, dashed line- watershed boundary, blue lines-river network).

3. The KVHK rainfall-runoff model

For modelling mean monthly runoff time series the KVHK conceptual water balance model developed at the Slovak University of Technology was used [6]. This model (see figure 2) simplifies a river basin into 2 nonlinear reservoirs, and it simulates water accumulation in the basin, snowmelt, evapotranspiration, runoff from impermeable areas in the basin, surface and subsurface runoff and baseflow.

The inputs required for the modelling of the water balance in a monthly time step are: the mean monthly precipitation for the basin, the mean monthly discharges in the outlet of the basin and the mean monthly potential evapotranspiration. For calculating the potential evapotranspiration, various methods can be used and accordingly additional climate data may be required (e.g. the long-term mean monthly hours of the duration of sunshine, the long-term mean monthly values of the relative air humidity, and the mean monthly air temperature values). The model contains a snowmelt reservoir, which enables the distinction between solid and liquid precipitation on the basis of the principle of threshold temperatures. The total runoff is calculated as the sum of the direct, surface, subsurface runoff components and of the baseflow. In this study a genetic algorithm was used to calibrate the 11 model parameters, and the traditional Nash-Sutcliffe criterion was employed as an objective function. We fitted the model on 32 years of monthly data from the period January 1961 - December 1992 (altogether 384 observations).

IOP Conf. Series: Earth and Environmental Science **92** (2017) 012052 doi:10.1088/1755-1315/92/1/012052

4. The KVHK model forecasting error time series analysis

Rainfall runoff model forecasting error time series data exhibit three major differences when compared to natural runoff data. First the time series takes all possible values, not only positive numbers. Second, the general property of the hydrological model error processes seems to be heteroscedascity. This behaviour present in the hydrological model error series can be explained by the fact, that the modelling and forecasting of flood flows (pulse like rising discharge) is a more demanding task than that of droughts (slowly decreasing flows). Therefore, unlike in natural data, the variance of the time series may not be considered as constant. The heteroscedasticity concept has not received much attention in hydrological literature so far, but it is gaining popularity as well, see e.g. [3,4].

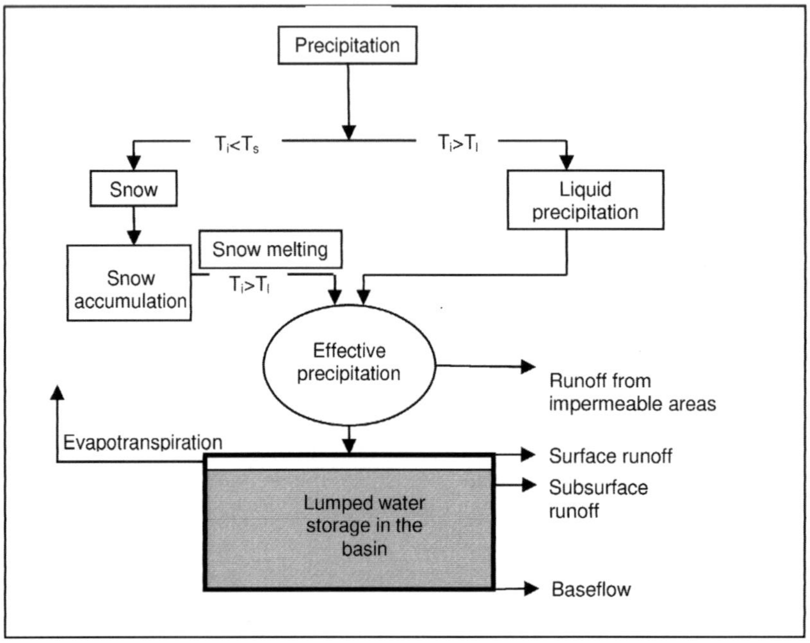

Figure 2. The flowchart of the KVHK monthly hydrological balance model T_i is the mean monthly air temperature, T_s and T_l are threshold temperatures for solid and liquid precipitation, $T_s<T_l$.

Third, unlike in natural discharge series, there is no physical justification for assuming seasonal and cyclical dependencies within the error time series. Model errors depend much more on the inadequate description of the modelled process within the conceptual model and the inputs. Large errors can occur anytime. Despite of this, for the sake of completeness, we analysed two cases - including and excluding the removal of the cyclical components from the time series (seasonality was not considered). We will denote the version of a model, where we removed the cyclical component with "-C".

Since in hybrid modelling schemes one of the oldest model classes the family of the autoregressive moving average models was usually applied [5], here these models were used as a basis for comparison to GARCH type models of the model error time series, too. The actual data analysis was arranged as follows (for details of the methodology see [2], where it was used on daily discharge data):

- We removed the deterministic components i.e. trend and cyclical components.
- We tested the stationarity using the Augmented Dickey -Fuller test (ADF) test, and fitted an appropriate ARMA model on the residuals. The choice of order of the ARMA model was based on the shape of the autocorrelation and partial autocorrelation functions (ACF and PACF). It may happen, that several alternative parameters of an ARMA model seem

acceptable. For this reason, the Akaike and Schwarz Bayes Information criteria (AIC, BIC) based on comparing the residuals from the different models were used and the model with the minimum information criterion value was preferred.

- Then we examined the residuals for possible heteroscedasticity. To test for existence of heteroscedasticity the Ljung-Box-Pierce Q-test was used on these to obtain the first estimate, and to verify it, we used the Engles's Langrange multiplier ARCH test (ARCH LM test).Subsequently we fitted a GARCH or EGARCH model and examined the fit and the residuals.
- We gave a comparison of predictions produced by the GARCH or EGARCH models against ARMA type models using the modified Diebold-Mariano (DM) statistics.

The Diebold-Mariano test comparisons between the distinct models were given in a form of a result table, the entries have 3 possible values: 0, when the models have statistically same performance, 1, when model A delivers statistically significantly better performance than B, and -1, when model A delivers statistically significantly worse performance than B (for more details on the test see e.g. [7]). The Nash-Sutcliffe coefficient, which is the most popular model performance criterion in hydrology was also used [6].

5. Results and discussion

The ARCH LM test rejected no heteroscedasticity in 3 out of 4 possible cases, the exception was the Banská Bystrica KVHK error series after the removal the cyclical components. The spectral analysis showed only one significant value in Brehy, corresponding to 1 year period and 2 values in Banská Bystrica - 1 year and 6 months. Surprisingly both data sets showed very little correlation, especially after the removal of the cyclical components.

We fitted an AR(1) in all four cases. However the GARCH model in its original form was not applicable for the same reasons as in [2]. We used EGARCH(1,1) on all three heteroscedastic model error series. Surprisingly the forecasting performance of the models was by far not as good as it was in [2] when applying these to daily data. The DM values are in table 1 and table 2, the Nash Sutcliffe coefficients in table 3.

Table 1. Performance comparison based on the DM test – Brehy.

	AR(1)	AR(1)-C	EGARCH(1,1)	EGARCH(1,1)-C
AR(1)	X	1	-1	-1
AR(1)-C	-1	X	-1	-1
EGARCH(1,1)	1	1	X	0
EGARCH(1,1)-C	1	1	0	X

Table 2. Performance comparison based on DM test – Banská Bystrica.

	AR(1)	AR(1)-C	EGARCH(1,1)
AR(1)	X	1	-1
AR(1)-C	-1	X	-1
EGARCH(1,1)	1	1	X

Table 3. Nash Sutcliffe S coefficients for Brehy and Banská Bystrica.

	NS Brehy	NS Banská Bystrica
AR(1)	-0.3452	0.1238
AR(1)-C	-0.6352	-0.1399
EGARCH(1,1)	-0.0442	0.2974
EGARCH(1,1)-C	0.0447	N/A

Comparing the Brehy models, the AR(1) and AR(1)-C models actually performed worse, than if we approximated the values by the time series mean, the AR(1)-C model being the poorest in performance altogether. The EGARCH(1,1) (-C) models did comparably against each other and outperformed the AR(1) (-C) models. Considering the Banská Bystrica models, again, the AR(1) -C delivered the worst performance. The EGARCH(1,1) model was slightly better than the simple AR(1) model. However, overall the EGARCH(1,1) models didn't offer much improvement compared to the already poorly performing AR models.

6. Discussion

The ARMA models may not properly reflect the structural properties of the heteroscedastic error series, but may outperform the GARCH models in practical performance. Indeed, as reported in the econometric literature, the EGARCH model showed better performance only on Banská Bystrica data, otherwise the forecasting performances were statistically equivalent with the respective ARMA models. Moreover, the forecasting performance of the hybrid framework did not matched its performance gained with daily data in [2] and was not good enough for practical purposes. Unlike in econometric time series, here the γ parameter was positive, which means that the data tend to react more strongly on positive changes, rather than the negative ones. It underpins the need of a non-mechanistic approach in the case based analysis of such data and the physical interpretation of statistical modelling results. In this particular case the general property of hydrological processes was manifested in the error series, that the rise of discharge is rainfall driven (a highly nonlinear chaotic intermittent process) and the decrease of discharge is ruled by the damping effects of the water storage in the driven system (catchment or river reach), which filters and smoothes the pulse like behaviour of the inputs. The fact, that this behaviour is present in the hydrological model error series, can be explained by the fact, that the modelling and forecasting of floods (pulse like rising discharge) is a more demanding task than that of droughts (slowly decreasing flows).

7. Conclusions

The hybrid modelling framework was used by analysing the structure of the one step ahead forecasting error series of deterministic hydrological model and forecasting these by time series models. So far only a few attempts to apply various nonlinear time series models were reported in the hydrological time series literature e.g. [1, 3, 4]. The fitting, forecasting and simulation performance of such models have to be explored case based [2].

The investigation of model properties and performances was thoroughly tested. In general, besides the one case, heteroscedasticity was present in the model error data. Considering the monthly error series, since all models delivered very poor performances, it is questionable, if the improvement offered by the EGARCH model is to be considered as significant at all. In these applications the manifestation of heteroscedasticity in the data was not so persuasive as with daily data and other type of models, mainly from the regime switching class (such as STAR, SETAR, LSTAR, [1, 5]) may be more appropriate for the modelling.

Acknowledgments

The authors gratefully acknowledge the VEGA Grant Agency support of the Project No. 1/0891/17.

References

[1] Komorníková M, Szolgay J, Svetlíková D, Szökeová, D, Jurčák S 2008 *J. Hydrol. Hydromech.* **56** 145-162
[2] Szolgayová E, Danačová M, Komorniková M, Szolgay J 2017 *Slovak J. Civ. Eng.* **25** 39-48
[3] Otache M, Ahaneku I, Mohammed A,Musa J. 2012 *Open J. Modern Hydrol.* **2** 79-90
[4] Wang W,Van Gelder P H A J M, Vrijling J K, Ma J 2005 *Nonlin. Processes Geophys.* **12** 55-66
[5] Valent P, Howden N, Szolgay J and Komorníková M 2011 *J. Hydrol. Hydromech.* **59** 157- 170
[6] Hlavčová K, Szolgay J, Kohnová S and Hlásny T 2008 *J. Hydrol. Hydromech.* **56** 163
[7] Komorník J, Komorníková, M, Mesiar, R, Szökeová D and Szolgay J 2006 *Phys.Chem.Earth* **18** 1127-45

IOP Conf. Series: Earth and Environmental Science **92** (2017) 012053 doi:10.1088/1755-1315/92/1/012053

Linking Benthic Macroinvertebrates and Physicochemical Variables for Water Quality Assessment in Saigon River and Its Tributaries, Vietnam

A D Pham

Faculty of Environment and Labour Safety, Ton Duc Thang University, Ho Chi Minh City, Vietnam

Email: phamanhduc@tdt.edu.vn

Abstract. The benthic macroinvertebrates living on the bottom channels are one of the most promising of the potential indicators of river health for the Saigon River and its tributaries with hydrochemistry playing a supporting role. An evaluation of the interrelationships within this approach deems necessary. This work identified and tested these relationships to improve the method for water quality assessment. Data from over 4,500 km^2 watershed were used as a representative example for the Saigon River and its tributaries. The data covered the period March and September, 2007, 2008, 2009, 2010 and 2015. To implement this evaluation, the analyses were based on accepted the methodology of Mekong River Commission and the studies of scientific group for the biological status assessment. For correlation analyses, the selected environmental variables were compared with the ecological indices, based on benthic macroinvertebrates. The results showed that the metrics of Species Richness, H', and 1-D_S had significant and strong relationships with the water quality variables of DO, BOD_5, T_N, and T_P (R^2 = 0.3751 – 0.8866; P << 0.05). While the metrics of Abundance of benthic macroinvertebrates did not have a statistically significant relationship with any water quality variables (R^2 = 0.0000 – 0.0744; P > 0.05). Additionally, the metrics of Species Richness, H', and 1-D_S had negatively correlated with the pH and TSS. Both univariate and multivariate analyses were used to examine the ecological quality of the Saigon River and its tributaries using benthic macroinvertebrates seems to be the most sensitive indicator to correlate with physicochemical variables. This demonstrated that it could be applied to describe the water quality in the Saigon River and its tributaries.

1. Introduction

The Saigon River is a river located in southern Vietnam that rises near Phum Daung in Southwestern Cambodia, flows South and South-Southeast for about 280 kilometers and empties into the Nhabe River, which its turn empties into the East Sea some 20 kilometers North-east of the Mekong Delta. The Saigon River is joined 29 kilometers Northeast of Ho Chi Minh City by the Dongnai River [1]. The Saigon River is very important to Ho Chi Minh City as it is a one of the main water supply as well as the host Saigon Port. The policies for socio-economic development and land-use change have impulsed the economic growth of Tay Ninh, Binh Phuoc, Binh Duong, and Ho Chi Minh City, but the water pollution have a tendency to increase. These consequences of environmental pollution are thus far from inevitable [2]. The high contents of organic pollutants degrade the water quality in receiving waters and threaten the aquatic ecosystems. In addition, the statistics show that the water sources has been, and continued to be, exploited speedily [3].

In order to contribute the water resources management and improve the water quality monitoring for the Saigon River and its tributaries, besides the physiochemical measurements, the aquatic organisms for the ecological health monitoring has been applied more and more because of many their advantages. In Vietnam, up to now, the application of these organisms is rather limited and not verified, especially, the relationships between benthic macroinvertebrates and physicochemical variables for the water quality evaluation [2,4].

According to this approach, the program "*Linking benthic macroinvertebrates and physicochemical variables for water quality assessment in Saigon River its tributaries, Vietnam*" will support for researchers and managers in field of ecology, resources and environment, who can applied these studies for the ecological health monitoring in the Saigon River and its tributaries. The overall objectives of the research were to: (1) Study on the relationships between benthic macroinvertebrates and physicochemical variables for the water quality assessment in the Saigon River and its tributaries; and, (2) Improve the biomonitoring method that serves for the water resources management and the environmental protection in the Saigon River and its tributaries.

2. Materials and methods

2.1. Study sites and sample collection

In the Saigon River and its tributaries, samples of qualitative and quantitative benthic macroinvertebrates, and water quality at 10 sites were collected in March and September 2007, 2008, 2009, 2010 and 2015 (see figure 1) [5].

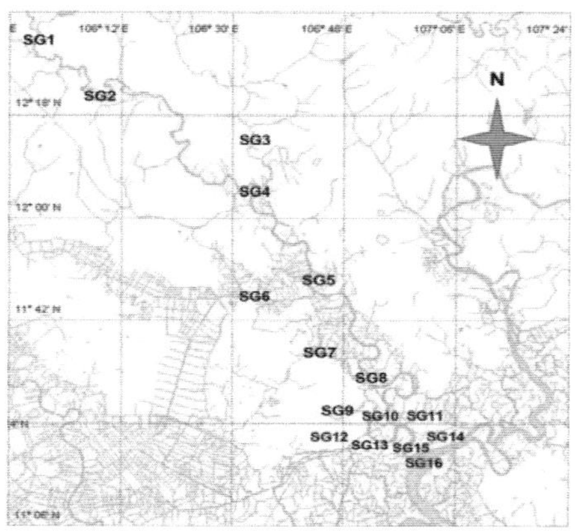

Figure 1. Map of sampling sites.

For benthic macro invertebrates, sample locations at each site were selected in each of the right and left parts of the river. Five locations were sampled at each of these parts of the river [2]. At each sampling location, a composite of four grabs was taken with a Petersen grab sampler, covering a total area of 0.1 m^2. If the sampler did not close properly because material such as wood, bamboo, large water-plants, or stones jammed its jaws, its contents were discarded and another grab was taken. The composite sample was washed through a sieve (0.3 mm) with care taken to be sure that macroinvertebrates did not escape. The contents of the sieve were then placed in jars and fixed with formaldehyde. Samples were sorted in the laboratory, because there was insufficient time at a site. The sample jar was labeled with the site location code, date, position within the river, and replicate number. The sampling location conditions, collector's name were recorded on a field sheet [2], [5]. All individuals collected were identified and counted under a compound microscope (with magnifications of 40 – 1200x) or a dissecting microscope (16 – 56x). Oligochaeta, Polychaeta, Gastropoda, Bivalvia, Ophiuroidea, and Crustacea were generally identified to species level. Insecta and Insecta larvae were classified only to genus level. The results were recorded on data sheets and specimens are kept at the Ton Duc Thang University, HCMC, Vietnam.

For water quality, the samples for environmental quality analysis in the field were collected according to the Operational Guide (3rd Ed.), UN Environment Programme (1992) [6]. The samples were taken in March and September of 2007, 2008, 2009, 2010, and 2015. Sample locations at each site were taken in the middle of the river with depth layer of surface water from 30 – 40 cm. The water samples were collected in 2 liter plastic bottles and kept at 2^0C temperature [6].

2.2. Physical, chemical and benthic macroinvertebrate analysis
The identification of benthic macroinvertebrates was based on morphology and taxonomic books such as Polychaeta [7,8,9,10]; Oligochaeta [11,12]; Gastropoda [12,13,14]; Bivalvia [12,13,14]; Crustacea [12,15,16,17,18,19,20,21]; Insecta [22,23,24,25,26,27].

The aquatic environmental parameters (pH, total suspended solid – TSS, dissolve oxygen – DO, biological oxygen demand – BOD_5, total nitrogen – T_N, and total phosphorus – T_P) were analyzed according to standard methods (APHA-AWWA-WEF, 1998) [28].

2.3. Data analysis
For all sites sampled in April and September 2007, 2008, 2009, 2010 and 2015 the following metrics were calculated (i) taxonomic richness (i.e. number of taxa); (ii) abundance (i.e. numbers of individuals per site); (iii) the Shannon-Wiener Diversity Index [29]; and, (iv) the Simpson Density Index [30]. The Pearson test (SPSS, version 16.0) was used for calculation on the correlation between benthic macroinvertebrates species number, abundance and biodiversity index) and environmental parameters in the Saigon River and its tributaries. The three metrics were tested for their potential as indicators of human impact by regressing values for two seasons of 2007, 2008, 2009, 2010 and 2015 (160 sampling events for 16 sites) against the water quality variables (pH, TSS, DO, BOD_5, T_N and T_P). For each metric examined against these variables, p values and r^2 values were calculated from regression analyses.

3. Research results

3.1. General characteristics of benthic macroinvertebrates
During the ten monitoring times, there were 49 taxa of benthic macroinvertebrates belonging to 6 groups of polychaetes, oligochaetes, gastropods, bivalves, ophiuroids, crustaceans, and insects. None organisms were dominant in the benthic macroinvertebrates communities. In each monitoring, species number of benthic macroinvertebrates ranged from 29 (March 2008) to 38 (September 2015) taxa.

The insects were the most species-rich group and occurred in almost sites. In addition, polychaets, oligochaetes, gastropods, bivalves and crustaceans also occurred widely in the studied areas. Taxon richness at a site ranged widely at the 16 sites sampled in March and September, 2007, 2008, 2009, 2010, and 2015. Richness ranged from 0 (a few canals inside of HCMC) to 15 (upper sites of Saigon River) taxa. The species of benthic macroinvertebrates recorded in the Saigon River and its tributaries were originated from the estuary or coastal region, including all polychaetes; *Melita* sp., *Grandidierella lignorum, Tachaea* sp., *Cyathura truncata, Apseudes vietnamensis, Alpheus bisincisus, Alpheus* sp. (Crustacea). A number of the species that were polluted tolerance such as *Nephthys polybranchia, Polydora* sp., *Eudistylia polymorpha* (Polychaeta); *Branchiodirus semperi, Limnodrilus hoffmeisteri, Branchiura sowerbyi* (Oligochaeta); *Melanoides tuberculatus* (Gastropoda); *Chironomus* sp., *Cryptochironomus* sp. (Insecta). While, there were few species these were sensitive, including *Neanthes meggitti* (Polychaeta), *Dromogomphus* sp., *Macronema* sp., *Hydropsyche* sp. (Insecta).

The number of individuals at sites was highly variable, ranging from 0 to 6,550 individuals/sample. The density of benthic macroinvertebrates tended to increase too high or to disappear all in near big cities or industrial areas. The oligochaete species of *Limnodrilus hoffmeisteri* was dominant in almost sites in the monitored area.

3.2. Bioindex analysis
The values of bio-indices for the water quality assessment for the Saigon River and its tributaries were presented in table 1.

3.3. Relationships of Benthic Macroinvertebrates and Physicochemical Variables

The metrics of Species Richness, H', and 1-D_S had significant and strong relationships with the water quality variables of DO, BOD_5, T_N, and T_P ($R^2 = 0.3751 – 0.8866$; $P << 0.05$) (figures 2 a, b, c).

Table 1. Bio-Indices of Benthic Macroinvertebrates for Water Quality Assessment in the Saigon River and its tributaries in March and September, 2007, 2008, 2009, 2010, and 2015.

Sites	H'	1-D_S	Ranking [2]
SG1	2.40 – 3.01	0.70 – 0.84	Light pollution
SG2	1.64 – 2.32	0.47 – 0.68	Light pollution – Low moderate pollution
SG3	1.21 – 1.54	0.40 – 0.46	Low moderate pollution
SG4	1.08 – 1.31	0.33 – 0.36	Low moderate pollution – High moderate pollution
SG5	0.83 – 1.09	0.21 – 0.30	High moderate pollution
SG6	0.32 – 0.51	0.08 – 0.15	Heavy pollution
SG7	0	0	Very heavy pollution
SG8	0.49 – 1.18	0.12 – 0.39	High moderate pollution – Heavy pollution
SG9	0	0	Heavy pollution
SG10	0.32 – 0.42	0.08 – 0.12	Heavy pollution
SG11	0.41 – 0.64	0.11 – 0.16	Heavy pollution
SG12	0	0	Heavy pollution – Very heavy pollution
SG13	0.48 – 0.72	0.15 – 0.19	Heavy pollution
SG14	0.85 – 1.59	0.22 – 0.45	High moderate pollution
SG15	1.48 – 2.77	0.44 – 0.84	Low moderate pollution
SG16	2.19 – 2.55	0.63 – 0.79	Light pollution

Notes: H' (Shannon-Wiener Diversity Index), and D_S (Simpson Dominance Index).

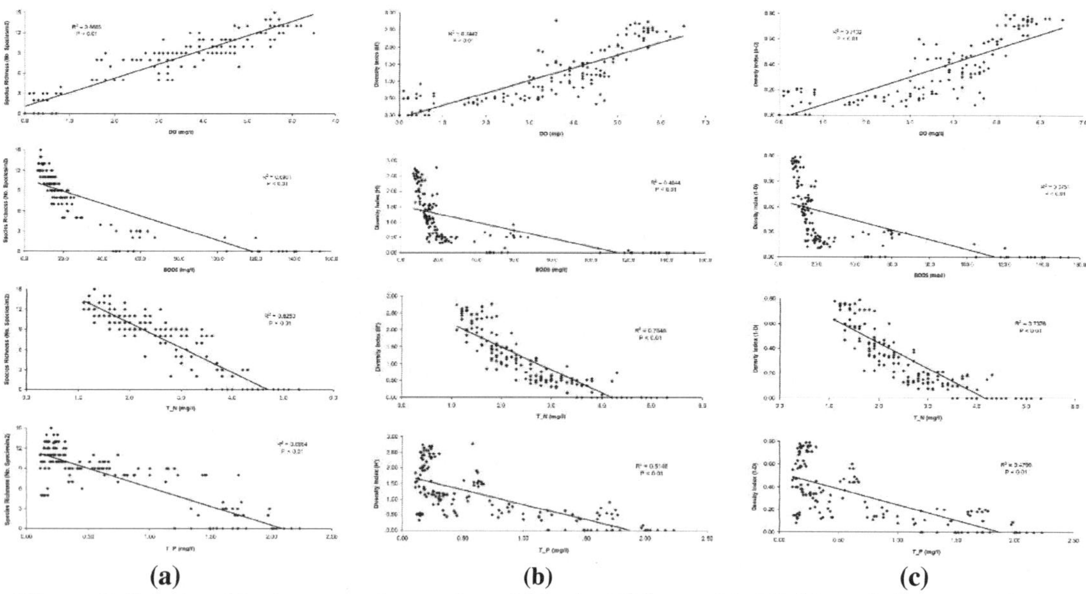

(a) **(b)** **(c)**

Figure 2. Relationships between the metrics of Species Richness (a); H' (b); and, 1-D (c) with the water quality variables for sites sampled in March and September, 2007, 2008, 2009, 2010, and 2015.

4. Conclusion

During the ten monitoring times in 2007, 2008, 2009, 2010 and 2015 at Saigon River and its tributaries, there were 49 taxa of benthic macroinvertebrates belonging to 6 groups of polychaetes, oligochaetes, gastropods, bivalves, ophiuroids, crustaceans, and insects, in the studied areas of which none organisms were dominant in the benthic macroinvertebrates communities. Generally, the changes of abundance and dominant species at 16 sites expressed clearly the environmental characteristics of the Saigon River and its tributaries, and the number of individuals tended to increase in urban and industrial sites, where more organic pollution. The bio-index values tended to decrease in the urban and industrial sites with high turbidity. These results were suitable for the analysis of benthic macroinvertebrates communities. The metrics of Species Richness, H', and 1-D_S had significant and strong relationships with the water quality variables of DO, BOD_5, T_N, and T_P. The metrics of Abundance of benthic macroinvertebrates did not have a statistically significant relationship with any water quality variables. Additionally, the metrics of Species Richness, H', and 1-D_S had negatively correlated with pH and TSS. Results of this study contributed the interesting information on benthic macroinvertebrates structure, their correlation with environmental parameters and ecological characteristics, which was quite limited in Saigon River and its tributaries. Besides, the results confirmed the advantage of using benthic macroinvertebrates and their indices as useful tools for environmental monitoring and ecological health assessment.

References

[1] Le T and Le Q H 2004 The Environmental Characteristics of the Saigon – Dongnai River Basin (Publisher of Science and Technology) pp 246
[2] Pham A D. 2014 Study on the Water Quality Assessment using Benthic Macroinvertebrates for Lower Dong Nai River Monitoirng (National University in HoChiMinh City) pp 159
[3] Nguyen V H 2009 Overview of Water Resources in Dongnai River Basin" pp 26–33
[4] Le P N, Le P Q and Pham A D 2012 Study of Ranking of Bio-Indices Using Benthic Macroinvertebrates for Lower Dongnai River System, Vietnam *J. Environ. Sci. and Eng.* **4** pp. 196 – 205
[5] Pham A D and Sangpradub N 2010 Benthic Macroinvertebrates in Biomonitoring Methods for the Lower Mekong Basin, MRC, Vientiane **8** pp 49-54
[6] UNWP. – GEMS/Water. 1992 Operational Guide UN Environment Programme pp 121
[7] Day J H 1967 Monograph on the Polychaeta of Southern Africa Part 2: Sedentaria, Trustees of the British Museum (Natural History) London pp 656
[8] Dejian Y and Raping S 1985 Polychaetous Annelids Commonly Seen from the Chinese Waters (Agriculture Publisher) pp 352
[9] Fauvel P 1953 The Fauna of India including Pakistan, Ceylon, Burma and Malaya. Annelida Polychaeta (Allahabad the Indian Press, LTD) pp 507
[10] Usakov P V 1955 URRS Fauna – Polychaeta (URRS Academy Publisher: Moscow) pp 218
[11] Thai T B 2007 Invertebrates (Publisher of Science and Technology: Hanoi) pp 382
[12] Dang N T, Thai T B and Pham V M 1980 Identification of Invertebrates in Nothr Vitnam (Publisher of Science and Technology: Hanoi) pp 573
[13] Brandt R A M 1974 The Non – Marine Aquatic Mollusca of Thailand, pp 423
[14] Dillon Jr R T 2004 The Ecology Freshwater Molluscs (Cambridge University Press: Cambridge) pp 509
[15] Dang N T and Ho T H 2012 Freshwater Decapoda in Vietnam: Palaemonidae, Atyidae, Parathelphusidae, Potamidae, (Publisher of Science and Technology: Hanoi) pp 264
[16] Gurianova E F 1951 Amphipoda of URRS Sea and Neighboring Waters (Amphipoda – Gammaridea), (URRS Academy Publisher: Moscow) pp 1029
[17] Hayward P J and Ryland J S 1990 The Marine Fauna of the British Isles and North-West Europe Introduction and Protozoans to Arthropods (Clarendon Press: Oxford) pp 416
[18] Holthuis L B 1950 Decapoda of the Siboga Expedition Part X: The Palaemonidae Collected by the Siboga and Snelluis Expeditions with Remarks on Other Species I, Subfamily Palaemininae (Leiden: Holand) pp 268
[19] Kensley B and Schotte M 1989 Guide to the Marine Isopod Crustaceans of the Caribbean (Smithsonian Institution Press: Washington, D.C. and London) pp 308
[20] Tiwari K K 1963 Alpheid Shrimps (Crustacea: Decapoda: Alpheidae) of Vietnam *Annual Faculty Science Saigon* 269–362.

[21] Bouvier E L 1925 Encyclopedie Entomologique IV: Recherches sur la Morphologie, les Variations et l Distribution systematique des Crevettes d'eau douce de la Families des Atyides (Pual Lechevalier: Editeur) pp 370
[22] Edmondson, W.T. 1976, Freshwater Biology (2nd Ed.), John Wiley & Sons, New York, 1248 pp.
[23] Lepneva S G 1964 URRS Fauna – Trichoptera (Science Publisher: Moscow) pp 38
[24] Sangpradub N and Boonsoong B 2006 Identification of Freshwater Invertebrates of the Mekong River and its Tributaries (Mekong River Commission: Vientiane) pp 267
[25] Thorp J H and Covich A B 2001 Ecology and Classification of North American Freshwater Invertebrates (Academic Press: California) pp 389
[26] Tsuda M 1987 Aquatic Entomology (Tokyo) pp 166
[27] Wiederholm T 1983 Chironomidae of the Holarctic Region. Key and Diagnoses. Part 1: Larva, Entomologica Scandinavica Supplement pp 457
[28] APHA-AWWA-WEF 1998 Standard Methods for Examination of Water and Wastewater (American Public Health Association: Washington, D.C.) pp 541
[29] Stiling P 2002 Ecology: Theories and Applications (Prentice-Hall of India Private Ltd: New Delhi) pp 403
[30] Mandaville S M 2002 Benthic Macroinvertebrates in Freshwaters: Taxa Tolerance Values, Metrics, and Protocols, Soil and Water Conservation Society of Metro Halifax pp 48

Radioactivity of buildings materials available in Slovakia

E Singovszka[1], A Estokova[1] and J Mitterpach[2]

[1]Technical University of Kosice, Faculty of Civil Engineering, Institute of Environmental Engineering, Kosice, Slovakia
[2]Technical University in Zvolen, Faculty of Ecology and Environmental Sciences, Department of Environmental Engineering, Zvolen, Slovakia

E-mail: eva.singovszka@tuke.sk

Abstract. In the last decades building materials, both of natural origin and containing industrial by-products, have been shown to significantly contribute to the exposure of the population to natural radioactivity. As a matter of fact, neither the absorbed dose rate in air due to gamma radiation nor the radon activity concentration are negligible in closed environments. The soil and rocks of the earth contains substances which are naturally radioactive and provide natural radiation exposures. The most important radioactive elements which occur in the soil and in rocks are the long lived primordial isotopes of potassium (40K), uranium (238U) and thorium (232Th). Therefore, additional exposures have to be measured and compared with respect to the natural radiation exposure. Further, it is important to estimate the potential risk from radiation from the environment. The paper presents the results of mass activities of ^{226}Ra, ^{232}Th a ^{40}K radionuclides in cement mortars with addition of silica fume. The gamma index was calculated as well.

1. Introduction

The most important radioactive elements which occur in the environment are the long lived primordial isotopes of potassium (^{40}K), uranium (^{238}U) and thorium (^{232}Th). These radioisotopes are occurring in almost all the building materials but in different concentration. About 85% of dose received by man is due to natural radioactivity, of these 55.8% of dose comes from ^{238}U series, 14% from ^{232}Th series and 13.8% from ^{40}K [1]. The radiation level due to natural radioactivity is about 2.4 mSv/y and the estimated worldwide average activity of ^{238}U, ^{232}Th and ^{40}K in the earth´s crust to be 32 B/kg and 45 Bq/kg respectively [2]. Additional exposure have to be measured and compared with respect to the natural radiation from environment. It is important to estimate the potential risk from radiation from environment. Man is continuously exposed to ionizing radiation from naturally occurring radioactive materials (NORM). The origin of these materials is the Earth's crust, but they find their way into building materials, air, water, food, and the human body itself. Measurement of activity concentrations of radionuclides in building materials is important in the assessment of population exposures, as most individuals spend 80% of their time indoors [3]. Building materials cause direct radiation exposure, because most of them contain naturally occurring radioactive materials, mainly radionuclides from the 226Ra and 232Th decay chains and 40K. The population-weighted average of indoor absorbed dose rate in air from terrestrial sources of radioactivity is estimated to be 84 nGy h^{-1} [4]. The worldwide average indoor effective dose due to gamma rays from building materials is estimated to be about 0.4 mSv per year [1]. Elevated indoor external dose rates may arise from high radionuclide content in building materials [5,6,7]. Generally, natural building materials reflect the geology of their site of origin. The average activity concentrations of ^{226}Ra, ^{232}Th, and ^{40}K in the Earth's crust are 35, 30, and 400 Bq kg–1, respectively. However, elevated levels of natural radionuclides causing annual doses of several mSv have been identified in some regions around the world, e.g. in Brazil, France, India, Nigeria, Iran [1,4]. This external radiation exposure, caused by gamma emitting radionuclides in building materials, can be assessed either by direct exposure measurements in the existing buildings or by radionuclide analyses of building materials with the dose rate modelling.

Moreover, the trend of incorporation of various wastes into cement composites may result in increased radiological activity of concretes and cement-based materials depending on the radioactivity of waste utilized. The aim of the paper is to study the radiological parameters (mass activity of radionuclides and gamma index) of cement mortars prepared with various portions of silica fume. The activity of natural radionuclides in building materials have to be in accordance with the legistalitve requiremernts in the Slovak republic [8,9,10]

2. Material and methods

2.1. Cement samples

Research on radioactivity of building materials was performed using 4 mixtures of cement mortars with different portions of silica fume as follows: 0 wt.% (KU 0 – reference sample), 10 wt.% (KU 10 sample), 20 wt.% (KU 20 sample) and 30 wt.% (KU 30 sample) as cement replacement. Table 1 describes the recipe of mixtures in more detail.

Table 1. Recipes of cement mortars per $1m^3$.

The components of mortar mix [kg]	Composition KU 0	KU 10	KU 20	KU 30
Cement CEM 42.5 R [kg]	450	405	360	315
Cement replacement [wt.%]	0%	10%	20%	30%
Silica fume [kg]	0	45	90	135
Fine aggregate [kg]	1350			
Water [L]	225			

Silica fume, also known as silica or microsilica is a by-product of manufacture of silicon or various silicon alloys by reducing quarz to silicon in an induction are furnace at temperature up to 2000°C. Gassified SiO at high temperatures condenses in the low-temperature zone to tiny spherical particles consisting of noncrystalline silica. The chemical composition of silica fume depends not only upon the raw materials used, but also upon the quality of the electrodes and the purity of the silicon product. Generally speaking, the impurities in condensed silica fume decrease as the amount of silicon increases in the final products. The by-products from silicon metal and ferrosilicon alloy industries, producing alloys with 75% or higher silicon content, contain up to 95% noncrystalline silica. Minor components in silica fume are aluminium oxide, iron (III) oxide, carbon, sulphur, calcium oxide, titanium (IV) oxide phosphorous(V) oxide and less than 1% alkalis [11]. The chemical composition of the silica fume, incorporated into the cement mortars in our experiments, is given in table 2.

Table 2. Chemical composition of silica fume.

Oxide	SiO_2	CaO	Al_2O_3	Fe_2O_3	MgO	K_2O	Na_2O	Cl	SO_3	C	LOI
[%]	71.65	18.17	0.70	1.19	1.95	1.14	0.52	0.14	0.10	2.39	1.21

It is well known that use of silica fume in concrete decreases or eliminates the free lime content, increases the strength and decreases or eliminates concrete. Silica fume showed an effect on the released lime 8 hour after the hydration of Portland cement, which is significantly faster than slag or fly ashes due to its small particle size. Laboratory results have indicated that the use of silica fume decreases the diffusion coefficient of contaminants very significantly [12].

2.2. Testing methods

The mass activities of radionuclides in cement mortars were measured using gamma ray spectrometry. Gamma ray spectrometry has been used for many years to measure the specific radionuclides activity in a given sample. This technique is an important tool in field of environmental radioactivity measurements due to its high resolution, large photo peak efficiency and it can measure different radionuclides in a single spectrum. Measurement of radioactivity was carried out using an EMS-1A SH (Empos, Prague, Czech Republic) detection system equipped with a NaI/Tl scintillation detection probe and a MC4K multichannel analyser with optimized resolution of 818 V, 4,096 channel and with 9 cm of lead shielding and internal lining of 2 mm tinned copper.

The background of the detection system plays a vital role in the measurement of low-level activity as typically found in construction materials. The counting system must have a background as low as attainable with a minimum number of spectral lines originating from natural radionuclides which may be present in the system components and in the surrounding environment of the counting facility. In the presented study, routine measurements of the background count rates for natural radionuclides were carried out before each set of measurements, each for a counting time of 43,200 s (*i.e.*, 12 h). The Cement specimens were crushed and milled to prepare the powder samples and consequently filled into Marinelli containers and weighed. The filled containers were sealed and kept for 30 days to achieve the radioactive equilibrium. The emphasis was on the determination of specific activity concentration of ^{226}Ra, ^{232}Th and ^{40}K. The radioactivity of ^{40}K was measured directly through its gamma ray energy peak at 1461 keV, while activities of ^{226}Ra and ^{232}Th were calculated based on the mean value of their respective decay products. Activity of ^{226}Ra was measured using the 351.9 keV gamma rays from ^{214}Pb and the activity of ^{232}Th was measured using the 238.6 keV gamma rays of ^{212}Pb. Every sample was measured for 86 400 s.

Radiation hazard index, called the gamma activity concentration index, I_γ, has been defined by the European Commission [13] and equation (1) is given below:

$$I_\gamma = A_U/300+A_{Th}/200+ A_K/3000 \quad (1)$$

The index I_γ is correlated with the annual dose rate due to the excess external gamma radiation caused by superficial material. The table 3 shows a dose rate criterion [14].

Table 3. Dose rate criterion for Gamma Index.

I_γ [-]	Dose rate criterion [mSv×y^{-1}]
$I_\gamma \leq 2$	0.3
$2 < I_\gamma \leq 6$	1
$I_\gamma > 6$	>1

3. Results and discussion

The table 4 presents the results of measured mass activities of radionuclides ^{226}Ra, ^{232}Th and ^{40}K. Comparing the radium activities measured, the limit value of 120 Bq.kg^{-1} was not exceed for any composite even silica fume itself. As presented in our previous work [15], the radionuclide activities in CEM I cements ranged from 3.69 – 36.8 Bq·kg^{-1}, 11.8 – 24.9 Bq·kg^{-1} and 36.98 – 331.4 Bq·kg^{-1} for ^{226}Ra, ^{232}Th and ^{40}K, respectively. According to Terpakova (2000) [16], the aggregates radioactivity strongly depends on the geological sources and ranges from 21.5 to 29.9 Bq·kg^{-1}for ^{226}Ra, 5.0 to 11.4 Bq·kg^{-1} for ^{232}Th and 106.5 to 270.6 Bq·kg^{-1} for ^{40}K. Cabanekova (1996) [17] presents a very wide range of radionuclides activities regarding concretes. As seen in Table 4, the silica fume´s radioactivity is comparable to the values presented for cement and aggregates as majority components of cement mortars. The highest values were measured for reference sample KU 0 without any silica fume. The hypothesis that the silica fume does not increase the radioactivity levels of concretes was confirmed. On the other hand, the hypothesis that silica fume can play a role of a dilution agent regarding the radioactivity levels in concretes was not quite confirmed when compared the mass activities of silica fume-based samples.

Table 4. Mass activities of radionuclides in studied cement mortars.

Sample	Mass activity [Bq.kg^{-1}]		
	^{226}Ra	^{232}Th	^{40}K
KU 0	5.72	20.74	80.45
KU 10	2.67	11.50	5.02
KU 20	2.18	12.74	18.17
KU 30	3.47	9.06	2.06
Silica fume	3.91	6.11	297.13

The mean values of index of natural radionuclide mass activity (gamma index, $I\gamma$) calculated from the measured activity concentration of ^{226}Ra, ^{232}Th and ^{40}K are presented in figure 1.

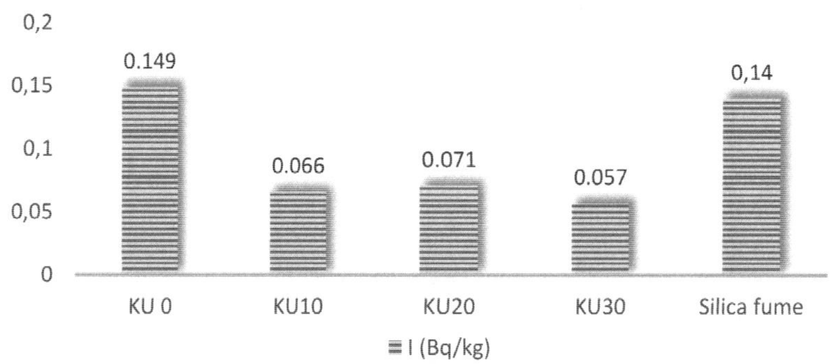

Figure 1. Gamma indexes of cement samples (I_γ).

Gamma index ranged from I = 0.149 to I = 0.057 for the sample with addition of silica fume. Silica fume had gamma index with value I = 0.14. The lowest value of was calculated for KU 30 cement sample with 30 wt.% silica fume. The highest value was found for sample KU 0 without any silica fume addition. The gamma index should also take into account typical ways and amounts in which the material is used in a building. The limit values depend on the dose criteria, the way and amount of the material and the manner in which it was used in a building and construction. The limit value of I = 1.0 for building materials was not exceeded in any cement mortar sample.

4. Conclusion

The natural level of radioactivity in building materials is one of the major causes of external exposure to γ-rays. By the determination of the radioactivity level in building materials, the indoor radiological hazard to human health can be assessed. Knowledge of basic radiological parameters and radioactive contents in the construction materials is important since it allows us to calculate the exposure of the population of radiation from natural sources. It plays an important role in the protection measurement, geoscientific research and guidelines for the use and management of these materials. During the last decades, there has been an increasing interest in the study of radioactivity in various building materials.

In this study, the commonly used composites based on silica fume addition as partial replacement of cement - in the amount of 0% (reference), 10%, 20% and 30% were investigated to assess the radiological hazards. The values obtained in the study are within the recommended safety limits, showing that these building materials do not pose any significant radiation hazard and hence the use of these materials in the construction for dwelling purpose can be considered to be safe for the inhabitants. This study can be used as a reference for more extensive studies of the same subject in future.

Acknowledgment
This work has been supported by the Slovak Grant Agency for Science (Grant No. 1/0648/17).

References

[1] UNSCEAR 1993 Sources and effects of ionizing radiation Report to General assembly, with scientific annexes United Nations New York
[2] UNSCEAR 2008 Sources and effects of ionizing radiation Report to General assembly, with scientific annexes (United Nations New York)
[3] Mustonen R 1984 Radioactivity in and Radon Exhalation from Finnish Building Materials *Health Phys.* **46** 1195–1203
[4] UNSCEAR 2000 Sources and Effects of Ionizing Radiation Report to the General Assembly with Annexes (United Nations NewYork)
[5] Chen C J and Lin Y M 1996 Assessment of building materials for compliance with regulations of ROC *Environ Int.* **22** (Suppl. 1) S221–S226
[6] Stoulos S Manolopoulou M and Papastefanou C 2003 Assessment of natural radiation exposure and radon exhalation from building materials in Greece *J. Environ. Radioact.* **69** (3) 225–240
[7] Richi S and Bruzzi L 2006 Natural radioactivity and radon exhalation in building materials used in Italian dwellings *J. Environ. Radioact* **88** 158–170
[8] 96/2017 Z. z. ktorým sa mení a dopĺňa zákon č. 541/2004 Z. z. o mierovom využívaní jadrovej energie (atómový zákon) a o zmene a doplnení niektorých zákonov v znení neskorších predpisov
[9] 103/2015 Úplné znenie zákona č. 355/2007 Z. z. ZÁKON o ochrane, podpore a rozvoji verejného zdravia
[10] Vyhláška č. 528/2007 Z. z.Vyhláška Ministerstva zdravotníctva Slovenskej republiky, ktorou sa ustanovujú podrobnosti o požiadavkách na obmedzenie ožiarenia z prírodného žiarenia
[11] Uchikawa H 1986 Influence of character of clinker on the early hydration process and rheological property of cement paste *Cem. and Conc. Res.* **15** (4) 561-572
[12] Feldman R.F and Huang C Y 1985 Properties of Portland Cement-Silica Fume Pastes I. Porosity and Surface Properties *Cem. and Conc.Res.* **15** 765-774
[13] EC (European Commission) 2000 Radiological Protection Principles Concerning the Natural Radioactivity of Building Materials. Directorate-General Environment, Nuclear Safety and Civil Protection (Luxembourg Belgium)
[14] ICRP 2002 Basic anatomical and physiological data for use in radiological protection: reference values (ICRP Publication 98 Oxford UK)
[15] Estokova A and Palascakova L 2013 Assessment of natural radioactivity levels of cements and cement composites in the Slovak *Rep.c Int. J. of Env. Res. and Pub. Heal.* **10** (12) 7165-7179
[16] Terpakova Eva 2000 Rádioaktivita stavebných materiálov Spišská Nová Ves: Stavebná fakulta Technická univerzita v Košiciach II. ročník konferencie *Rádioaktivita v životnom prostredí* 80-84
[17] Cabanekova H 1996 Concentration of natural radionuclides in various types of building materials in Slovakia Bratislava *J.of Rad. and Nucl. Chem.* **209** 301-306

IOP Conf. Series: Earth and Environmental Science **92** (2017) 012055 doi:10.1088/1755-1315/92/1/012055

Land management decisions in a carbonatic geo-environment

P Siska[1,3] and I-K Hung[2]

[1] Department of Ecochemistry and Radiology, UCM, Trnava, 91701 SVK
[2]College of Forestry and Agriculture, SFASU, Nacogdoches, TX 75961, USA

E-mail: peter.siska@ucm.sk

Abstract. Land is the uppermost territorial unit of the earth's surface that is quasi-homogeneous in its physical, natural, and also anthropogenic properties. The fundamental component of land is lithosphere. The focus of this work is on a carbonatic geo-environment that is dominantly characterized by Mesozoic rock complexes, significant chemical weathering, and a set of landforms that are unique to this type of a geological structures. In general, optimal land management is a composite of land sharing and land sparing practices; however, in order to answer the question: 'What is a parcel of land best suited for?' often requires well-organized spatial data. In this work, we have focused on developing a model that would evaluate the suitability of a carbonatic geo-environment for land management practices. Due to the potential hazards of some sinkhole infested areas, the risk of natural hazards must be first evaluated. In addition, the level of hazards depends on population pressure and the intensity of human impact on this particular environment. In this research, we have applied the principles of geostatistics to evaluate the probabilities for sinkhole hazards as well as fuzzy logic to evaluate the suitability of land sharing and land sparing management.

1. Introduction

Since the dawn of Homo Sapiens on this planet, approximately 100-200K BP, the interaction between humans and their land base has become increasingly more dynamic; as a result, new methods and procedures for land management have gradually been developed in order to cut down on expense and promote more sustainability. In the United States, nature conservation expenses have reached 80 billion USD annually, which is about five times lower than the actual need [1]. In the United Kingdom, the environmental land management costs already amounted to 2 billion GBP several years ago; most of this funding was dedicated to the conservation of biodiversity and mitigation of global warming [2]. The Institute for European Environmental Policy (IEEP) have estimated that the costs associated with environmentally benefitting land management will be 34 billion euros per year in 2020. In addition, the cost of keeping up environmentally sensitive farms, physical infrastructures, and various trainings is going to add nine billion euros, which raises the total cost to 43 billion euros per year with a margin of error +/- 8.5 billion [3]. These estimates include many practices associated with agriculture and forestry, i.e. about 18 billion euros of investments per year into arable land, and an additional ten billion per year for grassland management as well as the creation of woods and forests, etc.

Land management practices are frequently grouped into two categories: land sharing and land sparing practices [4]. In the first category, the land is shared by both people and the natural wildlife, resembling a rather symbiotic life style; in contrast, land sparing practices leave the land undisturbed to wildlife and natural processes. Due to the preservation of endemic flora, fauna, and the source of clean drinking water, the land sparing practices are more appropriate for some portion of carbonatic environment. In addition, the risk factors for humans are sometimes significantly higher due to the collapse of underground cavities. In an ideal case, the implementation of conjunctive land sparing and land sharing management practices appear to be the most effective from an economic point of view [4]. Examples of conjunctive land management in carbonatic geo-environment can be seen in a typical tourist environment where carbonate rocks usually create attractions (such as caves); the same land is used for tourism and the conservation of the original biosphere as well as its own special geological forms with their own wild life, i.e., such as species that have adapted to living without sunlight. The successful land management practices in a carbonate geo-environment require the assessment of potential risks associated with this land and its particular complex natural and human characteristics.

The world urbanization process has rapidly increased, and in some countries, it has almost climbed to 90%. The management of urban forestry, soils, air as well as industrial and historical complexes is a great challenge to society. Highly populated countries such as China face land management challenges that are associated with the development of megacities [5]. In this work, we will focus on data from the Ste. Genevieve and St. Louis limestones with high content of calcite mineral in Tennessee where urban areas and partially traditional agricultural land come together. The area also indicates some hazardous geo-environmental characteristics due to the infestation of sinkholes.

2. Model building

The purpose of this work is to build a two-step model that provides information about suitability of land for management practices. In the first block, indicator kriging method was used to assess the probability of hazards from collapsing sinkholes. In the second block, a fuzzy logic and GIS overlay are used to produce a land management propensity model (LAMP) that provides fundamental information about the potential of land for future development. The basis for this model is the three layers of information: a) probability of land hazards, b) the density of the population and c) landuse/landcover types. Technically, LAMP can be viewed as a set of models that use spatial data to evaluate a variety of land management aspects with the purpose to provide information on land management from different points of view. The fuzzy logic in GIS is an effective tool to evaluate the suitability question.

2.1 Hazard assessment

Sinkhole collapse is a hazard that occurs in carbonate areas when the structural roof of a void loses support and collapses. There are several parameters that lead to sinkhole collapse. For example, [6] identified the impact of four groups of parameters causing doline collapse: lithology, hydrology, topography and vegetation. In the Ebro Valley in Spain, six variables out of 27 were selected for model development [7].

For this model, two sinkhole morphometric parameters were identified in the field that are good indicators for sinkhole collapse: long to short sinkhole axes with a ratio equal and less than 1.8, and the angle of long sinkhole axis within 40-50, 70-80, or 320-330 degrees of azimuth [8]. The indicator kriging method (IK) was carried out to interpolate these parameters and produce a joint hazard probability map. In the first step, two indicators were created for each sinkhole geo-located by a vector of geographical coordinate's $\mathbf{u}_\alpha$:

$$i_1(\mathbf{u}_\alpha) = \begin{cases} 1 & \text{if } L/W \leq 1.8 \\ 0 & \text{otherwise} \end{cases} \tag{1}$$

$$i_2(\mathbf{u}_\alpha) = \begin{cases} 1 & \text{if azimuth } \subset [40,50] \cup [70,80] \cup [320,330] \\ 0 & \text{otherwise} \end{cases} \tag{2}$$

The spatial connectivity of these two sets of indicators was then quantified and modelled using the indicator variogram:

$$\gamma_I(h) = \frac{1}{2N(h)} \sum_{\alpha=1}^{N(h)} [i_k(\mathbf{u}_\alpha) - i_k(\mathbf{u}_\alpha + \mathbf{h})]^2 \qquad k=1,2 \tag{3}$$

where the indicator value of i_k (length-width ratio or alignment of long axis given in degrees of azimuth) at location $\boldsymbol{u}_\alpha$ is paired with another indicator value a lag distance $\boldsymbol{h}$ away. Last, the probability for the occurrence of each event was estimated at the nodes of a 200 m spacing grid by applying ordinary kriging to indicators (2) and (3). The IK estimate at node $\mathbf{u}_0$ was written as:

$$\hat{i}_k(\mathbf{u}_0) = \sum_{\alpha=1}^{n(\mathbf{u}_0)} \lambda_\alpha \, i_k(\mathbf{u}_\alpha) \tag{4}$$

where λ_α are kriging weights that are solutions of the following system of linear equations:

$$\begin{cases} \sum_{\beta=1}^{n(\mathbf{u}_0)} \lambda_\beta C_I(\mathbf{u}_\alpha - \mathbf{u}_\beta) + \mu(\mathbf{u}_0) = C_I(\mathbf{u}_\alpha - \mathbf{u}_0) \\ \sum_{\beta=1}^{n(\mathbf{u}_0)} \lambda_\beta = 1 \end{cases} \qquad \alpha = 1,\ldots,n(\mathbf{u}_0) \tag{5}$$

The assessment of hazards in a carbonatic environment based on natural processes is incomplete without population data and land use practices. If people are absent in a hazardous region, then the natural hazard itself is not harmful, yet it can still be harmful to animals or wildlife. In order to account for these variables, the 2010 National Census data and kernel density function were used to produce a population density raster layer with a 200 ft. resolution and State Plane coordinate system. The impact on the wild life and conservation was evaluated using the National Land Cover Database with a total of 15 classes found within the study area [9].

2.2. GIS fuzzy overlay

The fuzzy overlay is a frequently used method for suitability analysis. It combines fuzzy membership raster data. In the first step, the original data are transformed to a fuzzy membership based on its relationship to a specific set such as population density. These transformed values (from 0.0 to 1.0) define the possibility of a point or pixel to be in a specific set. The full membership receives the value 1, while the weak membership will be close to 0. In addition, fuzzy logic is a useful approach in regard to the imperfection of attribute values that do not have strict binary characters. Most of the variables in earth sciences are continuous rather than strictly defined categories. In other words, there is a natural transition zone between two different categories where variable properties change continuously from one type to another. For example, soil texture categories often change continuously from sandy soils to silty soils, or forest types from a hardwood to coniferous forest. These changes in nature occur slowly from one type to another with many transitional stages. Fuzzy logic accounts for this “fuzziness” of the earth’s natural-physical boundaries.

From the group of fuzzy overlays, the gamma overlay was selected due to its property to combine the fuzzy Sum and fuzzy Product characteristics. The three previous layers, a) probability of hazard as predicted by indicator kriging, b) population density, and c) landcover data were used to produce the LAMP model. The landcover data was also assigned a new weighting system that was based on the estimated residence time people spent in a specific landcover type. The weight for open water, woody wetlands, barren rock is the lowest (equal to one); shrub/scrub and forests is two; grassland, cultivated crop land, pastures is three; low intensity developed land and open land is four; and medium and high intensity land is five. The gamma fuzzy overlay was performed in a GIS environment to combine all three transformed raster data using the following equation:

$$\mu(x) = (\text{Fuzzy Sum})^{\gamma} * (\text{Fuzzy Product})^{1-\gamma} \tag{6}$$

IOP Conf. Series: Earth and Environmental Science **92** (2017) 012055 doi:10.1088/1755-1315/92/1/012055

where ‘μ’ is the LAMP index and the ‘γ’ value was determined to be 0.9; as it appears to be the optimal value between the fuzzy Sum and fuzzy Product for this dataset. The fuzzy Product decreases the final values while the fuzzy Sum increases the importance of values with high membership characteristics. Therefore, in the decision-making process regarding land management practices, the high membership values in the LAMP model warn managers regarding the safety of people and activities in this particular area.

3. Results

The results from indicator kriging that led to the evaluation of the spatial distribution of sinkhole collapse probabilities is presented in figure 1 (left). The spatial distribution model originated through a “joint” (combined) probability: $p = prob\ \{a_1 \cap a_2\}$ which is the probability that the two events a_1 = “length-to-width” sinkhole ratio is at or below the critical value of 1.8 and a_2 = “the long axis orientation” runs parallel to one of the three systematic joints sets occurring simultaneously. Hence, this joint probability was obtained by multiplying the two estimated probabilities of the two separate events that were estimated by indicator kriging. As figure 1 (left) indicates, the darker regions with a higher combined probability occur in a circular shape, mostly in the southwestern part running from Interstate Highway 24 in the southwest to the Montgomery County eastern boundary in the northeast.

Most of the current land management in this area focuses on land sharing activities between pastures and agriculture crops on one hand and the wildlife and conservation on the other. The higher risk also continued along the Red River where parks and reservations such as the Port Royal State Park with a high historical significance are located. The area marked as “C” is especially worth noticing; it indicates that the intersection area is highly prone to sinkhole collapse. The right side of figure 1 shows the results of the fuzzy overlay.

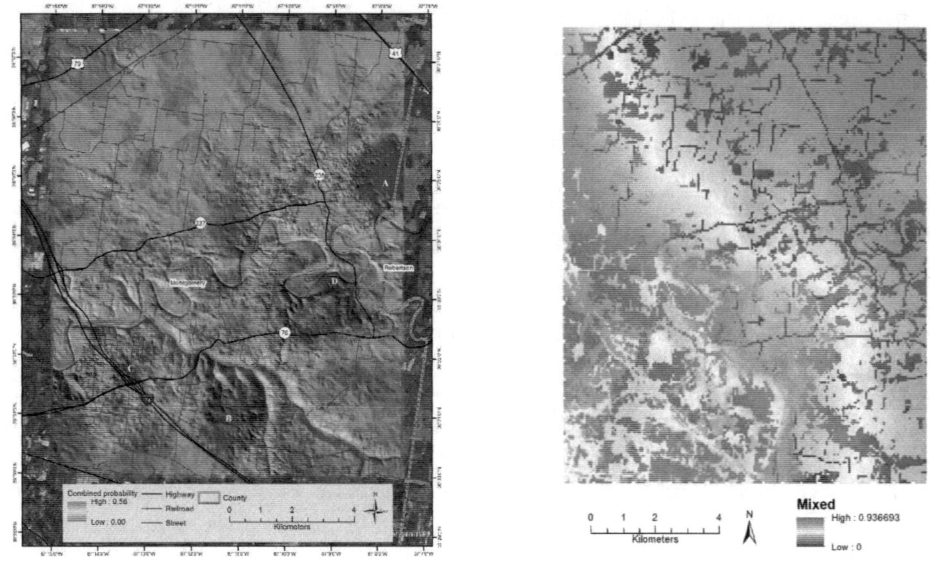

Figure 1. Spatial models indicating sinkhole collapse probabilities (left) and LAMP model (right).

In order to evaluate this area properly, one must consider the population density and the type of landuse. Therefore, some of the highly probable areas for sinkhole collapse disappeared in the LAMP model, especially the area marked with the letter “A.” There is also no imminent potential threat to people based on the estimated residence time. Hence, the potential collapse of sinkholes can be hazardous only if a high number of people reside in this area and/or other properties that are located in this type of land. In our opinion, this part of land is suitable for land sparing practices. On the other hand, the combination of low (blue) sinkhole collapse probability values from the left map and low hazard potential (green) values from the right map of the same figure can be considered for further development. Therefore, in order to make sound decisions about the future of this land, the LAMP model, together with the potentiality for sinkhole collapse model provide valuable information for the land decision-making process.

4. Conclusion

Land is our prime resource. Due to an increasing population and economic developments, land is becoming more expensive and the land management processes more complex. There are six major criteria that must be considered regarding future land management:1) the production of food to satisfy even the most populated areas in the world, 2) the conservation of the natural environment, 3) urban development, 4) the detoxification of polluted regions, 5) economic and technological advancements and 6) safety factors that guarantee life without conflict or threat. The idea of sustainable development is closely related to the goals of land management. This study focuses on developing a new perspective on land management; it inspires better solutions for future decisions regarding the management of land resources and land space for human development. In spite of a great complexity of socio-economic, demographic, and conservation problems, the fundamental question is very simple: How are we going to survive on this planet and "What can this land offer?" These are basically the same questions humans have been asking themselves since the beginning of our existence. The purpose of developing the LAMP model is to produce a tool for planners, developers, practitioners, local governments and public offices for making informed decisions about the future of land and its components: natural-physical and anthropogenic. The LAMP model also provides information about the risks or vulnerability factors. People make decisions to live in places where natural hazards are of a genuine concern (e.g. active tectonic zones) or where wild elements pose a significant hazard or even where people become a hazard to endangered species.

We demonstrated the development of the LAMP model on the example of the Red and Cumberland Rivers in Tennessee where three major criteria were considered: 1) the propensity to existing land hazards, 2) a high population density, 3) and the amount of time people are spending in hazardous areas. As can be seen in figure 1 (right), the LAMP index, which is close to the 0.9 value is mostly in the southwestern corner of the study area where all of the three mentioned criteria have a high probability of occurring; on the other hand, values near zero occur in the northeastern area.. Some of this region has the potential for implementing land sparing management practices; it is a safe area where wild life and land aesthetics can be developed. Therefore, each value in the LAMP model is an impetus for corresponding land management decisions and practices. The areas with high values (near one) are suitable for conjunctive land sharing and land sparing practices, while the low coefficient values near zero are attractive for land sparing management. Therefore, land management practices must be well-planned and geared towards people and their needs.

References

[1] Shaffer M L, Scott J M and Casey F 2002 *Bioscience* **52** 439-43

[2] Cao Y, Elliott J, McCracken D, Rowe K, Whitehead J and Wilson L 2009 Estimating the scale of future environmental land management requirements for the UK *Land Use Policy Group Report* 59

[3] Hart K, Baldock D, Tucke G, Allen B, Calatrava J, Black H, Newman S, Baulcomb C, McCraken D and Ganitioner S 2011 London *DG Environment Report contract No ENV.F.I/ETU/2010/0019r*

[4] Legras S, Martin E and Piguet V 2018 *Ecol. Econ* **143** 170-87

[5] Zeng C, Yang, L and Dong J 2017 *J Clean Prod* **153** 68437-647

[6] Yilmaz I 2007 *Eng. Geol.* **90** 89-103

[7] Galve J P, Bonachea J, Remondo J, Gutierrez F, Guerrero J, Lucha P, Cendrero A, Gutierrez M and Sanchez J A 2008 *Eng Geol* **99** 185–97

[8] Siska P, Goovaerts P and Hung I 2016 *Prog. in Phys. Geog.* **40** 579-97

[9] Homer C G, Dewitz, J A, Yang L, Jin S, Danielson P, Xian G, Coulston J, Herold N.D, Wickham J D and Megown K 2015 *Photogramm Eng. Rem. S.* **81** 345-54

IOP Conf. Series: Earth and Environmental Science **92** (2017) 012056 doi:10.1088/1755-1315/92/1/012056

Monitoring based localisation of pollution sources

M Sokac[1] and Y Velisková[2]

[1]Department of Sanitary and Environmental Engineering, Faculty of Civil Engineering, Slovak University of Technology in Bratislava, Slovak Republic
[2]Institute of Hydrology, Slovak Academy of Sciences, Bratislava, Slovak Republic

Email: marek.sokac@stuba.sk

Abstract. Water quality modelling is currently very effective and important tool in context of the task to ensure the required quality of water resources, respectively, to achieve (maintain) good water status according the Water Framework Directive (2000/60/EC). This paper analyses the current status in numerical modelling of pollution dispersion in streams and use of some modelling approaches for the inverse task. Inverse task means a modelling technique, which is focused on the localisation of unknown pollution source (typical common models or equations are rather focused on the pollution spreading simulation, whereas the pollution source location is known). Paper offers an idea of such inverse task solution. It is based on the known pollution concentration time courses or it can be based on the results of the on-line monitoring of the specific water quality parameters as well. For the application of inverse tasks in conditions of real streams and rivers a large number of various requirements and conditions in specific river should be considered, i. a. the non-prismatic river bed, occurrence of dead zones, dispersion rate etc. Paper also describes the first version of the software for solving inverse tasks and preliminary experiences of using this software.

1. **Introduction**

Currently, in water quality modelling practice are prevailing tasks, for which initial and boundary conditions are known and the model is solving pollution concentration in downstream direction. It is typical for the tasks of this type, that the quantity, the time course of discharge and concentration of emitted pollutant, as well as the hydraulic conditions in the stream are known parameters. The result of such modelling tasks is a time course of pollutant concentration in the particular stream locations. The overwhelming majority of existing simulation models is conceived exactly for that purpose - their focus is on pollution modelling, i.e. such models are able to simulate the pollution spreading (concentration) only downstream and these models require to enter strictly all the initial and boundary conditions (discharges, concentrations of the pollutant).

However, in practice converse problems may occur: pollution concentration time courses in specific locations are known (e.g. based on the on-line monitoring), but pollution discharge location, as well as the pollution time course are unknown. The objective of this task is to determine the location of pollution outlet (localization task) as well as the overall mass and time course of pollution concentration. This type of task can occur in case of accidental pollution, if the origin of the pollution is not known, or even with intentional act that can be defined as an offense or crime (illegal business - production of various substances, illegal wastewater outlets etc.).

2. The theoretical part - inverse tasks

As mentioned above, most of the current models are focused on modelling the pollution spreading in downstream direction. The inverse task – pollution outlet localization on the basis of known monitoring data is referred as a "inverse problem". It should be noted, that such inverse task has not always an unequivocal solution, but the result is usually few plausible solutions. Inverse problem is currently far more understood as a math problem, verified practical applications or commercial software that would deal with the inverse problem are not currently known. In the scientific literature a number of works dealing with these issues can be found, e.g. [1,2,3], but in general it can be stated, that the presented methods and solutions are relatively complex and complicated, and in the practice there is a large number of limitations and restrictions.

Based on such definition, we believe that the general and exact analytical solution of inverse problem is not possible: every solution, using any method will generate only the approximate solution or estimate, or will generate a particular - most probable solution among number of plausible solutions (simulations), the results of which are in best coincidence with monitored data.

Overall, we can divide the solving methods of inverse problems into the following groups:

- Empirical methods, e.g. [4,5].
- Simple methods- based on simplified analytical solutions of the advection – dispersion equation. These methods focus on the simulation of the individual (potential) pollution incidents, using for simulation simple analytical solutions and consequently comparing of simulation results with measured (monitored) data. The advantage of these methods is their simplicity, the possibility of governing equations modifications for the specific conditions (e.g. occurrence of dead zones, etc.). Their disadvantage is the necessity of large amounts of numerical simulations, thereby long computation time.
- More complex procedures focused on the inverse numerical simulation in upstream direction. From a mathematical point of view these methods are more accurate, but also more complicated and more difficult to code [1,2].

For practical solution of an inverse problem following assumptions are often taken:

- The pollution is conservative.
- There is a simple pollution source, which means, the stream is defined as a simple river section without tributaries and lateral inflows.

Pollution conservatism means that examined pollution is not subject to any physical, chemical or biological (biochemical) processes that take place in the stream. It is obvious that absolutely conservative substance in reality does not exist, so it is realistic to require that ongoing changes are not random and fast, but rather slow and steady (slow changes - assume that the concentration changes are reflected in the order of days or in a longer time period).

In case of a not conservative pollutant (polluting substance is a subject of physical, chemical or biological processes), inverse problem could be solved by similar processes, but decomposition (concentration decrease) of the substance should be considered in the simulation.

Assuming single pollution source means, that in the investigated river section is only one, spatially invariant source of pollution, which emits pollution directly into the river (not into tributary). Another assumption is a single river section, i.e. the deployment of monitoring sections will clearly determine the river section, where the pollution enters into the river. This simplifies the inverse task to localise the source of pollution in a particular river section without tributaries. The monitoring results in downstream end of the examined river section are the basis for the source localisation.

For proper operation of such localization tool the key task will be the determination of the exact hydraulic parameters (discharges, cross sectional area, velocity) in the stream. The task will be even more complicated by the fact, that the hydraulic parameters may vary depending on the hydrological situation (discharge) in the river.

3. Proposed method of pollution sources localization

The project VEGA Nr. 1/0805/16 focuses on methods for practical solutions of inverse problems, i.e. localisation of pollution sources. The following section describes proposed method, which will be further developed and improved.

The general solution method is based on the computation of simulations for (all possible) boundary conditions:

- Different distances between the source and the monitoring profile.
- For the various alternatives of pollution concentrations time courses.

Simulation results of each of these alternatives will be compared with data monitoring - simulation that will best match the monitored data can be considered as probable solution. The simulation itself is based on simplified analytical solution, which (as discussed further in this paper) should be further modified and developed.

In principle, it as a "brute force" method. The risk of such method is in the computer time, necessary to carry out a huge number of numeric simulations. This computational time will depend on the length and time steps, which implies the total number of simulations needed. Under the term 'adjustable length interval' can be a distance between points understood, where the pollution outlet is assumed (discretisation of stream section length), but also number of river sub-sections, where individual definition of constant hydraulic parameters is assumed. Simulation time step is used for time discretization of the pollution inflow. Decreasing the time step dt increases the number of possible pollution inflow (time courses) variations and thus the simulations count. A key factor in the choice of the time step dt is variability of pollution inflow. The choice of time step will be one of the key tasks and we assume to perform huge amount of numerical tests to obtain the optimal value of the time step *dt*.

Another assumption is the time invariability, proportionality principle and the superposition principle. Time invariability means that the pollution input causes always the same time course of the pollution concentration (pollutogram) independently of pollution entry in other time steps. In other words, if the pollution entry will shift the timeline with time step dt, then the resulting pollutogram will also move the time dt, and does not change its shape or size.

The proportionality principle means that the system (river) response is linearly proportional to the input, i.e. if the concentration will increase twice, then the output (concentration pollutogram) will be also two times bigger.

The last assumption is that the output concentrations, obtained by transformation of unit inputs, can be added together (superposition principle). This assumption is generally valid, if the concentrations are low, whereas in high concentrations, close to the saturation concentration, this assumption is not valid. However, if we assume that the monitored concentrations are sufficiently low, it is possible to apply the superposition principle of full concentration range. A summation of transformed particular pollution time courses then determines the final resulting concentration time course (pollutogram).

The definition of the parameters needed for the pollution transport simulation in a river has to reflect the fact, that the basic hydraulic parameters, characterizing the flow in a stream, are variable in time, depending on the current hydrological situation. It is therefore necessary to define the flow area, the average flow speed, eventually the discharge characterized as a function of the water depth. This principle is common in hydrological measurement profiles and hydrology service is generally able to provide such data. Inverse task should be then solved with such specific hydrological parameters, which were monitored during event (accident).

Another problem is the spatial variation of hydraulic parameters. In solving inverse problems, we assume spatial discretization - the distribution of the investigated river branch into several sections in which we can assume constant hydraulic conditions (flow velocity, depth, cross-sectional area...). The resulting pollution pollutogram will be then simply the result of the partial transformation of the input pollutogram in particular sub-sections of the investigated river branch.

Risk of the proposed inverse problem solution procedure is the sensitivity of the method to various influences, such as hydraulic stream parameters and the precision of the numerical description of the dispersion process in real streams, especially the presence of dead zones, singularities, etc. This results from the fact that the solution of the given inverse task is determined simply by the similarity of measured and modelled pollutograms. This clearly demonstrates the need for further research and improvement of modelling procedures focused on dispersion processes in conditions of real streams.

4. Results

For project implementation a SW tool was developed (working title Locator Of Pollution Source - LOPS), according to the principles described in the previous section. The basic input parameters of the SW are the longitudinal length step dx, the time step dt, sensitivity of the pollution entry (number of intervals), the searching procedure starting length (ending length is the river branch length). Besides this it is necessary to enter the monitored pollutogram, as well as the basic hydraulic parameters of the river section (currently only one section is implemented) – average flow velocity, cross sectional area and dispersion coefficient.

The first version of the LOPS successfully passed the tests with virtual (modelled) data and was preliminary tested also on real data. The test with real data and various input pollutograms were performed on a small mountain creek. Beside this was the LOPS also tested on real data from earlier tracer experiments on lowland stream, but only with instantaneous pollution injection, thus the tests should be extended with different pollution input conditions on different streams types.

The test results show, that for each field experiment can be achieved at least one good pollution input localisation, with the minimal best fit length error up to 10 percent. The problem is, that this best result was achieved with various and heterogeneous combination of SW tool computation parameters and it is not possible to determine, which parameter combination is the best one. Decrease of time or length steps, as well as increase of the SW tool input sensitivity generally does not guarantee better localisation results. A very big results heterogeneity was achieved for the pollutions source localisation length. Average localisation error (about 30%) is typically 10-30 times bigger, compared to the minimal error (less than 10%), the maximal localisation error is as well very high, typically higher the 50%.

Tests of sensitivity confirmed two important factors, that significantly improves the results of the LOPS computational procedure – pollution input sensitivity parameter and the average flow velocity. The first parameter is in fact the number of intervals, in which is pollution input unit divided to. Higher numbers of intervals lead to better results, but on the other hand significantly extend the computational time. For the second parameter – the average flow velocity - almost linear dependency of the LOPS results was confirmed and this parameter seems to be the crucial one.

During the LOPS SW test was another issue found: the localization results are based on the search for the best fit between the monitored and modelled data, but the current (first) version of the LOPS routine is not able to represent some special particular problems, e.g. occurrence of dead zones in the investigated stream, so there is significant difference between the pollutograms. Therefore, one of the main tasks in the future should be to use (or to develop) simple computational routines, based on analytical solutions, which incorporate modelling of particular dispersion problems and irregularities.

5. Conclusions

Systems for water quality on-line monitoring in rivers have a great potential for the control, eventually management of water quality in rivers. They enable operational monitoring and control of discharges of specific substances. In case of water quality accidents (acute and extraordinary deterioration of water quality) such systems can provide the information necessary for operational measures, as well as to provide data for models to simulate a prediction of water quality in real time. Specific problem related to the on-line monitoring is the polluters identification. This type of task can occur in case of accidental pollution, where the polluter is not known, as well as cases of operational or explorative monitoring of water bodies. As noted above, general and accurate analytical solutions of inverse problems are currently not known, using any method always gives us only the approximate solution or an estimate, eventually the choice of a particular solution – the most probable case from a number of possible simulations, the results of which fits best with monitored data. In this paper we also describe the pollution source localization SW – LOPS. It represents one of possible approaches how to solve the inverse problem, which is universal and simple. The preliminary tests of the LOPS show various SW tool accuracy and overall results heterogeneity. Sensitivity analysis of the SW tool has shown high correlation between the SW tool results and the velocity input value, so it means that the river flow velocity is the crucial parameter for the SW localisation tool accuracy.

Another crucial issue will be to use (or to develop) simple computational routines, based on analytical solutions, which are able to simulate particular dispersion problems and irregularities to achieve better coincidence of measured and modelled pollutograms and thus better localisation results.

Acknowledgment
This work was supported by the Scientific Grant Agency VEGA [grant number VEGA 1/0805/16, title "Localisation of accidental point sources of pollution in watercourses based on-line monitoring data"] and by the project ITMS 26240120004 "Centre of excellence for integrated flood protection of land" supported by the Research & Development Operational Programme funded by the ERDF.

References
[1] Verdière N, Joly-Blanchard G and Denis-Vidal L 2013 Identifiability and identification of a pollution source in a river by using a semi-discretized model *App. Math. and Com.* **221** 1-9
[2] El Badia A and Hamdi A 2007 Inverse source problem in an advection–dispersion– reaction system: application to water pollution *Inv. Prob.* **23** 2103–2120
[3] Andrle M and El Badia A 2012 Identification of multiple moving pollution sources in surface waters or atmospheric media with boundary observations *Inv. Prob.* **28 7**
[4] Šajer J 2014 Input time estimation *Vodní hospodářství* **6** 12-15
[5] Šajer J 2015 Example of tracer experiment use *Vodní hospodářství* **4** 1-7
[6] https://en.wikipedia.org/wiki/Molecularly_imprinted_polymer
[7] SOVVA Slovakia in the Seventh Framework Programme (FP7) — Project Overview & Results 2014 Project Goldfish 114-115 http://www.sovva.sk/files/attachments/SOVVA_SLOVAK%20FP7%20SUCCESS%20STORIES_v00.pdf 114-115

IOP Conf. Series: Earth and Environmental Science **92** (2017) 012057 doi:10.1088/1755-1315/92/1/012057

Comparative study of antimicrobial efficiency of metallurgical slags suitable for construction applications

J Strigac[1], N Stevulova[2], J Mikusinec[1], L Varecka[3] and D Hudecova[3]

[1]Povazska cementaren, a.s. Ladce (Povazska cement plant Ladce), Ladce, Slovakia
[2]Technical University of Kosice, Faculty of Civil Engineering, Institute of Environmental Engineering, Kosice, Slovakia
[3]Slovak University of Technology in Bratislava, Faculty of Chemical and Food Technology, Institute of Biochemistry and Microbiology, Bratislava, Slovakia

Email: julius.strigac@gmail.com

Abstract. The article deals with comparative study of antimicrobial efficiency of metallurgical slags suitable for construction applications. The tested slags were as follows: granulated blast-furnace slag (GS1), air cooled blast-furnace slag (AS2), demetallized steel slag (DS3), calcerous ladle slag (LS4), slag from copper refining (CS5). The antimicrobial activity was tested on selected representatives of bacteria, yeasts, and filamentous fungi. The highest antibacterial activity possessed LS4, which intensely inhibited growth of bacteria with the lowest concentration of slag (10%) in the growth media. 100% inhibition of growth of some bacteria was observed only in slags LS4, DS3 and AS2 in concentrations 20% - 60% of slag. Antibacterial activity of slag samples was decreasing in the order: LS4 > DS3 > AS2 > GS1 > CS5. Growth of all model yeasts was 100% inhibited at as low concentration as 20% of slag GS1 and DS3, and 10% of slag LS4. Antiyeast activity of slags was decreasing in the order: LS4 > GS1 = DS3 > AS2 > CS5. Regarding that filamentous fungi were selectively sensitive to presence of slags, it is possible to determine only approximate order of inhibition effectiveness of slags to fungi: LS4 > GS1 = DS3 > AS2 = CS5.

1. Introduction

Microbially influenced concrete biocorrosion is a serious problem wherever the suitable conditions for a life of the particular microorganisms, such as temperature, pH and moisture of the environment, presence of oxygen, radiation, input of nutrients and others, have come to existence. Biodeterioration reduces the utility efficiencies of concretes, mortars and their useful life. Biocorrosion, as a specific type of chemical corrosion, is caused by various biogenic organic acids, mineral acids (sulphuric acid H_2SO_4, nitric acid HNO_3) as well as by corrosive hydrogen sulphide H_2S and ammonia NH_3, which result from metabolic activity of microorganisms [1]. These aggressive metabolites react mainly with calcerous components of concrete and mortar stone forming their non-binding calcareous salts, some of them cause sulphate degradation, resulting in extreme expansion in the concrete and leads to a complete destruction of concrete structural elements. It was found out in the previous study [2], that the fine ground granulated blastfurnace slag possesses fungistatic properties. Since metallurgical slags are utilized in concretes as artificial, manufactured aggregates, as fine ground inorganic type I addition – filler aggregates and as fine ground inorganic type II addition – latent hydraulic material, this article aims to study the antimicrobial activity of all types of metallurgical slags used in concretes, possibly mortars, plasters and to compare them mutually.

2. Materials and methods

2.1. Materials

Following kinds of metallurgical slags were tested: granulated blast-furnace slag (GS1), air cooled blast-furnace slag (AS2), demetallized steel slag (DS3), calcerous ladle slag (LS4), slag from copper refining (CS5). All of them were ground to a specific surface area of 400 m^2/kg. The chemical composition of the slags was analysed by X-ray fluorescence analysis (XRF) according to EN 196-2 using apparatus SPECTRO X-LAB 2000. The semi-quantitative chemical composition of the slags is given in table 1.

Table 1. The chemical composition of metallurgical slags in wt.%.

Slag	Unit	GS1	AS2	DS3	LS4	CS5
L.O.I*	[wt.%]	0.95	0.09	6.02	5.32	+ 4.30**
SiO_2	[wt.%]	42.17	40.57	12.81	13.97	27.26
Al_2O_3	[wt.%]	6.87	8.12	1.64	17.77	7.01
Fe_2O_3	[wt.%]	0.32	2.81	29.78	1.90	46.64
CaO	[wt.%]	41.92	41.73	52.30	58.97	7.48
TiO_2	[wt.%]	0.42	0.11	0.34	0.14	0.21
MgO	[wt.%]	10.39	8.44	2.54	3.30	1.90
K_2O	[wt.%]	0.60	0.72	0.04	0.06	0.40
Na_2O	[wt.%]	0.17	0.19	0.07	0.07	1.07
SO_3	[wt.%]	1.84	2.39	0.28	1.98	0.15
MnO	[wt.%]	0.68	2.31	3.54	0.38	0.61
P_2O_5	[wt.%]	0.05	0.14	0.48	0.05	1.26
Cl^-	[wt.%]	0.0173	0.0112	0.0138	0.0017	0.0012

*L.O.I: Loss on ignition

**Increment on ignition (caused by e.g. oxidation of metallic particles, sulphides, disulphides, sulphites etc.)

The mineralogical composition of the samples was determined by XRD technique using BRUKER AXS D8 Advance device. The identified mineralogical composition of the slags is as follows: GS1 – melilite (solid solution of gehlenite C_2AS and akermanite C_2MS_2); AS2 – melilite, brownmillerite C_4AF, quartz SiO_2; DS3 – wüstite FeO, brownmillerite C_4AF, free lime CaO, portlandite $Ca(OH)_2$, β-C_2S, quartz SiO_2; LS4 – free lime CaO, β-C_2S, shanonite γ-C_2S, gehlenite C_2AS, tricalcium aluminate C_3A, gypsum $CaSO_4.2H_2O$, quartz SiO_2 and CS5 – fayalite $2FeO.SiO_2$, anortite CAS_2, pyroxene type $CaAlAlO_6$ (abbreviations in the compounds mean C = CaO, M = MgO, S = SiO_2, A = Al_2O_3, F = Fe_2O_3).

2.2. Methods of testing

The antimicrobial activity of metallurgical slags was tested on selected representatives of bacteria, yeasts, and filamentous fungi (moulds). Following microorganisms were used: G^+ bacteria - *Bacillus subtilis*, *Staphylococcus aureus*, *Micrococcus luteus*; G^- bacteria - *Escherichia coli*, *Pseudomonas aeruginosa*, *Serratia marcescens*; yeasts - *Candida albicans*, *Rhodotorula glutinis*; microscopic filamentous fungi (moulds) - *Aspergillus niger*, *Penicillium funiculosum*, *Chaetomium globosum*, *Alternaria alternata*, *Trichoderma viride*, *Cladosporium herbarum*. As growth media, meat-peptone bouillon for inoculation of bacteria, meat-peptone agar for cultivation of bacteria, Sabouraud glucose bouillon for inoculation of yeasts and malt agar for cultivation of yeasts and mycelial fungi were applied. For diluting of inoculum of bacteria and yeasts, saline was used, for preparation of spore suspension (moulds), water solution of Tween 80 was used. The tests were performed in the laboratory incubator at temperatures 30 °C for bacteria, 28 °C for yeasts, 25 °C for filamentous fungi and at relative humidity 95%.

Antimicrobial activity was determined by dilution methods in agar growth media so that resulting concentration of tested slags in growth media was 10, 20, 40 and 60% by weight. The pH value of growth media with addition of slags was strongly basic (pH 11), thus half of samples of each slag was tested at this pH, and the second half of samples was tested at modified pH (bacteria pH 7.2, yeasts and filamentous fungi pH 6.6). The first half of samples with original pH represented real conditions for growth of microorganisms in concrete, the second half of the samples with modified pH represented optimal conditions for growth of microorganisms *in vitro*.

The microorganisms growth at slags presence was compared with the microorganisms growth in control growth media without slags in dependence on time. Inhibiting effect of slags was inspected: microbistatic (microorganisms are not growing; after transfering microorganisms into growth media without inhibitor, they are growing broader) and microbicidal - lethal (microorganisms are dead; after transfering microorganisms into growth media without inhibitor, they are not growing).

3. Results and discussion

The highest antibacterial activity was achieved at slag LS4, which intensely inhibited growth of G^+ as well as G^- bacteria, what was proved also at the lowest concentration of the slag LS4 (10%) in growth media. Bacteria, except *M. luteus* did not grow at higher concentrations of slag LS4. The growth of G^+ bacteria *S. aureus* and *B. subtilis* was inhibited by slag DS3 in growth media as early as the concentration 10% and at higher concentrations, the growth was completely (100%) inhibited. 100% inhibition of growth of some bacteria was observed only in the samples of slags LS4, DS3 and AS2 in concentration range 20% - 60% of slag. Antibacterial activity of slag samples was decreasing in the order: LS4 > DS3 > AS2 > GS1 > CS5.

The growth of all model yeasts was 100% inhibited at as low concentration as 20% of slag GS1 and DS3, and 10% of LS4, at basic as well as modified pH of growth media. Growth of model yeasts was intensively reduced at the presence of slag AS2 in concentrations 40% and 60%. Slag CS5 partially inhibited the growth of model yeasts without 100% inhibitions at the highest 60% concentration. Antiyeast activity of slags was decreasing in the order: LS4 > GS1 = DS3 > AS2 > CS5.

The tested model filamentous fungi were sensitive to presence of the slag samples in various manners. As it is apparent from the results of inhibition, all slags inhibited the growth of filamentous fungi by 100% - 40% in concentration 60% of slag. The most sensitive to presence of all slags were *A. niger* and *T. viride*, the growth of which in concentrations 20% - 60% of all slag samples at basic as well as modified pH value, was inhibited on 100% with fungistatic effect (they stopped growth of fungi), and in concentration 60% of slag LS4 with fungicide effect (killed fungi) at *T. viride*. The most resistant was *A. alternata*. Its growth was the most intensely inhibited by slag LS4, 100% inhibition was observed at 40% - 60% of slag, with mostly fungistatic effect. The most inhibiting activity for all fungi was disposed by slag LS4, which 100% inhibited growth of almost all model filamentous fungi in concentrations 20% - 60% at basic pH of growth media, with mostly fungistatic effect. Only the *T. viride* growth was inhibited on 100% with fungicide effect at basic pH of growth media in concentration 60% of slag LS4. At modified pH of growth media, the growth of model filamentous fungi was inhibited on 100% in concentration range 20% - 40% of slag LS4 with fungistatic effect and at concentration 60% of slag LS4 was inhibited on 100% with fungicide effect on *A. alternata, A. niger, T. viride, Ch. globosum*. Lower inhibiting effect on filamentous fungi growth was observed at slags GS1 and DS3, but they inhibited the growth of all tested filamentous fungi on 40% - 100% at concentration 40%. The lowest impact on the growth of model filamentous fungi was observed at the presence of slag CS5 in growth media, which affects only the growth of the most sensitive fungi *A. niger* and *T. viride*. The most resistant was *A. alternata*. Its growth was the most intensely inhibited by slag LS4, 100% inhibition was observed at 40% - 60% of slag, with mostly fungistatic effect.

Regarding that model filamentous fungi were selectively sensitive to the presence of tested slags, it is possible to determine only approximate order of inhibition efficiency of slags to filamentous micromycetes: LS4 > GS1 = DS3 > AS2 = CS5.

The pH values of growth media did not significantly affect the intensity of inhibition of model microorganisms growth.

4. Conclusion

According to these results, metallurgical slags possess the great potential for using as antimicrobially active agents for building materials against biodeterioration. The highest antimicrobial efficiency had calcerous ladle slag (LS4), the medium efficiency granulated blast-furnace slag (GS1) and demetallized steel slag (DS3), lower efficiency air cooled blast-furnace slag (AS2) and the lowest efficiency slag from copper refining (CS5), respectively.

References

[1] George R P, Vishwakarma V, Samal S S and Mudali U K 2012 Current understanding and future approaches for controlling microbially influenced concrete corrosion: a review *Concr. Res. Lett.* **3** 491–506
[2] Strigac J and Martauz P 2016 Fungistatic properties of granulated blastfurnace slag and related slag-containing cements *Ceram.-Silik.* **60** 19–26

IOP Conf. Series: Earth and Environmental Science **92** (2017) 012058 doi:10.1088/1755-1315/92/1/012058

The antifungal efficiency of carbide lime slurry compared with the commercial lime efficiency

J Strigac[1], J Mikusinec[1], J Strigacova[2] and N Stevulova[3]

[1]Povazska cementaren, a.s. Ladce (Povazska cement plant Ladce), Ladce, Slovakia
[2]University Hospital in Trencin, Department of Clinical Biochemistry, Hematology and Microbiology, Trencin, Slovakia
[3]Technical University of Kosice, Faculty of Civil Engineering, Institute of Environmental Engineering, Kosice, Slovakia

E-mail: strigac.j@pcla.sk

Abstract. The article deals with studying the antifungal efficiency of carbide lime slurry compared to industrially manufactured commercial lime. Antifungal efficiency expressed as mould proofness properties was tested on the fungi using the procedure given in standard CSN 72 4310. A mixture of fungi *Aspergillus niger*, *Chaetomium globosum*, *Penicillium funiculosum*, *Paecilomyces variotii* and *Gliocladium virens* was utilized for testing. The scale for evaluating mould proofness properties according to CSN 72 4310 is from 0 to 5 in degree of fungi growth, where 0 means that no fungi growth occurs and the building products and materials possess fungistatic properties. The study confirms the fungistatic propeties of carbide lime slurry as well as industrially manufactured commercial lime. However, carbide lime slurry and industrially manufactured commercial lime possess no fungicidal effect.

1. Introduction

Calcium carbide CaC_2 is industrially produced in an electric arc furnace with graphite electrodes from a mixture of lime and coke at approximately 2000°C by reaction (1) [1, 2]. The carbide product produced generally contains around 80wt.% CaC_2. Calcium carbide is generally used for the generation of acetylene gas (C_2H_2) by the reaction (2).

$$CaO + 3\,C \rightarrow CaC_2 + CO \tag{1}$$

$$CaC_2 + 2\,H_2O \rightarrow C_2H_2\,(g) + Ca(OH)_2 \tag{2}$$

Carbide lime (CL) is high quality hydrated lime slurry produced as a by-product of acetylene production, composed mainly of calcium hydroxide $Ca(OH)_2$ with minor portions of $CaCO_3$, and the presence of carbon as graphite particles and amorphous films in the waste [3, 4]. Carbide lime has been used for decades as an alternative source for hydrated lime in such applications as municipal and industrial wastewater treatment, flu-gas desulfurization, sludge conditioning, road stabilization and other applications where normal commercial lime is used. The $Ca(OH)_2$ content of the dry solids is in the 90% range. $Ca(OH)_2$ (hydrated lime) slurry produced as, and delivered, in a slurry form. The study [5] was aimed at the possibility of using the carbide lime waste as an alternative material to the conventional lime used for cement-lime mortar. Specimens were tested for mechanical properties and the results have shown that the carbide lime waste can be readily used for masonry cement mortar work.

Calcium carbide lime is practically used for various chemical and industrial uses such as pH control, industrial water and as sewage treatment, stabilization and agricultural purposes [6]. It is solid or liquid with white-gray colour. Lime is non-toxic, however, it may cause skin and eye irritation and burns. The irritant effects of lime are due primarily to its alkalinity, but dehydrating and thermal effects may be contributing factors. Due to the alkalinity of lime it may be subject to different regulations in different locations.

Monopoly producer of calcium carbide CaC_2 is company Fortischem a.s. Novaky (former NCHZ a.s. Novaky) in Slovakia [7]. The manufactured CaC_2 is ground, sorted and supplied in number of qualitative types. It is used as a raw material for the production of acetylene, calcium cyanamide, in metallurgy for metal desulphurization. Lime fine powder, lime powder and lime hydrates are supplied in various qualities. Lime powders and dust regulate soil pH. They are also used to adjust water pH in waste water treatment plants [7]. The industrial wastes including carbide lime slurry $Ca(OH)_2$ are separately deposited at the waste dump - landfill site in the vicinity of Novaky. However, several hundred thousand - millions tons of carbide lime slurry $Ca(OH)_2$ are landfilled till now, without any utilization, which is serious environmental load and burden.

The ordinary commercial limes are extensively utilized throughout the world according to the technical standards and regulations. They can be applied as building lime EN 459-1, as a part for masonry cement EN 413-1, for rendering and plastering mortar EN 998-1, for masonry mortar EN 998-2, for unbound and hydraulically bound mixtures - soil treated by lime, for hydraulically bound mixtures - hydraulically stabilized soils EN 14227-11, for calcium silicate masonry units EN 711-2 and for autoclaved aerated concrete masonry units EN 711-4, respectively.

The disinfective effect of lime at hygienisation is very good known for centuries and it is very well utilized at sanitisation (hygienisation) till now, as well. Disinfection by lime in the wastewater reclamation plant reduced number of microorganisms extensively when operated at pH 11.2 [8]. Lime enables the destruction of all pathogens due to the pH effect, which provides alkaline hydrolysis, combined with the temperature increase - thermolysis, that quicklime hydration brings [9]. Of course, the thermolysis effect can be decreased in the case of slaked, hydrated lime $Ca(OH)_2$, however, the alkaline hydrolysis can be considered as the main antimicrobial potential of lime. Alkaline hydrolysis (OH^-) results in the destruction of protein-based cellular walls and of enzymes, the destruction of proteins at polypeptide bonds to amino acids and oligopeptides, the destruction of nucleic acids (RNA), the destruction of carbohydrate cell constituents and lipids and the denaturing of enzymes [9]. Therefore, this article aims at the studying the antifungal effect of carbide lime slurry in comparisom to the industrially manufactured commercial lime with emphasis on increasing its potential application.

2. Materials and methods

2.1. Materials

Following lime samples were tested: carbide lime slurry (CLS) and comparative industrially manufactured commercial lime (IMCL) as the reference material. The chemical composition of tested samples was performed by XRF according to EN 196-2 using apparatus SPECTRO X-LAB 2000. The semi-quantitative chemical composition of the samples is given in table 1.

The mineralogical composition of the samples was determined by XRD technique using BRUKER AXS D8 Advance device and it is given in table 2.

Table 1. The chemical composition of tested carbide lime slurry (CLS) and industrially manufactured commercial lime (IMCL) in wt.%.

Comp.	Unit	CLS	IMCL	Comp.	Unit	CLS	IMCL
L.O.I*	[wt.%]	27.70	22.22	V_2O_5	[ppm]	88.0	96.0
SiO_2	[wt.%]	2.243	1.079	Cr_2O_3	[ppm]	30.96	25.49
Al_2O_3	[wt.%]	1.317	0.424	CoO	[ppm]	3.30	3.41
Fe_2O_3	[wt.%]	0.1503	0.262	NiO	[ppm]	6.85	7.68
CaO	[wt.%]	66.642	65.278	CuO	[ppm]	0.51	7.07
TiO_2	[wt.%]	0.0501	0.023	ZnO	[ppm]	27.3	11.79
MgO	[wt.%]	0.375	9.883	As_2O_3	[ppm]	3.04	5.15
K_2O	[wt.%]	0.0812	0.143	Cd	[ppm]	1.69	0.79
Na_2O	[wt.%]	0.926	0.406	SnO_2	[ppm]	28.0	20.69
SO_3	[wt.%]	0.4177	0.127	Sb_2O_5	[ppm]	7.27	6.64
MnO	[wt.%]	0.0043	0.0065	Hg	[ppm]	1.18	1.83
P_2O_5	[wt.%]	0.0239	0.029	Tl	[ppm]	2.37	2.88
Cl^-	[wt.%]	0.0101	0.0058	PbO	[ppm]	0.93	13.01

*L.O.I: Loss on ignition

Table 2. The mineralogical composition of tested carbide lime slurry (CLS) and industrially manufactured commercial lime (IMCL) in wt.%.

[wt.%]	portlandite $Ca(OH)_2$	calcite $CaCO_3$	quartz SiO_2	periclase MgO	graphite C	hydrocalumite Friedel's salt♦
IMCL	+	+	+	+	-	-
CLS•	70.10	22.40	-	-	2.37	5.13

♦ – the hydrocalumite phase $Ca_4Al_2(OH)_{12}(Cl,OH)_2.4H_2O$ ($Ca_2Al(OH)_6Cl.2H_2O$, β-$Ca_4Al_2O_6Cl_2.10H_2O$, or $Ca_4Al_2(OH)_{12}(Cl,CO_3,OH)_2.4H_2O$) is structurally like as Friedel's salt $3CaO.Al_2O_3.CaCl_2.10H_2O$ and so it is adequately attributed to Friedel's salt phase, as well.

• – the quantitative mineralogical composition determined by XRD analysis connected with Rietveld computation of quantitative analysis.

2.2. Method of testing the resistance of building products and materials to filamentous fungi (moulds) according to standard CSN 72 4310

The antifungal activity of carbide lime slurry (CLS) and comparative industrially manufactured commercial lime (IMCL) was tested using the procedure given in the Czech standard CSN 72 4310 – The testing of mould proofness of building products and materials [10]. The mould proofness properties expressed as the intensity of fungi growth on building products and materials are determined by both artificial and natural contamination, with exposure to the selected testing moulds under the prescribed conditions presented in the standard [10]. The testing method is fully described [11], in detail. The scale of evaluation of the mould proofness properties of building products and materials according to CSN 72 4310 [10] is given in table 3. The degree of the fungi growth is expressed by a value from 0 to 5, where the value 0 is attributed to no occuring fungi growth and the building products and materials possess fungistatic properties; in some cases, fungicidal properties also occured after the formation of an inhibiting zone in the broth around the sample. The final evaluation was carried out by microscopic examination of the intensity of mould growth on the sample surface. The intensity of mould growth was evaluated according to the scale from 1 to 5 given in table 3. The mould proofness properties of the building products and materials were evaluated according to CSN 72 4310 [10] as follows:

- the building products and materials are fungicidal – mould does not develop on the sample. An inhibiting zone forms on the agar around the sample;
- the building products and materials are fungistatic – mould does not develop on the sample. Mould growth on the agar is not affected;
- the building products and materials are not mould-proof (the intensity of mould growth on the sample surface itself is from 1 to 5).

Table 3. The scale of evaluation of the mould proofness properties of building products and materials according to standard CSN 72 4310 [10].

Degree of fungi growth	Description
0	No growth of fungi
1	The growth of fungi is negligible (the colonies of fungi are dispersed)
2	The growth of fungi is gradual (numerous small fungi colonies which cover 25% of the sample surface)
3	The growth of fungi is intensive (fungi colonies cover up to 50% of the sample surface)
4	The growth of fungi is very intensive (fungi colonies cover up to 75% of the sample surface)
5	The sample surface is fully covered by fungi (fungi colonies cover 100% of the sample surface)

The testing was realized in the accredited microbiological laboratory of Testing Institute for Textiles (Textilni Zkusebni Ustav, s.p.) in Brno, Czech Republic. The test conditions are as follows: sample size Ø 5.5 cm; temperature in the incubator 28±1°C; rel. humidity in the incubator 95%; incubation period 3 months; standard broth media.

A mixture of fungi *Aspergillus niger* (CCM 8155), *Chaetomium globosum* (CCM 8156), *Penicillium funiculosum* (CCM F-161), *Paecilomyces variotii* (CCM F-566) and *Gliocladium virens* (CCM 8042) was applied for the testing. The tested samples were exposed to the contamination by the chosen microorganisms in the broth medium according to exactly defined conditions given in CSN 72 4310 [10] during the incubation period of 3 months, after which their mould proofness was evaluated.

3. Results and discussion

After the 3 months incubation period according to standard CSN 72 4310 [10], the mould proofness properties of carbide lime slurry (CLS) and comparative industrially manufactured commercial lime (IMCL) were evaluated as the degree of fungi growth.

Generally speaking, the CLS and IMCL (reference material) are suitable for reaching 0 degree of fungi growth according to CSN 72 4310 [10]. On the basis of the experimental results, it can thus be stated that the CLS and IMCL are defined as the fungistatic building materials according to CSN 72 4310 [10]. The degree of fungi growth 0 means that no growth of fungi occurs and the building products and materials possess fungistatic properties.

The inhibiting zone did not form on the agar around all of the tested samples. This means that CLS and IMCL (reference material) possess no fungicidal effect. However, the results unambiguously showed that CLS and IMCL have a fungistatic effect. Moulds did not develop on the samples, but the mould growth on the agar was not affected.

The fungistatic effect of CLS and IMCL is declared on the basis of reference testing, which was carried out in an accredited microbiological laboratory at the Textile Testing Institute (Textilni Zkusebni Ustav, s.p.) in Brno, Czech Republic, which is an approved certification body, as well. CLS and IMCL complying with the requirements of CSN 72 4310 [10] can be used for the production of fungistatic building materials and products. An application of CLS and IMCL into the building materials and products increase their mould proofness and antimicrobial properties.

According to the results, the addition of fungistatic CLS and IMCL (reference material) into the building products and materials is suitable for increasing the antimicrobial effects and for achieving fungistatic protection, as well. From a practical point of view, fungistatic CLS and IMCL are suitable for achieving a mould-free environment. CLS and IMCL potentially offer wide application possibilities in the building industry from preventive use to the purposes of repairing and reconstruction. However, the mechanism of the antimicrobial effects of CLS as well as IMCL should be the subject of further study.

4. Conclusion

The main conclusions that can be drawn from this experimental study are summarized as follows:

- carbide lime slurry (CLS) and comparative industrially manufactured commercial lime (IMCL) is a fungistatic building material according to the requirements of standard CSN 72 4310 with 0 degree of fungi growth;
- no growth of fungi is observed in CLS and IMCL, with mould proofness property 0;
- the inhibiting zone did not form on the agar around the tested samples, which means that CLS and comparative IMCL possess no fungicidal effect;
- fungistatic CLS as well as IMCL (reference material) are suitable for use in a wide scope of applications aimed at eliminating fungal growth, starting with preventive purposes and ending with repairing and reconstruction works.

References

[1] Orsula I, Lehocky M and Steltenpohl P 2015 Simulation of calcium acetylide and acetylene production *Acta Chimica Slovaca* **8** 91-96

[2] Mineral produkt industry 1995 Part 11.4 Calcium Carbide Manufacturing 11.4-1-11.4-6, https://www3.epa.gov/ttnchie1/ap42/ch11/final/c11s04.pdf

[3] Materials of Gilmour & Company Inc. Washington Corp. including lime slurry products, carbide lime slurry and hydrated lime slurry 2017 (http://64.234.209.72/lime_products.html; http://64.234.209.72/carbide_lime.html; http://64.234.209.72/hydrated_lime.html)

[4] Cardoso A F, Fernandes H C, Pillegi R G, Cincotto, M A and John V M 2009 Carbide lime and industrial hydrated lime characterization *Powder Technology* **195** 143-149

[5] Waheeb A Al Khaja 1992 Potential Use of carbide lime waste as an alternative material to conventional hydrated lime of cement-lime mortars *Eng Jour of Qatar University* **5** 57-67

[6] Intern. Industrial Gases Ltd. 2006 Material Safety Data Sheet for Hydrated lime slurry, carbide lime slurry (Calcium hydroxide slurry) http://www.iigas.com/carbide_limeMSDS.htm

[7] Fortischem a. s. Novaky, Slovakia 2017 Products http://www.fortischem.sk/en/our-products

[8] Grabow W O K, Middendorf I G and Basson N C 1978 Role of lime treatment in the removal of bacteria, enteric viruses and coliphages in a wastewater reclamation plant *Applied and Environmental Microbiology* **35** 663-669

[9] Remy M 2009 Scientific evidence on the effectiveness of lime *Conference Animal health & welfare: a sustainable solution from industry*, 18th February 2009, Brussels, Belgium, ed. European Lime Association (EuLA), EuLA c/o IMA-Europe 11-21

[10] Czech standard CSN 72 4310: 1977 *The testing of mould proofness of building products and materials*

[11] Strigac J and Martauz P 2016 Fungistatic properties of granulated blastfurnace slag and related slag-containing cements *Ceram.-Silik.* **60** 19–26

Efficiency of innovative technology in construction industry

H Stverkova[1] and V Vaclavik[2]

[1]Department of Business Administration, Faculty of Economics, VSB – Technical University of Ostrava, Czech Republic
[2]Institute of Environmental Engineering, Faculty of Mining and Geology, VSB-Technical University of Ostrava, Czech Republic,

E-mail: hana.stverkova@vsb.cz

Abstract. The need for sustainability increasingly influences the development of new technologies, business processes and working practices. Innovations are an important part of all business processes. The aim of innovation is, in particular, to reduce the burden on the environment. The current trend in the construction industry is diamond rope cutting. The aim of the paper is to evaluate the most advanced technology for cutting and removing concrete structures in terms of efficiency.

1. Introduction

The concept of sustainable development is associated with environmental problems, particularly with the issue of natural resource management, economic challenges and global climate changes or the transformation of post-industrial society. The basic aspects of achieving or maintaining the competitiveness of the economy and sustainable development include the advanced knowledge industry, especially the adaptation to the changes brought by the so-called Fourth Industrial Revolution. The cornerstone of this period is the knowledge society and the innovation. The sustainable economic development has been chosen as one of the key areas for the competitiveness of the national economy. The main microeconomic priorities lie in the improvement of science and research, innovation, the business environment, the modernization and the development of information technologies and also the development of technical infrastructure and sustainable consumption of natural resources.

The principles of sustainable development are based on the belief that the needs of the present generation must be fulfilled in such a way that the possibilities of future generations are not weakened. Their part is the promotion of life in its diversity, based on the principles of democracy, the gender equality, the solidarity, the respect for the rule of law and human rights, including the peace and equal opportunities for all.

The need for sustainability increasingly influences the development of new technologies, business processes, business organization and working practices. The innovation is an important part of all business activities. The innovation is characterized as a transfer of the idea of a new or improved business process used in industry or in business. The technical innovation is characterized by the emergence of new products, processes and significant technical changes in manufacturing processes.

The aim of the innovation is especially the reduction of the burden on the environment.

Nowadays, the new technologies are at the forefront of ecology, economy, and efficiency. The current trend in the construction industry is the diamond rope cutting. The construction structures, whether administrative buildings, bridge constructions, hall buildings, multifunctional units and others, are built mainly of reinforced concrete. The aim of this paper is to evaluate the most advanced technology for the cutting and the removing of the concrete structures in terms of efficiency.

2. Innovation and the sustainable development

The innovation represents a key element from the point of view of the enterprise for its further development and the increased competitiveness within the current globalized market. In compliance with the international terminology, these types of the innovation are distinguished [5]:

- The innovation of the product, which mainly involves the introduction of new or significantly improved products or services, especially in the field of technical parameters or user-friendliness;
- The process innovations that are aimed at introducing new or significantly changed production, process change, or supply methods;
- The marketing innovations that are focused on new marketing methods in product support, valuation, design and other areas;
- The organizational innovations which are focused on new organizational methods, changes in organizational arrangements or external relations.

In the manufacturing industry, there prevails the product and the process innovation, or the technical innovations; in the services, there is a higher share of the innovative activities in terms of marketing and organizational innovation. Non-technical innovations make it possible to create the new and effective ways to promote new products and services, or introduce the flexible changes in organization management to respond to new market trends and customer demands. The ecological innovation, according to Eurostat methodology [5], is not considered as a separate innovation. It is a group of environmental benefits of innovations that are considered to be the by-products of other named innovation groups (especially in product and in process).

Since 2008, the overall innovation activity in the Czech Republic has been decreasing, mainly due to the lower intensity of the innovation efforts in the area of non-technical innovation. From the international point of view, enterprises in the Czech Republic innovate less than the average of the EU28 countries, but in the area of the technical innovation, the Czech Republic is close to the EU average. The business ownership and the size of an enterprise affect the intensity of innovation activities. The most intensively innovating are the large enterprises with more than 250 employees [5,3,7].

The business entities in the business sector are innovating in their products, internal processes, or business strategies and try to accommodate the existing market or technological conditions or anticipate their development. The successful innovation activities are a key prerequisite for the business growth and survival in a globalizing market. The sustainable development involves not only an environmental protection and the efficient use of natural resources, but also aspects of the economic growth and social cohesion. There is a predominant view that the principles of sustainable development should also be reflected in decision-making processes at all levels. A capacity for mobilizing and using science and technology (S&T) is increasingly recognized as an essential component of strategies for promoting sustainable development [6,8].

The accuracy of the fulfillment of sub-parts of the construction phase is the key to an effective outcome. The diamond cutting and the drilling is currently a highly efficient and fast way to dismantle reinforced concrete structures, concrete products, masonry and other natural and man-made building materials. The unique technology of cutting concrete using a rope with a diamond surface allows for dismantling even under the most demanding conditions: the cutting walls with a thickness of more than 70 centimeters, dismantling of monolithic foundations and structures, the work on coastal objects (pier, docks, and girders), and the arc cutting and cutting with tilting in hard-to-reach places. Also, the diamond drilling technology is widely used in construction for drilling technological holes in hard materials such as concrete, reinforced concrete, bricks, stone and other [2,4]. Thanks to this no-break method, the gap drilling takes place without a noise and a vibration, and the process technology allows the drilling gaps of ideal shapes and exact dimensions in any position and at different angles. In the case of the core drilling, it is mainly the production of passages for water, hot water, air-conditioning, etc. Unlike the manual cutting using the drill hammers and pneumatic hammers, it is possible to achieve the core drilling of concrete and reinforced concrete, 10-20 times more work efficiency is achieved thanks to the diamond segments ensuring an easy penetration into the structure. This method is used, for example, for drilling gaps for piping, anchoring, drilling wells and perforation boreholes for the removal of concrete and the reinforced concrete.

3. Diamond cutting efficiency

The benefits of the diamond cutting have been analyzed with 15 experts from the special field, using the QFD methodology, more in [1] and the results are demonstrated in the following table 1.

Table 1. The benefits of the diamond cutting.

	Importance			
	little	medium	much	frequency
Very precise gaps without the need for further structural modifications		2	13	123
Fast, accurate and efficient work		4	11	111
Saving the time and financial cost vs. classical cutting	10	4	1	31
Possible cutting even in heavily armored constructions		1	14	129
Work without vibrations and so without damage of other constructions			15	135
The possibility of carving large openings, which are divided into manipulated pieces		2	13	123
Reduction of a noise and dust to a minimum		13	2	57
Maximum environmental friendliness	4	8	3	55

Source: own elaboration

From the point of view of specialists, this is an advantage in terms of statics - work without vibrations and so without damage of the other constructions, the second most advantageous is the suitability for use in heavily reinforced constructions, and the last two most important aspects are the possibility of cutting large gaps, which are divided into manipulated pieces as well as precise gaps without the need for further structural modifications. The professionals also evaluate the diamond cutting as a benefit in terms of accuracy and efficiency. The weakest aspect is the saving the time and financial costs versus the classic cutting. In terms of time, it would be possible to talk about the time saving, but on the economic aspect – not the financial costs. The use of diamond technology is costlier than the classic cutting.

For the evaluation of the effectiveness of the most advanced technology for cutting and removing concrete structures, is given an example of practical experiences.

The diamond rope can cut the building gaps from 0.5m thick the concrete, the reinforced concrete, the concrete pipes, and the brick masonry. The diamond rope also commonly cuts a wall of 4m thickness. At the highest performance, the engagement of the diamond rope is within 10m. The other use of the diamond rope is cutting the small square gaps, and for measurements below 0.6 x 0.6m of different thickness is a suitable diamond rope system.

The concrete and the reinforced concrete are possible to cut by diamond wheel to 1m of the masonry or the monolith. This process can only be used for long cuts (over 10m of the cut length). In the case of smaller gaps cutting, this technology cannot be used as the masonry cuts of 1m in length - more half the diameter of the disk. This presents a problem in terms of pillars, pipe bells and beams, there may come their disturbance.

The work with the diamond rope has a slightly lower performance than the diamond wheel cutting. The sections of diamond rope in the steel reinforcements are not as accurate as the diamond disks. The explanation is in the diamond rope slip, which are at an angle of up to about 60 °, there is a grinding of the steel reinforcement and not the pruning (instead of 1cm cut is there 3 cm cut, the rope vibrates gently and there is a curl of the cut). The diamond disk cuts the steel reinforcement continuously. With the reinforcing concrete or the steel reinforced brick masonry, the large vibrations occur, the steel reinforcements are released and thus their adhesion and the reduction of the strength of the reinforced concrete are impaired. The use of diamond rope appears to be problematic, especially for poorly made steel reinforcements (curl or bend when assembling), this causes the cut to be cut even in a place that cannot be seen from the outside. Because of this, it is impossible to carry out any manipulation with the concrete scraper (carved block) and the diamond rope system is then inefficient.

Nowadays, the so-called banked diamonds are used, which have better work performance and are of better quality. The diamond ropes are driven mainly by a hydraulic saw, but there are also diamond rope self-tensioning saws.

The disadvantage of the diamond rope system cutting at higher cutting thicknesses is the flatting down the rope. Also, due to lack of space in the rope stack when performing longer cuts, the rope must be shortened. Therefore the rope is re-routed to the desired length by cardan joints or fixed ones, which - when are new - have a larger diameter than the diameter of the worn rope, and this creates a problem by tearing the rope often and re-connecting.

The diamond rope is now able to cut round gaps. The diamond rope usually cuts from the diameter of 1.2m, up to 1.2m diameter is usually used a core drilling with crowns.

4. Conclusion

While diamond rope cutting technology has been known to demolition of large structures for decades, its use in other sector is growing slowly. The process has many advantages including minimal vibration or damage to any structural element, low noise or high accuracy of cuts and large depth of cuts. The diamond rope cutting, diamond wheel and core drilling are gradually being discovered in the construction industry and despite the higher price; it is being used more often in practice.

References

[1] Blecharz P and Štverková H 2011 Product Quality and Customer Benefit *App. Econ., Busin. and Devel. Inter. Symp.* p 382 - 388
[2] Cash D W, et Al. 2003 Knowledge systems for sustainable development *Natl. Acad. Sci.* **100** pp 8086–8091
[3] Chytilova L 2013 Rating prediction using neural networks *VI. mezinárodní vědecká konference doktorandů a mladých vědeckých pracovníků* (Karviná: Slezská univerzita v Opavě)
[4] RYLES D Q 2017 *Diamantové řezání*
[5] EUROSTAT 2012 *The Oslo Manual is one of the central documents providing an overview of the basic definitions and practices in measuring innovative business activities*
[6] Ruttan V W 2001 Technology, Growth, and Development. *NY Oxf. Univ. P.*
[7] Vymetal D, Hucka M, Hunka F and Kasik J 2008 Production planning model using REA ontology *E & M EKONOMIE A MANAGEMENT* **11** p 93-102
[8] World Bank 1999 *World Development Report*

IOP Conf. Series: Earth and Environmental Science **92** (2017) 012060 doi:10.1088/1755-1315/92/1/012060

Monitoring of Thermal and Gas Activities in Mining Dump Hedvika, Czech Republic

D Surovka[1], E Pertile[1], V Dombek[1], M Vastyl[1] and V Leher[1]

[1] VSB – Technical University of Ostrava, Faculty of Mining and Geology, Institute of Environmental Engineering, Ostrava-Poruba, Czech Republic

Email: dalibor.surovka@vsb.cz

Abstract. The negative consequences of mining of the black coal is occurrence of extractive waste storage locations - mining dumps. The mining activities carried out within the area of Ostrava are responsible for at least six mine dumps of loose materials arising as wastes from mining of mineral resources, many of which show presence of thermal processes. The thermal activity in dumps is responsible for many hazardous substances that pollute the environment and harm human health in the surroundings. This paper deals with the results of the first phase of project CZ.11.4.120/0.0/0.0/15_006/0000074 TERDUMP, on exploration of thermally active mining dumps are published in the article. As a first studied thermally active dump was a Hedvika dump. To localize of hot spots with hot gas emission was used a thermovision scanning by drone. The place with high temperature (49.8 °C) identified natural gas emission through natural cracks. Analysing the occurring pollutants in Hedvika Dump using the GC-MS or HPLC, respectively and the inert gases (CO_2, CO and SO_2) were determined by ion chromatography. The pollutants were determined in five sampling points during two measurements executed from July to August 2017.

1. Introduction

Intensive mining of black coal in the underground mines in the Ostrava-Karviná district (OKR) has left behind a large number of dumps that have resulted from the weighing of the produced carboniferous tailings as mining waste. In OKR, mining from deep mines was carried out on a hail, and extracted carbon tailings from the sorting process, preparatory works, and the treatment of hard coal was necessary to be placed on the surface [1]. As a consequence of the deposition of the tailings after the surface treatment of coal, dumps were basically found at all mines from the beginning, which can be considered as a characteristic feature of the mining landscape. About 50 dumps of 600 ha were created within the region of Ostrava and its surroundings.

The tailing deposited on the dumps of the OKR consists mainly of the admixtures of the rocks of Petřkovice and Hrušín layers, which consist mainly of sandstone, dust, claystone and also coal residues. Just these residues of black coal of the Upper Silesian Basin create also the main flammable component of the dump [2]. As mentioned above, due to the imperfect sorting of the tailing from the mineral, a considerable amount of coal substance remained in the rolling stock. Then they were susceptible to self-ignition under certain conditions. This phenomenon can occur primarily by aerating a carbon-rich coal tar. Oxygen and water vapor are therefore critical factors for this self-inflammation. The greatest amount of heat is released primarily through the oxidation of unsaturated organic substances and is supported by exothermic reactions. The humidity of the environment is mainly due to the reactions of water molecules with oxygen and coal. This accelerates oxidation and water vapor condenses on the surface of the coal. The result is sensory changes not only on the surface but also in the body of the rock. If the released energy is not continuously removed, occurs desorption of coal gas when the temperature is below 160 °C. On the contrary, oxidative complexes are formed at higher temperatures, and burning occurs above 300 °C [3]. So this may result in radiant heat fires. In the aeration zones, the gases can both ignite and reduce the temperature to the surface air. This generation of gaseous substances is nothing special and occurs on a shallow surface. Substances that emerge at these sites can be crystallized in the effluent and react with the tailing material. Conversely, endothermic reactions occur in the part of the dump body without air access. Thermal decomposition of the carbonaceous matter starts and at a temperature above 1 000 °C there is a high temperature carbonization and consequently a number of secondary products are formed [4].

The arising risks from the deposited carbon sinks are permanent. Pollutants as well as gaseous combustion products can be released into the water. Due to carbonization, emissions of volatile and partially volatile organic compounds are released into the atmosphere. Their condensation then produces PAHs. In case of inadequate oxygen supply to the combustion zone, burning downstream producers produce mainly CO2, CO, NOx, SOx, HCl, HF, hazardous metals and many other substances [5]. In general, the composition of gaseous substances which are released from the thermally active dumps into the atmosphere differs not only in relation to the composition of the deposited material but also to the presence or absence of soil cover. Thermically active dumps, due to processes with varying intensities in their bodies, they produce large amounts of pollutants into the air and, due to unfavorable diffusion conditions in the atmosphere, they can further concentrate and subsequently negatively affect the health of the population living near the dumps [6].
Thermically active dumps can also cause falls in dumps, surrounding forest fires, and health hazards due to airborne dust emissions, gaseous pollutants, hazardous elements and organic compounds. Unchecked coal fires can cause property destruction and ecosystem degradation. An important aspect of the thermodynamic activity is the composition of the heap. Dumps degrades aesthetic values of the landscape and its ecological functions [7].

Thermically active dump Hedvika was monitored within the framework of screening. The exact date of establishment of the Hedvika dump is unknown. It is supposed that the rock was weighed here since 1903, originating from the excavation and opening works. There was probably deposited materiál from the digging of the pit and the openings of the mine of the Hedvika (the original name the Albrecht mine). Surface was spread from the original area of the Hedvika mine's own area to the surroundings, especially in the northwest direction, directly to Michalkovice. At the turn of the 1960s and 1970s there was a period of intense widening of the dump to roughly today's surface area. The deposition of tailings in the cesspool was completed in 1998, when the disposal of part of the tailings production from the nearby mine Pokrok [8].

Dump is located on the eastern outskirts of cadastral areas of Radvanice and Michálkovice (Ostrava district) in the local parts of Chotebuz and Zaryje, on the border of the former districts of Ostrava and Karvina. The eastern part of the area under Karvina is part of the area of Petrvald near Karvina. It is a massive, powerful and vastly formed formation of tailings that follow the northern and western edges of the Hedvika plant of the former J. Fucik Mine. Its own dump is an extensive area that occupies an area of about 32 hectares. In the northeast, it is bordered by the Michalkovice-Petrvald road and in the southwest adjoins the mine site. With increasing volumes of mining, it was spreading northwest to Michalkovice. The transport of tailings was done by mines. This well is reclaimed and partially biologically stabilized with progressive revitalization (partly rainforest, partially set up by recultivating tree species). The part of the dump at the northern edges is formed by relatively fresh embankments [9].

It is a massive area-wide formation of tailings with an area of approximately 40.6 ha and an average height of the ridges of 12-15 m at present. The maximum height of the dump to the surrounding natural environment is up to 32 m. Thus the morphologically shaped complex of tailings with a high proportion of the residual coal substance is predisposed to the formation of fumes and the development of endogenous fires. Thermal processes have been registered in various parts of the dump since the 1950s. According to preserved documentation, however, they were locally eliminated, in particular by extraction, extinguishing, and above all by covering with insulating layers. These interventions, however, failed to prevent the massive development of thermal activities, especially in the central part of the dump during the 1990s, when the temperatures in the focus of the fire exceeded 500 °C. However, the last experimental remediation of thermal activity in 2006 by flooding the hotbeds with liquid clay slurry was not effective enough and so these experiments were completed. After the initial multi-year stabilization of the outbreak, dynamic temperature migrations have been observed since 2010, also in the direction of the original area of the Hedvika mine, which is built up by building objects of various uses [10].

2. Material and methods

2.1. *Characteristic of locality and sampling points*

Two sampling points were selected for the thermically active dump so as to objectively include, if possible, all intensively thermally active sites. Their initial choice was based on the conclusions of thermovision, but with regard to the real situation in the field (gas leakage, vents, etc.) was specified directly on the site. The first sampling point 1F (GPS: 49° 49'36.2, N 18° 21'36.6 E) is located on the left side of the forest road embankment. On the right side of the forest road there is a takeoff point 2F (GPS: 49° 49'35.6, N 18° 21'37.6 E). More thermically-active point 2F is located on a significantly fallen surface over the surrounding terrain. It is clear that thermal activities occur here, due to the presence of burning residues of higher plants.

2.2. *Thermovision*

Identification of the most thermically active locations on Hedvika's dump was performed by an air thermovision using a drone realized by Fly Vision (Czech Republic). The measurement was performed on a 100×50 m early in the morning (temperature-avoidance), with a view to the representativeness of the results, repeatedly to eliminate the possible accidental effects (28th March 2017 and 3rd June 2017).

2.3. *Analytical Methods*

Due to the main objective of the project is to compare and subsequently unify the methodologies of sampling and adaptation of individual samples and used analytical methodologies in the framework of close cooperation between VSB-TUO and GIG Katowice with the aim of their mutual compatibility and the most objective evaluation. Initial screening was also attended by experts from both sites for gas emissions.

Sampling was carried out by employees of the Health Institute with headquarters in Ostrava (hereinafter referred to as ZU OVA). For sampling (July 19, 2017), specially pretreated PUF cartridges SKC, 22×60 mm, were used in conjunction with a quartz fiber filter. Subsequent extraction was carried out in a Tecator device using dichloromethane for 8 h. The extracts were then concentrated with nitrogen in a TurboVap and subsequently transferred to a solvent compatible with the mobile phase. For the chromatographic analysis, the extracts were diluted so that the analyte responses were in the range of the calibration curve used and the linear response of the detector used. For quantification, a certified calibration standard, Absolute Standard, was used. Custom analysis, data acquisition and evaluation were performed using Empover software.

Determination of PAHs was performed at VSB-TUO by high performance liquid chromatography (HPLC) on a Waters Acquity UPLC I-Class device with dual detection. Separation was performed on a Waters PAH C column 18.5 μm 150×2.1 mm i.d. by gradient elution with acetonitrile/water.

3. Results and discussion

It was important to find those sampling points that exhibited the highest thermal activity for primary screening. For this purpose, infrared thermography was used. When there is enough air that enters the drain, it initiates the oxygen contained by the spontaneous combustion of the stored material. During the burning of coal residues a crack is formed on the surface, through which it escapes together with flue gases and heat. It can be seen from figure 1 that it practically burns the entire surface of the Hedvika dump. Under the surface of the dump there is a huge amount of coal that has accumulated here as a result of mining, and not quite effective treatment processes in the past. The maximum surface temperature measured by thermovision imaging was 71.6 °C. Clearer places to represent a temperature difference of about 50 °C. The ambient temperature in this case was 3.4 °C.

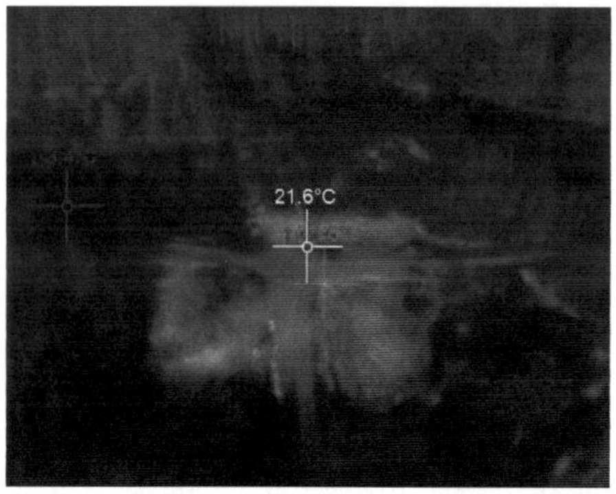

Figure 1. Overall view of thermically active dump Hedvika (June 2017).

As part of the initial joint screening carried out on July 19, 2017, an analysis of PAHs representatives was carried out, captured in the concentrate of the GIG Katowice and captured on the PUF (VSB-TUO). The results of individual PAHs at the most thermically active sampling point 1F for the initial screening are shown in table 1.

Table 1. PAHs identified in the most thermically active point of dump Hedvika.

Analyte	Sampling point 1F			Sampling point 2F		
µg m^3	Concentrate	PUF	Σ	Concentrate	PUF	Σ
Naphthalene	110.00	0.35	110.35	27.67	0.25	27.92
Acenaphthylene	*	<0.01	<0.01	*	<0.01	<0.01
Acenaphthene	<0.05	<0.05	<0.05	<0.05	1.10	1.10
Fluorene	13.17	0.72	13.89	24.83	0.59	25.42
Phenanthrene	26.67	4.00	30.67	92.17	5.06	97.23
Anthracene	0.83	0.52	1.36	12.83	0.50	13.33
Fluoranthene	13.83	0.44	14.27	36.00	0.38	36.38
Pyrene	<0.05	0.32	0.32	30.50	0.31	30.81
Benzo[*a*]anthracene	<0.05	0.18	0.18	<0.05	0.14	0.14
Chrysene	<0.05	0.19	0.19	<0.05	0.16	0.16
Benzo[*b*]fluoranthene	<0.05	0.09	0.09	<0.05	0.10	0.10
Benzo[*k*]fluoranthene	<0.05	0.03	0.03	<0.05	0.03	0.03
Benzo[*a*]pyrene	<0.05	0.07	0.07	<0.05	0.07	0.07
Dibenzo[*ah*]anthracene	<0.05	0.01	0.01	<0.05	0.03	0.03
Benzo[*ghi*]perylene	<0.05	0.04	0.04	8.17	0.07	8.24
Indeno[1,2,3-*cd*]pyrene	<0.05	0.58	0.58	<0.05	0.03	0.03
Σ16 PAHs	164.50	7.53	172.03	232.17	8.81	240.98

Note to the table: * not determined

Based on the screening analysis, it is not yet possible to reach overall conclusions, but some general assumptions have already been confirmed. For example, some concentrations of PAHs, whether measured in the concentrate or captured in PUF, are worth mentioning. It is true that low molecular weight PAHs (2-3 benzene nucleus) are contained in the atmosphere only in the gas phase as diffusion-moving molecules, whereas higher molecular weight polyaromatic hydrocarbons are bound to aerosol particles or sorbed on organic matter. In the concentrate, the presence of mainly multi-agent representatives of PAHs, with the exception of Benzo[ghi]perylene (8.17 μg m^3), was detected below the limit of determination. Higher concentrations of Phenanthrene and Pyrene, which can be considered as indicators of high temperature pyrogenic processes, have also been confirmed. At the most thermically active sampling point 2F, Phenanthrene (97.23 μg m^3) and Pyrene (30.81 μg m^3) were found at a much higher concentration than at the sampling point 1F, which no longer exhibits such thermal activity. Thermically active dump will be regularly monitored, at intervals of 3 months.

4. Conclusion

Based on a screening analysis of PAHs representatives on a thermically active dump Hedvika, was found that endogenous fire causes release of gaseous emissions, the dominant component of which is Phenanthrene. The composition of emissions as well as the concentration of individual pollutants will be affected by many factors, most likely the most significant being the seasons and the current meteorological conditions.

Taking into account to the heterogeneous nature of the material of the dump, which affects the resulting concentrations of pollutants during thermal activity, it is necessary to implement long-term monitoring of the monitored pollution indicators directly in the individual vents. In order to better understand the ongoing thermal processes, for regular quarterly monitoring of PAHs, VOCs and inorganic gases (CO, CO_2, H_2S, SO_2, etc.) there will be drilled boreholes for permanent probes on thermically active dump Hedvika.

Acknowledgments

This work was supported within the European Regional Development Fund under the grant no. CZ.11.4.120/0.0/0.0/15_006/0000074.

References

[1] Schwarzerová I 2011 Analýza rizik odvalů zasažených endogenní činností p 1–2
[2] Zástěrová P, Marschalko M, Durďák J and Niemiec D 2015 Nature Conditions of Waste Dumps in the Czech Parts of the Upper Silesian Coal Basin *Proc.Ear. and Plan. Sci.* **15** 395–400
[3] Tvrdý J and Sejkora J 1999 Hořící uhelné haldy a redepozice toxických látek při samovolném termickém rozkladu uhelné hmoty p 5
[4] Gazdová O and Jelínek P 2009 Jak dál s největší ostravskou doutnající haldou? *Diamo občasník* **14** (2) p 4
[5] Smolińnski A, Drobek L, Dombek V and Bak A 2016 Modeling of experimental data on trace elements and organic compounds content in industrial waste dumps *Chemosphere* **162** 189-198
[6] Pertile E, Surovka D, Sarčáková E and Božoň A 2017 Monitoring of Pollutants in an Active Mining Dump Ema *Inzy. Min.* **1** p 45-50
[7] Pertile E, Surovka D and Božoň A 2016 The study of occurrences of selected PAHs adsorbed on PM_{10} particles in coal mine waste dumps Hermanice and Hrabuvka (The Czech Republic) *GeoConference-SGEM* p 161-168
[8] Jochymková L, Surovka D and Sarčáková L 2014 Determination of PAHs in the dust particles from the burning slag heap EMA *SGEM 2014* **2** (5) p 179-184
[9] Sendlerová L and Dombek V 2015 Possibilities of reclamation the tank for further treatment of water from mine Fucik in the village Petřvald *SGEM 2015* **2** (5) p 427-432
[10] Sarčáková E, Pertile E, Surovka D and Jochymková L 2014 Sample Preparation Procedure For The Determination Of Polycyclic Aromatic Hydrocarbons In Dust Particles *SGEM* 2014 **2** (5) 619-624

Separation of gaseous air pollutants using membrane contactors

T Sverak[1,2], P Bulejko[1], J Ostrezi[2], O Kristof[1], J Kalivoda[2], P Kejik[1], K Mayerova[1] and M Adamcik[1]

[1]Heat Transfer and Fluid Flow Laboratory, Faculty of Mechanical Engineering, Brno University of Technology, Technicka 2896/2, 616 00 Brno, Czech Republic
[2]Institute of Materials Chemistry, Faculty of Chemistry, Brno University of Technology, Purkynova 464/118, 612 00 Brno, Czech Republic

E-mail: sverak@fch.vut.cz

Abstract. This work deals with the separation of CO2 gaseous pollutant from gas mixtures to a water solution using the laboratory contactor. The laboratory set process parameters showed the rate of carbon dioxide transition through the interface in a so promising level the contactor separators can be considered as a very promising pathway to reduce the content of this greenhouse gas from the air.

1. Introduction

Adsorption, membrane, condensation and absorption processes are basic separation processes which have widely been applied to a large industrial scale for separation of gaseous pollutants from the air. Absorption is widely used diffusion separation process where gas/liquid contact occurs. With the use of membrane contactors these phases are in contact through pores of a membrane which provides very high contact surface. In this work we demonstrate separation efficiency of a membrane contactor with an example of absorption of CO_2 into aqueous solution.

2. Absorption rate

Absorption rate can be expressed using equation describing the rate of mass transfer between phases. Using partial mass transfer coefficients of the substance in the liquid and gaseous phase, tha absorption rate is given by the following equation (1):

$$\frac{\dot{n}}{A} = k_{gP} \cdot \left(P_g - P_f\right) = k_{lC} \cdot \left(C_f - C_l\right) \tag{1}$$

Where $\dot{n}$ is the constant flow of molar amount of absorbed substance, A is the surface area of the interface k_{gP} and k_{lC} are the partial coefficient of the mass transfer in the gaseous and liquid phase, respectively. P_g and P_f, are the partial pressure of the absorbed substance in the gaseous phase and at the interface, respectively, C_l and C_f are the molar concentration of absorbed substance in the liquid and at the interface, respectively.

3. Membrane contactors basic principles

Membrane contactors are mostly used as a bundle of hollow fibre membranes (HFMs) with very small fibre diameter. The membrane pores provides contact between the two phases. The fibres are mostly made of hydrophobic materials such polyethylene, polypropylene, Teflon, or polyvinylidenfluoride. The HFM's function is to separate the two phases and to prevent their mixing. The interface is kept in the pore surface of the HFM which is filled with one of the phases. The direct contact of the phases is thus through the pores and the contactor works as device of a continual mass transfer similar to an absorber [1].

The operation of a membrane contactor is based on a presence of capillary forces which drive one of the two phases on the membrane surface. If the gas phase is inside the hollow fibres and submerged into a liquid the liquid will not pass through the membrane until the liquid pressure ΔP_C reaches a certain value given by the Young-Laplace equation (2) [2,3]:

$$\Delta P_C = -\frac{4\lambda\cos\theta}{d} = P_K - P_P \qquad (2)$$

Where λ is the liquid surface tension, θ is the contact angle between liquid and the membrane surface, d is the diameter of the membrane pore, P_K is the liquid pressure outside of the membrane and P_P is the gas pressure inside of the membrane.

4. Materials and methods

4.1. Experimental device

The testing equipment consists of two independent circuits i.e. gas and liquid circuit. The scheme of the device is shown in figure 1 and the real view of the device in figure 2. For further detail see work of Ostrezi [4].

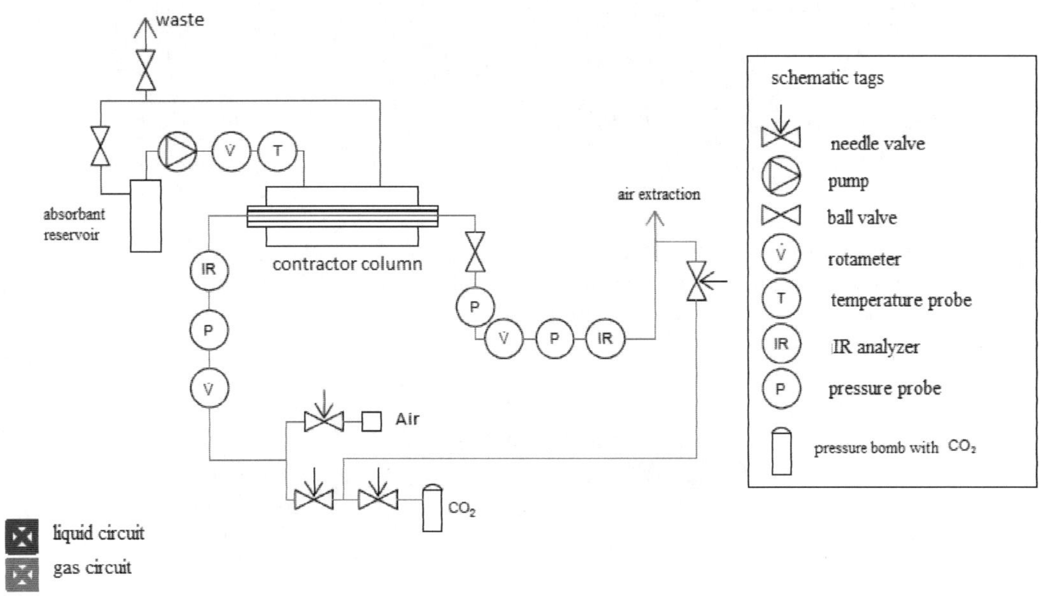

Figure 1. Technologic scheme of the experimental device.

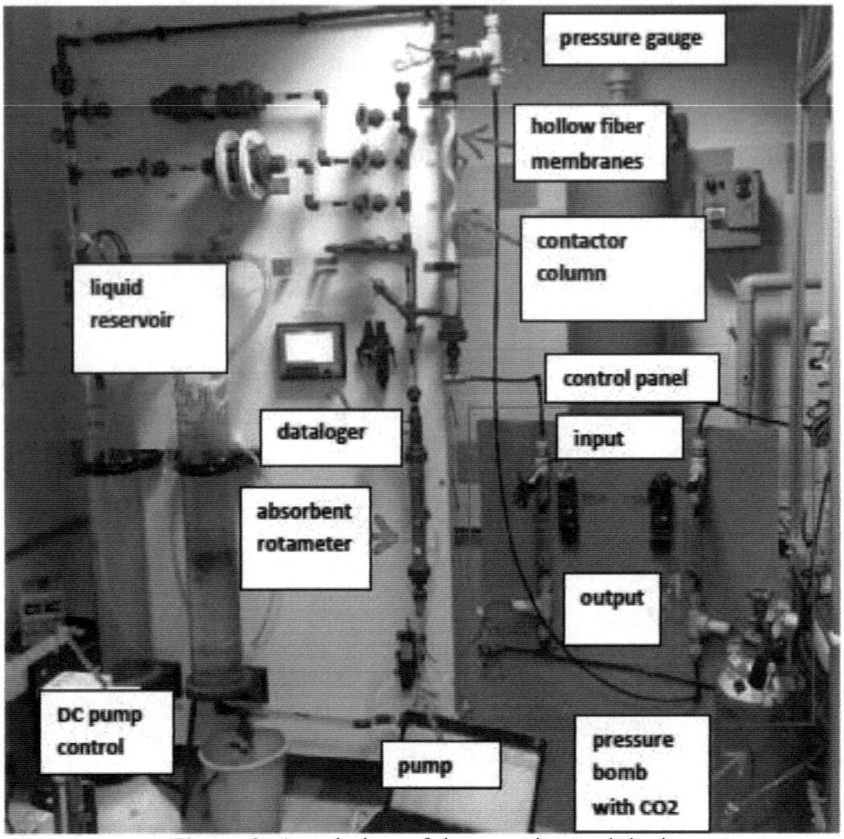

Figure 2. A real view of the experimental device.

4.2. Hollow fibre membrane (HFM) module

The HFM made of polypropylene by Zena Membrane [5] were used. The HFM's parameter are shown in table 1, a real view of the membrane is in figure 3.

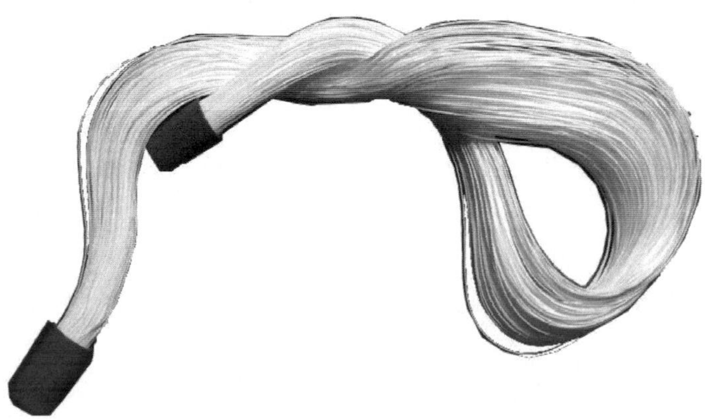

Figure 3. A real view of the HFM

Table 1. Parameters of the hollow fibre membrane used in the experiments.

Fiber diameter [μm]	Surface treatment	Surface area [m^2]	Number of fibers
620	hydrophobic	0.4	300

4.3. Calculation of the efficiency of the absorption process

The process efficiency η was calculated as follows (equation (3)):

$$\eta = \frac{\Delta \dot{n}_{CO2}}{\dot{n}_{CO2\,IN}} \cdot 100\% = \frac{\left(\dot{n}_{CO2\,IN} - \dot{n}_{CO2\,OUT}\right)}{\dot{n}_{CO2\,IN}} \cdot 100\% \tag{3}$$

where $\dot{n}_{CO_2IN}$ is the amount of CO_2 transferred to liquid phase and $\dot{n}_{CO_2OUT}$ is the amount of CO_2 in the gas phase.

5. Results and discussion

The rate of the CO_2 transfer per unit of area and the concentration gradient is shown in figure 4.

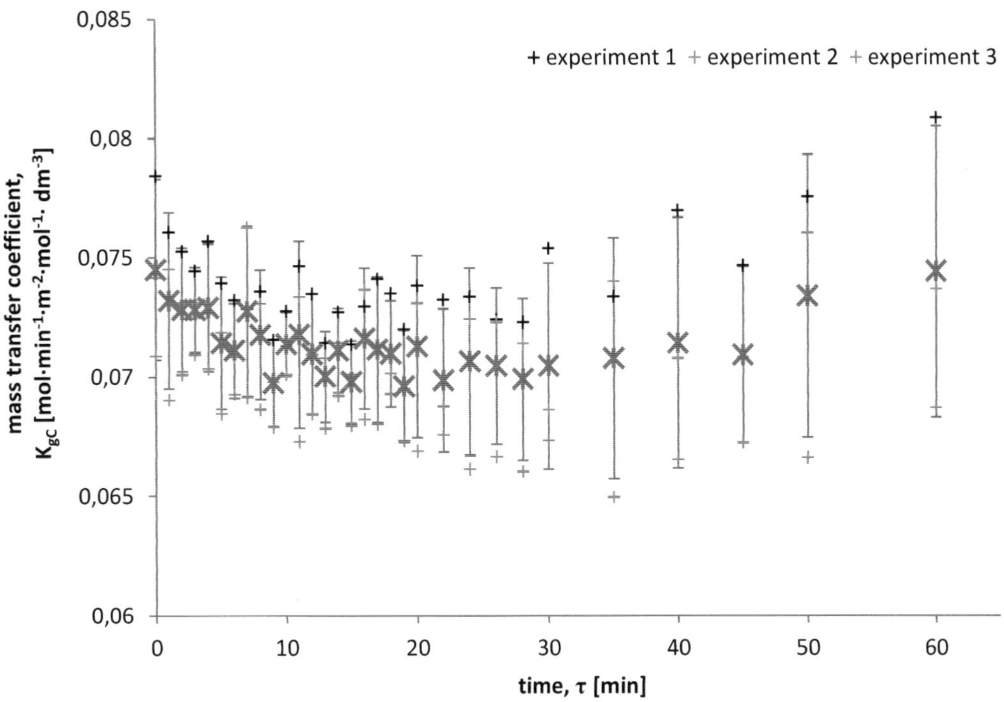

Figure 4. Mass transfer coefficient as a function of time.

6. Conclusion

Using the membrane contactor developed in our laboratory, the CO_2 mass transfer rate from air to aqueous environment was $1.9 \cdot 10^{-4}$ mol/min. Expressed as the mass transfer coefficient at given conditions it was 0.072 ± 0.005 mol·min^{-1}·m^{-2}·mol·dm^{-3}.

Acknowledgments
The research leading to these results has received funding from the Ministry of Education, Youth and Sports under the National Sustainability Program I (Project LO1202).

References

[1] Mansourizadeh A A A F Ismail 2009 Hollow fiber gas–liquid membrane contactors for acid gas capture: A review *J. of Hazard.Mater.* **171** 38-53
[2] Neužil L Mika V *Chemické inženýrství I.* Vysoká škola chemicko-technologická
[3] Pabby Anil Kumar S S H. Rizvi A Ana Maria Sastre 2009 *Handbook of membrane separations: chemical, pharmaceutical, food, and biotechnological applications* (Boca Raton: CRC Press)
[4] Ostrezi J 2017 *Separation of gaseous pollutants on membrane contactors* Diploma Thesis (VUT Brno)
[5] Zena-membranes http://www.zena-membranes.cz/

The design of flood protection in Kobeřice municipality

J Svoboda[1], V Václavík[1], T Dvorský[1] and L Klus[1]

[1]Institute of Environmental Engineering, Faculty of Mining and Geology, VSB-Technical University of Ostrava, Czech Republic

Email: jakub.svoboda@vsb.cz

Abstract. This paper describes the current situation of flood protection of the municipality of Kobeřice and the method to be used to increase it, which is the construction of a flood pool – detention basin intended for the reduction of the negative impact of flash floods within the territory of the municipality. The actual design of the detention basin is based on hydraulic calculations and the prepared drawing documentation.

1. Introduction

Water is one of the main prerequisites for life on the Earth and it forms an important part of the environment as a renewable natural resource of significant importance. At present, the obvious rise of climate changes in connection with the global warming of the Earth's surface also help to understand the importance of water for the life and development of the society. It is evident that there will be more inconsistence in annual precipitation in very near future. This will result in frequent and large floods, or a period of low precipitation volume and low humidity, which will have adverse affects on agriculture and water supply.

Flood protection means measures that prevent and minimize the damage of floods with respect to the lives and property of citizens, companies, and the environment. The purpose of flood protection is the highest possible reduction of damage due to the flooding, especially of urbanized areas. It is a fact that absolute flood protection is impossible to achieve. We can partially limit and influence the peak flood flow rates or transform the flood wave, thus affecting the time course of the floods. This allows us to take more effective measures to save lives and property.

2. Description of the current situation

The municipality of Kobeřice is situated in an area where it is not endangered by any larger watercourse (figure 1). The Oldřišovský Stream runs through the centre of the municipality and it mostly passes through a built up area. In municipality of Kobeřice are 3 301 permanent residents (as of January 1, 2016). At present, Kobeřice does not have any flood protection measures. The so-called storm rainfalls that can cause flash floods are the biggest threat to the municipality. The location of the municipality from the point of view of the geomorphology of the terrain makes it even worse. The last flood in the municipality was recorded in June 2010. The Oldřišovský Stream drainage basin is mainly formed by cultivated agricultural land, the slopes located in this area are not provided with any landscaping elements (e.g. alleys, balks, etc.) and there are no anti-erosion measures either. As a result, the retention capacity of the landscape is very limited, the rate of rainwater drainage from the landscape is high, and there is also land erosion. This causes a direct threat to the lands, residential buildings and the inhabitants of the municipality in the form of storm rainfalls.

IOP Conf. Series: Earth and Environmental Science **92** (2017) 012062 doi:10.1088/1755-1315/92/1/012062

The territory of Kobeřice municipality is drained by watercourses of the second and third order. The largest watercourse in the area under consideration is the Oldřišovský Stream, which runs to the territory of the Republic of Poland as the Bílá Voda watercourse, where it subsequently flows into the Odra River. The average specific outflow in the area is 3 l/s^1km^2. There are small water reservoirs located on the Oldřišovský Stream, which do not serve as the flood protection of the municipality [1,2].

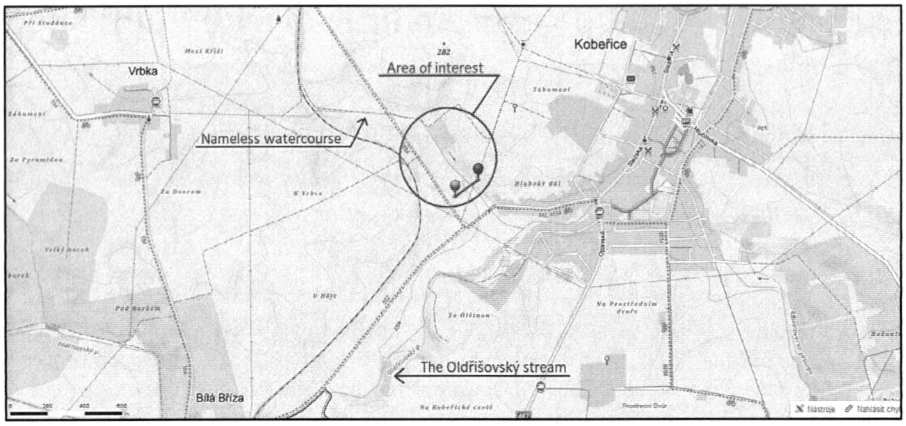

Figure 1. Location of the area of interest in Kobeřice municipality [3].

3. Design of flood protection

With respect to the local, morphological and agricultural conditions, the concept of a small water reservoir with a retention function is proposed as the most suitable flood protection of the municipality of Kobeřice. There is a nameless watercourse running through the site of the designed flood pool which is a left-side tributary of the Oldřišovský Stream (figure 1). The total length of this stream is 2.64 km and its hydrological characteristics are shown in table 1.

Table 1. Hydrological data of the left-side tributary of the Oldřišovský Stream.

Watercourse	Left-side tributary of the Oldřišovský Stream (HOZ - IDVT 10217069)
Hydrological order number	2-04-01-0070
Profile	r. km app. 0.370 – Tributary from Vrbka village
Long - term average annual precipitation level in the basin of P_a	628 mm
Long-term average flow rate	9.1 l/s

A similar concept of flood protection was dealt with in the cadastre of Bolatice municipality, which is referred to in the article called The Design of Flood Protection in the Cadastre of Bolatice Municipality [5].

3.1. Construction of flood-protection dam

The design of the flood pool includes a homogeneous earth dam and the type of soil to be used for the construction of this dam is selected on the basis of engineering-geological survey recommendations. The most widespread types of suitable soils found in the area in question are brown and luvic soils [6,7].

The dam is designed with the height of 4 m and the length of the dam crest is 141.50 m. The width of the dam crest is 4 m and it will be at an altitude of 253.26 m above sea level. The bottom of the flood pool is located at an altitude of 249.26 m above sea level, while topsoil removal reaching the depth of 0.6 m will be made at the site of the future dam. The removed topsoil will be used for the final fertilization of the windward and waterward slopes. The dam crest is designed as horizontal and movable. There is a 3 m wide service paved road. Quarry rubble with the fraction of 8/16 mm in the thickness of 100 mm and with the fraction of 63/125 mm in the thickness of 300 mm is to be used for the road reinforcement. The road design is carried out in compliance with ČSN 75 2410 [8]. A sample cross section of the dam body is shown in the figure (figure 2). The following table presents the basic dam technical data.

Table 2. Technical parameters of the dam.

Dam parameters	Description / Value
Dam type	Homogenous earth, frontal, flow through
Dam crest length	141.5 m
Dam height	4 m
Dam crest width	4 m
Service road width	3 m
Crest elevation	253.26 m above sea level
Dam bottom elevation	249.26 m above sea level

A safety crest spillway with a capacity of Q_{100} is designed as part of the dam profile to transfer high water. The safety spillway is connected to a drainage channel with a depth of 0.8 m, a bottom width of 4 m and a gradient of slopes of 1: 1. The drainage channel will be led and discharged into the left-side tributary of the Oldřišovský Stream at an angle of 45 °. A 3 m wide cutoff is designed according to ČSN 75 2410 [8] in order to stabilize the dam. The depth of the cutoff is designed to 0.8 m. The designed waterward slope gradient is 1: 3 and quarry stone along the entire length of the slope is designed as the toe of dam at its end. The windward slope is designed with a gradient of 1: 2.5. The reinforcement of the windward and waterward slope is designed using humus with sowing in the min. thickness of 100 mm across the whole range. A toe drain at the windward toe of the dam is designed to drain away the water leaking through the dam body and its subsoil.

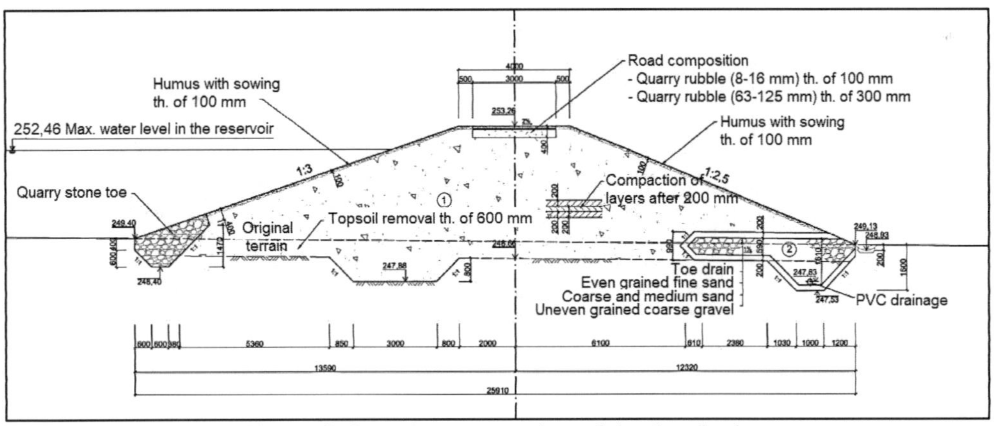

Figure 2. Sample cross section of the dam body.

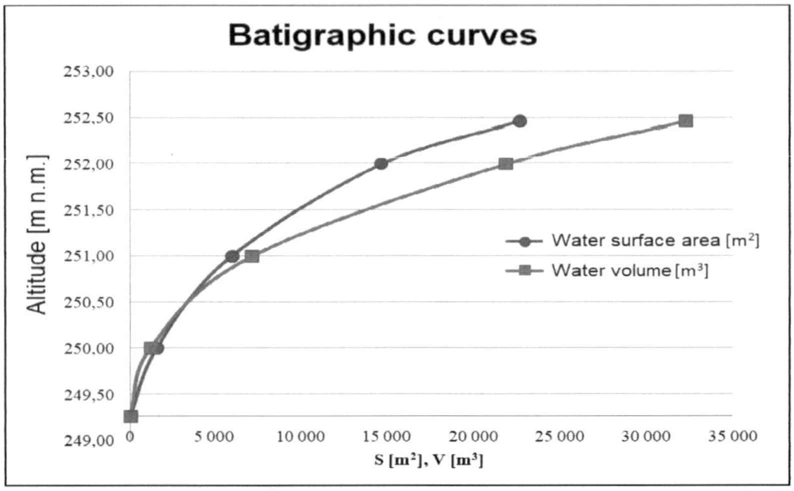

Figure 3. Batigraphic curves.

The characteristic data of the reservoir (water level, flood area and reservoir volume) are expressed by the so-called batigraphic curves. They represent the dependence of the volume of accumulated water in the reservoir and the flood area on the water level in the reservoir. Fig. 3 shows the batigraphic curves of the reservoir.

3.2. Flood pool outflow regulation

Gullet type of draining device has been chosen to regulate the drainage of accumulated water from the flood pool. The location of the gullet is designed in the waterward slope at a distance of 8.5 m from the pool axis and 5.1 m from the toe of the slope. The water inlet to the gullet will be provided by a reinforced channel to the toe of the waterward slope. In addition, a concrete trough at a slope of 10 ‰ with concrete wings and a total length of 4.6 m is designed through the dam body. The normal flow rate in the stream-bed of the left-side tributary of the Oldřišovský Stream will be taken away by a circular culvert DN 350 located at the bottom part of the gullet. A coarse screen is designed in front of the circular culvert. At peak flow and full flood pool filling, the water will drain through the gullet stop log wall. A reinforced concrete pipeline DN 600 running through the dam body is designed to drain the water from the gullet.

In order to ensure the safety of the dam, a safety spillway is designed to transfer the peak flow rate of Q_{100}, whose value is 5.54 m^3/s. It is possible to take part of the flow rate away by means of the gullet, thus reducing the designed flow rate of the safety spillway. Due to the safety of the dam (e.g. sewer pipe blockage, etc.), the Q_{100} flow rate has been chosen as the designed one. The spillway edge of the safety spillway is designed for the length of b = 7.0 m, the overflowing jet height was designed at h = 0.4 m and the slopes of the safety spillway are designed at a gradient of 1: 5. The safety spillway will be reinforced by means of quarry stone pavement in the concrete bed with a thickness of 0.40 m.

3.3. Flood wave transformation

In order to capture and transform the flood volume through the retention area of the flood pool, it is necessary to know the volumetric flow of the flood over time, the so-called flood hydrograph. This graph (see figure 4) represents the course and transformation of the flood wave, which shows the relation between the inflow and outflow of water from the flood pool per unit of time. From the point of view of the protection against the adverse effects of flood, it is especially important to reduce the peak flow rate and shift the critical flood flow time. The main issue is to ensure the safety of the waterwork itself and the safety of the area below the flood pool. The drainage of the water from the designed flood pool is simulated during a calculation of the flood wave transformation to the capacity flow rate of the drainage pipe DN 600 Q_{kap} = 0.816 m^3/s. The controlled drainage of the water from the flood pool is made by means of a circular culvert DN 350 controlled by a slide-valve and by a spillway through the gullet stop log wall. In case of Kobeřice flood pool, the flood wave simulations and transformation of their peak flow rates are performed for the flow rates of Q_2, Q_{20} a Q_{100}.

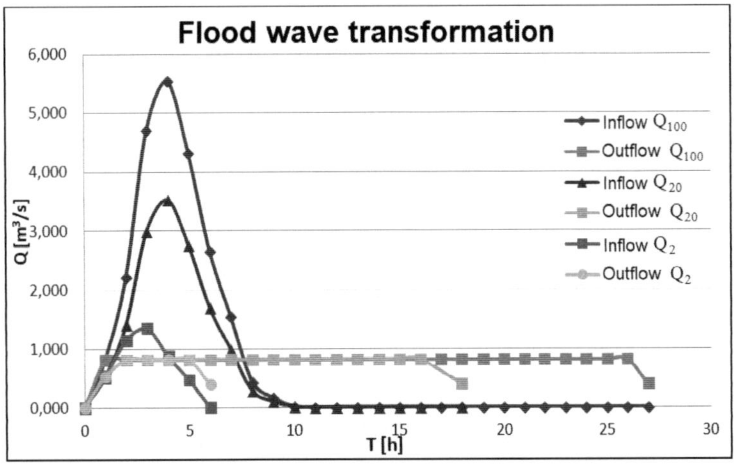

Figure 4. Flood wave transformation.

4. Conclusion

A small water reservoir with a retention function has been designed in order to protect the municipality of Kobeřice, whose task is the retention of a flood flow and its controlled drainage. The water reservoir has a designed earth-fill homogeneous dam with the height of 4 meters and the length of 141.5 m. This dam determines the controllable area of the reservoir with the volume of V_{max} = 32 258 m^3 and a flood area of S_{max} = 22 651 m^2. The safety of the dam is ensured by a safety spillway designed to transfer the peak flow rate of Q_{100}. A gullet outlet, with a lower circular culvert DN 350 equipped with a reinforced concrete waste pipe DN 600, is used to drain the accumulated water from the reservoir.

References

[1] *Obec Kobeřice* 2009 http://www.koberice.cz/
[2] HALUZA J 2011 *Vyhodnocení vlivu územního plánu Kobeřice na udržitelný rozvoj území*
[3] *Mapy.cz* 2015 http://www.mapy.cz
[4] *Český hydrometeorologický ústav* 2017 http://portal.chmi.cz/
[5] Gola L, Václavík V, Dvorský T and Valíček J 2014 Flood protection concept in the register of commune Bolatice *Aktuální vodohospodářské problémy* Ostrava pp 65-72
[6] Balatka B, Demek J and Mackovčin P 2006 *Zeměpisný lexikon ČR* (Brno: AOPK ČR)
[7] Narita K 2000 *Design and Construction of Embankment Dams* Aichi http://aitech.ac.jp/~narita/tembankmentdam1.pdf
[8] ČSN 75 2410: 2013 *Small water reservoirs*

IOP Conf. Series: Earth and Environmental Science **92** (2017) 012063 doi:10.1088/1755-1315/92/1/012063

Impact of mine dumps on transport the invasive plant species to Upper Silesia

N Sotkova[1], B Lokajickova[1], J Mec[1], H Svehlakova[1] and B Stalmachova[1]

[1]VŠB – Technical Univerzity of Ostrava, Faculty of Mining and Geology, Institute of Environmental Engineering, Czech Republic

E-mail:jan.mec.st@vsb.cz

Abstract. Human activities significantly change the species composition in the area. The main factor of change was the mining industry, which changed the natural conditions on Upper Silesia. The anthropogenic relief of mine dumps are the main centre of alien plant in an industrial landscape. The poster deals with the state of the invasive plant species by the phyto-sociological surveys on Upper Silesia.

1. Introduction

The investigated areas used in this research are those of Upper Silesia. Upper Silesia is located in the temperate belt of Central Europe. This highly urbanized landscape has been marked by ongoing mining extraction since the 18th century to the present. There are still preserved remains of mine dumps which are the subject of our research. These are mining depots consisting of tailing which have a specific microclimate. Uncovered tailing offers specific conditions for primary succession of species. This situation is used by species in our conditions considered as non-original.

Figure 1. Map of mine dumps-subject of our research (google.cz/maps, edited by authors).

Intensive coal mining has been involved in the relief formation resulting in both concave and convex shapes. The mine dumps belong to the most significant and the most numerous ones. An indirectly caused anthropogenic shapes are declining areas, often flooded with water [1]. These lands have become valuable habitats providing optimal conditions for many plant and animal species.
The monitored area is situated in the river basin of Odra, which originates in the mountain range of Oderské vrchy and together with rivers Olše, Ostravice, Opava and Moravice they form a network, converging towards the Ostrava basin [2]. Rivers are an important transport medium for plant seeds.

This way, the seeds of the invasive species *Reynoutria Sp.* and *Impatiens glandulifera* easily root in riverbeds. Soils in the Ostrava region are significantly affected by the contamination of ecotoxic elements and compounds mainly from the mining and chemical industry not only from Ostrava but also from the whole of Upper Silesia.

According to the Regional Phytogeographical Division of the Czechoslovak Republic [3]. The Ostrava Basin is included in the phytogeographical district No. 83 known as Carpathian mesophysic. These areas are slightly warm with typical Central European flora. The climatic conditions of the Ostrava Basin are the factors influencing species composition. The Ostrava Basin is influenced from the north by the Polon province and from the south by semi-thermo-thermophilic to thermophilic plants from the Pannonian thermophilic [2]. The Polon province is represented mainly by Eurasian species [4]. Forest is the most important connecting element. Together with the territory It forms a compact factor that is harmonious and makes the main protective factor of other environmental components [5].

We included in the areas of our research following mine dumps: Urx, Ema, František, Barbora, Lazy, Heřmanice, the Darkov sea and the reference area was the forest near Šenov.

2. Invasive species

The Ostrava Basin is taken as an area with a relatively monotonous vegetation and the landscape is mostly anthropogenic in nature [6]. Due to anthropogenic burdens and urbanization the biotopes are degraded. The species with a wide ecological valency assert there and the landscape is being fragmented and ruderalized [7]. Into such modified communities invasive species are introduced more easily and expansive species are spread.

Non-native species are those transported outside its original area. Invasive plant species are those that endanger the biodiversity of the original ecosystem [8]. These species come to us deliberately (botanical gardens, seed transport, etc.) or unintentionally (with cargo, animals, etc.). Our research areas include: *Reynoutria Sp., Impatiens glandulifera, Robinia pseudoacacia, Solidago canadensis.* Among the potentially dangerous invasive species *include Quercus rubra* and *Larix decidua.*
Well-known iInvasive species *Reynoutria Sp.* were imported from Southeast Asia. They reached our continent in the first half of the 19th century. Nowadays mostly northern and central Europe is swamped with it. Here it was introduced as a decorative plant [9]. It was used in this way, but it quickly spread from the gardens into the open country. It is now a highly invasive species. It grows mainly along roads, streams and rubble and it is one of the most popular plants in the waste dumps of Upper Silesia.

Impatiens glandulifera was introduced here from the Western Himalayas region. It spread in the first half of the 19th century as an ornamental plant. It has adapted to wetlands, river banks and wetter places. It forms continuous vegetation particularly along river banks, where it is very successful. We can also find it in anthropogenically disrupted habitats [9].

The issue of non-native species also applies to *Robinia pseudoacacia.* It is located on steep slopes and unpaved spruces, where its roots strengthen the soil. In the countryside, it benefits as a honey tree (melliferous), but its wood is not suitable for economic use. On the whole, its presence is disturbing because it depauperates the soil of nutrients [5].

The original home of *Solidago canadiensis* is the USA and southern Canada. It introduces to ruderal habitats and anthropogenic affected areas, which include plants, railways and roads. It reproduces vegetatively using a large number of achenes. It is highly competitive to other plants and displaces native species. However, this highly invasive plant is melliferous and it also has healing properties [9].

Quercus rubra is a non-native tree that was introduced here from North America in the 17th century. Now it is widely introduced into the country. It is widely used in parks, urban greenery and sometimes also forests. On our continent, it penetrates into the oak, oak-hornbeam, pine-oak, fir and beech forests. It is very resistant to dirty environment. They also survive in bigger shade than our original oak species. Its fruits are not edible for our birds because of the size, and therefore the seeds remain on the sites and root [9].

3. Methodology

As a basis for this research the methodology by Moravec et al. [10] was chosen. A retrospective field survey was conducted in the spring months. This way an approximate inventory of species composition was obtained. For selection of habitats, a subjective method was chosen [10]. There have been chosen appropriate areas with homogeneity of vegetation throughout their area.

The size and shape of the study area is determined by the methodology. Forest communities were mapped on 225 m^2 in the square shape. On selected sites, we conducted research based on the species composition (floristic composition of the vegetation cover), stratification (vertical structure), numerosity and coverage of individual species. The Braun-Blanquet 1951 scale was used to evaluate stratification and coverage of vegetation [11].

Table 1. An example of a phytosociological reléve.

Serial number of reléve	34	35	36	37	38
Job number of reléve	1	10	16	17	26
Plot size (m^2)	225				
The number of species in a reléve	16	20	15	19	26
Cover of E3 (%)	-	50	60	60	70
Robinia pseudoacacia	.	.	.	1	2
Acer pseudoplatanus	.	3	1	+	r
Fagus sylvatica	.	.	.	1	.
Quercus rubra	.	.	.	.	+
Cover of E1 (%)	60	70	60	30	90
Reynoutria japonica	2	.	.	.	.
Galeobdolon luteum	1	.	.	+	1
Impatiens parviflora	2	1	+	1	1
Impatiens noli-tangere	.	r	1	.	.
Urtica urens	4	2	3	.	+
Acer campestre juv.	.	.	.	.	+

4. Result and discussion

In our research we focused on the mapping of invasive species on mine dumps in Upper Silesia. It is clear from the results that mine dumps of Upper Silesia are a great place to begin the expansion of invasive non-native species. Invasive species naturalize here, which means-repeated successful introduction of the reproductive population (Assessment and management of alien species that threaten ecosystems, habitats and species, 2001).

We included in the areas of our research following mine dumps: Urx, Ema, František, Barbora, Lazy, Heřmanice, the Darkov sea and the reference area was the forest near Šenov. Selected dumps were either left to natural succession, or reclaimed in the past. The results show that the most widespread invasive species is: Reynoutria Sp. Poster also contains a graph including the percentage of invasive species on mine dumps. From these results, it is clear that the most affected area is the Urx mine. In conclusion, it can be said that invasive species have a dangerous effect on our original biocenose. However, their overall impact cannot be predicted in advance. Research on dumps will continue in the coming years.

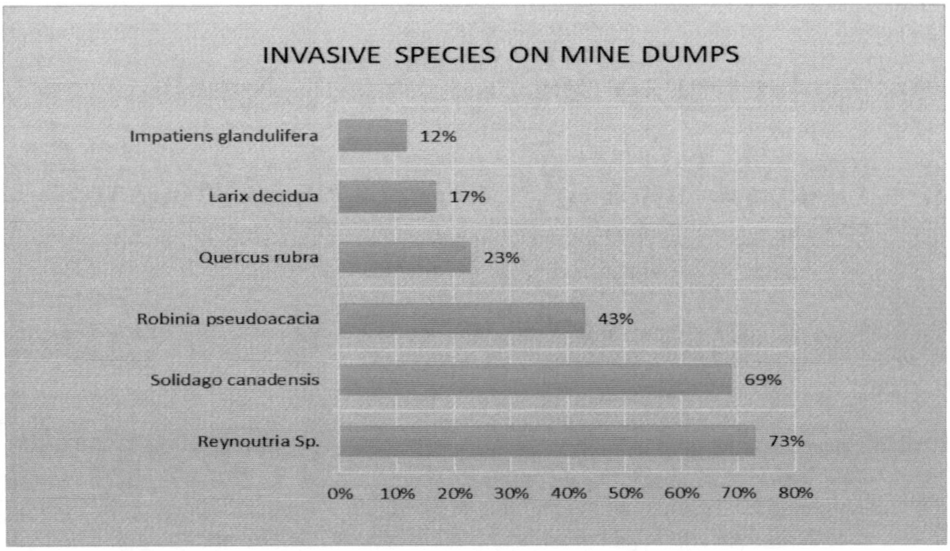

Figure 2. Invasive species on mine dumps-subject of our research.

Acknowledgments
This article and poster were funded from sgs projects.

References

[1] Hruban R Ostravská pánev http://moravske-karpaty.cz/prirodni-pomery/geomorfologie/ostravska-panev/
[2] Weissmann H 2004 *Ostravsko* (Praha: Agentura ochrany přírody a krajiny ČR) p 454
[3] Skalický V 1988 Regionálně fytogeografické členění Hejný S. and Slavík B. *Květena ČSR I* Academia Praha p 103-121
[4] Neuhäuslová Z et al. 1998 *Mapa potencionální přirozené vegetace České republiky: te tová část* (Praha: Academia) p 341
[5] Vokřínek P Nadace Landek 2003 Landek: Svědek dávné minulosti (Český Těšín: FINIDR) p 120
[6] Ševčík J Pečínka A and Plášek 2005 *Rostliny Národní přírodní památky LANDEK* (Ostrava: Biologické-ekologický klub)
[7] Stalmach J and Stalmachová B 1997 Sanace odvalu Ur v Ostravě-Petřkovicích: Biologické hodnocení vlivu záměru na biocenózu p 27
[8] https://www.cbd.int/doc/publications/cbd-ts-01.pdf
[9] Aichele D and Golte-Bechtle M 2007 *Co tu kvete? Kvetoucí rostliny střední Evropy ve volné přírodě* (Praha: Knižní klub, Euromedia Group) p 430
[10] Moravec J 1994 Fytocenologie: nauka o vegetaci (Praha: Academia) p 403
[11] Šotková N 2015 Mapování reálné vegetace na území NPP Landek Master Thesis (Ostrava: VŠB-TU)

IOP Conf. Series: Earth and Environmental Science **92** (2017) 012064 doi:10.1088/1755-1315/92/1/012064

Environmental Engineering in the Slovak Republic

N Stevulova[1], M Balintova[1], M Zelenakova[1], A Estokova[1] and S Vilcekova[1]

[1]Technical University of Kosice, Faculty of Civil Engineering, Institute of Environmental Engineering, Kosice, Slovakia

E-mail: nadezda.stevulova@tuke.sk

Abstract. The fundamental role of environmental engineering is to protect human population and environment from impacts of human activities and to ensure environmental quality. It relates to achieving the environmental sustainability goals through advanced technologies for pollutants removing from air, water and soil in order to minimize risk in ecosystem and ensuring favourable conditions for life of humans and organisms. Nowadays, a critical analysis of the environment quality and innovative approaches to problem solving in order to achieve sustainability in environmental engineering, are necessary. This article presents an overview of the quality of the environment and progress in environmental engineering in Slovakia and gives information regarding the environmental engineering education at Faculty of Civil Engineering at Technical University in Kosice.

1. Introduction

The fundamental role of environmental engineering is to protect human population and environment from impacts of human activities and to ensure environmental quality. It relates to achieving the environmental sustainability goals through advanced technologies for pollutants removing from air, water and soil in order to minimize risk in ecosystems and ensuring favourable conditions for life of humans and organisms [1,2,3,4]. Sustainability in environment focuses on practical experience of a wide range of research areas within the general field of environment which covers mainly the following areas: waste water management, air and soil pollution control, waste disposal and recycling strategies, radiation protection, industrial hygiene, environmental technologies, processes linking environments, urban planning and development, environmental economics, policy and legislation [5,6,7,8]. It also includes management of environmental risk and safety, mainly assessment of environmental risks from climate changes and anthropogenic activities in the catchments of water bodies. Studying the impact of construction systems in buildings in terms of harmful substances emissions release into indoor environment is very important. Environmental engineering realizes a wide variety of ecological activities for creating appropriate living and technological conditions to maintain the right parameters with regard to the human environment; provides environmental technologies for cleaner production allowing to reduce emissions and waste production, the usage of non renewable resources, recycling the waste, mitigates the effects of natural disasters (floods, droughts, pollution in water, air and soil) [7,8,9].

Nowadays, a critical analysis of the environment quality and innovative approaches to problem solving in order to achieve sustainability in environmental engineering, are necessary. The aim of this paper is to give the information of the quality of the environment in a clear and concise form, to show progress of the Slovak Republic (SR) in the transition to a low-carbon and green economy. Ensuring improved environmental sustainability is linked to the education of environmental experts having a broad knowledge base and skills to solving the actual environmental problems. Further part of this article brings information of environmental engineering education at Faculty of Civil Engineering of Technical University of Kosice.

2. The environment quality of the Slovak Republic

A cross-cutting source of information on the status of the environment in the SR presents the environmental regionalization of Slovakia (ERS). Over 20 year's period, environmental protection and risk factors: air, water, rock environment, soil, biota and landscape, wastes, as well as criteria for their mutual evaluation have been gradually profiled. The Slovakia territory is in terms of the environmental regionalization divided into three categories. About 51% of its territory is included into the 1st category – the highest environmental quality where classified regions with undistorted environment were. A decrease of 18 % of the regions with heavily distorted environment was recorded in 2015 in comparison with 1992.

The decline in the size of agricultural land and increase in the size of forest and urban areas have been long-recorded. Agricultural land is of high quality from point of view of its contamination; more than 99% is hygienically suited. Soils monitoring has been showed that the most apparent is the physical degradation of soils – especially by erosion and compaction. Regarding the chemical degradation of soils there has been observed the decrease of accessible nutrients - particularly phosphorus and potassium. The area of acidic soils has been decreasing together with the decrease of acidic air pollutants. Average organic carbon content in arable land is significantly lower than on permanent grassland. The soils contaminated in the past are contaminated also at present. The evaluation of agricultural lands according to quality level in 2014 confirmed that 35% share is of the highest quality (1-5 levels). Improving the air quality and reducing atmospheric deposition have led to improved forest health, but it is still assessed as negative.

The water resources are distributed unequally, not only regarding the quantity but also quality.

A very good and good environmental status of surface water bodies was recorded at 56.2% with a length of 8,073.43 km. The average environmental status was 34.8% of water bodies, which represents 7,565.46 km. The bad and very bad state was set in 9% of water bodies with a length of 2,159.41 km. The assessment of chemical status of surface water bodies in the period 2009 - 2012 was carried out in 1,510 water bodies. Good chemical status reached 1,473 (97.6%) of water bodies of the SR and 37 (2.4%) of water bodies did not achieve good chemical status.

The contamination of territory by harmful pollutants coming from human activities poses a significant risk to human health or environment, groundwater and soil. In Slovakia 902 probable and 281 contaminated sites were recorded. It includes a wide range of areas contaminated by industrial, military, mining, transport and agricultural activities, but also by inappropriate waste handling. Despite a lower production of municipal waste per capita than the average in the European Union countries, low poor separation of its components and recycling still persists and land filling in high share prevails. Ratio of Gross Domestic Product (GDP) to quantity of waste generated (excluding municipal waste) shows a positive trend, the so-called absolute decoupling.

The energy consumption of the Slovak economy is declining; however, when compared to the EU average, it is still high. The share of renewable sources in gross final energy consumption of energy is increasing. In 2014, it amounted to 23.0% in electricity production, 10.4% in heat and cold production, and 6.6% in transport.

Emissions of air pollutants such as NO_x, SO_2, CO, $PM_{2.5}$, PM_{10} and non methane volatile organic compounds (NMVOC) have been declining in the long term, but after 2000 the rate of their decline slowed down significantly. The total volume of emissions of persistent organic pollutants (POPs) is increasing. A positive trend in produced greenhouse gas emissions quantity was recorded. The 45% decline in 2014 has been found in comparison to 1990. Problematic sectors where the effective regulation of greenhouse gas emissions increase is not successful are the road transport and energy production by the burning of fossil fuels in households. The limit values for selected pollutants have been exceeded in some air quality monitoring stations, but air pollutant limit values to protect vegetation were not exceeded. However, ground level ozone remains a serious problem, where limit values for the protection of vegetation and forest are long term and regularly exceeded.

As people spend more than 2/3 of days in buildings, focusing on monitoring indoor air quality is very important from the point of view of ensuring healthy housing. The assessment of indoor air quality in schools in Europe and ambient air in the vicinity of schools (projects SEARCH a SINPHONIE) resulted in worsening the indoor air quality related to outdoor air pollution, design errors of buildings and poor building maintenance, poor cleaning and lack of ventilation. The proposed recommendations were aimed at preventing and improving indoor air quality in schools.

Drinking water quality has been showing high levels for a long time, the domestic consumption has been declining in the long term. The number of inhabitants supplied with water from public water supplies is increasing. Amount of waste water and water pollution has been declining in the longer term. The connection of population to public sewerage system has been also increasing in the long term; however, it lags far behind the public water supply.

Weather in the Slovak Republic has become more extreme in recent decades. One of the most significant negative effects of climate change is the more frequent occurrence of very strong local storms with high total rainfall causing unexpected floods and huge flood damages. On the other hand, it is the occurrence of long-lasting heat and droughts. The alternation of these extreme phenomena has negative impacts on ecosystems, economic and social spheres what is the most pressing environmental problems at present.

The GDP of the Slovak Republic since 2000, with the exception of 2009, records an annual year-on-year increase. Industrial production has the greatest share (about one-fifth) in its creating among others of industrial sectors. Resource productivity in the Slovak economy, measured as the ratio of GDP at constant prices as at 2010 to the domestic material consumption (DMC), increased by 42% compared to 2000 but Slovakia still significantly lags behind the average resource productivity in the EU.

CO_2 productivity expressed as a ratio of GDP at constant prices to the total quantity of CO_2 emissions shows a long-term positive development. On the other hand, thus the carbon intensity of the SR economy is decreasing.

The most important voluntary tools of the care for environment include environmental labelling, the application of environmental management systems, green public procurement and environmental innovation. The Slovak Republic is among the EU countries with low eco-innovation performance. Within the „Eco-innovation Scoreboard“ in 2013, Slovakia ranked 25th from all EU countries (Slovakia is followed only by Cyprus, Poland and Bulgaria). The most important instruments of financial support of the care for the environment are the Environmental fund and programs for receiving the aid from the EU funds - for the years 2007–2013 the Operational Programme Environment (OPE), and for the years 2014-2020 the Operational Programme Quality of Environment [10].

3. Education in Environmental Engineering

Environmental engineering programmes are offered at either faculties/departments of civil engineering or the departments of chemical engineering at the technical universities. Significant efforts in shaping the environmental science field and application of the principles of sustainable development in education at the Faculty of Civil Engineering of the Technical University in Košice (FCE TUK) resulted from the global trend in higher education in the field of ecology and environmental science in the early 1990s. There was a need to solve the interactions of technical activities with the environment in terms of the complex protection of human and environment and to live in a healthy environment. This was linked with the need for education and training of environmental experts. Education in Environmental Engineering as a discipline in the field of technical sciences is directed to utilization of engineering methods for preserving, rational shaping and using external natural environment (e.g. water resources, waste management, air protection, soil protection), for preserving and creating indoor environment in rooms, buildings and constructions. The first accredited master study branch Ecology in Construction (1993) was followed the accreditation of bachelor, master and doctoral studies in the field of Environmentalistics (2001). The three level study in environmental engineering was provided within the programme Environmental Structures since 2004. Nowadays, the FCE TUK provides 4 interdisciplinary study programmes accredited in the combination of two field of study - Environmental Engineering and Construction/Structural Engineering at bachelor and master level of the university study. Doctoral study program is provided in the field of study of Environmental Engineering. The study programmes are created in accordance with the concept of sustainable development in environmental engineering and construction and it is defined by the needs for industrial, engineering, design, municipal and state spheres associated with the design and making of environmental structures, with the assessment and solution building interactive relationships with the environment.

Bachelor study programme: Environmental Structures
Programme graduates are proficient in using environmental and civil engineering knowledge to develop technologies of air, water and soil protection, waste processing and recycling. The graduated students are able to assess environmental impacts of activities and buildings, and to solve simple engineering projects in the field of landscape design, protection and revitalisation based on new legal regulations and state environmental policy.

Bachelor study programme: Structures for Sustainable Water Management in the Landscape
Graduated students will acquire basic knowledge of water management in urbanized areas and in the landscape; monitoring and analyzing the state of pollution of environmental components; design, construction and operating of water structures; protection and restoration of the landscape and legislation in force in the addressing areas. The graduated students are able to use the acquired knowledge, practical procedures and tools for solving simple engineering projects from the field of environmental protection, water management, and design of simple constructions and fragments of complex constructions and are able to assess the impacts of activities and constructions on the environment.

Master study programme: Environmental Structures
Graduates can work in the design, operation and management of civil engineering and environmental structures with emphasis on the protection, creation and restoration of the environment and can find applying in state administration and in the area of environmental impact assessment and evaluation of environmental performance of materials and buildings during their whole life cycle. Graduates are also able to carry out all the activities in the preparation, implementation, reconstruction, rehabilitation and use of all types of structures.

Master study programme: Structures for Sustainable Water Management in the Landscape
Graduated students gain comprehensive knowledge and can solve engineering problems of water management, water structures and environmental protection. They will acquire modern methods of experimental research, computerized data processing, environmental monitoring of the quality of environmental components with an emphasis on the aquatic environment and evaluation of extreme hydro-climatological events - floods and droughts. Graduates can analyze, design, operate, as well as prepare and manage the construction of various types of water structures. They are able to implement new technical solutions and the best available technologies to the operating systems, to introduce new legislative regulations and standards. They can perform research with a high degree of creativity and autonomy.

Doctoral study programme: Environmental Engineering
The programme is suitable for engineers who wish to prepare for their future scientific research career. Programme graduates are able to clearly formulate scientific research problems, target objects and

goals of research and progress in the field of environmental engineering; they are able to solve formulated problem with respect to social, ethical, legal and economic impact of their research.

4. Conclusion

This paper presents information on the status of the environment in the SR. Innovative approaches to achieve sustainability in environmental engineering are still necessary nowadays. Solving the interactions of technical activities with the environment in terms of the complex protection of human and environment is linked with the need for education and training of environmental experts. Study programmes in the field of the environmental engineering at FCE TUK are oriented on education of multidisciplinary graduates - environmental "civil" engineers focusing on environmental chemistry, advanced air, soil and water treatment technologies, bioremediation, separation processes, recycling, water treatment plant design, hydrology, water resources management and environmental assessment of technological activities, building products as well as environment components.

Acknowledgments

The paper was created as part of the NFP 26220120037 project solution.

References

[1] Engh E D and Smith B F 2002 *Environmental Science: A study of interrelationships* Mc Graw-Hill Higher Education
[2] Chiras D D 2006 *Environmental Science*. Jones and Bartlett Publishers
[3] Singh Y K 2006 *Environmental Science*. New Age International Publishers
[4] Corbitt R A 1989 *Standard Handbook of Environmental Engineering* McGraw-Hill Publishing
[5] Duggal K N 2008 *Elements of Environmental Engineering* S. Chand Publishing
[6] Mackenzie L D and Masten S 2003 *Principles of Environmental Engineering and Science.* McGraw-Hill Higher Education
[7] Mihelcic J R and Zimmerman J B 2014 *Environmental Engineering: Fundamentals, Sustainability, Design* Wiley
[8] Mines R O 2014 *Environmental Engineering:Principles and Practice.* John Wiley & Sons
[9] Gilbert M M and Wendell E 2008 *Introduction to Environmental Engineering and Science.* Prentice Hall
[10] Lieskovska Z 2017 *Environment of the Slovak Republic in Focus*. Ministry of Environment of the SR; Slovak Environment Agency

Identification of air pollution by macro-tracers for source apportionment

B Svedova[1], J Ruzickova[1], M Kucbel[1,2], H Raclavska[1,3] and K Raclavsky[1]

[1]ENET Centre-Energy Units for Utilization of Non-traditional Energy Sources, VSB-Technical University of Ostrava, 17. listopadu 15, 708 33 Ostrava, Czech Republic
[2]Institute of Environmental Engineering, Faculty of Mining and Geology, VSB-Technical University of Ostrava, 17. listopadu 15, 708 33 Ostrava, Czech Republic
[3]Institute of Geological Engineering, Faculty of Mining and Geology, VSB-Technical University of Ostrava, 17. listopadu 15, 708 33 Ostrava, Czech Republic

Abstract. Identification of air pollution sources in the city of Olomouc was performed during the heating and non-heating season 2016. The contributions of air pollution sources were determined by ratios and equations for organic carbon (OC), elemental carbon (EC), sulphates, nitrates, levoglucosan, etc. During the heating season, the concentration of EC was six times higher compared to the non-heating season.

1. Introduction
Atmospheric particles are generated through a variety of physical and chemical mechanisms from different sources, affecting the occurrence and representation of individual PM components [1]. Organic carbon (OC) and elemental carbon (EC) are products of incomplete combustion of fossil fuels, biomass and biofuels. Elemental carbon is only a primary product of imperfect combustion, whereas OC can also be of secondary origin, i.e. photochemical reactions of volatile organic compounds in the atmosphere [2]. OC and EC cannot be considered as typical "markers" but they can be successfully utilized to identify the type of combustion material [3]. The significance of EC increases during the winter months due to the greater extent of incineration in local furnaces [4]. Sulphates and nitrates can account for 30 to 60% of $PM_{2.5}$. By oxidizing precursor gases (SO_2) and nitrogen oxides (NO_x) in the atmosphere, sulphates and nitrates are generated as a product of some anthropogenic activities. The presence of sulphates in the aerosol is mostly the result of long-distance transmission, nitrates occur in local or regional sources [5]. Potassium is a water-soluble ion commonly used as a marker for the identification of biomass combustion [6], and it is present mainly in fine grain particles of diameter ≤ 1 µm [7]. Levoglucosan and mannosan are the main products of thermal decomposition of structural polysaccharides contained in biomass [8]. Levoglucosan is used as an indicator for wood burning emissions [9]. The ratio between the K^+/EC concentration is used as an indicator for biomass combustion. High K^+/EC ratios (range: 0.21-0.46) are typical for biomass combustion and low ratios (range: 0.025-0.09) for fossil fuel emissions [10]. The OC/levoglucosan ratio can be used to evaluate the contribution of OC emissions from biomass combustion [11]. According *to Frey et al., 2006* [12], this ratio is 9.8 in the case of normal combustion of biomass, and 24.6 in the case of smouldering. The mean OC/levoglucosan ratio for biomass combustion is 14.2, but the ratio values for different types of wood ranged from 3.0 to 100 [13].

The EC/OC ratio for the exhaust gases from the combustion of diesel is around 1.2 and the ratio of EC/OC is 0.45 for the exhaust gas emissions from the combustion of petrol in cars with the catalyst.

Coal combustion has an EC/OC ratio of about 0.3, for wood combustion, the values range from 0.06 to 0.25, and ratio value for biogenic emissions is 0.03 [14].

The aim of the contribution was to research the chemical character of PM_{10} and to assess the sources of air pollution at the localities of Olomouc Hodolany and Klášterní Hradisko during the summer and winter season 2016 based on information on the content of organic, elemental carbon and other components contained in PM.

2. Sampling site and methods

Dust particles (PM_{10}) were collected on quartz fibre filters using a high-volume sampler. PM sampling took place from 27th to 29th July 2016 (summer) and from 5th to 7th December 2016 (winter) at both localities at the same time. The sampling site Olomouc Klášterní Hradisko (1) is in front of a house; there is a field in the immediate surroundings. The site is in an area without direct industrial influence. The sampling site Olomouc Hodolany (2) is situated in a garden in front of a house, adjacent to a field and in vicinity there is a built-up area with houses, see figure 1.

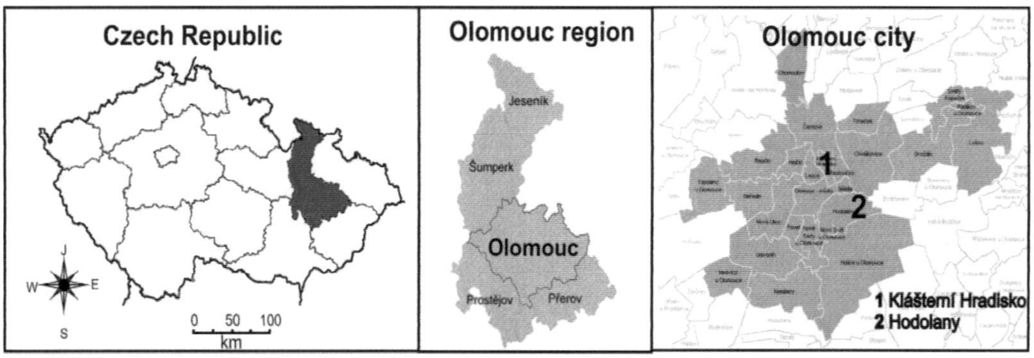

Figure 1. Sampling sites Klášterní Hradisko and Hodolany.

For assessment of the impact of anthropogenic contamination, OC and EC measurements were made by thermo-optical analysis on an OC/EC analyser (SunSet Lab) using the EUSAAR 2 temperature protocol. Determination of water soluble sulphates, nitrates and potassium in PM was done by dissolving PM_{10} particles in deionized water with subsequent determination of ions using the ion chromatography method on the Metrohm 850 Professional IC instrument [15] according to ISO 10304 Water quality - Determination of dissolved anions by liquid chromatography of ions and ISO 14911 Water quality - Determination of dissolved Li^+, Na^+, NH_4^+, K^+, Mn^{2+}, Ca^{2+}, Mg^{2+}, Sr^{2+} and Ba^{2+} using ion chromatography. Determination of anhydrosaccharides and organic compounds was performed using pyrolysis gas chromatography mass spectrometry (Py/GC/MS).

3. Results

The OC, EC, OM and PM_{10} concentrations for both localities during summer and winter sampling in 2016 are shown in table 1. During the heating season (winter season), the PM_{10} daily limit (50 $\mu g/m^3$ according to the Act on Air Protection 201/2012 Coll.) was slightly exceeded at the Hodolany locality.

During the heating season, the PM_{10} concentration was about twice as high at both localities. In the heating season, there was a 2.6 times higher increase in OC concentrations at both localities and a nearly 6-fold increase in EC concentrations compared to the non-heating season. The increase in OC and EC concentrations is due to a higher intensity of incineration processes during the winter months (combustion of fossil fuel and biofuels).

Table 1. OC, EC, and PM_{10} concentrations in heating and non-heating season.

Locality	PM_{10}	OC	EC	OM*	OC in PM_{10}	EC in PM_{10}	OM in PM_{10}
	(µg/m³)				(%)		
Hodolany- summer	28.46	8.22	0.80	11.51	41.83	8.80	58.57
Hodolany - winter	50.93	21.31	4.48	29.83	28.90	2.80	40.50
Klášterní Hradisko - summer	22.77	6.23	0.77	8.72	36.02	9.92	50.43
Klášterní Hradisko - winter	45.06	16.23	4.47	22.72	27.30	3.40	38.30

OM value (organic matter) = OC concentration 1.4. The value 1.4 is mostly used for both urban and rural areas [4].

In the winter season, 288 organic compounds were identified in the PM_{10} samples, which is about 3 times more than in the summer season (97 organic compounds). Sources of organic pollution include both anthropogenic activities and natural sources of biogenic compounds. The dominant group of organic compounds in the summer season were monocyclic arenes (aromatic hydrocarbons containing only one benzene nucleus) that are part of emissions from the transport. During the heating season, concentrations of all organic compounds with exception of monocyclic arenes (especially benzene compounds and BTEX group compounds) increased. Carboxylic acids (organic acids containing the carboxyl group -COOH) represented a dominant group in PM_{10} in the heating season at both locations. Anthropogenic sources of carboxylic acids are primarily the combustion of biomass and solid fossil fuels, or road transport. Also, polycyclic aromatic hydrocarbons (PAHs) showed significantly higher concentrations at both localities during the winter than other organic compounds. Sources of carboxylic acids, PAHs, sulphur-containing compounds, alkenes are primarily combustion processes (fossil fuel combustion and biofuels).

3.1. Identification of the origin of dust particles using diagnostic ratios

The value of the OC/EC ratio for Hodolany was 10.27 in the summer and 4.76 in the winter, and for Klášterní Hradisko the value was 8.09 during the summer campaign and 3.63 during the winter season. The values of the OC/EC concentration ratio for Hodolany and Klášterní Hradisko during the summer period correspond to emissions from biomass combustion. During the winter season, there is a significant decrease in emissions from biomass combustion and the effect of local heating is evident. EC/OC values were 0.1 for Hodolany and 0.12 for Klášterní Hradisko during the summer period and 0.21 (Hodolany) and 0.28 (Klášterní Hradisko) for the winter season, see figure 2. The EC and OC emissions in PM_{10} during the summer correspond mainly to biomass combustion, and in the winter period, the EC/OC ratio corresponds to coal and wood combustion (the EC/OC ratio for coal combustion is 0.3 and for wood combustion 0.06-0.25) [14]. OC also includes biogenic matter, therefore, the difference between the heating and non-heating season is not as significant as in the case of elemental carbon (EC), which is a direct indicator of combustion processes. The K^+/EC ratio in Hodolany was 0.28 and in Kláštěrní Hradisko, the value was 0.29 during the non-heating season. These values indicate that both localities are affected by biomass burning or by biogenic mass during the summer period. In the winter season, the value of K^+/EC= 0.13 (identical for Hodolany and Klášterní Hradisko) indicates fossil fuel combustion, which was also confirmed by chemical analysis carried out at these localities, see figure 2.

Nitrates are a major component of PM_{10} at both sites. In the winter season, nitrate concentrations in Hodolany increased about 10 times and in Klášterní Hradisko about 12 times compared to the summer, see table 2. The increase in nitrate concentrations in the winter season can be influenced by combustion processes (fossil fuels and some types of biomass), but also by environmental conditions affecting nitrate formation (low precipitation, humidity and temperature). The concentrations of sulphates in the winter period are almost identical for both localities. Sulphate concentrations in Olomouc correspond to the range of average sulphate concentrations for Central Europe (3 to 7 µg/m³ in *Schaap et al., 2004* [16]). The sources of an increase in sulphate concentrations in the winter season are combustion processes. Some diagnostic ratios can be used to determine the origin of dust particles. For example, the $(SO_4)^{2-}/(NO_3)^- < 1$ ratio indicates the origin from transport sources, while $(SO_4)^{2-}/(NO_3)^- > 1$ refers to stationary sources of combustion, especially of coal [17]. On the basis of this ratio, surprisingly during the winter season, the influence of transport sources was demonstrated, and during the summer, the value of the ratio indicates partial influence from industrial sources.

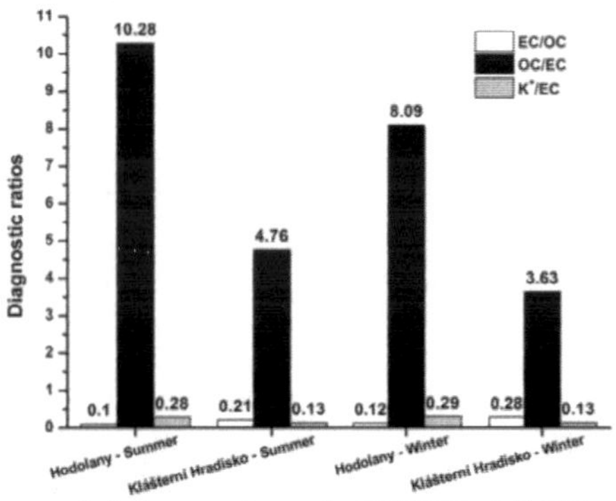

Figure 2. Diagnostic ratio EC/OC, OC/EC and K^+/EC.

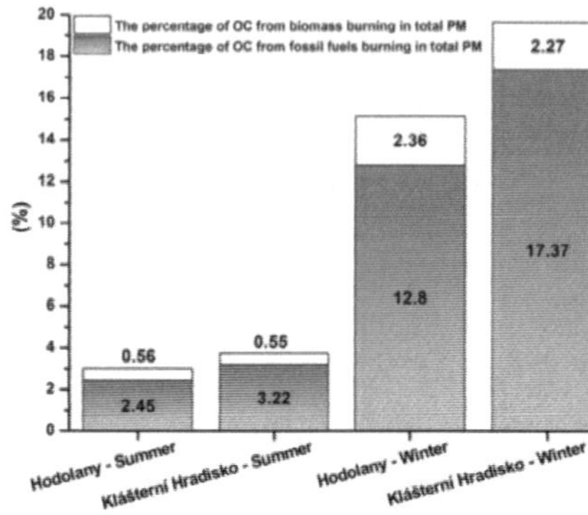

Figure 3. Percentage contribution of individual sources.

Table 2 lists the concentrations of anhydrosaccharides (L - levoglucosan and M - mannosan) during both the summer and winter seasons. The L/M ratios are comparable during the summer and winter months at both localities. The L/M diagnostic ratio makes it possible to identify the type of combustion biomass. Based on the L/M ratio, the dominant species of combustion biomass is coniferous wood. The ratio L/M for coniferous wood ranges from 3.6 to 3.9 [18]. The study by *Caseiro et al., 2009* [19] states that the L/K^+ ratio in Austria ranged from 0.9 to 1.7 during the winter months and from 0.22 to 0.56 during the summer months, while the L/K^+ ratios in Olomouc were lower. The L/K^+ ratio ranged from 0.11 to 0.14 in the summer, and from 0.36 to 0.40 in the winter.

Table 2. Concentrations of selected water-soluble ions in the summer and winter season.

Locality	$(NO_3)^-$	$(SO_4)^{2-}$	K^+	L	M	$(SO_4)^{2-}/(NO_3)^-$	L/M	L/K^+
	(µg/m^3)					-		
Hodolany- summer	0.799	3.289	0.222	0.03	0.01	4.12	3.6	0.14
Hodolany - winter	7.669	4.552	0.595	0.24	0.07	0.59	3.6	0.4
Klášterní Hradisko - summer	0.642	3.603	0.221	0.03	0.01	5.61	3.7	0.11
Klášterní Hradisko - winter	7.818	4.566	0.591	0.21	0.06	0.58	3.4	0.36

3.2. Contribution of OC from biomass and fossil fuel combustion using equations

The study by *Caisero et al. 2009* [19] reports that the L/M ratio for hardwood is 14.4-14.8 and for softwood 3.6-3.9. Using the L/M ratio, it is possible to determine the type of combustion wood using the following equation (1), and to determine the contribution of OC and PM from biomass combustion (equations (2), (3)):

$$\%\ spruce\ wood = \frac{\left(14.8 - R_{L/M}\right)}{0.112} \tag{1}$$

Information on the share of soft and hardwood (according to equation (1)) was used to calculate the conversion factor OC/levoglucosan. Foreign literature states the value of 7.35 as the general conversion factor OC/levoglucosan for the common European background environment [20]. The values of the OC/levoglucosan factor for both localities in Olomouc were 5. This calculated value of conversion factor (5) was used to calculate the contribution of the OC concentration from biomass combustion (OC_{BB}) using equation (2):

$$OC_{BB} = Levoglucosan \cdot conversion\, factor \tag{2}$$

For calculation of dust particles contribution from biomass combustion (PM_{BB}) to total PM_{10}, equation (3) with PM/levoglucosan conversion factor 11.2 (the recommended value for Central Europe) was used [20]:

$$PM_{BB} = Levoglucosan \cdot conversion\ factor \qquad (3)$$

The OC contribution from fossil fuel combustion (OC_{FF}) was found using equation (4) [21]:

$$OC_{FF} = 1.1 \cdot (EC - 4.2 \cdot Levoglucosan) \qquad (4)$$

The average percentage contribution of PM from biomass combustion to total PM_{10} was 1.2% (Klášterní Hradisko) and 1.3% (Hodolany) during the summer season and 5.2% (Klášterní Hradisko) and 5.3% (Hodolany) during the winter. The percentage contribution of OC from biomass and fossil fuel combustion is shown in figure 3.

4. Conclusion

The proportion of organic matter in PM_{10} in the winter season was 58.6% for Hodolany and 50.4% for Klášterní Hradisko. During the summer months, the OM percentage in PM_{10} was about 0.7 times lower for both Hodolany and Klášterní Hradisko. The OC/EC and K^+/EC ratios show that during the winter months the air quality at both localities is influenced by fossil fuel combustion, and by biomass combustion or transport during the summer months. The OC contribution from biomass combustion was 0.6% of PM for both localities during the non-heating season, and 2.3% (Klášterní Hradisko) and 2.4% (Hodolany) from PM during the heating season. Contribution of OC from fossil fuel combustion from PM was 2.5% (Hodolany) and 3.2% (Klášterní Hradisko) during the summer season, and 12.8% (Hodolany) and 17.4% (Klášterní Hradisko) during the winter season. In future, more attention will be paid to the EC and OC contributions caused by different kinds of fuel.

Acknowledgement

This work was supported by research projects of the Ministry of Education, Youth and Sports of the Czech Republic: National Sustainability Program LO1404 - TUCENET, SGS 2017/35 Research in Selected Areas of "Smart Energy" in the 21st Century and SP2017 /185 Characteristics of Unburned Carbon from Local Furnaces.

References

[1] Pachauri T, Singla V, Satsangi A et al. 2013 *Aerosol Air Qual. Res.* **13** 523-536
[2] Zhu CH S, Cao J J, Tsai CH J et al. 2014 *Sci. of the tot. Environ.* **466–467** 203-209
[3] Oros D R, Abas M R , Omar M J N Y et al. 2006 *Appl. Geochem.* **21** 919-940
[4] Pio C, Cerqueira M, Harrison R M et al. 2011 *Atmos. Environ.* **45** 6121-6132
[5] Qin M, Wang X, Hu Y et al. 2015 *Atmos. Environ.* **112** 81- 89
[6] Zhang T, Claeys M, Cachier H et al. 2008 *Atmos. Environ.* **42** 7013-7021
[7] Duan F, Liu X, Yu T and Cachier H 2004 *Atmos. Environ.* **38** 1275-1282
[8] Caseiro A, Marr I L, Claeys M et al. 2007 *J. Chromatogr.* **1171** 37-45
[9] Simoneit B R T, Schauer J J, Nolte CG et al. 1999 *Atmos. Environ.* **33** 173-182
[10] Andreae M O 1983 *Science* **220**1148-1151
[11] Zdrahal Z, Oliveira J, Vermeylen R et al. 2002 *Environ. Sci. Technol.* **36** 747-753
[12] Frey A, Virkkula A, Saarnio K et al. 2006 *Rep. Ser. Aero. Sci.* **83** 325-329
[13] Fine P M, Cass G R and Simoneit B R T 2002 *Environ. Sci. Technol.* **36** 1442-1451
[14] Zheng M, Hagler G S W, Ke L et al. 2006 *J. Geophys. Res.* **111** D20313
[15] Rogula-Kozłowska W, Sówka I, Mathews B et al. 2013 *J. of Environ. Protec.* **4** 371-379
[16] Schaap M, van Loon M, Brink H M et al. 2004 *Atmos. Chem. Phys.* **4** 857-874
[17] Xiu G, Zhang D, Chen J, Huang X, Chen Z, Guo H and Pan J 2004 *Atmos. Environ.* **38** 227-236
[18] Ward T J, Hamilton R F, Dixon R W et al. 2006 *Atmos. Environ.* **40** 7005-7017
[19] Caseiro A, Bauer H, Schmidl C, Pio CA, Puxbaum H 2009 *Atmos. Environ.* **43** 2186-2195
[20] Cheng Y, Engling G, He K-B et al. 2013 *Atmos. Chem. Phys.* **13** 7765-7781
[21] Pio C A, Legrand M, Alves C A et al. 2008 *Atmos. Environ.* **42** 7530-7543

IOP Conf. Series: Earth and Environmental Science **92** (2017) 012066 doi:10.1088/1755-1315/92/1/012066

Possibilities of the management of *Helianthus tuberosus* species in Poodri PLA (Czech Republic)

H Svehlakova[1], A Janikova[1], J Kupka[1], N Sotkova[1] and T Rajdus[1]

[1]Institute of environmental engineering, Faculty of mining and geology, VSB – TU Ostrava, 17.listopadu 15, 708 33 Ostrava – Poruba, Czech Republic

E-mail: hana.svehlakova@vsb.cz

Abstract. This article deals with the possibilities of population management of invasive and in Czech Republic no - native species *Helianthus tuberosus*. We have chosen study areas in Poodri PLA, it is located near the industrial city Ostrava (Moravian-Silesian region). Invasive *Helianthus tuberosus* forms monodominant vegetation,excludes the original and often rare plant species and reduces biodiversity. It is a risk for the further development of the landscape of Poodri PLA. There is no known effective way to dispose of invasive *Helianthus tuberosus* so far. We have applied the proposed management works on 15 monitoring areas and then evaluated their effectiveness. We have used chemical (herbicides) and mechanical (mowing, digging up) methods and their combinations. The most effective was the combination of 3 interventions: manual mowing - spraying – mowing by mechanization, just mowing itself rather promotes the spread of the population. We can say this is a pilot study from a long-term project dealing with the effectiveness of different ways of invasive plants management.

1. Introduction

The Poodri Protected Landscape Area (PLA) is located in Czech Republic in the Moravian-Silesian Region close to the Ostrava agglomeration. It is exceptional especially due to the preservation of the natural water regime of the Odra river [1]. Most parts of the area is formed by floodplains, there are biotopes closely related to water: alluvial meadows, floodplain forests and permanent or periodic pools [2]. Also relatively dense system of ponds and drainage channels is typical for this area [1,2]. A major problem in Poodri PLA is the spread of invasive plant species. One of the non-native introduced species is Jerusalem artichoke (*Helianthus tuberosus*), a perennial tuberous herb of the *Asteraceae* family, originally from North America. Jerusalem artichoke belongs to an unpretentious plants capable of forming a dense, monodominant, successful phytocoenosis, which suppresses other species and can reduces biodiversity of territory [3]. *Helianthus tuberosus* is capable of rapid and successful vegetative reproduction, using especially tubers, rhizomes, but can spread also seeds [4,5,6].

The dynamics of the Ondrejnice watercourse contribute significantly to *Helianthus tuberosus* distribution, tubers and rhizomes being transported by annual floods. Litoral and alluvial meadows are the most threatened biotopes from this point of view. The most invaded communities by *Helianthus tuberosus* belong to the phytocoenological associations *Arhenatherion*, *Bidention tripartitae*, *Salicion albae* and *Alnion glutinosae*. For this reason, we proposed in cooperation with experts of Nature Conservation Agency of the Czech Republic (NCA CR) experimental management on selected areas with Jerusalem artichoke. Management of experimental areas with the occurrence of Jerusalem artichoke includes mechanical methods (mowing and digging up), chemical methods (application of herbicides), or combinations thereof.

2. Materials and methods

The selected areas are located in the Poodri PLA, cadastral territory Proskovice (see figure1). It is an alluvial meadow, which the river Ondrejnice flows through the and which, due to the floods, is heavily invaded by *Helianthus tuberosus*. These invasive plants are located mainly along the banks of Ondrejnice, in the adjacent meadow, in the remains of floodplain forests and in new forest planting.

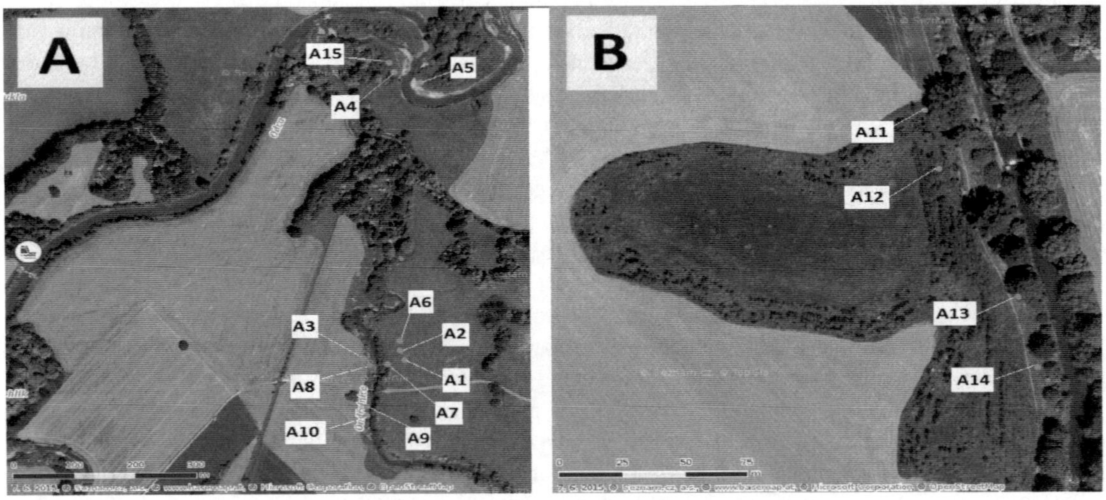

Figure 1. Selected monitoring areas of *Helianthus tuberosus.*

In 2014, we selected 4 experimental areas (A1-A4) with various managements and 1 control area (A5) where no intervention has taken place and serves to observing standard lifecycle and development of Jerusalem artichoke. It also serves to compare the effectiveness of the methods used. In 2016, we have expanded number of study areas with *Helianthus tuberosus* by another 9 experimental areas (A6-A14) and 1 new control area (A15). Thus, we have monitored population of *Helianthus tuberosus* on a total of 13 experimental and 2 control areas. Areas have about 100 m^2 and are square in shape. The location of areas corresponds to the typical habitats of *Helianthus tuberosus* - sunny banks (A1, A2, A9, A14), shaded banks (A3, A5, A7, A8, A10, A13, A14), meadow (A4)), tree planting (A11, A12). Control areas are located in shaded banks (A5) and in the shaded meander (A15).

The following interventions were performed on the individual monitoring areas:

- the mowing - mechanical mowing in the regular management of meadows twice per season, supplemented by manual mowing or brush cutting in places where regular mowing did not take place;
- digging tubers up - removal of *Helianthus tuberosus* tubers by spade at the end of the growing season;
- application of herbicides - spraying plants by herbicides ROUNDAP, GARLON NEW, BOFIX, HERBISTOP. ROUNDAP is a systemic non-selective herbicide that belongs to the so-called total herbicides that destroy weeds from leaves to roots [7]. The active ingredient of the herbicide is glyphosate. GARLON NEW - a selective herbicide, it is intended for the control of unwanted tree species and dicotyledonous plants on meadows, pastures, temporarily uncultivated soils. The active substances are fluroxypyr and triclopyr [8]. HERBISTOP is a total natural herbicide based on pelargonic acid, intended for the disposal of monocotyledonous and dicotyledonous weeds. Pelargonic acid is a naturally occurring fatty acid that leaves no residue and is environmentally friendly [9]. BOFIX is a selective herbicide able to kill resistant dicotyledonous weeds in lawns. The active substances are clopyralid, fluroxypyr and MCPA [10].

- Management combinations - mowing + digging up, mowing + spraying, spraying and digging up, mowing, spraying and digging up.

3. Results and discussion

We have performed management interventions of *Helianthus tuberosus* on 5 monitoring areas (A1-A4) in 2014 - 2016, one area was control (A5). The management ran from June 2014 to October 2015, and in the 2016 season, we have conducted a field survey to determine the state of populations *Helianthus tuberosus* after interventions. The meadows were mowed always twice in June and August 2014 and 2015. The application of herbicides was at the turn of August and September 2014 (ROUNDUP) and 2015 (GARLON NEW). Tuber harvesting was performed in October 2015. In the autumn months of 2016, we have supplemented the network of experimental areas (A6-A14) and control areas (A15). We have expanded also a range of interventions. At the end of June and July 2017, and in August 2017 we have performed mowing, in early August 2017 application of herbicides (BOFIX, HERBISTOP) was carried out. Digging tubers up is scheduled for October 2017.

Mowing. Areas A2, A6. We also include the area A14 here, which is mowed and after this season we plan digging up of tubers. We can say plant compensates removing of terminal buds after mowing by their doubling and also rapid regeneration from roots. So it optically and actually increases the population. This increase is evident in the spring aspect of the following season and only slightly declines in autumn. Due the mowing weakening of plants occurs, which is evident in the size of the tubers. They are very small compared to the area without mowing. We can say exclusive mowing in *Helianthus tuberosus* does not cause the decreasing of population of *Helianthus tuberosus*; however, it negatively affects the height of the plant, which does not exceed 30-40 cm on the area of A2 and A6. The area of A14 was massively rejuvenated by young individuals and the plants reached a height of 60 cm in 2017. Since none of areas has created flowers, mowing also negatively affects generative reproduction.

Digging tubers up. Area A8. It will be done in autumn 2017.

Mowing and digging up. Areas A1, A14. Combination of mowing and harvesting of tubers is more effective in the elimination of the Jerusalem artichoke. Mowing weakens plants and harvesting of tubers removes the source of vegetative propagation. The abundance of plants on area A1 decreased by about 40%. We have to note that mowing took place twice in 2014 and twice in 2015 by mechanization, and in September 2016 it was supplemented by manual mowing. Area A14 was mowed in 2017 (but young plants quickly regenerate), digging of tubers up will be done in October.

Application of herbicide. Area A4 (GARLON NEW), A11 (HERBISTOP), A9, A12 (BOFIX).The application of GARLON NEW is ideal at the time of flowering. Jerusalem artichoke´s flowering on A4 began in a very dense vegetation formation, when defoliation of plants´stems has reached the height 1.4 m as a result of competition for light. This herbicide acts through the leaf, so it was not effective. We can say that its application is not suitable for dense plant stands, or it is necessary to apply it early spring before flowering. The application of HERBISTOP on A11 was not easy due to the high growing of the Jerusalem artichoke individuals. It was not possible to hit top of stems of all individuals. After application of the herbicide, assimilation organs of plants have been damaged (wilting, necrotisation). Lower individuals were completely destroyed and did not come into flower. Higher individuals over 2.5 m show leaf damage up to 2/3 height, but tops of stems remain undamaged, some had flowers. So we recommend using herbicide earlier, preferably at the turn of spring and summer. The BOFIX application to *Helianthus tuberosus* on areas A9 and A12 was successful. There was damage of most of plants assimilation organs (wilting, necrotisation), tops of stems were deformed and without flowers also in individuals over 2.5 m. On A9, the tubers will be dig up in October.

The application of herbicide and digging up. Area A3 (ROUNDUP). An interesting result was the application of the ROUNDAP herbicide to area A3 followed by digging tubers of survive individuals up. The spraying was carried out in the autumn months and, after its application, the invasive population almost did not recover. From the original abundance about 90%, *Helianthus tuberosus* was eliminated for a spot occurrence - in the following season we found only 6 individuals. Plants that survived, however, reached a height of 2.8 m and a diameter of stem up to 7 cm and were very dominant in the stand. In August, they created up to tens of flowers per one plant. In October 2015, these individuals were harvested, their tubers were dig up and we have found that the root system of the plants was more potent and the tubers were 5 times larger than on other areas. Digging tubers up after application of ROUNDAP, we consider it very important, because of the high viability of individuals that survived the spraying.

Mowing and herbicide application. Areas A7 (BOFIX), A10 (HERBISTOP), A13 (HERBISTOP). Area A7 was manual mowed (July 2017) and treated with the BOFIX (August 2017), then it was mechanically mowed (August 2017). We have checked area A7 in September 2017 and have found the area is completely free of rejuvenated individuals. Plants of *Helianthus tuberosus* on the border of area, which have not been affected by mowing, were very damaged, without functional photosynthethic organs. The A10 area was mowed once and then HERBISTOP was applied to the regenerated plants. After the herbicide application, most of *Helianthus tuberosus* died, but lot of young individuals appeared in September. As a result of receiving more light, also the *Colchicum autumnale* bloomed on this habitat. A13 area was manually mowed in July 2017, the HERBISTOP was applied on gradually regenerating individuals. The herbicide-affected individuals are completely destroyed, but the new ones are again being rejuvenalised. In the autumn 2017, digging tubers up is planned. We terminated the use of herbicides ROUNDAP and GARLON NEW in 2015. We excluded the ROUNDAP because of the high toxicity of glyphosate and GARLON NEW was inappropriate for much closed plant formations with defoliation of stalks up to 1m.

HARTMANN et al [11] recommends mechanical methods twice a season (June, August). Our results show that the most effective method to the elimination of *Helianthus tuberosus* from alluvial meadow is the combination of manual mowing → application of BOFIX herbicide → mowing by mechanization (area A7). With this intervention, *Helianthus tuberosus* was completely removed without regeneration. Only BOFIX spraying (areas A9, A12) is also effective and also affects plants over 2.5 m high, but a part of plants survives and is able to be regenerated. BOFIX can be used even in mild winds and temperatures up to max. 25°C. Rain can reduce efficiency within 2 hours after application. The price for spraying of 100 m^2 is 80 CZK, which is favourable. It is also very effective to spraying the ROUNDUP herbicide and subsequently digging tubers and surviving plants up (A3). These individuals are generally very high, stable and viable and could become the basis of a rapidly spreading invasive population. The application of ROUNDAP was, however, stopped in the meadows of Poodri PLA due to glyphosate toxicity. Using herbicide HERBISTOP, the plant assimilation organs were very damaged. Generative organs were damaged in lower individuals (up to 2m), higher individuals were able to form flowers and expand populations. HERBISTOP is therefore not suitable for high vegetation, or it is necessary to apply it early in the spring, but in the sunny, windless weather, which is rare in the alluvial meadows of Poodri PLA in the spring months. An advantage is the degradation of the herbicide without residue in soil and water. The least suitable herbicide is GARLON NEW (A4), which can be applied to stem leaves. However leafs often absent in densely-growing formations of *Helianthus tuberosus* or are too high (above 1 m). The application can be done optimally early in the spring in sunny weather without rain. Mowing (A2, A6, A14) is less effective if it is not followed by another intervention of management. FEHÉR and KONČEKOVÁ [12] came to similar conclusion. They recorded decreasing of *Helianthus tuberosus* population of about 44 % after 3 years mowing. However, it can be said that mowing negatively affects the development of the plant's generative organs and the size of the tubers and it is extremely effective in combination with other interventions (especially herbicides). The digging tubers up themselves (area A8) will be carried out in autumn 2017, but results show that, like mowing, it will be more efficient combined with other management methods.

4. Conclusions

The article focuses on the effectiveness of various landscape management methods for the removal of invasive species *Helianthus tuberosus* from the alluvial communities of Poodri PLA. It summarizes results of the pilot research launched at the instigation of the NCA Czech Republic in 2014. We have chosen 4 basic management methods - mowing, digging tubers up, application of herbicides ROUNDUP, GARLON NEW, BOFIX, HERBISTOP and their combination - mowing and digging up, mowing and application of herbicide, application of herbicide and digging up. We can say that the most effective management method is a combination of twice to mow and in the meantime to apply a suitable herbicide. The most suitable herbicide is BOFIX. ROUNDUP also has excellent results, but its use with respect to glyphosate content is not possible in the long term. Relatively effective and environmentally friendly is HERBISTOP, but it is necessary to use it in the spring before the first mowing, higher plants survive the spraying and they flourish. So it is not possible to follow the optimal scheme of mowing - spraying - mowing. The research will be supplemented by the digging tubers up, because we can say on example ROUNDAP application, that surviving individuals have strong viability and removing of tubers from the soil can be prevent them from expanding.

References

[1] Neuschlová Š 1999 *Současné výsledky výzkumů v chráněné krajinné oblasti Poodří: Poodří chráněná krajinná obast* (Společnost přátel poodří v Ostravě: Ostrava)
[2] Weissmannová H et al. 2004 *Ostravsko, svazek x – chráněná území ČR 2004* Nature Conservation Agency of the Czech Republic and EkoCentrum
[3] Mlíkovský J and Stýblo P 2006 *Nepůvodní druhy fauny a flóry české republiky* ČSOP
[4] Dostál J 1989 *Nová květena čssr* (Academia)
[5] Honsová H *Topinambur lze využít k mnoha účelům* http://biom.cz/cz/odborne-clanky/topinambur-lze-vyuzit-k-mnoha-ucelum
[6] Lubojacka A 2016 The Invasive Helianthus tuberosus in phytocoenosis of Poodří PLA and possibilities of its management (VŠB – TUO: Ostrava)
[7] Roundup a jeho použití http://www.randap.cz/
[8] Garlon New *Vše o přípravcích na ochranu rostlin* http://www.agromanual.cz/cz/pripravky/herbicidy/herbicid/garlon-new.html
[9] Herbistop *Herbistop koncentrát ničitel plevele 500 ml* http://herbistop.cz/?2,herbistop-koncentrat-br-nicitel-plevele-500-ml
[10] Bofix *Vše o přípravcích na ochranu rostlin* https://www.agromanual.cz/cz/pripravky/herbicidy/herbicid/bofix
[11] Hartmann E, Schuldes H, Kubler D and Konold W 1995 *Neophyten. Biologie, Verbreitung und Kontrolle ausgewählter Arten* (Ecomed: Landsberg)
[12] Fehér A and Končeková L 2009 Evaluation of mechanical regulation of invasive Helianthus tuberosus populations in agricultural landscape *J. of Centr. Europ. Agr.* **3** 245 -250

IOP Conf. Series: Earth and Environmental Science **92** (2017) 012067 doi:10.1088/1755-1315/92/1/012067

Pedotransfer function for calculating the potential of volume changes in soils

A Tall[1], M Gomboš[1], B Kandra[1] and D Pavelková[1]

[1]Institute of Hydrology, Slovak Academy of Sciences, Dúbravská cesta 9, 841 04 Bratislava, Slovakia

Email: tall@uh.savba.sk

Abstract. The volume changes in the soils are manifested both by the formation of cracks and by the vertical movement of the soil surface. Both of these could be generally negative for engineering constructions. The Coefficient of Linear Extensibility (COLE) is used to quantify volume changes. In this work a pedotransfer function was designed to calculate the COLE potential. Calculation is based on the texture of the soil. This function was designed on the basis of 169 samples of soil samples taken at 11 locations in the East Slovak lowland.

1. Introduction

Volume changes of soil are caused by clay minerals. These minerals can change their volume when they are wetting (soil is swelling) or drying (soil is shrinking). The more content of clay minerals, the more significant volume changes. Content of clay minerals in soil is close related to the content of particles with diameter < 0.002 mm. Therefore, when soil texture is known, it is possible to predict a potential of volume changes of soil. Volume change of soil is three-dimensional process, which in vertical direction results in vertical movement of terrain and in horizontal direction results in creation of soil cracks. Both, vertical movement and cracks can affect engineering constructions (buildings, pavements, pipelines etc.) almost in negative way. The principle and mechanism of soil´s volume changes is relatively well described in literature [1-11].

Most of present research in soil physics is oriented to mathematical methods used for obtaining hydrophysical characteristics of soil from data, which can be easily measured (in most cases from the particle size distribution of soil). Outputs are called "pedotransfer functions" (PTF), which are useful for calculation hydrophysical characteristics of soils. This paper presents PTF for computing potential of volume changes of soils from soil texture. Potential of soil´s volume changes is expressed in form of Coefficient Of Linear Extensibility (COLE).

2. Materials and methods

2.1. Sample collection and preparation

The soil sampling site was situated in the East Slovak Lowland (ESL) (figure 1). This lowland is characterized by the high textural diversity of its soils. This is caused by complicated tectonic development of this area. The soils of ESL are mostly agricultural used. For soil sampling were selected 11 localities of ESL. Together, 169 undisturbed soil samples were taken into the Kopecky cylinders with a volume of 100 cm^3. At the same time, disturbed soil samples were also taken from each site for grain analysis. The parameters of sampling sites are listed in table 1. All undisturbed soil samples were saturated by water in the laboratory to reach the full water capacity.

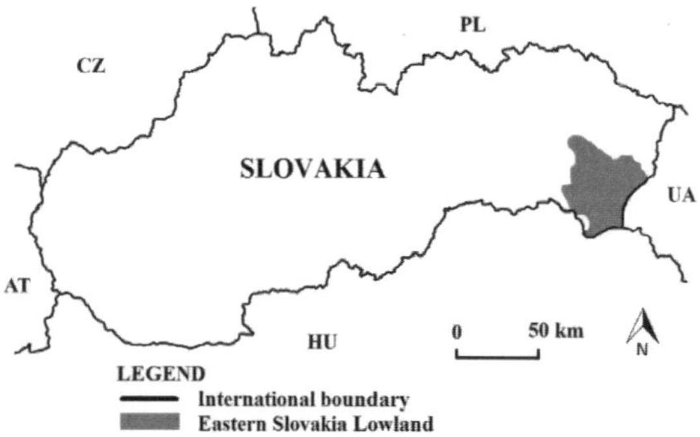

Figure 1. Map of the study area (Eastern Slovakian Lowland).

Table 1. Parameters of sampling sites.

Locality	Coordinates	Sampling depth [m]	No. of samples
Michalovce	N48° 44,255′ E21° 56,664′	0 – 0.8	24
Milhostov	N48° 40,185′ E21° 44,248′	0 – 1.6	32
Pribeník	N48° 23,688′ E21° 59,547′	0 – 1.0	6
Senné	N48° 39,802′ E22° 02,892′	0 – 0.8	42
Sírnik	N48° 30,538′ E21° 48,830′	0 – 0.8	12
Somotor 1	N48° 23,748′ E21° 48,471′	0 – 0.8	23
Somotor 2	N48° 23,173′ E21° 48,237′	0 – 1.0	2
Veľký Horeš	N48° 22,540′ E21° 53,907′	0 – 1.0	6
Veľký Kamenec	N48° 21,048′ E21° 48,877′	0 – 1.0	4
Vysoká nad Uhom	N48° 36,796′ E22° 06,898′	0 – 0.6	16
Zatín	N48° 28,725′ E21° 54,918′	0 – 1.0	2
			∑169

2.2. Laboratory measurements

The soil texture analysis was determined from disturbed soil samples by hydrometer method. According to the grain size analysis, the soils were classified according to the USDA classification (sand: 2-0.05 mm, silt: 0.05-0.002 mm and clay: <0.002 mm). The undisturbed soil samples were weighed after saturation and then dried in a laboratory dryer at 105 °C to a constant weight. After drying, the volumes of the cylinders were measured. Volume measurements consisted in measuring the height and diameter of the cylinder by a caliper. Each dimension was measured 2x (in perpendicular directions). From the average values of heights and diameters were calculated volumes of the soil samples.

2.3. Data analysis

To quantify the volume changes of soil in this work, the coefficient COLE was used. This parameter is used to quantify the shrink – swell potential of the soil. Equation (1) was established by Grossman [12]:

$$COLE = \sqrt[r_s]{\frac{V_{wet}}{V_{dry}}} - 1 \qquad [-] \qquad (1)$$

V_{wet} means soil volume in the moist state (i.e. saturated soil sample) and V_{dry} means soil volume in the dry state (i.e. owen dried sample) and r_s is the shrinkage geometric factor. If the value of r_s factor is three, it means that soil is isotropic shrinking. In this study was assumed isotropic shrinkage of soils, so r_s factor had assumed to be three. Simple classification of soils based on COLE was introduced by Parker [13] (table 2).

Table 2. Soil classification by COLE.

Shrink-swell potential	COLE
Low	<0.03
Medium	0.03-0.06
High	0.06-0.09
Very high	>0.09

By multiplying the COLE value by the thickness of the soil layer, we get the shrinkage potential of this layer expressed in the length unit (for example, if a soil layer of 100 cm has a potential of COLE 0.10, this soil layer can potentially shrink by 10 cm).

Based on measured COLE values for 169 texturally diverse soils, a PTF was designed to calculate the COLE parameter from the texture of the soil (from the clay particles (<0.002 mm)).

3. Results and discussion

The soil grain analysis results are shown in figure 2. In the examined data set, the content of the clay component ranged from min. 9.89% to max. 69.13%.

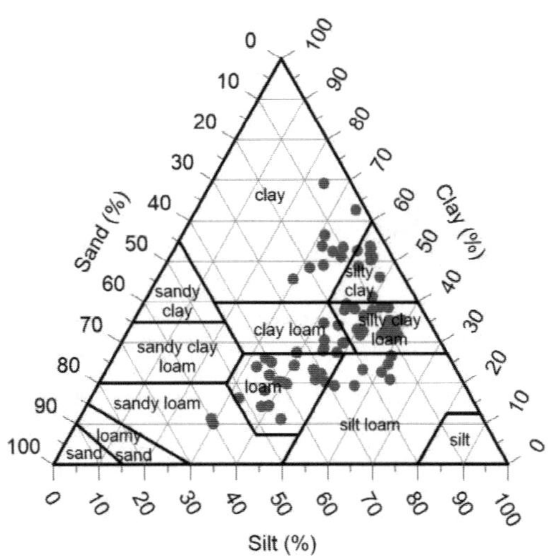

Figure 2. Distribution of soil textures of the 169 soil samples in the USDA textural triangle.

Values of V_{dry} used in relation (1) ranged from 54.5 cm^3 (maximum shrinkage) to 100 cm^3 (no shrinkage). The values of V_{wet} had a constant value of 100 cm^3. Calculated COLE values varied in interval from 0.00 to 0.22.

The dependence of COLE on clay content (particles <0.002 mm) based on 169 soil samples is in figure 3. Pedotransfer function was obtained in the form of polynomial function of second degree:

$$COLE = 0.00004\,(clay)^2 + 0.0006\,(clay) - 0.0056 \quad [-] \tag{2}$$

A very high level of reliability of PTF is confirmed by a high value of coefficient of determination (r^2 = 0.937) and relatively low value of mean-absolute error (MAE = 0.010%) (figure 4).
Figure 5 shows a comparison of obtained PTF with previously published studies [13-15]. As seen in figure 5, the previous studies expected the linear relationship between COLE values and clay content.

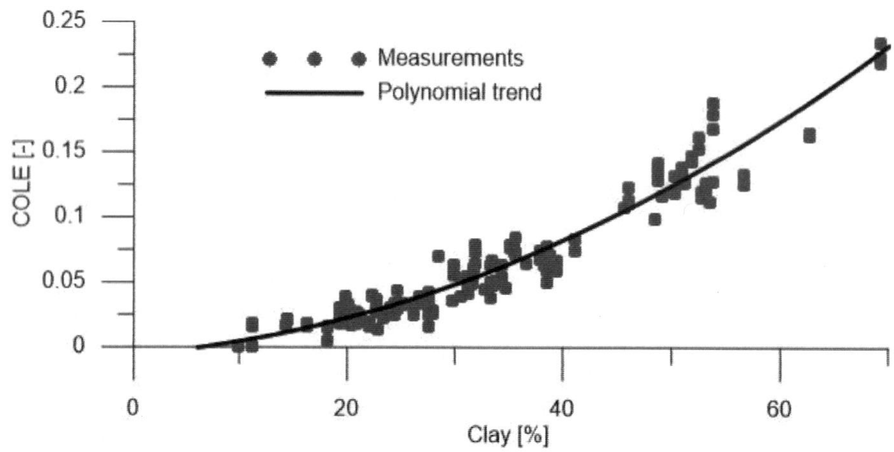

Figure 3. Relationship between COLE values and clay content of the soil.

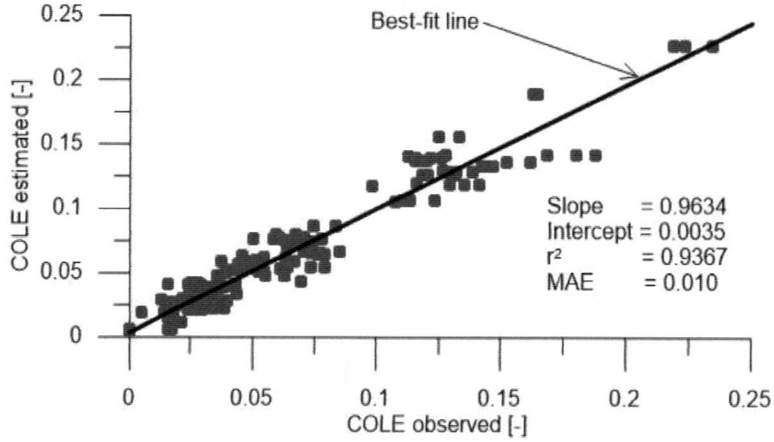

Figure 4. COLE values calculated vs. COLE values measured.

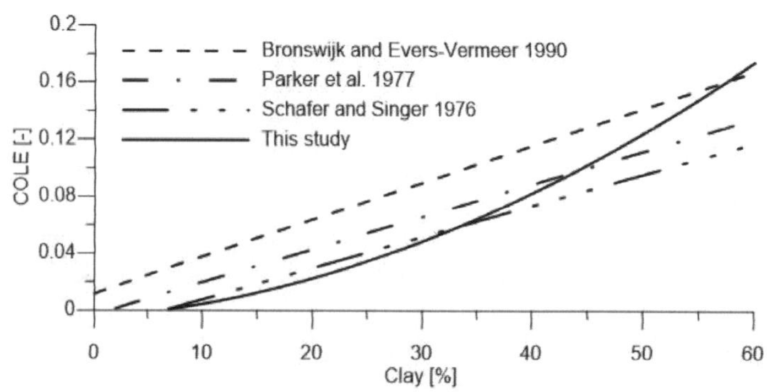

Figure 5. Obtained pedotransfer function vs. some similar functions published in world´s literature.

This study introduces nonlinear relationship between COLE values and clay content, which seems to be more accurate. The magnitude of volume changes is null till clay content is smaller than 7%. Process of volume changes in soil starts when clay content in soil is greater than 7%.

4. Conclusions

The occurrence of volumetric changes in the soils can be a potential risk for engineering constructions. In order to determine the potential for volumetric changes, the shrink-swell potential of COLE is used. Direct determination of COLE is relatively laborious and time-consuming. Indirectly, the potential of COLE can be estimated by PTF, based on the texture of the soil. Based on 169 undisturbed soil samples from 11 locations on ESL, the PTF was submitted to calculate the COLE potential. COLE values are calculated from the clay content in the soil.

Acknowledgments

This study is the result of the project implementation: Centre of excellence for the integrated river basin management in the changing environmental conditions, ITMS code 26220120062; supported by the Research & Development Operational Programme funded by the ERDF.
Authors are also grateful to Scientific Grant Agency of the Ministry of education, science, research and sport of the Slovak Republic: (project VEGA: 2/0062/16).

References

[1] Bronswijk J J B 1989 Prediction of actual cracking and subsidence in clay soils *Soil Sci.* **148** 87
[2] Bronswijk J J B 1990 Shrinkage geometry of a heavy clay soil at various stresses *Soil Sci. Soc. Am. J.* **54** 1500
[3] Bronswijk J J B 1991 Relation between vertical soil movements and water-content changes in cracking clays *Soil Sci. Soc. Am. J.* **55** 1220
[4] Gebhardt S, Fleige H and Horn R 2010 Shrinkage processes of a drained riparian peatland with subsidence morphology *J. Soils Sediments* **10** 484
[5] Gregory A S, Bird N R A, Whalley W R, Matthews G P and Young M 2010 Deformation and shrinkage effects on the soil water release characteristic *Soil Sci. Soc. Am. J.* **74** 1104
[6] Chertkov V Y 2003 Modelling the shrinkage curve of a soil clay pastes *Geoderma* **112** 71
[7] Chertkov V Y, Ravina I and Zadoenko V 2004 An approach for estimating the shrinkage geometry factor at a moisture content *Soil Sci. Soc. Am. J.* **68** 1807
[8] Chertkov V Y 2005 The shrinkage geometry factor of a soil layer *Soil Sci. Soc. Am. J.* **69** 1671
[9] Lu D, Shao M, Horton R and Liu Ch 2004 Effect of changing bulk density during water desorption measurement on soil hydraulic properties *Soil Sci.* **169** 319
[10] Chertkov V Y 2012 An integrated approach to soil structure, shrinkage, and cracking in samples and layers *Geoderma* **173** 258
[11] Chertkov V Y 2007 The reference shrinkage curve of clay soil *Theoretical and Applied Fracture Mechanics* **48** 50
[12] Grossman R B, Brasher B R, Franzmeier D P and Walker J L 1968 Linear extensibility as calculated from natural-clod bulk density measurements *Soil Sci. Soc. Am. J.* **32** 570
[13] Parker J C, Amos D F and Kaster D L 1977 An evaluation of several methods of estimating soil volume change *Soil Sci. Soc. Am. J.* **41** 1059
[14] Bronswijk J J B and Evers-Vermeer J J 1990 Shrinkage of Dutch clay soil aggregates *Neth. J. Agr. Sci.* **38** 175
[15] Schafer W M and Singer M J 1976 Influence of physical and mineralogical properties on swelling of soils in Yolo County California *Soil Sci. Soc. Am. J.* **40** 557

Energy Management of Historic Buildings

A Tokarčík[1] and M Rovňák[1]

[1]Department of Environmental management, Faculty of Management, PU, Konštantínova 16, 080 01 Prešov, Slovakia

E-mail: martin.rovnak@unipo.sk

Abstract. The paper is focused on solving problems of design of energy management model in the context of historical buildings and monuments. The main goal of the energy management is to find solutions that preserve the historical heritage while not damaging the uniqueness of the buildings in question and their surroundings. The paper proposes a scheme of activities of energy management of historical monuments in the implementation of energy-efficient measures.

1. Introduction

Building an energy management system using energy management in historic buildings and monuments is a strategic tool for an efficient energy management system. In essence, it is the introduction of a control activity that relies on data from extensive monitoring of energy consumption and ways of losing it [1].

The subsequent estimate of energy consumption, based on data from previous periods, then clearly defines the requirements for managers. Restricting the analysis to specific places where building elements of historic architecture are sufficiently homogeneous is necessary for the development of technical guidelines for energy efficiency [2].

2. Specification of basic tasks of energy management

To optimize the use of energy in historic buildings in a particular area, managers must also look for solutions that preserve historical heritage while not damaging the uniqueness of the buildings in question and their surroundings. Only a well-designed grouping of professionals who know the specifics of the environment is a guarantee of introducing an alternative energy management solution to increase energy efficiency at all levels of consumption while maintaining historical values for future generations. If there is to be a creative environment that brings innovations in the energy management environment while taking into account the environmental demands of society, then it is a necessary requirement to build an energy management that can improve the energy efficiency of the environment and increase its energy efficiency on the basis of its own know-how.

However, such an energy management must find a suitable form of cooperation between experts from both the internal and the external environment, which, through a suitable combination of the performance of individual group members, can optimize the technical, technological and management activities and at the same time bring the requirements of the landowners into conformity [4].

The analysis of historical buildings and monuments consists of the audit and standardization of heat consumption data according to standardized weather conditions, structural damage to historic buildings according to normal structural damage, preliminary assessment of technical condition and detailed assessment of damage of the building covering [3].

The basic prerequisite for the conscious protection of the monuments is not only their awareness but, above all, their preservation for other generations. An indicator that best describes the care for preserving the cultural monuments and historic buildings is the observation of their construction and technical condition. The biggest challenge in building an energy management system is (in the framework of the priorities) to bring energy-saving rescuing of monuments, restoration of monuments and securing the financing of their renewal. Because of the specific building and restoration procedures required to restore historic monuments, this process is costly. The recovery priority is saving the historical original, not replacing it with a new, often cheaper elements. However, long periods of neglected maintenance or inappropriate pre-recovery procedures require complete and fundamental solutions. Through in-depth energy audits, simulations, and performance enhancement applications, representative constructions can also be defined as design models useful for identifying energy efficiency measures that are customizable for the particular building [2].

The restoration of the cultural monument is any intervention in the monument, which also includes activities that are otherwise called maintenance or repair - new interior painting or painting of facades, replacement of windows and doors, plaster repair, realization of new wiring, etc. It follows that not only the total renovation and reconstruction of the monuments, but also those partial activities, are subject to a decision or a binding opinion by the competent monuments office.

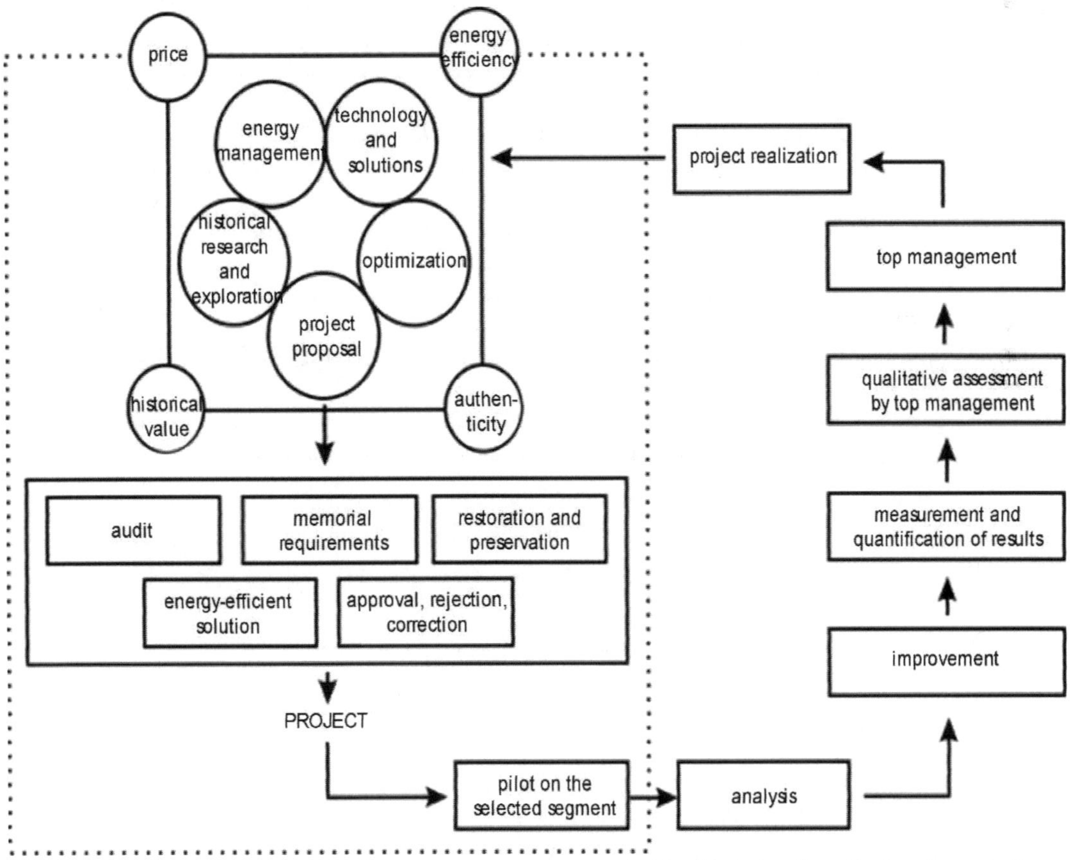

Figure 1. The proposed scheme of energy management activities historic sites, the implementation of energy efficiency measures.

The recovery process also includes the processing of specialized preparatory and project documentation and the execution of construction or restoration work. The main prerequisite for the preparation of professional and quality renovation of the building cultural monument is its knowledge. For buildings that are several centuries old, it is necessary to specify the construction and development stages, to evaluate the suitability of previous building interventions and modifications. Lastly, it is necessary to process the Framework Restoration Proposal itself, which takes into account the energy intensity of its operation and long-term sustainability. From the point of view of energy management, it is therefore necessary to have experts who are able to carry out a probing monumental architectural-historical and artistic-historical research. Another team of experts who coordinate the energy management system is the restoration team. Their task is to focus on defining the original color and visual solution of the exterior and interior plaster surfaces, specifying their technical state and proposing ways of their renovation, taking into account new technical solutions and results of the research, development and innovation. This expert background can identify unrecognized paintings and surface details that have long been covered with layers of secondary plasters or paints. The subsequent group of experts is represented by managers in the field of energy, technology, projections, audits, energy purchasers and other related professions, including the academic community. All these activities should be carried out by a group of experts, which we can call the energy management of historical monuments (figure 1). If we can activate the required group of market professionals, we will build the core framework of the operating system.

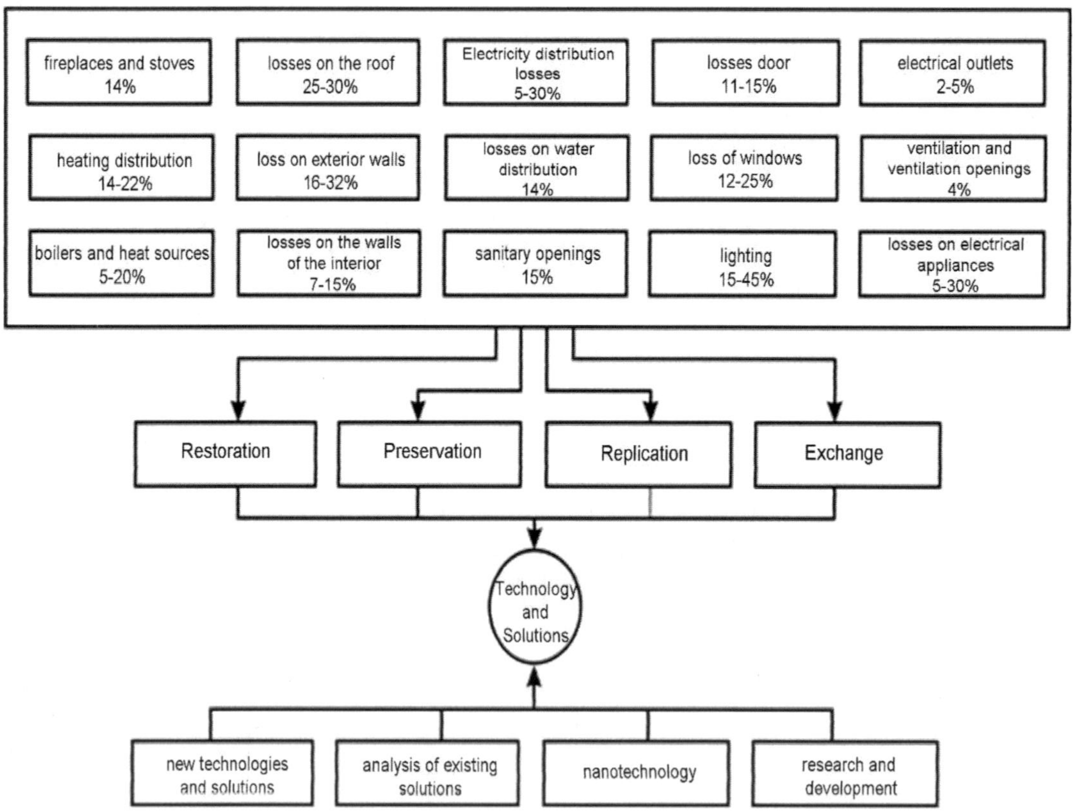

Figure 2. The proposed scheme for the implementation of energy management requirements for the selection of technological solutions for monuments and historical buildings.

From an energy point of view, energy management of historic sites should concentrate on the amount of energy needed to meet all the energy needs associated with the normal use of the building, in particular the amount of energy needed for heating, hot water, cooling, ventilation and lighting [1].

When determining it, it relies on calculations or calculations using measured energy consumption. In terms of numerical indicators, its demands express the need for energy and primary energy in the historic building. In order to collect heat and hydrometer data on traditional materials and techniques of the architectural heritage analyzed, the calculations based on typical packaging compositions are compared with laboratory results as well as the on-site test data [2].

3. Possibilities to increase the energy efficiency of buildings using the energy management of historic buildings

The potential sources of energy leakage in buildings are: roof, foundations, walls, windows, doors, floors, thermal bridges. For a complex solution of the building, it is also necessary to focus on insulating the floors, windows, doors, pipes, external joints, thermal bridges inside the building, in the areas of the contact between the roof and the building elements of the masonry and removing visible cracks or holes. Figure 2 shows the proposed scheme for the implementation of energy management requirements for the selection of technological solutions for monuments and historic buildings.

With exemplary design solutions of the constructions, it is possible to thoroughly analyze the relevant aspects such as, for example, the construction and conditions of its protection, including its impact on energy improvements. In addition, further relevant research developments in the area concerned, namely the economic feasibility and environmental impact of measures to modernize the historic building through representative buildings, need to be examined. In monumental architecture, the significance of particular features of individual constructions limits the suitability of the proposed approach compared to standard buildings where the repeating characteristics are more pronounced. Nevertheless, there are serious difficulties in balancing the energy efficiency target with the need to preserve the heritage [2].

There is a relationship between the technical state of buildings and energy consumption in buildings for space heating. Although this relationship is unambiguous, it appears that the technical state of buildings must be taken into account when assessing the potential of buildings for their modernization [3].

4. Conclusions

The introduced models accumulate people who have the technical ability and high working knowledge of historic buildings, a prerequisite for proper system maintenance and adaptation in energy efficiency processes. Their advice can thus prevent damage, inappropriate costs and defects while preserving the historical heritage. Only an appropriate balance in the design process and the subsequent qualitative consultation between the interested parties is the guarantee of improving the energy conservation of the historic buildings and monuments.

Acknowledgments

The paper was elaborated in the framework of the project "V4EaP Visegrad University Studies Grant No. 61500079 ".

References

[1] Colla C and Gabrielli E 2016 Integrated diagnostic approach for the structural and energetic evaluation of historic buildings *Ener. Effic. and Comf. of Hist. Build.* pp 10-16

[2] Genova E, Giovanni F and Calogero V 2016 The recurrent characteristics of historic buildings as a support to improve their energy performances: the case study of Palermo *Ener. Proc.* **113** pp 452-461

[3] Kass K et al. 2016 *Pre*-assessment method for historic building stock renovation evaluation. *Ener. Proc.* **113** pp 346-353

[4] Tokarčík A et al. 2015 *Analytická štúdia efektívnosti pri používaní energií v zariadeniach v zriaďovateľskej pôsobnosti Prešovského samosprávneho kraja* (Prešov: Prešovský samosprávny kraj) p 263

[5] Tokarčík A, Rovňák M and Adamkovič J 2015 Energy management in the public sector according to ISO 50001 *Proc. Int. Conf. SGEM* pp 199-206

[6] Troi A 2016 Comfort and energy efficiency in historic buildings – the 3ENCULT experience *Ener. Effic. and Comf. of Hist. Build.* pp 10-16

The influence of kind of coating additive on the compressive strength of RCA-based concrete prepared by triple-mixing method

K Urban[1] and A Sicakova[1]

[1]Technical University of Kosice, Faculty of Civil Engineering, 042 00, Kosice, Slovakia

Email: karol.urban@tuke.sk

Abstract. The paper deals with the use of alternative powder additives (fly ash and fine fraction of recycled concrete) to improve the recycled concrete aggregate and this occurs directly in the concrete mixing process. Specific mixing process (triple mixing method) is applied as it is favourable for this goal. Results of compressive strength after 2 and 28 days of hardening are given. Generally, using powder additives for coating the coarse recycled concrete aggregate in the first stage of triple mixing resulted in decrease of compressive strength, comparing the cement. There is no very important difference between samples based on recycled concrete aggregate and those based on natural aggregate as far as the cement is used for coating. When using both the fly ash and recycled concrete powder, the kind of aggregate causes more significant differences in compressive strength, with the values of those based on the recycled concrete aggregate being worse.

1. Introduction

There are whole a lot of construction and demolition waste being generated worldwide. Therefore, recycling of the concrete, bricks and masonry rubble as concrete aggregates is an important way to contribute to a sustainable material flow. Conserving natural raw materials such as natural aggregates while reusing construction and industrial wastes for this purpose is the most immediate and obvious way to achieve more sustainable construction. Unfortunately, the composition of these aggregates can vary substantially and consequently the varying properties have a significant influence on the properties of the concrete. In fact, the properties of concrete made with recycled concrete aggregates are inferior to those made with natural aggregates [1].

Recycled aggregates are composed of original aggregates and adhered mortar [2]. The major problems with the use of recycled aggregates in structural concrete are their high water absorption capacity, porosity and lower strength. The presence of RCA and the porous nature of the old cement mortar affect the bond between the RCA and cement paste when used in new concrete. That is why the poorer quality of RCA often limits its utilization [3]. The influence of using recycled aggregate as a replacement of natural aggregate, on the properties of recycled aggregate concrete, has been topic of research by many, from the results of which it was observed that the properties of the RAC were lower than those of the traditional concrete, which has been attributed primarily to the adhered (porous) mortar. With respect to compressive strength, concrete made with 100% of recycled coarse aggregate with lower w/c ratio than the traditional concrete can have a larger compressive strength. When the w/c ratio is the same the compressive strength of concrete made with 100% of recycled aggregate is lower than that of traditional concrete [2]. Chai [4] reported that the compressive strength of RAC decreased by 4 to 37% with increase in percentage replacement of coarse RCA. Jian Zhuang Xiaoa et al. also concluded in 2004 [5] that the compressive strength of recycled aggregate concrete generally decreases with increasing recycled aggregate contents. According to Patel et al. concrete made with recycled aggregate has a compressive strength (28 days) about 27% - 30% less than that of made with natural aggregates [6].

Different mixing methods (double mixing, triple mixing) are presented as a way how to improve the RCA properties directly during mixing in order to improve the behaviour of the whole concrete by [7, 8, 9, 10, 11, 12].The principle of these methods lies in dividing the mixing process in two/three steps, differing in the order and timing of addition of concrete components. This in principle results in coating of aggregate, thus improving its surface character.

In this study, the recycled coarse aggregate´s surface was coated with fly ash and fine fraction of recycled concrete to improve the quality of concrete. The triple mixing technology was applied for this purpose while RCA was coated in the first stage. The effect of kind of coating material on the compressive strength of concrete after 2 and 28 days of hardening is evaluated.

2. Materials and methods

The triple mixing method (TM) was presented and proved by [9] with the aim of enhancement of interfacial transition zone (ITZ) quality between RCA and binder paste. The principle of this mixing is given in figure 1. The purpose of this mixing is to achieve the coating of the aggregate with a thin layer of paste in the first phase, the paste being made of pozzolanic materials such as fly ash, granular blast furnace slag or silica fume. Thus, the triple mixing method is a way how to improve the surface quality of RCA directly during mixing the concrete.

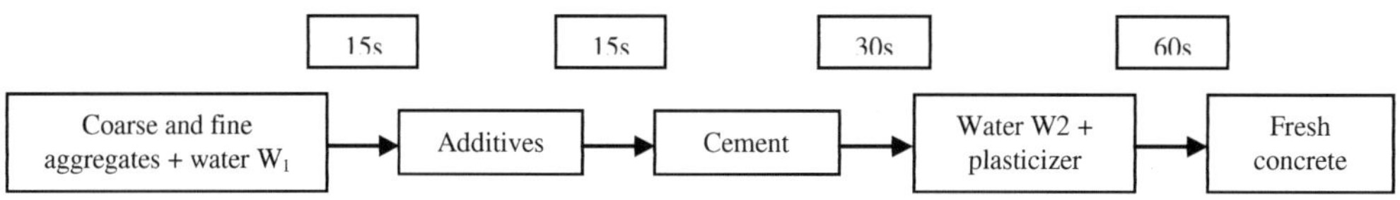

Figure 1. The principle of triple mixing method according to Kong [9].

In presented experiment, TM method given by Kong was slightly modified to be useful for intended purposes. The mixing procedure mainly differs in sequence of adding aggregate (only coarse portions of RCA are of subject for testing and therefore they only will be coated in the first stage of mixing), method of calculation of coating layer volume as well as method for the calculation of W_1 and W_2 (see figure 2). The amount of water for the first stage of mixing is increased by a value corresponding to the absorption capacity of RCA, i.e. the "effective w/b ratio" is applied.

The calculation of a particular recipe is based on the proposed thickness of the coating layer in the first mixing step: $\delta = 0.150$ μm. The volume of the paste for the coating as well as the volume of the paste to fill the gap between the grains is calculated by the Kennedy method. Subsequent weight of both the coating additives and cement is calculated on the base of their density. The mixing procedure for this experiment is given in figure 2.

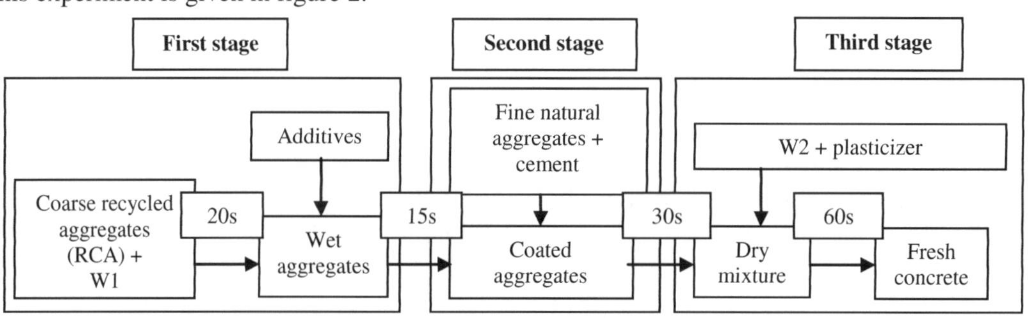

Figure 2. Experimental triple mixing procedure.

Two kinds of additives are tested for the coating of RCA: fly ash (FA) and recycled concrete powder (RCP). The characteristics of the materials used are as follows:

- Aggregates:
  - Fraction 0/4: natural aggregate (NA) is used for all recipes.
  - Fraction 4/8 and 8/16: recycled concrete aggregate (RCA) is used in this study. The 0/32 fraction of RCA was obtained from a company dealing with C&DW treatment (ENVIRONCENTRUM Ltd., Slovakia). Within the experiment, it was crushed and sorted to standard fractions. NA is also used as reference aggregate.
- Coating additives:
  - Fly ash (FA): coming from the energy segment of the steel-making factory from Eastern Slovakia. The original grain size of fly ash is $d_{(0.9)} = 95$ μm.
  - Recycled concrete powder (RCP): this material was prepared with the idea of using fine portion of RCA which is otherwise difficult to recycle in concrete production. Particles under 125 μm were separated by sieving from the material that remained after the sorting of above mentioned fractions 4/8 and 8/16.
- Cement: type CEM I 42.5 R.
- Admixture: polycarboxylate type of plasticizer.

Six recipes were tested (see table 1), which varied from each other by the kind of coarse aggregate (RCA and NA) and by the kind of coating additives (FA and RCP). Coating of coarse aggregates by cement was performed for comparison. Fraction 0/4 was the same in all tested mixtures (NA). The samples of cubes shape with dimensions 100x100x100 mm were prepared with standard methods. Samples were demoulded after one day and consequently cured in water under standard conditions until the test execution. The standard test of compressive strength was executed after 2 and 28 days of setting and hardening, according to [13].

Table 1. Concrete mixture proportions.

Component			RCA - coarse aggregate			NA - coarse aggregate		
			Coating material					
			CEM	FA	RCP	CEM	FA	RCP
			3-RCA_{CEM}	3-RCA_{FA}	3-RCA_{RCP}	3-NA_{CEM}	3-NA_{FA}	3-NA_{RCP}
CEM I 42.5 R	[kg]		310	310	310	336	336	336
NA	[kg]	0/4	898	898	898	896	896	896
		4/8	-	-	-	269	269	269
		8/16	-	-	-	627	627	627
RCA	[kg]	4/8	224	224	224	-	-	-
		8/16	545	545	545	-	-	-
Coating material	[kg]		80	68	68	55	47	47
Admixture	[kg]		2.5	2.5	2.5	2.7	2.7	2.7
Water content	[kg]	W_1	84	78	78	36	33	33
		W_2	166	166	166	179	179	179

3. Results

Compressive strength of concrete mixes which were produced by substitution of natural aggregates with recycled concrete aggregates is shown in table 2. No differences in type of failure were witnessed. It is evident that all of the tested samples had the same compressive strength time development (increase) while the rate of increase is quite significant. The increase is expressed in % and is also given in table 1. Although the values of all samples are within a narrow range (48%-54%),

clear positive effect is visible in the case of using FA and RCP for coating, comparing CEM. The well-known long-term strength development of pozzolanic materials has also appeared here.
The highest compressive strength after two days of curing obtained sample 3-NA_{CEM} (19.4 MPa), while after 28 days it was 3-NA_{FA} (38.2 MPa) – it corresponds with strength class C30/37, following the [14]. The lowest compressive strength reached mixture 3-RCA_{RCP}; 11.7 MPa after 2 days and 25.6 MPa after 28 days - it corresponds with class C20/25.

There is no very important difference between samples based on RCA and corresponding samples based on NA as far as the cement only is used for coating the aggregate (3-RCA_{CEM} and 3-NA_{CEM}). Once the additives (FA and RCP) are used for coating, the difference between the RCA and NA based samples is more significant with the values of those based on RCA are worse. Comparing the kind of coating material, FA and RCP additives give worse results comparing the cement and FA gives better results than RCP. The influence of coating material change is seen in the table 2, where differences between values of compressive strength are given for particular combinations of coating materials (CEM/FA, CEM/RCP and FA/RCP). Results are grouped by the kind of aggregate.

Table 2. Compressive strength of samples and differences in values resulting from kind of coating material.

Coarse aggregate	Time	f_c - compressive strength [MPa]			Differences of values by the change of coating material [%]		
		CEM	FA	RCP	CEM/FA	CEM/RCP	FA/RCP
RCA	2 days	18.3	13.4	11.7	-27	-36	-13
	28 days	37.0	27.7	25.6	-25	-31	-8
Increase in time [%]		51	52	54	-	-	-
NA	2 days	19.4	17.6	15.7	-9	-19	-11
	28 days	37.0	38.2	34.1	+3	-8	-11
Increase in time [%]		48	54	54	-	-	-

4. Conclusions

The triple mixing technology was applied for improvement the quality of concrete based on RCA, while effect of different kinds of coating additives are evaluated. Both the fly ash and the fine fraction of recycled concrete were used for coating in the first stage of triple mixing, as well as cement as the reference material. For the same reason, samples with natural aggregate instead of RCA were tested as well. Compressive strength is evaluated based on the results obtained after 2 and 28 days of setting and hardening. Following conclusions can be formulated:

- all samples show positive time development of compressive strength;
- FA and RCP have better effect on the rate of increase the compressive strength in time than the cement;
- comparing the kind of coating material, FA gives higher values of compressive strength than RCP;
- there is no very important difference between samples based on RCA and NA as far as the cement is used for coating the aggregate;
- in case that both the FA and RCP are used for coating, the kind of aggregate causes the more significant differences in compressive strength, with the values of those based on RCA being worse.

Generally, testing the alternative powder additives (FA and RCP) for coating the coarse RCA in the first stage of triple mixing resulted in decrease in compressive strength comparing the cement.

However, the results that have been achieved (27.7 MPa and 25.6 MPa) are at practically useful values. This approach is open for next optimization of presented principle and is very promising for saving cement and supporting the RCA consumption, while keeping the technical limits for concrete production and application.

Acknowledgments
This work was supported by the project ITMS Center of excellent integrated research of progressive building constructions, materials and technologies [grant number 26220120037] and the project VEGA - 1/0677/14 „Research of construction efficiency improvement through MMC technologies".

References

[1] Achtemichuk S, Hubbard J, Sluce R and Shehata M H 2009 The utilization of recycled concrete aggregate to produce controlled low - strength materials without using Portland cement *Cement and Concrete Composites* **31** (8) p 564-569
[2] Etxeberria M, Vázquez E, Marí A and Barra M 2005 Influence of amount of recycled coarse aggregates and production process on properties of recycled aggregate concrete *Cement and Concrete Research* **37** p 735-742
[3] Spaeth V and Tegguer A D 2013 Improvement of recycled concrete aggregate properties by polymer treatments *International Journal of Sustainable Built Environment* **2** p 143-152
[4] Chai N S NGO 2004 High-Strength Structural Concrete with Recycled Aggregates B.E Thesis. Report submitted to University of Southern Queensland Faculty of Engineering and Surveying (Queensland, Australia) p 112
[5] Jian Zhuang Xiaoa C, Jiabin L and Zhangb CH 2004 Mechanical property of recycled aggregates concrete under uniaxial loading a Department of Building Engineering, Tongji University, Shanghai 200092, China Department of Civil Engineering, University of Applied Sciences Zittau/Gorlitz, D-02763 (Zittau: Germany) p 1-112
[6] Patel M K, Nayak CH and Patel S Use of recycled aggregates in concrete: a critical review *International journal for technical education and research innovations* **1** (6) p 1-6
[7] Tam V, Tam C M, Wang J 2007 Optimization on proportion for recycled aggregate in concrete using two-stage mixing approach *Construction and Building Materials* **21** p 1928-1939
[8] Otsuki N, Miyazato S I and Yodsudjai W 2003 Influence of Recycled Aggregate on Interfacial Transition Zone Strength Chloride Penetration and Carbonation of Concrete *Journal of material and civil engineering* **15** p 443-445
[9] Kong D, Lei T, Zheng J and Jiang J 2010 Effect and mechanism of surface coating pozzalanic materials around aggregate on properties and ITZ microstructure of recycled aggregate concrete *Construction and Building Materials* **24** (5) p 701-708
[10] Otoko G R 2014 A solution to the problem of recycled concrete aggregates *Interntional Journal of Engineering and Technology Research* **2** (4) p 1-6
[11] Kou S C, Poon C S and Etxeberria M 2011 Influence of recycled aggregate on long term mechanical properties and pore size distribution of concrete *Cement and Concrete Composites* vol **33** p 286-291
[12] Tam V W Y, Gao X F and Tam C M 2005 Microstructural analysis of recycled aggregate concrete produced from two-stage mixing approach *Cement and Concrete Research* vol **35** p 1195-1203
[13] STN EN 12390-3:2010 Testing hardened concrete Compressive strength of test specimens
[14] STN EN 206:2015 Concrete. Part 1: Specification, performance, production and conformity

IOP Conf. Series: Earth and Environmental Science **92** (2017) 012070 doi:10.1088/1755-1315/92/1/012070

Utilization of a new optical sensor unit to monitor the electrochemical elimination of selected dyes in water

M Valica[1], T Černá[1] and S Hostin[1]

[1]Department of Ecochemistry and Radioecology, Faculty of Natural Sciences, University of SS. Cyril and Methodius in Trnava, Nám. J. Herdu 2, Trnava, SK-917 01, Slovak Republic

Email: martin.valica.mv@gmail.com

Abstract. This paper presents results obtained by developed optical sensor, which consist from multi-wavelength LED light source and two photodetectors capable of measuring the change in optical signal along two different optical paths (absorbance and reflectance measurements). Arduino microcomputer was used for light source management and optical signal data measuring and recording. Analytical validation of developed optical sensor is presented in this paper. The performance of the system has been tested with varying water solution of dyes (malachite green, methyl orange, trypan red). These results show strong correlations between the optical signal response and colour change from the dyes. Sensor was used for continual in-situ monitoring of electrochemical elimination of selected dyes (current density 15.7 mA cm^{-2}, electrolyte volume 4 L and NaCl concentration 2 g L^{-1}). Maximum decolorization level varies with each dye. For malachite green was obtain 92,7 % decolorization (25 min); methyl orange 90,8% (8,5 min) and trypan red 84,7% decolorization after 33 min of electrochemical treatment.

1. Introduction

Environmental pollution is a global issue. Synthetic dyes are used in many fields of industry. Textile and printing industries are regarded as the most polluting sectors, due to high discharge volume of dyes-contained wastewater. Dyes are highly soluble and persistent in polluted water, because they have long degradation time in the environment [1]. Pre-treatment has been found important for significant decrease of organic content and colour of dye polluted water. In recent years, electrochemical treatment processes, especially electrochemical oxidation and electrocoagulation, have been studied as alternatives for degradation of various types of organic pollutants in wastewater [2,3]. Electrochemical oxidation is the process of oxidizing pollutants on the electrode surface, through the formation of disinfectants such as oxygen species from water itself as in equation (1-5):

$$2\,H_2O - 4e^- \rightarrow O_2 + 4\,H^+ \quad (1)$$

$$2\,H_2O + 2\,e^- \rightarrow H_2 + 2\,OH^- \quad (2)$$

$$O_2 + 2\,H^+ + 2\,e^- \rightarrow H_2O_2 \quad (3)$$

$$2\,H_2O \rightarrow OH\bullet + e^- + H^+ \quad (4)$$

$$3\,H_2O \rightarrow O_3 + 6\,e^- + 6\,H^+ \quad (5)$$

If suitable electrolyte is presence in solution (e.g. NaCl), it lead to formation of active chlorine [4] as in equation (6-8):

$$Cl^- - 2\,e^- \rightarrow Cl_2 \quad (6)$$
$$Cl_2 + H_2O \leftrightarrow HOCl + Cl^- + H^+ \quad (7)$$
$$HOCl \leftrightarrow ClO^- + H^+ \quad (8)$$

Even low concentration of chlorides in water (0,004 mol L^{-1}) is sufficient for electrochemical elimination undesired pollution and production of active chlorine in effective level. This level of chlorides is naturally presence in surface waters [4].

Common used UV/VIS spectrophotometric methods for monitoring of water quality are based on specific absorption of light with various wavelengths [5]. The current trend is to simplify and increase the effectiveness of pollution monitoring in real time. Recent development in optoelectronics allows to constructs a low cost in-situ sensor for continuously monitoring. Optical sensors are based on detection of physical changes of light and can be used for qualitative and quantitative analysis of pollution substances. The low cost components allow to connected sensors into monitoring networks and performed real time, continuous *in situ* water pollution analysis [6].

The aim of paper is to verify the possibility of using a low-cost multi-wavelengths spectrophotometric sensor for monitoring the electrochemical purification of water contaminated with selected dyes. The dyes were selected based on the maximum absorption band, which was in the wavelength range corresponding to the radiation of light source inside spectrophotometric sensor. In natural waters these dyes generate toxicological risk for biological system.

2. Materials and methods

Schematic representation of electrochemical treatment device based on previous work [7] with spectrophotometric unit controlled by microprocessor platform Arduino is shown in the figure. 1.

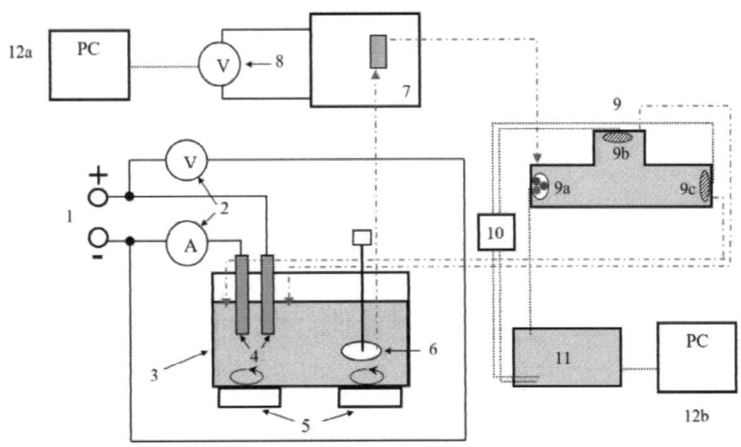

Figure 1. Scheme of the experimental equipment for the electrochemical elimination of dyes in water (1 – DC power supply, 2 – voltmeter and amperemeter with USB, 3 – electrolysis reactor, 4 – carbon electrodes, 5 – magnetic stirrers, 6 – pump, 7 – VIS spectrophotometer with flow cuvette, 8 – voltmeter with USB, 9 - optical sensor unit, 9a – RGB LED light source, 9b – resistance detector of absorbance, 9c – resistance detector of reflection, 10 – A/D converter, 11 - Arduino microcomputer, 12a,b - data logger computers).

The electrolytic cell contained two vertically positioned graphite electrodes (Elektrokarbon, Slovak Republic) connected to a regulated DC power supply (Elektrolyser, Veb Mlw Labortechnik Ilmenau, Germany) and submerged into electrolyte solution (4 L) with 2g L^{-1} content of NaCl. The resulting electrolyte was used to monitor the process of elimination of selected water-soluble dyes (malachite green, methyl orange, trypan red). All degradation processes were carried out under the galvanostatic condition at current density 15,7 mA cm^{-2} at constant laboratory temperature 25°C.

During electrolysis, the values of electric current and voltage transferred through the electrolyte were recorded on a computer in time intervals of two seconds. For this measurement, two separate digital multimeters (METEX M-3850D with software ScopeView ver. 1.06; and AXIO MET AX-18B with software PC Link soft) were used. Electrochemical water treatment was finished, when the lowest and constant values of the absorbance of wastewater were obtained.

Continual measurement and proper circulations into laboratory spectrophotometer (Spekol 11- Carl Zeiss JENA) in conjunction with measuring cell of spectrophotometric sensor was performed by pumping device. Data from laboratory spectrophotometer was recorded into PC by digital VOLTCRAFT VC820 with software Digiscop for Windows ver. 2.06. Because output values of spectrophotometer were in electric current, it was necessary to convert these values to absorbance using the calibration curve between these two variables.

The optical sensor is devised to give data on bulk water property changes, particularly opacity and colour changes. For record values gained by spectrophotometric sensor was used the Measurino software with connection to Arduino microcomputer. Detectors in spectrophometric unit are capable of measuring the transmission of light emitted by RGB LED (Super Flux RGB LED SFL RGB 2200/80° with red LED 620-630 nm, green LED 515-525 nm and blue LED 460-470nm) light source through water, while simultaneously measuring the side-scattering of the light measured at right angles to the transmission path. Arduino measured change in the frequency values on the absorbance photoresistor detector and the reflectance detector (Vactrols VT83N4) after converting via an analog transducer from resistance units.

Elimination efficiency was calculated from ratio of content dye at the initial C_0 and the given C_t time in model waste water. The remaining concentration of dye in wastewater was determined from the equation of the linear dependence between the concentration of dye and absorbance, for solutions with the known concentration of dye.

3. Results and discussion

The results obtained from the laboratory tests on the food dyes are represented here. The accuracy of spectrophotometric sensor response on pollution presence in water can be evaluating against to laboratory spectrophotometer in figure 2c.

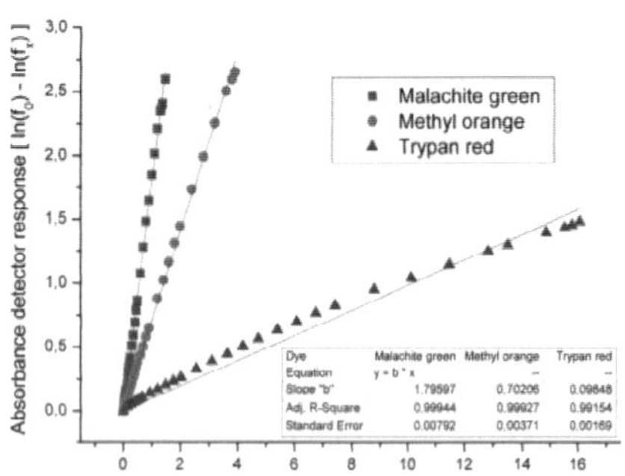

(a)

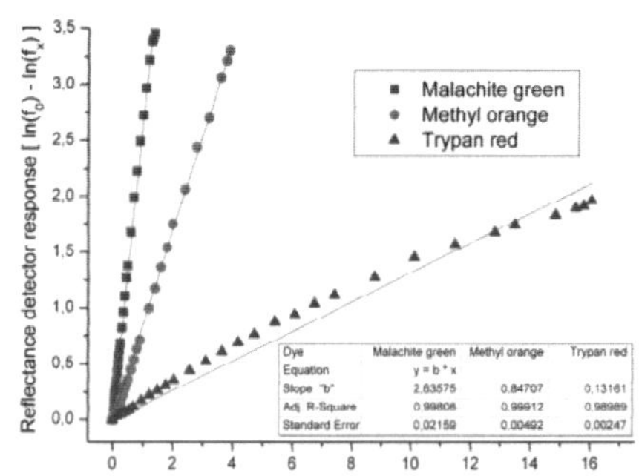

(b)

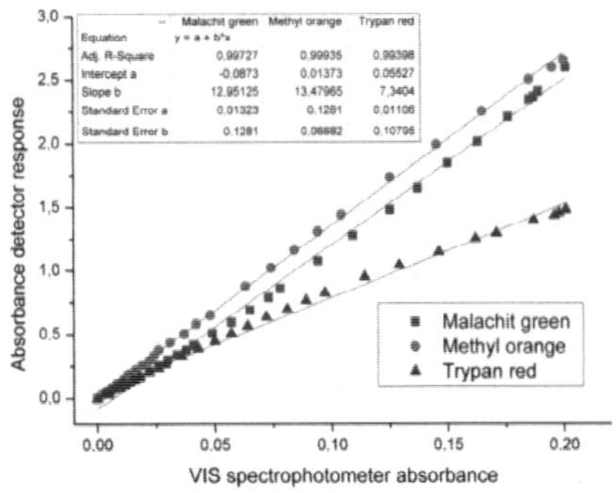

(c)

Figure 2. Dependence of absorbance (a) and reflectance (b) detectors response to content of the dyes in water and correlation between the absorbance from VIS spectrophotometer and the absorbance from detector of optical sensor unit (c).

Figure 2 a,b shows the absorbance and reflectance detector response on change in the optical signal caused by increasing concentration of selected dyes in tested water. Response of detector is shown to be a linear, for absorbance detector with $R^2 > 0.99$. Response of reflectance detector was less accuracy due to very low irradiation of photoresistor close to detection limit. Dyes solutions were clear without signs of turbidity. Response recorded by reflectance detector was caused by my measuring reflected light on sensor cover and its absorption by dye. In fact reflectance detector during calibration test measure absorption and trend of calibration curves from absorbance and reflectance detector are very similar.

Table 1. Evaluation of electrochemical elimination of dyes by absorption detector.

Dye	Initial content (mg/L)	Electrolyses time (min)	Elimination efficiency (%)	Rate constant (min^{-1})		
				Value	Adj. R-Square	St. Err.
Malachite green	1.37	25	92.7	-0.05562	0.97237	2.35E-04
Methyl orange	3.7	8.5	90.8	-0.43209	0.98846	2.09E-03
Trypan red	14.4	33	84.7	-0.40839	0.95231	2.13E-03

Treatment time needed for maximum level of decolorization varied for each dye (see table 1). Methyl orange and malachite green showed good degradability by electrochemical treatment with elimination efficiency > 90%. In case of electrochemical elimination of trypan red was maximum elimination level only 84.7% gained after 33 min. Electrolysis continued after this time, but elimination efficiency showed no increase. Differences in elimination efficiency were probably caused by different initial content of dye in water, because initial concentration was modified on colour change capability of each dye.

Figure 3 showed continuous recording of electrochemical decolorization dyes from absorbance and reflectance detector.

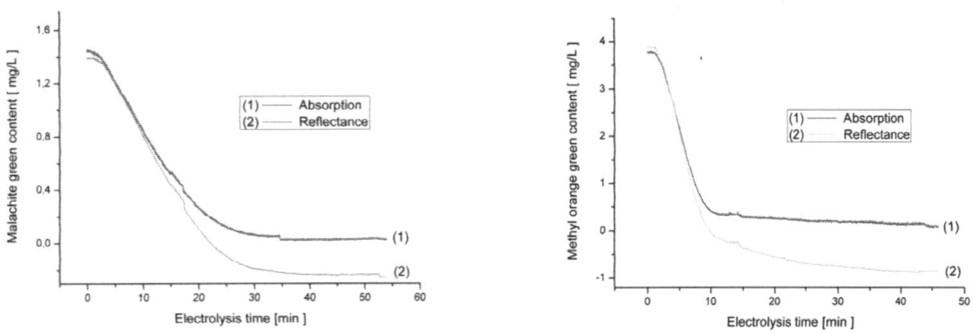

(a) (b)

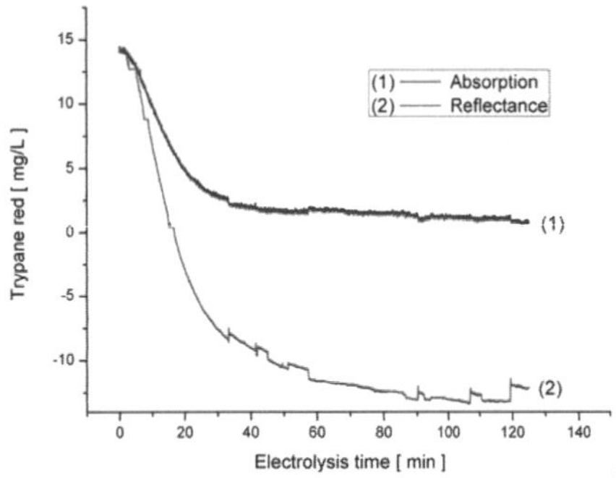

(c)

Figure 3. Decrease of content Malachite green (a), Methyl orange (b), Trypan red (c) during the electrolysis, determined by the absorption and reflectance detectors (content NaCl in electrolyte 2 g/L, current density 15,7 mA cm^{-2}).

The response of detectors depends on many factors such as light reflection from sensor walls, light source radiation, absorption of radiation by the analyzed substance, radiation of scattering, refractive index. During electrolysis are generated gas bubbles (equations 1-5), which circulate with solution to the sensor measuring cell. Different refractive index of water and air causes increase in light diffusion. This explains the negative absorbance values measured by the reflectance detector, because with treatment time increase saturation of solution, which lead to more irradiation of detector than was measured at the beginning of the experiment. After completion of electrolysis and washing sensor, absorbance measured by at reflectance detector stabilized at the original/blank value. Small deviation observed on both elimination curves for each dyes were cause by attaching gas bubbles nearby detectors and by their removal with temporal increased liquid flow. Response reflectance detector on dispersion confirmed usage for measuring turbidity, which is important fact for next experimental work.

4. Conclusions

Results obtained by developed low cost optical sensor consist from multi-wavelength LED light source and two photodetectors capable for absorbance and reflectance measurements. The performance of the system has been tested with varying water solution of dyes compared with results from laboratory spectrophotometer. These results show strong correlations between the optical signal response changes from the dyes. The results of these experiments suggest, that spectrophotometric unit with new optical sensors can be used to monitoring pollution during electrochemical water treatment.

Acknowledgments

We would like to acknowledge support of this work provided by Department of Ecochemistry and Radioecology, Faculty of Natural Sciences, University of SS Cyril and Methodius in Trnava. This paper was supported by grants FPPV-18-2015 and APVV-14-0716.

References

[1] Azami M, Bahram M, Nouri S and Naseri A 2011 *J. Serb. Chem. Soc.* **76** 1–21
[2] Panizza M, Bocca andC Cerisola G 2000 *Water Res.* **34** 2601–05
[3] Santos E V, Sáez C, Martínez-Huitle C A, Cañizares P and Rodrigo M A 2015 *Electrochem. Commun.* **55** 26–9
[4] Wei L, Guoa S, Yana G, Chena C and Jiangb X 2010 *Electrochim. Acta.* **55** 8615–20
[5] Korostynska O, Mason A and Al-Shamma'a A 2012 International Journal on Smart Sensing and Inteligent System **5** 147-71
[6] O'Toole M and Diamond. D 2008 *Sensors* **8** 2453–79
[7] Valica M, Hostin S 2016 *Nova Biotechnologica et Chimica* **15** 55–64

Reduction of mercury emissions from coal combustion in the Czech Republic

Z Vávrová[1], O Palička[1], P Dvořák[1], R Fojtů[1], Z. Szeliga[1] and L Pilař[2]

[1]Centrum ENET, Technical University of Ostrava, Czech Republic
[2]ÚJV Řež, a. s., Husinec - Řež, Czech Republic

E-mail: zuzana.vavrova@vsb.cz

Abstract. Under the new legislation, in force since 2021, mercury emission limits for large combustion plants (LCP) are determined. This paper describes formation of mercury during coal combustion process and present methods of removing mercury from flue gases. This paper also reviews the fuel base of the Czech Republic and current measurement results of mercury emissions from some important LCPs in the Czech Republic. Measurement shows that most LCPs in the Czech Republic do not meet these emission limits and it leads to find out technical and economic solution.

1. Introduction

In terms of proven ecotoxicity and accumulation, global mercury (Hg) issues are addressed in recent years. The researches relate the harmfulness of individual forms of Hg to human health, sources of Hg that discharge it into the food chain as well as into the air. One of the most significant sources of Hg emissions are fossil fuel burning units. For these reasons, the EU has also newly revised the emission limits for Hg concentrations for LCP, it means for units with a thermal input higher than 50 MWt. New legislation will come into force in the second half of 2021. In the Czech Republic Hg emissions have not been legislatively established. Only once a year, the Hg measurement was carried out according to Czech technical standard (ČSN). However, this has changed with the new legislation. Every LCP has to solve in a relatively short time several issues. It means how much Hg emissions are emitting from LCPs and in case of failure to meet the required limit, find measures to meet the emission limit. Another sector linked to the new legislation is the measurement of Hg emissions. The actual determination of Hg emissions is both specific and novelty. Hg emissions are nowadays a highly discussed topic at all levels in the Czech Republic.

2. New emission limits valid from 2021

Commission implementing decision (EU) 2017/1442 of 31 July 2017 establishing best available techniques (BAT) conclusions, under Directive 2010/75/EU of the European Parliament and of the Council, for large combustion plants. There are also new emission limits for mercury. See table 1. Hg emission limits are divided by type of fuel for new and existing sources and by heat input. Sources of thermal input below 300 MWt will show mercury emissions four times a year from the average of samples obtained over one year. Sources with thermal inputs more than 300 MWt will then report an annual average of continuous measurement.

Table 1. Mercury Hg emission limits.

Yearly average or average of samples obtained during one year Hg [ηg/Nm^3]				
Heat input [MWt]	New plant		Existing plant	
	Hg (coal)	Hg (lignite)	Hg (coal)	Hg (lignite)
< 300	< 1 - 3	< 1 - 5	< 1 - 9	< 1- 10
≥ 300	< 1 - 2	< 1 - 4	< 1 - 4	< 1- 7

3. Mercury in the process of coal combustion

During the coal combustion process, mercury is released into flue gases in three forms.

- Gaseous oxidized mercury (Hg^{2+});
- Gaseous elemental mercury (Hg^0);
- Mercury bound to particles (Hg_p).

Each of Hg forms has its specific properties that influence their formation and reduction during the combustion process. The amount of Hg emissions emitted from individual large combustion plants is then dependent on the type of fuel and the emission control devices. Gaseous Hg^{2+} is water soluble. Therefore in the atmosphere it can be removed by rainfall or moisture. Hg^{2+} presented in flue gases can be removed by wet flue gas desulphurization systems (WFGD). It all results in that the Hg^{2+} represents only 3-5% of the total mercury in the atmosphere. Gaseous Hg^0 stays in the atmosphere in average of 0.5 - 2 years. It is long enough for transporting around the world. First it is necessary to oxidize Hg^0 to Hg^{2+}. This oxidization is important for reducing Hg^0. For that reason Hg^0 is major problem for large combustion plants. Mercury bound to particles Hg_p is adsorbed onto surfaces – slag, fly ash. This form can be easily captured by dust control devices (precipitators or baghouses). Figure 1 shows the formation of individual forms of mercury in the flue gases from the combustion process.

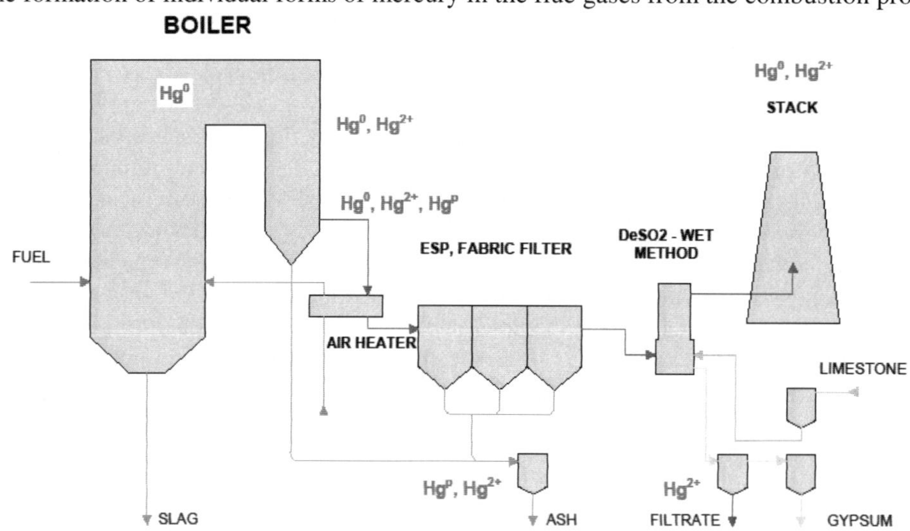

Figure 1. Forms of mercury emissions during coal combustion.

4. Fuel base of the Czech Republic

Mercury is in trace amounts already in the coal. The amount of Hg varies depending on the type of coal (black, brown) and the origin of the coal. Another important factor is the content of halogens in the coal, especially the content of chlorine (Cl). Chlorine is very important for the oxidation of Hg^0 to Hg^{2+} during the combustion processes.

Knowledge of fuel is the basic input information. In LCPs in the Czech Republic, mainly lignite is burned. Table 2 shows the lignite sample analyses (content of Hg and Cl), which are the most used in the Czech Republic.

Table 2. Analysis of lignite samples.

Lignite sample from quarry	Hg mg/kg$_{dry}$	Cl %$_{dry}$
Bílina	0.22	0.0074
Nástup Tušimice	0.203	0.0088
Jiří	0.482	0.0095
Vršany	0.262	0.0061

For comparison, a sample of black coal was also analysed. The results of the analysis shows that Hg content is 0.516 mg/kg$_{dry}$ and Cl content is 0.19 %$_{dry}$.

It is obvious from the analysis of coal burned in the Czech Republic that the content of Hg in lignite is lower than in black coal. Similarly, the content of Cl is significantly higher for black coal than for lignite. However, the resulting value of Hg emissions from LCP does not affect only the amount of Hg in coal. Another factor is the combustion process itself and the choice of the flue gas cleaning method, especially the choice of denitrification system and desulfurization system.

5. Methods of removing mercury from flue gas

For reducing mercury concentrations in flue gases from solid fossil fuel combustion, the development is focused on two main areas. The first area is the development of techniques and technologies designed to promote the oxidation of Hg^0 itself to Hg^{2+} in the flue gas and the subsequent removal of the Hg^{2+} by means of scrubbing in the FGD systems. The second area is the development of sorbent and associated systems for the sorption of Hg on sorbent surfaces.
The following technologies and methods are available to control mercury Hg emissions.

5.1. Sorbents

- Activated carbon – by adding active carbon, it is possible to reduce the Hg concentration by as much as 90%. Activated carbon is added in front of the dust control device.
- Addition of sorbents based on halides, bromides, iodides. Sorbents are added both to the combustion chamber and then as the activated carbon before the dust control device. It is also possible to use zeolite-based sorbents for capturing mercury.

5.2. Oxidation on the catalyst layer

- The oxidation occurs on a catalyst layer primarily designed to reduce the NO_x emissions. Selective catalytic reduction (SCR) technology must be combined with the dust control device and FGD system using a wet method. SCR technology oxidizes the Hg^0 to Hg^{2+}. This fact allows us to subsequently removed Hg in the WFGD reactor. It is important that flue gas contains sufficient amount of Cl to oxidize mercury in the presence catalyst.

Mercury emission control system must be carried out in various combinations. The following table 3 and table 4 shows technological possibilities of reducing Hg emissions.

Table 3. Technologies used to reduce pollutant emissions and their effect on Hg reduction.

BAT technology	Primary use	Reduction of Hg
Bag filters	To remove dust	Reduction of only Hg_p
ESP	To remove dust	Reduction of only Hg_p
SCR	To reduce NO_x emissions	It is necessary to combine SCR with the wet flue gas desulfurization (WFGD)
Dry or semi-dry desulfurization	Especially for removal of SOx, HCl and HF in flue gas	-
WFGD		Reducing Hg^{2+}

Table 4. Technologies used to reduce Hg emissions.

BAT technology	Reduction of Hg	Comment
Fuel choice	Choosing a fuel with low ash content and / or Hg	Limited by local conditions, availability of other fuels and energy policies
Sorbent - Halides	Oxidation of elemental Hg^0 to the oxidized form	Adding into combustion chamber or in front of dust control device. It is necessary in combination with WFGD
Sorbent – activated carbon	Mercury capture on sorbent surface	Dosing in front of dust control device in combination with flue gas desulphurization using a semi-dry method
Fuel pre-treatment	Washing, forming mixtures to reduce the Hg content of the fuel or increasing the efficiency of Hg separation from flue gas	Applicability is highly dependent on specific conditions

6. Measuring and status of current mercury emission in the Czech Republic

6.1. Black coal-fired combustion plants

Selected LCP burning black coal equipped with pulverized boilers, followed by ESP and WFGD system. Part of the source is ecologized and 2 of 4 boilers are equipped with SCR technology.

Measurement was carried out on the boiler without SCR and then on the SCR boiler. Without SCR was average Hg concentration in the flue gas 5.92 μg/Nm^3. With SCR was average Hg concentration 5.44 μg/Nm^3. However, the emission limit value for Hg on a black coal-fired block with a heat input more than 300 MWt is 4 μg/Nm^3. The emission limit was exceeded in both cases. The Cl concentration was around 50-80 ppm for both measurements.

6.2. Lignite coal-fired combustion plants

Mercury emissions were measured on the following selected LCP with thermal input more than 300 MWt. [3] For these energy sources new legislation sets an emission limit of 7 μg / Nm^3.
LCP number:

- 4 pulverized boilers, with ESP and WFGD system.
- 2 pulverized boilers and 2 grid boilers with ESP and FGD by a semi-dry method.
- Lignite in combination with biomass in 2 boilers with circulating fluidized bed with a fabric filter. FGD system is based on the dry additive method.

Results of Hg concentration measurements are shown in the following table 5.

Table 5. Concentration of Hg in flue gas – lignite.

LCP n.	Fuel	$DeSO_x$ technologies	Hg μg/Nm^3	Emission limit μg/Nm^3
1.	Lignite	WFGD	15.0 – 24.0	7.0
2.	Lignite	Semi-dry FGD	18.0 – 26.0	7.0
3.	Lignite/biomass	Dry additive method – fluidized bed boiler	0.2 – 2.0	7.0

The values in table 5 shows a high exceedance of approved emission limit. Emission values differ is based on fuel, type of boiler (combustion process) and use of emission control systems.

7. Conclusion

Based on commission implementing decision (EU) 2017/1442 for LCPs are set the Hg emission limits. Mercury emission is currently the most discussed topic in the Czech Republic's energy sector.

Measurement results from some LCPs in the Czech Republic shows that most of them do not meet the emission limits. These LCPs will have to implement measures that will reduce mercury emissions to the desired value. For each LCP, the method leading to the reduction of Hg emissions will vary according to fuel, combustion process, emission control device and investment and operating costs of selected method. LCPs of the Czech Republic have a relatively short period of time to install measures to reduce mercury emissions.

These measures are associated with substantial investment costs that may be critical to the continued operation of the facility. Research and measurement in the field of mercury is an important part of further developments leading to the reduction of mercury emissions.

Acknowledgments

This paper is supported by the following project: LO1404: Sustainable development of ENET Centre; CZ.1.05/2.1.00/19.0389 Development of the ENET centre research infrastructure; SP2017/96 Students Grant Competition, Czech Republic.

References

[1] Frolka P 2016 *BREF Large combustion plants* XVII Forum for information exchange about BAT: MPO
[2] Vávrová Z 2015 *Design of Treatment Facilities for Research in Mercury Emissions from Coal Combustion* (Ostrava: Technical University of Ostrava)
[3] Technical reports - Accredited workplace for diagnostics and operation of power generation equipment Centrum ENET

IOP Conf. Series: Earth and Environmental Science **92** (2017) 012071 doi:10.1088/1755-1315/92/1/012071

[4] Best Available Techniques (BAT) 2016 Reference Document for Large Combustion Plants: Industrial Emissions Directive 2010/75/EU Bureau

[5] Pilař L, Szeliga Z and Zbieg R Monitoring conversions of Hg0/Hg2+ in flue gas from energy sources

Contribution to the Study of Ammonia Removal from Digestate by Struvite Precipitation

P Vidlarova[1], S Heviankova[1] and M Kyncl[1]

[1]Institute of Environmental Engineering, VŠB - Technical University of Ostrava, Czech Republic

Email: petra.vidlarova@centrum.cz

Abstract. This study deals with an issue of ammonia removal from anaerobic digestion's by-product - digestate. The aim of the research is ammonia removal by chemical precipitation using the combination of Mg2+, NH4+ and PO43- ions in struvite (MgNH4PO4 . 6 H2O). Laboratory tests were carried out using the sample of digestate liquor. The final product of the treatment may have the potential to be used as a fertilizer.

1. Introduction

Anaerobic digestion is a combination of processes leading to breaking down biodegradable materials in the absence of oxygen. It is a combination of physical, physicochemical and biological processes caused by functional groups of anaerobic microorganisms of transformation into final products – biogas and digestate [1]. Because residual biomass is generated mostly in agriculture, the most widely spread type of biogas plants are the agricultural ones. Such biogas plants process wastes from livestock production, crop residues or farm animals' excrements; recently also purpose-grown plants are used as feedstocks [2]. As it is forbidden to process wastes in the agricultural biogas plants according to Czech national waste legislation Act 185/2001 Coll., and other materials according to Regulation (EC) 1774/2002 of the European Parliament and of the Council, these biogas plants' digestates are the most suitable for their subsequent use. According to the Czech and also European legislation, digestates are considered to be organic fertilizers in most cases. The digestate's content of nutrients and also their subsequent availability for plants is affected by the input substrate; properties may also vary over time within one biogas plant. However, a common feature is the low dry matter content (for agricultural biogas plant between 2.5 and 8.5 %). C:N ratio is often between 10:1 and 2:1.

The issue is that the chemical composition of digestate is closer to a mineral fertiliser rather than an organic one, and almost all easily degradable organic material has already decomposed during anaerobic digestion resulting in only minimum of beneficial substances usable by plants. Nitrogen content in fresh digestate is about 0.25-0.75 %, of which 25 to 75 % are in the form of ammoniacal nitrogen (i.e. $N\text{-}NH_4^+$) causing a weak alkalinity of digestate (pH = 7-8) [3]. Thus, land spreading of digestates can cause a discharge of large quantities of nutrients into the environment, which contributes to eutrophication and depletion of dissolved oxygen in water bodies. Owing to the expansion of biogas plants and the associated quantity of digestate production, it is necessary to look for other means of its application or properties enhancement. One of the options is the chemical precipitation of $N\text{-}NH_4^+$ in the form of struvite, which is a mineral that can be reused as a slow-release fertilizer [4]. Frequently precipitation of struvite is an undesirable accompanying phenomenon at urban waste water treatment plants. In many countries, commercially available reactors are used in large waste water treatment plants for phosphorus recovery and it is further used as a certified fertilizer [5]. These days there is an increasing interest in the application of this method. According to [6] there are already published studies in combined wastewater from bovine and leather tanning factories, swine waste, anaerobic supernatant from a centrifugation section of a civil biological nutrient removal plant, effluent from the biologically treated opium alkaloid wastewater, anaerobic treatment of the baker's yeast industry, domestic wastewaters, landfill leachate, etc. In addition, this study deals with the precipitation of ammoniacal nitrogen ($N\text{-}NH_4^+$) in struvite from the liquid phase of digestate, digestate liquor, from an agricultural biogas plant.

2. Materials and methods

2.1. Materials

All laboratory experiments were conducted in real digestate samples provided by Stonava Farm. The Stonava Farm's Biogas Plant is a purely agricultural. The input material for anaerobic digestion is pig slurry in combination with maize silage and crushed corn grain. Two fermentation stages operate at 42-45 °C.

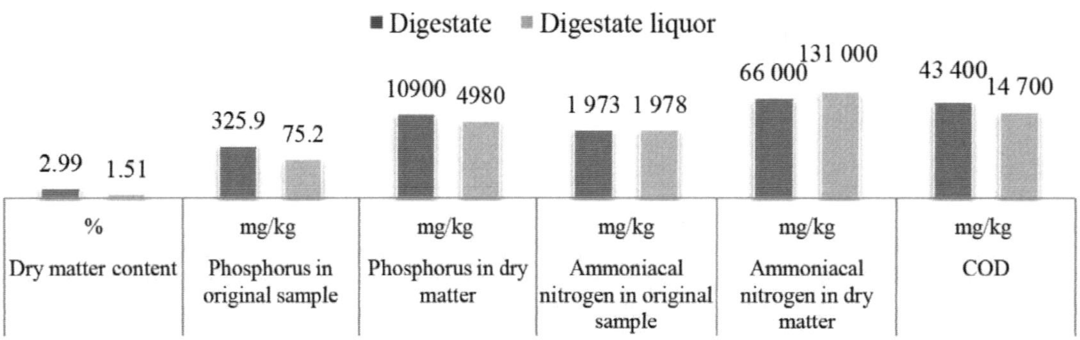

Figure 1. Model chemical composition of digestate and digestate liquor from Stonava Farm. The phosphorus and ammoniacal nitrogen concentrations were converted to the original weight of the sample depending on the dry matter content.

As digestate can be dewatered there are two components: the liquid phase – digestate liquor and the solid phase – digestate fibre. These two phases differ in their properties. In particular, they differ in the dry matter content and amounts of ammoniacal nitrogen, which can be separated by precipitation into the form of the magnesium ammonium phosphate – struvite. The amount of ammoniacal nitrogen in dry matter also differs in digestate and digestate liquor as can be seen in figure 1. For this reason, the experiments were conducted only with digestate liquor.

2.2. Experimental

Struvite precipitation begins by exceeding the equilibrium concentrations of the reaction ions Mg^{2+}, NH_4^+ and PO_4^{3-} and proceeds according to the simplified equation (1).

$$Mg^{2+} + NH_4^+ + PO_4^{3-} + 6\,H_2O \leftrightarrow MgNH_4PO_4 \,.\, 6\,H_2O \quad (1)$$

The solubility of struvite thus depends on the concentration of magnesium, nitrogen and phosphorus and usually decreases along with increasing pH when the minimum is at pH = 9-11 [7]. In order to guarantee that the product of solubility of the potential struvite precipitate as in equation (1) would be exceeded, Mg^{2+} and PO_4^{3-} were added to the sample of the centrifuged digestate. In the first part of the research, summarized in this contribution, Magnesium sulphate anhydrous – $MgSO_4$ (*Penta s.r.o*) and Disodium hydrogen phosphate dodecahydrate $Na_2HPO_4 \,.\, 12\,H_2O$ (*Penta s.r.o*) were used for this purpose. Chemical reagents were dosed according to the chemical analysis carried out in the laboratory of the *Institute of Environment Engineering at HGF, VSB-TUO*. The results of the digestate liquor analysis are shown in Table 1.

Table 1. Input parameters of digestate liquor after centrifugation in centrifuge *BEHO CHC 61A*

Parameter	Dry matter content	PO_4^{3-}	NH_4^+	$N\text{-}NH_4^+$	Ca^{2+}	Mg^{2+}	Na^+	COD
Unit	%	[$mg.l^{-1}$]	[$mg.l^{-1}$]	[$mg.l^{-1}$]	[$mg.l^{-1}$]	[$mg.l^{-1}$]	[$mg.l^{-1}$]	[$mg.l^{-1}$]
Value	2.17	640	2 138	1 660	279	136	215	34 000

Experimental procedure was drawn up based on experiments already carried out according to expert articles [8, 9]. Initial sample temperature was ±20 °C. The pH value was adjusted using 30% NaOH. The pH value = 8.90-9.25 was tolerated, because in accordance with [10] struvite has the lowest solubility at this interval. Also samples were tested without pH adjustment. Stirring was carried out on a *Lovibond* 4-position mixing column at 250 and 150 RPM. Volume of the samples was V = 150 ml. First, PO_4^{3-} ions were added, stirring was done, pH was measured, and Mg^{2+} reagent was added, and the sample was homogenized. pH was repeatedly measured or adjusted by NaOH. Subsequently, RPM were reduced and the time of precipitation was measured from this moment. For samples at the temperature of t = 35 °C, the *Lavat Chotutice* magnetic stirrer *MM4* with the heating function was used and the temperature was measured by pH meter *WTW pH 330i* with a temperature measurement function. Upon expiration of the precipitation time the samples were subjected to separation. The tested separation methods were:

- Vacuum filtration on a Vacc-Space 30 apparatus through a KA-2 quantitative analysis paper (high speed) at a pressure p = -0.2 MPa (-3 psi) according to recommendations of [8].
- Centrifugal centrifugation of *Merci* type *NF 800* for 10 minutes at a frequency of 4 100 RPM according to recommendations of [6].

3. Results and conclusions

To find a suitable ratio of precipitating agents $Na_2HPO_4 . 12 H_2O$ and $MgSO_4$ in various stoichiometric ratios were dosed. Tested ratios of the reactants are given in Table 2.

Table 2. Tested weight ratios and appropriate concentration of ions.

Tested weight ratio	Mg^{2+} [$mg.l^{-1}$]	NH_4^+ [$mg.l^{-1}$]	PO_4^{3-} [$mg.l^{-1}$]
3.2 : 1.0 : 1.7	9 363	2 138	18 725
2.0 : 1.0 : 0.5	5 618	2 138	5 618
3.2 : 1.0 : 0.8	9 363	2 138	9 363
1.3 : 1.0 : 0,8	3 745	2 138	9 363

After the first experiments, we found that the samples could not be separated by vacuum filtration through a KA-2 filter paper for quantitative analysis. Therefore, the solid phase was separated using the second method. The first tested ratio was Mg^{2+}:NH_4^+:PO_4^{3-} = 3.2:1.0:1.7. Results obtained were: The NH_4^+ removal efficiency was approximately identical in case of 15 min. precipitation at a) pH = 6.9 (without pH adjustment), and b) pH = 9.0 (with pH adjustment). For the first case a) was NH_4^+ removal 77.3 % and 81.3 % for the second case b). However, for the sample b) with pH adjustment there were obtained lower residual concentrations of PO_4^{3-} and Mg^{2+} ions compared to sample a). These ions could be precipitated in the form of $Mg_3(PO_4)_2$. Residual PO_4^{3-} concentration was very high (equals 3 700 $mg.l^{-1}$). At weight ratio of 2.0:1.0:0.5 there was the influence of pH adjustment (mostly pH = 9.0), duration and temperature on ammoniacal nitrogen removal tested (see figure 2). Ions were removed from these samples in average of 56.4 %. In the samples where precipitation was using increased temperatures at t = 35 °C, no significant increase in ammonium ion removal efficiency was observed. On the contrary, prolongation of the precipitation time to t = 60 min. led to the slightly increased efficiency. Because the ammonia removals are noticeably lower due to small amounts of PO_4^{3-} ions added another essential ions ratio was tested. At ratio 3.2:1.0:0.8 the maximum NH_4^+ removal efficiency (82.5 %) was obtained in a sample after 15 minutes precipitation with pH adjustment to pH = 9 and raised temperature to t = 35 °C. According to the calculations of removed ions ratio, N-NH_4^+ should be removed from this sample from up to 80.0 % in struvite. For total NH_4^+ ions removal, see in figure 3.

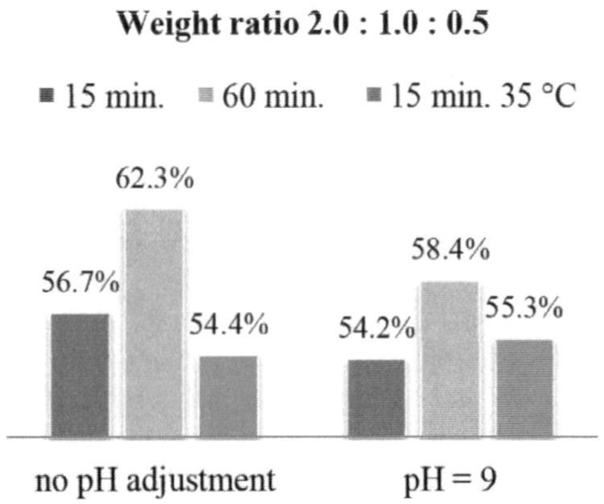

Figure 2: Total NH_4^+ ions removal. Tested weight ratio 2.0 : 1.0 : 0.5.

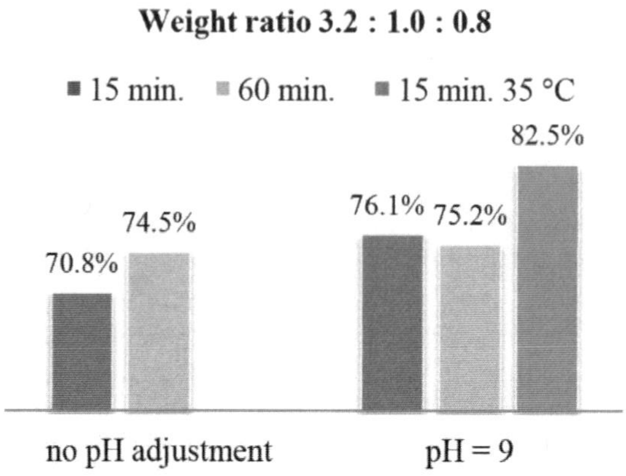

Figure 3: Total NH_4^+ ions removal. Tested weight ratio 3.2 : 1.0 : 0.8.

Due to the high residual concentrations of magnesium in acquired digestate liquor - c(Mg^{2+}) = 3 200 mg.l^{-1}, lower amount of Mg^{2+} ions added was tested along with the last tested amount of PO_4^{3-} ions. It means the ratio 1.3: 1.0: 0.8. Over 75.9 % of ammonium ions were removed in spite of low Mg^{2+} doses. Ammonium ions could be from 72.0% removed as struvite precipitate according to the calculations.

According to all the obtained results, the list of best removals in certain conditions was assembled (see table 3). One experiment was chosen from the list to be examined in detail.

Table 3. The best chosen removals of NH_4^+ ions and residual concentrations of ions.

Tested weight ratio	NH_4^+ Removal [%]	t [min]	pH [-]	t [°C]	Mg^{2+} [mg.l^{-1}]	PO_4^{3-} [mg.l^{-1}]	NH_4^+ [mg.l^{-1}]
3.2 : 1.0 : 1.7	81.3	15	9.0	20	1 980	3 700	400
2.0 : 1.0 : 0.5	62.3	60	no	20	2 320	160	805
3.2 : 1.0 : 0.8	82.5	15	9.0	35	3 220	192	375
3.2 : 1.0 : 0.8	74.5	60	no	20	3 101	700	545
3.2 : 1.0 : 0.8	76.1	15	9.0	20	3 252	280	510
1.3 : 1.0 : 0.8	77.8	15	no	20	129	603	475
1.3 : 1.0 : 0.8	78.5	60	9.0	20	768	525	460

Chosen dose of reagents was in a ratio of 1.3: 1.0: 0.8. The precipitation lasted for 15 minutes and the pH value was not adjusted in advance. Input concentrations of used ions corresponded to 3 745 mg.l^{-1} Mg^{2+} ions and 9 363 mg.l^{-1} $PO4^{3-}$ ions. The precipitated sludge after separation was analysed for the presence of struvite by X-Ray Powder Diffraction (XRD), for the dry matter content and sum of SO_4^{2-} ions. Total dry matter was 15.76 % and the sum of SO_4^{2-} in the sludge was 35 076 mg.kg^{-1} of dry matter. According to XRD the sludge contained the most the precipitating agent as a mineral in its hydrated form - $MgSO_4 \cdot 6\ H_2O$ (Precisely 37.29 %). Even though the struvite was represented in this sample less than was expected, namely 33 %, N-NH_4^+ was removed from the sample with an efficiency of η = 77.8 %. N-NH_4^+, which was not detected as struvite, may be present in an amorphous form in which cannot be distinguished by XRD analysis. Other minerals monitored in a sludge were:

Quartz - SiO_2 (15.96 %), Calcite - $CaCO_3$ (10.26 %), Brucite - $Mg(OH)_2$ (3.01%) and Sylvite - KCl (0.49 %). Parameters of supernatant liquor gained after precipitation and centrifugation are in table 4.

Table 4. Parameters of supernatant liquor gained after precipitation.

Parameter	COD [mg.l^{-1}]	NH_4^+ [mg.l^{-1}]	N-NH_4^+ [mg.l^{-1}]	PO_4^{3-} [mg.l^{-1}]	Mg^{2+} [mg.l^{-1}]	Ca^{2+} [mg.l^{-1}]	Na^+ [mg.l^{-1}]	SO_4^{2-} [mg.l^{-1}]
Value	14 200	475	369	603	129	13	2 184	9 275
Removal	-58.2 %	-77.8 %	-77.8 %	-5.78 %	-5.14 %	-95.3 %	+90.2 %	-34.7 %

4. Conclusions

The aim of the research was to verify the possibilities of struvite precipitation in the liquid phase of the digestate using $MgSO_4$ reagent, which is not a commonly used reagent according to summary in [6]. The experiments proved that struvite was able to precipitate at various Mg^{2+}:NH_4^+:PO_4^{3-} ratios, whereby the stoichiometric ratio of 1.0:1.0:1.0 is not necessary to keep up, since the full use of the introduced ions will never be achieved. Furthermore, it was found that the combination of increased pH to a level of 9.0 and increased temperature can achieve the maximum removal of NH_4^+ ions. However, almost identical results under lower energy demands can be achieved by appropriate ion ratio. In general, it is not advisable to use $MgSO_4$ as a source of Mg^{2+} ions, because the need of high concentrations of SO_4^{2-} permits a subsequent use only with further treatment.

Acknowledgments
The paper was elaborated with the support of the Ministry of Agriculture within the Project QJ1320159 “Research in processing, use and disposal of waste products from biogas stations”.

Reference list

[1] Dohányos M 2000 *Anaerobní čistírenské technologie* **1** (Brno: NOEL)
[2] Mužík O and Slejška A 2003 Možnosti využití anaerobní fermentace pro zpracování zbytkové biomasy
[3] Varga L and Lošák T Použitie digestátov (biokalu) z BPS pro hnojení poľnohospodárskych plodin a možnosti ovplyvnenia rozkladných procesov pomocou bakteriálnych preparátov *Naše pole* **17** 32-33
[4] Siciliano A and De Rosa S 2014 Recovery of ammonia in digestates of calf manure through a struvite precipitation process using unconventional reagents *Environmental Technology* **35** 841-850
[5] Evans T D 2007 Recovering ammonium and struvite fertilisers from digested sludge dewatering liquors *Proc. IWA Specialist Conf. Moving Forward–Wastewater biosolids sustainability*
[6] Uludag-Demirer S, Demirer GN and Chen S 2005 Ammonia removal from anaerobically digested diary manure by struvite precipitation *Process Biochem* **40** 3667-3674
[7] Pitter P 2009 *Hydrochemie* **4** (Praha: VŠCHT)
[8] Lhotský O, Zápotocký L, Šváb M and Hudcová T 2013 Srážení struvitu pro recyklaci nutrietů z odpadních vod: Vliv pH a kinetika procesu v modelových systémech *Wasteforum* **40** 120-126
[9] Sýkorová E, Wanner J and Beneš O 2014 Analysis of Phosphorus Recovery by Struvite Precipitation from Sludge Water in Selected Wastewater Treatment Plants *Chemické listy* **108** 610-614
[10] Nelson N O, Mikkelsen R L and Hesterberg D L 2003 Struvite precipitation in anaerobic swine lagoon liquid: effect of pH and Mg:P ratio and determination of rate constant *Bioresource Technology* **89** 229-236
[11] STUMM W and MORGAN J 1996 *Aquatic chemistry: Chemical Equilibria and Rates in Natural Waters* **3** (New York: John Wiley & Sons)

Algae and their biodegradation effects on building materials in the Ostrava industrial agglomeration

H Vojtková[1]

[1]VŠB – Technical University of Ostrava, Institute of Environmental Engineering, Czech Republic

Email: hana.vojtkova@vsb.cz

Abstract. Microorganisms cause changes in the building stone, which reduce its usable life and reliability. Microalgae make important parts of the biodegradation consortia of microorganisms on the surface of building materials. Via their metabolites, microalgae affect the stability of mineral components and thus lead to the material destruction. The aim of the paper was to identify aerophytic microalgae on the surface of engineering structures in the Ostrava agglomeration, and to describe the basic interactions between such microorganisms and the building materials, which may lead to the destruction of the materials.

1. Introduction

The investigations of microflora abundant on the building materials show that microbial communities of bacteria, algae, blue-green algae and microscopic fungi develop on the surfaces of structures, which may form more complex consortia (biofilms) and form symbionts of higher orders (lichens). The colonisation of structures by such organisms is in opposition with the human efforts to maintain structures free of damage, either from the constructional or aesthetic point of view. It has been established that the action of microorganisms with the building materials leads to biodeterioration changes that reduce their lifespan and reliability [1,2].

Microalgae represent polyphyletic microorganisms that have adapted to wide spectra of the environments, in which in dependence on various conditions these alter their morphology and physiological strategy of survival. Despite the fact algae are primarily aquatic organisms, they have managed to well adapt to terrestrial conditions, and nowadays are considered to be cosmopolitan organisms as they colonise all types of environments [3]. The dominant percentage of aerophytic species of microalgae belongs among *Chlorophyta*; the algae have adapted to grow in aerobic conditions on the morphological as well as physiological levels [4]. Aerophytic algae are able to grow in different environments, for example in sites with extreme pH values or temperatures – algae have been documented for their capacities to tolerate temperature changes in the range from 0 to 85 °C [5], [6]. Urban agglomerations with built-up areas may be, to a certain extent, considered extreme environments where algae find habitats with extreme temperature conditions resulting from large artificial structures, especially in the summer. However, the pH value is also important for the occurrence of algae on manmade structures. In dependence on the species, its optimal value may range from 3.5 to 9 [7]. In general, algae find pH over 12 difficult to tolerate. In case of accelerated carbonation of the building materials due to surplus of CO_2 in the atmosphere, the pH value falls below 9, and the onset of algae on the structure surface is very probable. As a result, green algae may be found particularly on older, wet facades with pH around 7, while cyanobacteria grow better on alkaline surfaces.

The major danger for the building materials caused by the abundance of algae lies in their photosynthetic production of carbon dioxide that chemically affects the components of the building material. At increased amounts of aggressive carbon dioxide in the environment due to the industrial emissions, CO_2 reacts with calcium hydroxide from cement sealant under the formation of $CaCO_3$ and water; the action of water causes the dissolution of $CaCO_3$ and in the reaction with other molecules, CO_2 forms hydrogencarbonate complexes [8]. It has been established that microalgae also produce organic acids that disturb the building substrate and thus enable the development of other microorganisms responsible for biodegradation – bacteria and fungi. These are often sources of changes in the aesthetic and mechanical properties of building materials. Algae as organisms that combine water cause physical corrosion of such materials via penetrating into the porous systems of building materials and contribute to the formation of micro-fissures. Under higher moistures, microalgae increase their volumes and erode the surrounding material by swelling pressure. Along with the impurities from the atmosphere in towns (soot, dust particles, microorganisms, etc.) they form a mucous bio-layer of the material surface that further supports the retention of water. In addition, algae may also participate in the formation of a crust as they produce the so-called extracellular polymer substances (EPS), which have been identified on the surfaces of a range of urban structures [9,10]. These substances significantly affect the physicochemical properties of materials, e.g. building stone, and support the bacterial growth and activity leading to the release of inorganic substances useful for organisms in the same environment.

2. Methods and experimental

The samples were repeatedly drawn using sterile Pasteur pipettes (Fisher Scientific Inc., USA) from the surfaces of building materials of residential buildings within the City of Ostrava. The basic identification and determination were made using an Olympus CX41 light microscope and expert literature [11].

Table 1. Microalgae isolated of materials from building surfaces

Phylum *Chlorophytae* * species – name is currently accepted taxonomically [14]	Samples					
	1	2	3	4	5	6
Apatococcus lobatus (Chodat) Petersen		+			+	
Chlorella vulgaris Beyerinck [Beijerinck]	+	+	+	+	+	+
Chlorococcum infusionum (Schrank) Meneghini	+	+	+	+	+	+
Cosmarium undulatum Corda ex Ralfs				+		
Desmococcus vulgaris Brand	+		+			
Klebsormidium flaccidum (Kützing) Silva, Mattox & Blackwell		+	+			+
Monoraphidium griffithii (Berkeley) Komárková-Legnerová						+
Pleurococcus vulgaris Meneghini	+	+			+	+
Protococcus nivalis (Bauer) Agardh	+		+			+
Scenedesmus quadricauda (Turpin) Brébisson						+
Scotiellopsis terrestris (Reisigl) Puncochárová & Kalina		+				
Stichococcus bacillaris Nägeli	+	+		+	+	+
Trebouxia decolorans Ahmadjian			+			+
Trentepohlia umbrina (Kützing) Bornet	+	+	+	+	+	+
Ulothrix tenuissima Kützing	+				+	+

The samples come from the surface of the facades (plaster) of the panel buildings in the city districts of Ostrava (Ostrava-Poruba, Ostrava-Zábřeh, Moravská Ostrava.) The monitored buildings underwent revitalization in 2010-2012; within the framework of their modernization, the defects of the sealing

between the panels were removed, the damage of the panel parts and the thermal insulation of the perimeter skin was remedied. Although the external thermal insulating system (ETICS) with the use of facade polystyrene, mineral fiber boards and plasters on based of silicates and acrylate, at present, there is already an apparent growth of cyanobacteria and algae on the plaster of these buildings.
Algae microorganisms were identified using genomic DNA based on the DNA of a target DNA region of 18S rRNA gene [12] using eukaryotic standard primers 20F (5'-GTAGTCATATGCTTGTCTC-3') and 18L (5'-CACCTACGGAAACCTTGTTACGACTT-3'); according to the methodology published in Hamby et al. [13].

3. Results and Discussion

In the four samples of materials from building surfaces, there were identified 27 taxa, namely 8 taxa of *Cyanobacteria*, 15 taxa of *Chlorophyta*, and 4 taxa of *Chrysophyta*. The green algae *Chlorella vulgaris*, *Chlorococcum infusionum* and *Trentepohlia umbrina* were identified in all sampling sites (table 1).

The genus *Chlorella* was the most widespread, occurring on all buildings reported, represented by two subspecies and occurring on four different substrata.

The members of *Chlorococcum* genus were identified in all examined building substrata, including the newly insulated concrete-panel houses. Other abundant algae were the representatives of *Stichococcus* genus, and thus it may be assumed that *Chlorella*, *Chlorococcum* and *Stichococcus* are green algae that colonise the buildings in the Ostrava Region the most. This is in agreement with Ortega-Calvo et al. (1995), who state that the occurrences of *Chlorella*, *Chlorococcus*, *Klebsormidium*, *Pleurococcus* and *Trentepohlia* (figure 1) may be observed in the monuments in Europe, America and Asia, but due to their cosmopolitan distribution it is not possible to correlate the genera and specific substrata or climates [15,16].

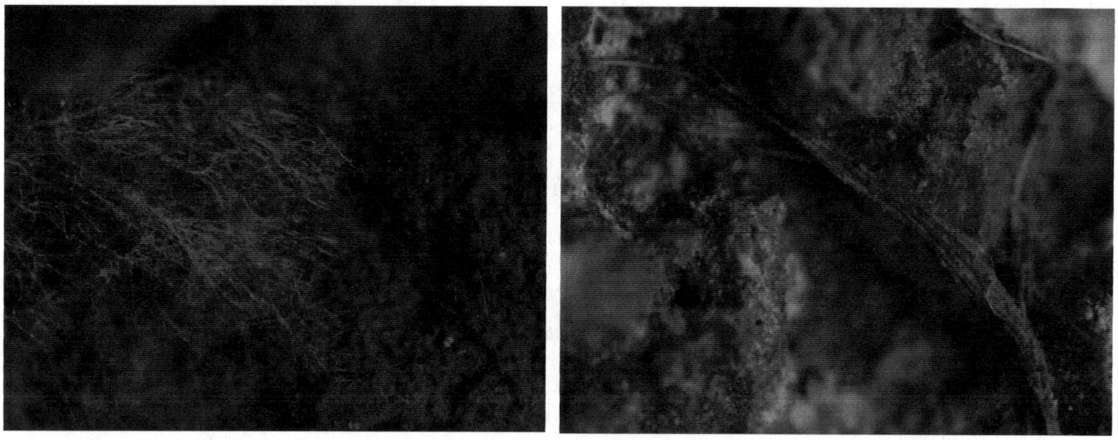

Figure 1. The progressive biodeterioration processes on the surface of building materials: algae constitute the medium for the growth of fungi that pass through the building material by means of their hyphae. The hyphae of microscopic fungi on the base comprising of *Trentepohlia* sp. algae (figure on the left); the figure on the right shows the hyphae penetrating the eroded building material (SEM coloured microphotography, photo by author).

Microalgae of the genera Chlorella and Trentepholia are also represented among the Chlorophyte genera that colonise stone substrata in the Mediterranean Basin. The occurrence of Trebouxia and Trentepohlia indicates that these microalgae may be involved in the lichenisation leading to colonisation by lichens [17]. In fact, the genus Trebouxia occurs in approximately 20% of all lichens and has rarely been found free-living. Regarding the endolithic growth of green algae, Trentepohlia, Chorella and Klebsormidium, they were found growing in monuments of Portugal [18].

Cryptoendolithic growth of Stichococcus bacillaris was also observed in granite of the Cathedral of Toledo, Spain [19].

4. Conclusion

Microalgae make significant parts of the biodegradation microbial consortia on the surfaces of building materials, which influence the stability of mineral components via their metabolites, and thus reduce their lifespan and reliability. Among the major determinants of the type and extent of microbial colonisation there are the mineral composition of the building substrata and their physicochemical properties. Nevertheless, the surrounding atmosphere contains high amounts of pollutants of different origins (especially air pollutants related to the industry and transport), which largely influence the course of the biodegradation process. The investigations of the microflora on the surface of building materials have shown that even in the conditions of industrial agglomeration communities of aerophytic microorganisms thrive, particularly bacteria, algae, blue-green algae and microscopic fungi that form biofilms and crusts on the surfaces.

In the environment of the Ostrava industrial agglomeration, there were identified microalgae on the building material surfaces, among which the species *Chlorella vulgaris*, *Chlorococcum infusionum*, *Stichococcus bacillaris*, *Pleurococcus vulgaris* and *Trentepohlia umbrina* may be considered to be ubiquitous colonisators of building materials. The industrial environment of Ostrava City, with its increased level of pollutants, is not a factor that determines algal growth on the structure surfaces. On the contrary, it may be a factor inhibiting their diversity, which needs to be confirmed by further expert studies.

Acknowledgments

This research was financially supported by Project for Specific University Research (SGS) No. SP2017/8 from the Ministry of Education, Youth and Sports of the Czech Republic & Faculty of Mining and Geology of VŠB – Technical University of Ostrava.

References

[1] Walters A H and Elphick J J 1968 *Biodeterioration of Materials* (London: Elsevier)
[2] Gaylarde C, Ribas Silva M and Warscheid T 2003 *Mat. and Struc.* **36** pp 342–352
[3] Karsten U, Schumann R and Mostaert A 2007 Aeroterrestrial Algae Growing on Man-Made Surfaces *Alg. and Cyan. in Ext. Env.* ed J Seckbach (Dordrecht: Springer) pp 583–597
[4] Hoffmann L 1989 *Bot. Rev.* **55** pp 77–105
[5] Tripathi S N, Tiwari B S and Talpasayi E R S 1990 *Ener. and Build.* **15** pp 499–505
[6] Benson E, Harding K and Day J G 2007 Algae at Extreme Low Temperatures *Alg. and Cyan. in Ext. Env.* ed J Seckbach (Dordrecht: Springer) pp 365–383
[7] Seckbach J, Chapman D J, Garbary D, Oren A and Reisser W 2007 Algae and Cyanobacteria Under Environmental Extremes *Alg. and Cyan. in Ext. Env.* ed J Seckbach (Dordrecht: Springer) pp 781–786
[8] Rovnaníková P, Rovnaník P and Křístek R 2004 *Stavební chemie: Modul 3 – Degradace stavebních materiálů a chemie kovů* (Brno: VUT – FAST)
[9] Kiessl K 1989 *Bauphysik* **11** pp 44–49
[10] El-Sheekh M M, Khairy H M and El-Shenody R 2012 *Ira. J. of Env. Heal. Sci. & Eng.* **9** 10
[11] Poulíčková A and Jurčák J 2001 *Malý obrazový atlas našich sinic a řas* (Olomouc: UPa)
[12] Haddad R, Alemzadeh E, Ahmadi A R, Hosseini R and Moezzi M 2014 *Ira. J. of Mic.* **6** pp 437–442
[13] Hamby R K, Sims L, Issel L and Zimmer E 1988 *Pla. Mol. Bio. Rep.* **6** pp 179–197
[14] Guiry M D and Guiry G M 2017 *AlgaeBase: World-wide electronic publication* (Galway: National University of Ireland) http://www.algaebase.org
[15] Scheerer S, Ortega-Morales O, Gaylarde CH 2009 Chapter 5 Microbial Deterioration of Stone Monuments—An Updated Overview *Adv. in App. Mic.* **66** pp 97–139

[16] Macedo M F, Miller A Z, Dionísio A and Saiz-Jimenez C 2009 *Microbiology* **155** pp 3476–90
[17] Chapman R L and Waters D A 2002 Lichenization of the Trentepohliales *Sym.: Mech. and Mod. Sys.* ed J Seckbach (Dordrecht: Springer) pp 359–371
[18] Pereira de Oliveira B, Miller A, Sequeira Braga M A, Macedo M F, Dionisio A and Silveira T 2008 *14th Int. Bio. and Bio. Sym. IBBS-14* (Messina: International Biodeterioration and Biodegradation Society) p 72
[19] Ortega-Calvo J J, Ariño X, Hernandez-Marine M and Saiz-Jimenez C 1995 *Sci. of The Tot. Env.* **167**, pp 329–341

IOP Conf. Series: Earth and Environmental Science **92** (2017) 012074 doi:10.1088/1755-1315/92/1/012074

Landscape management in an area affected by surface brown coal mining

J Vráblíková[1], E Wildová[1], P Vráblík[1] and M Blažková[1]

[1]Department of Natural Sciences, Faculty of Environment, J. E. Purkyně University in Ústí nad Labem

Abstract. The contribution summarizes results of a project concentrated on landscape management of an area affected by brown coal mining located in northern Bohemia (The Most basin) focusing on restoration and reclamation processes. It describes in particular the shares of individual types of reclamations in the area of interest. A strategic document that also supports landscape restoration in anthropogenically burdened regions was written within the project called "Restart" and the second part of the contribution is focused on its chapters which address this issue.

1. Introduction

The area of interest is the Most basin of 1420 km^2, located in the northern part of the Czech Republic. Intensive extraction of brown coal has influenced a wide area, first of all the Chomutov, Most, Teplice and Ústí nad Labem districts. Those districts were the most burdened regions in Central Europe as part of the so called “Black Triangle”. The greatest problem in the entire “Black Triangle” area was air pollution, caused by thermal power stations, heating plants, open surface coal mines and chemical plants. The long-term burden on the environment in the past was the cause of its deteriorated condition, with some of its components also reflected in the poor health condition of the inhabitants. The area is severely damaged, not only from an environmental point of view (landscape degradation, air pollution etc.), but economic and social problems also occur there. Landscape management of the area is based on restoration and reclamation processes, which ensure integration of the damaged landscape back to the environment. The article is generally focused on the shares of individual landscape recovery methods, which help with the resocialization of the area. [1].

The contribution also describes the “Restart” project, which is concentrated on financial and systematic support for regions affected by coal mining, and on its connection with restoration processes. Official title is “ The project for restructuring of the Ústí nad Labem (ÚR), Moravian-Silesian (MSR) and Karlovy Vary (KVR) regions “ and is being implemented in collaboration with the Office of the Government Commissioner, the Ministry of Industry and Trade and the Ministry of Regional Development, under the patronage of the Government of the Czech Republic.

2. Methods

Methods of research consisted firstly in studying available materials that are connected with landscape recovery in the model area. The main documents were Mining Yearbooks that are published by Czech Mining Authority and the Mining and Oil Industry Employers' Union [2].

Individual companies participating in the restoration of the landscape after mining in the Most Basin describe here the amount of completed reclamations, but not always with a distinction into individual types of reclamations (agriculture, forest, hydrological and other). Professional experience and results of several researches done by Faculty of Environment of the J. E. Purkyně University provided necessary information to analyze acreage of completed and planned reclamations in the model area. Second important document which had to be studied was "Overall Action Plan of the Restructuring Strategy for the Ústí nad Labem, Moravian-Silesian and Karlovy Vary Regions 2017 – 2018*"* [3].

Programmes concerning reclamation of the landscape affected by mining were chosen for further description.

3. Results

3.1. Landscape restoration in the model area

The reclamation of the landscape must be understood as the managed regeneration process of a landscape affected by mining and possibly also other human activity. Its aim is the restoration of the landscape's natural balance. It includes work of a technical nature (landscaping, stabilization measures, hydro-technical measures, etc.) and also of a biological nature (creation of agro-ecosystems, agricultural use, forest plantation, silvicultural care etc.). It is necessary to continue supporting revitalisation, i.e. functional integration into the landscape or a final treatment of the devastated area that will ensure the creation of an aesthetic landscape phenomenon and the restoration of the ecosystem's natural functions, while at the same time enabling the full utilisation of the area in accordance with the zoning plan [4].

The largest share within the scope of reclamation completed by 2016 is represented by the reclamation of 6,433.49 ha (46%) of forest. Next in line is agricultural reclamation at 4,150.88 ha (29%); however, land fertilisation during this process has been complicated. The area involving other forms of reclamation is constantly increasing (2,600.3 ha, or 18%), as the areas created adapt to society very well. That is why a significant expansion of these other areas is expected in the future. For now, the smallest share is represented by hydrological reclamation (986.56 ha, or 7%). However, with the end of open-cast coal mining, their share will increase thanks to the possibilities of utilising residual quarries (figure 1). The entire strategy for reclaiming the landscape below the Ore Mountains is based, in the long term, on the real possibilities in this area for creating conditions for this region to become, in its post-mining phase, a valuable recreational area, equipped not only for leisure-time activities, but also for a production capacity that is environmentally sustainable and undisruptive.

Within the scope landscape regeneration, it is necessary to apply the principle of integrated regional management, which ensures the involvement of all interested subjects, both from the state and private sectors, as well as the local inhabitants, who ensure that the newly created areas are subsequently utilised within the scope of reclamation and restoration. If the concept of sustainable development is adhered to within the scope of reclamation, whereby all of its pillars are taken into consideration, the newly created landscape will also be prepared for the possible consequences of global warming. Sufficient water and forests will create suitable conditions for housing, recreation and regional agriculture.

An important concept that links the ecological interests of an anthropogenically burdened landscape with economic and social aspects is its so-called resocialization [5] whose main aim is the return of people to the reclaimed and restored area.

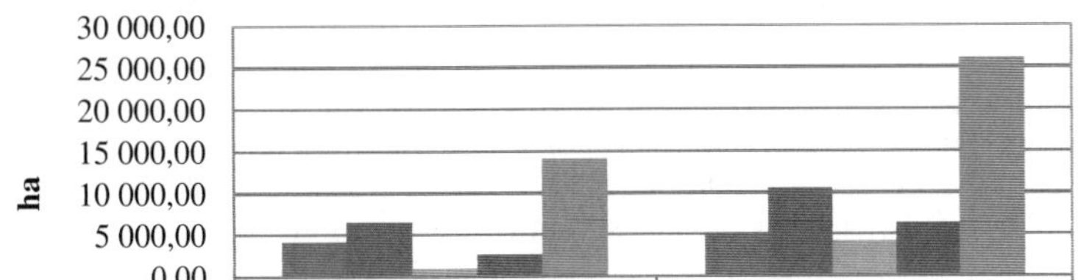

	Reclamations completed in 2016 [ha]	Area after termination of mining [ha]
agriculture	4 150,88	5 129,08
forest	6 433,49	10 546,81
hydrological	986,56	4 153,41
other	2 600,30	6 263,91
total	14 171,23	26 093,21

Figure 1. Reclamations completed in 2016 in the model area [ha][2].

3.2. Programmes supporting landscape management within the "Overall Action Plan of the Restructuring Strategy for the Ústí nad Labem, Moravian-Silesian and Karlovy Vary Regions 2017 – 2018"

The government decided, via Resolution no. 826 of the October 19th 2015, to economically restructure the Ústí nad Labem, Moravian-Silesian and Karlovy Vary regions, thereby showing that it is aware of the problems in these structurally affected areas, and committed to stop them lagging behind, and to kick-start their prosperity in 7 different areas: Entrepreneurship and Innovation (Objective: growing businesses capable of dealing with changes in global markets), Direct Foreign Investment (Objective: more direct foreign investment with higher added value), Research and Development (Objective: research and development with greater benefits for the economy), Human Resources (Objective: competent people for industry, services and public administration), Environment (Objective: revitalised and regenerated areas for better entrepreneurship and healthier lives of the inhabitants), Social Stabilisation (Objective: removal of barriers to development connected with social instability) and Infrastructure and Public Administration (Objective: higher quality infrastructure and public services). Via the Office of the Government, the Ústí nad Labem, Moravian-Silesian and Karlovy Vary regions requested financial and systemic support for specific measures which will help to restart these regions' economies. What specifically will happen in the Moravian-Silesian, Ústí nad Labem and Karlovy Vary regions is formulated in the so-called Action Plans. These programmes are based on the strategic framework and relevant discussions, and include the specific measures proposed in the particular localities. The first Restart Action Plan focuses on support for the creation of new activities in the areas after mining ceases. The objective is a regenerated landscape which is ready for further use and resettlement, including the maximum utilisation of its new potential, whether it is business opportunities, new houses, lakes, the option of pumped-storage power stations etc. [6].

The Action Plan for 2017 - 2018 is the first implementation document of the economic restructuring process of MSR, ÚR and KVR. Therefore, the draft Action Plans are the result of broad discussion at many levels, and among many participants from various sectors. The measures represent consensus among these participants on activities and programmes which will begin to improve the situation in structurally affected regions, and which at the same time can soon begin to be implemented. The Action Plans are divided according to the 7 pillars of the Strategic Economic Restructuring Framework. Within every pillar, the measures are divided according to the strategic objectives to which they contribute. Some measures can contribute to multiple strategic objectives or, in exceptional cases, also to the strategic objectives of multiple pillars. Each pillar sets forth so-called "supra-regional" measures, i.e. measures which are common for all three structurally affected regions, and for both Action Plans. As a rule, these measures take the form of programmes in which regional players will apply for support. In some cases, these programme measures take the form of specific invitations (within the scope of existing across-the-board programmes), which will be announced for structurally affected regions, or will be focused on their specific problems. Furthermore, some pillars also set forth so-called "regional" measures, which are specific for the relevant region or take into account its particular needs. These are either significant actions with a region-wide impact, or a mutually (factually, time) interconnected set of sub-interventions/projects, where only the implementation of the entire integrated set of interventions linked to a certain area will bring about significant change in the region.

The Action Plans also contain measures which are not expected to commence this year or in the first half of next year, but whose implementation will be prepared so that they can be put forward for implementation in one of the subsequent Action Plan updates. These measures are desirable in the regions, or are regarded as beneficial for the restructuring, but their preparation requires more time, more complex negotiations or the reaching of a consensus on their content. Therefore, they are included in the Action Plans as important, but the Office of the Government Commissioner will be entrusted with their verification and possible further development, in concord with regional players and players at a national level. The revitalisation measures in the Ústí nad Labem region, which relate to the regeneration of the area after mining, are addressed by the Environment pillar. In the Overall Action Plan, a chapter is devoted to this issue in more detail. The sub-chapter "Revitalise and regenerate an area heavily affected by mining and industrial activity" is directly related to support for the regeneration of the model area. The sub-chapter includes several points.

3.3. Acceleration of a programme to resolve the remediation of ecological damage - obligations arising from privatisation contracts (competence of the Ministry of Finance of the Czech Republic)

This measure represents an appeal to accelerate the fulfilment of obligations from privatisation contracts. Its objective is to reduce the extent of the ecological damage in MSR, ÚR and KVR, which the state committed to remove at the start of the 90s. This ecological damage restricts regional development, endangers the environment, and prevents companies from utilising their assets. In the model area, this measure relates to the remediation of the landfill area in the Litvínov locality which belongs to UNIPETROL, a.s., ecological contract number 14/94 concluded between UNIPETROL, a.s., and the Ministry of Finance of the Czech Republic.

3.4. Reconfiguration and strengthening of the programme for the remediation of ecological damage financed at the Operational Programme Environment level, which is designated for towns and municipalities - obligations beyond the scope of privatisation contracts (competence of the Ministry of the Environment).

The objective of the programme is to reduce the environmental risks connected with the existence of ecological burdens at the level of the population and individual components of the environment. At the same time, conditions will be created for the reduction of economic and developmental disparities in the region, in which ecological burdens play a part. The interventions will be linked primarily with the Operational Programme Environment, whose objective is:

- inventorying of contaminated and potentially contaminated locations, categorisation of the priorities of the contaminated locations according to severity;
- implementation of research works (including additional researches), risk analyses, and remediation of severely contaminated localities.

3.5. Prepare a draft programme for the revitalisation/resocialization of areas and regions which were previously recultivated after mining.

The objective is to improve the image of the regions, the quality of the environment and economic and social conditions, particularly in the industrial agglomerations of the Moravian-Silesian, Ústí nad Labem and Karlovy Vary regions. The prioritisation of natural successions during reclamations is suitable for smaller localities, and cannot be applied to already approved general restorations and reclamations, where it is necessary to proceed in accordance with the provision of "Mining Legislation", which is not familiar with the concept of natural succession. Individual activities will be assessed in terms of respect for the protection of nature and the landscape. The entire implementation period of this programme, 2019 - 2025, is expected to cost 2.7 billion crowns.

3.6. Analyse existing programmes in the area of the environment, and existing measures for the protection of the environment.

A healthy environment is one of the basic conditions for the economic restructuring of all three regions. At the same time, it is an indirect condition for the implementation of all measures aimed at hiring qualified workers, attracting investors to the region, and stabilising the social situation. The task is to analyse existing programmes in the area of the environment, and existing measures for the protection of the environment, on the basis of the following breakdown:

- for these programmes, propose ways of taking the specifics of structurally affected regions into account;
- for problem areas which will not be sufficiently covered by existing programmes and measures, prepare specific target-oriented programmes and measures to improve the air quality in towns and municipalities [3].

4. Conclusion

An area affected by coal mining has its own specifics, and its reintegration into the surrounding environment is a complicated process. The model area of northern Bohemia is a good example of successfully integrated anthropogenic creations into natural landscape. Landscape management is now concentrated on forest reclamations that are supplemented mainly by agriculture reclamations. In the future, hydrological and other reclamations will be expanded, as it will allow resocialization of the area. In this respect, the "Restart" project is an umbrella programme for regions which are dealing with this issue, and helps them to improve their situation from a social, economic and ecological perspective. An Action plan written within the project is an important document as it serves as a conceptual tool that supports sustainable development of the area.

Acknowledgement

This article was supported by project QJ1520307, entitled "Sustainable Forms of Management in an Anthropogenically Burdened Region." This project was carried out with financial support from state budget resources through the KUS program, the Ministry of Agriculture of the Czech Republic.

References

[1] Vráblíková, J et al. 2016 *Pro. Rep. C 004 „Area Analysis in Podkrušnohoří landscape"* A study made within the project QJ1520307 Sustainable forms of management in an anthropogenically affected landscape J. E. Purkyně University in Ústí nad Labem p 295

[2] Czech Mining Authority and the Mining and Oil Industry Employers' Union *Mining year books 2005–2016* Ostrava Czech Republic

[3] Ministry for regional development and Ministry for industry and trade 2017 *Overall Action Plan of the Restructuring Strategy for the Ústí nad Labem Moravian-Silesian and Karlovy Vary Regions 2017 – 2018* p 237

[4] Dejmal, I 2007 *Demographical and social limits of development of Podkrušnohoří basin* p 24

[5] Vráblíková J, Vráblík P and Zoubková L 2014 *Landscape creation and protection* J. E. Purkyne University in Ústí nad Labem p 150

[6] RESTART 2016 *Restructuring of Ústecký, Moravskoslezský and Karlovarský region* https://www.restartregionu.cz/

Analysis of change of retention capacity of a small water reservoir

R Výleta[1], M Danáčová[1] and P Valent[1]

[1]Department of Land and Water Resources Management, Faculty of Civil Engineering, Slovak University of Technology, Radlinského 11, 810 05 Bratislava, Slovakia

Email: roman.vyleta@stuba.sk

Abstract. This study is focused on the analysis of the changes of retention capacity of a small water reservoir induced by intensive erosion and sedimentation processes. The water reservoir is situated near the village of Vrbovce in the Western part of Slovakia, and the analysis is carried out for a period 2008-2017. The data used to build a digital elevation model (DEM) of the reservoir's bed came from a terrain measurement, utilizing an acoustic Doppler current profiler (ADCP) to measure the water depth in the reservoir. The DEM was used to quantify the soil loss from agricultural land situated within the basin of the reservoir. The ability of the water reservoir to transform a design flood with a return period of 100 years is evaluated for both design (2008) and current conditions (2017). The results show that the small water reservoir is a subject to siltation, with sediments comprised of fine soil particles transported from nearby agricultural land. The ability of the water reservoir to transform a 100-year flood has not changed significantly. The reduction of the reservoir's retention capacity should be systematically and regularly monitored in order to adjust its operational manual and improve its efficiency.

1. Introduction

One of the biggest challenges in the operation of small water reservoirs is the problem associated with their siltation with sediments coming from agricultural land as a result of erosion and transport processes. The siltation of water reservoirs is a worldwide problem that significantly reduces their retention capacity, deteriorates their efficiency, and shortens the time of their operational use. In order to increase the lifespan of water reservoirs, to maximize their efficiency in terms of retention capacity, and their ability to maintain good quality of water, it is necessary to know the current state and the dynamics of the changes of their bathymetry. The quantification of sedimentation processes, and a regular monitoring of reservoir's capacity and the volume of bed sediments plays an important role in the process of estimating the dynamics of their siltation, and their ability to meet the tasks for which they have been built. From the practical point of view the monitoring of this dynamics could be used in designing various mitigation measures aiming to reduce the amount of sediments eroded from the basin and deposited in the reservoir. The realisation of various management measures in the basin, as well as optimizing operation of water reservoir it is also possible to reduce the increasing costs of reservoir operation and maintenance. The problem of siltation of small water reservoirs is a very active research task both worldwide [1,2,3] and in Slovakia [4,5].

The main objective of this study was to monitor the bathymetry of a small water reservoir using an ultrasonic Acoustic Doppler Current Profiler (ADCP), and to compare the results with the survey from

2008 in order to estimate the amount of bed sediments, and the intensity of its siltation. The results were also used to recalculate its actual retention capacity, and its ability to accommodate a 100-year design flood. Within the study the mean annual volume of soil loss from the reservoir's basin was also evaluated.

2. Methodology

The area of the interest is situated in the western part of Slovakia in a cadastre of the Vrbovce village. A multipurpose water reservoir Vrbovce was built at the Zápasečník Creek in the southeast part of the village in 1966. The area of the basin contributing to the water reservoir is 3.24 km^2 with an altitude range between 342 and 587 m a.s.l. The parameters of a 100-year design flood were estimated using a SCS-CN method with a peak flow of 4.3 m^3s^{-1} and a total volume of 17926.1 m^3 [6,7,8]. The basin of the reservoir is predominantly covered by arable land (42%), which is followed by meadows and pastures (31%), and broad-leaved forests (25%) (see figure 1). The predominant type of soil texture is loam with as much as 85% of the basin area belonging to this category. The area of interest is situated at the border between lowland and highland climate. The mean annual temperature in this region is 7.5 °C, with January and July being the coldest and warmest months respectively. The mean annual precipitation depth in this area ranges between 650 and 850 mm.

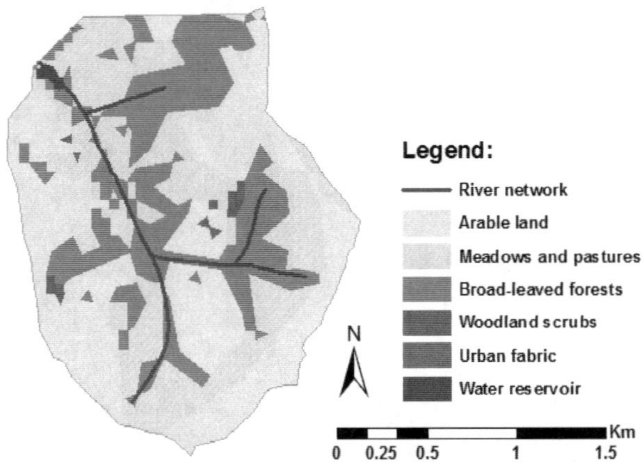

Figure 1. Land utilization of the Zápasečník Creek basin.

2.1. Determination of soil erosion

The effect of water erosion on the agricultural land is quantified by soil loss, which was estimated by the universal soil loss equation using the USLE2D model integrated in GIS environment. The method is based on the following equation proposed by [9]:

$$G = R \cdot K \cdot LS \cdot C \cdot P \tag{1}$$

Each member of the equation represents a factor influencing the soil loss. For each factor a 10x10 m raster was created in GIS to account for its spatial variability. The area of interest is located near the city of Senica, for which the value of the rainfall-runoff erosivity factor (*R*) of 28 $MJ.ha^{-1}.cm.h^{-1}$ has already been estimated [10]. The values of the soil erodibility factor (*K*) were based on the soil texture classification (see [10]). The *LS* factor, which takes into account the impact of slope length and steepness, was calculated as a mean value of the *LS* factor estimated by three methods of these authors: Wischmeier-Smith, McCool – moderate, Nearing and McCool by the USLE2D model [11]. The cover-management factor (*C*) was estimated for various types of land use including: arable land ($C = 0.18$ for winter wheat), meadows and pastures ($C = 0.005$), and broad-leaved forests ($C = 0.001$), [12]. The support practice factor (*P*) was determined according to the steepness of the ploughed slopes (contour ploughing) [9].

The Williams' model was used to estimate the quantity of transported sediments (soil yield), which is generally given in our physical-geographic conditions by the relationship:

$$SDR = 1,366.10^{-11} F^{-0,0998} RP^{0,3629} CN^{5,444} \quad (2)$$

where: *SDR* is the sediment delivery ratio, *F* is the basin area (km^2), *RP* is the relief-length ratio (m/km) calculated as a ratio of the difference between mean and minimum elevations and the longest flow path, and *CN* is the weighted value of the SCS curve number (*CN* = 67.5).

2.2. Monitoring of reservoir and data collection

In order to analyse the changes in the retention capacity of the water reservoir two monitoring surveys were conducted. The first one took place in 2008 when the reservoir was emptied and the sediments were removed. The second survey took place in 2017 during the full operation of the reservoir.

During the first monitoring in June 2008 a thorough geodetic survey of the reservoir and its near vicinity was conducted. The survey was focused on the obtaining of a precise location and elevation of the reservoir's dam, bed, surrounding terrain, technical objects and buildings. The planimetric survey (see figure 2) was performed using the method of precise tachymetry utilizing the Leica's TC 600 electronic total station. The crest of the reservoir's dam is situated at an altitude of 349.85 m a.s.l. and the emergency spillway at 348.25 m a.s.l.

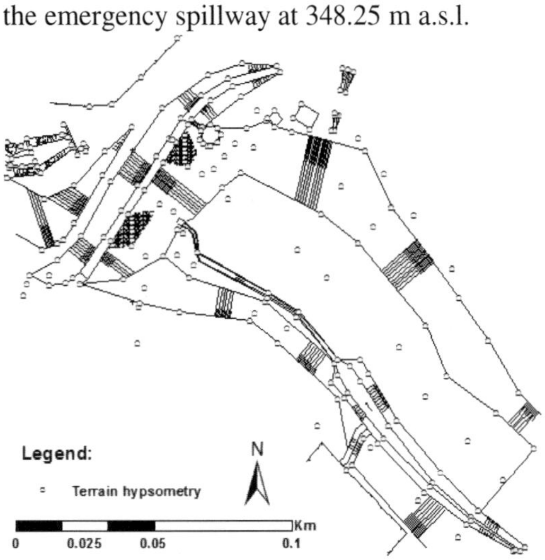

Figure 2. Planimetric and hypsometric survey of the water reservoir in 2008.

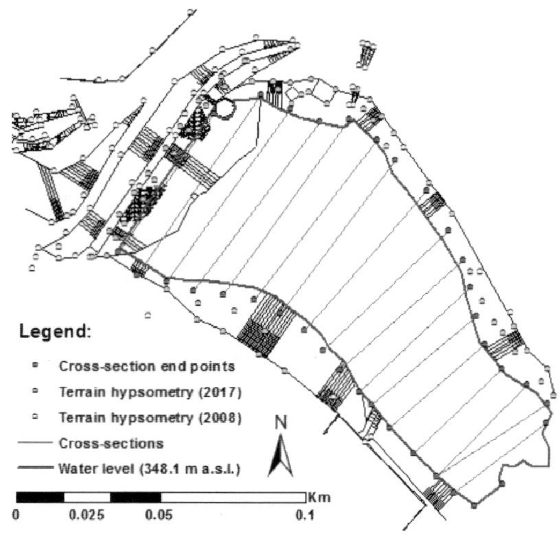

Figure 3. Planimetric and hypsometric survey of the water reservoir in 2017.

The second survey took place in April 2017 and was focused on the monitoring of the reservoir's bed elevation and surrounding terrain. The geodetic survey of the planimetry and hypsometry of the terrain (green points) and water level (blue line) was conducted using a GPS station (see figure 3). In order to monitor and evaluate the extent of the reservoir's siltation, 18 monitoring cross-sections were selected, each approximately 10 m apart. Within each cross-section an ADCP StreamPro, with a measuring range of 0.1 - 6 m, was towed from one bank of the reservoir to another to measure bathymetry. The interval of the measurements within each cross-section was set to 0.5 m. In order to reduce the measurement error two measurements were conducted for each cross-section. The end points of each cross-section (see figure 3 – red points) were fixed and geodetically surveyed using a GPS station (see figure 3). The elevation of the reservoir's bed was calculated by subtracting the bathymetry data from the elevation of the water level at the time of measurement (348.1 m a.s.l.).

The geodetic data from both surveys were used to create a DEM with a raster resolution of 0.5 m, which were further used in the analysis of the changes in the reservoir's retention capacity.

3. Results

The results of the calculation of the long-term annual soil loss and the corresponding volume of soil particles transported from the basin are shown in table 1. When converting the estimated soil loss from $t.year^{-1}$ to $m^3.year^{-1}$ the bulk density of soil of 1800 $kg.m^{-3}$ was used. The volume of sediments which could be deposited in the reservoir were estimated by multiplying the annual soil loss by *SDR*.

Table 1. Mean annual soil loss and a total volume of sediments washed out from the basin.

Object	SDR [-]	Area [km^2]	Average of soil loss [t.ha^{-1}.rok^{-1}]	Volume of soil loss [m^3]	Sediment loss [m^3]	Degree of Erosion
Basin	0.425	3.24	9.22	1657.5	705.4	moderate

The analysis of the changes in reservoir's bed bathymetry between years 2008 and 2017 is shown in figure 4. The results show that most of the reservoir's bed is a subject of deposition of eroded material (see figure 4 - red and yellow areas). The monitoring showed that the maximum depth of the sediments in was 2.25 m and was situated in the vicinity of the reservoir's dam, as well as near the inflow to the reservoir. At some places it was also possible to observe that the reservoir's bed was eroded with the maximum difference between its state in 2008 and 2017 of 1.5 m.

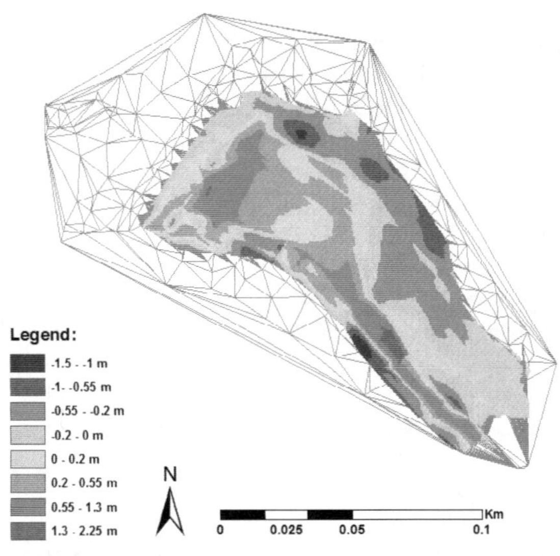

Figure 4. The depth of deposited and eroded particles in the reservoir's bed.

The impact of the reservoir siltation on its retention capacity was also evaluated for various values of water level (see table 2).

Table 2. Changes in the retention capacity of the water reservoir under various values of water level.

Year	Volume V (m^3)		
	Water level (348.1 m a.s.l.)	Emergency spillway (348.25 m a.s.l.)	Dam crest (349.5 m a.s.l.)
2008	17022.4	18309.1	34515.8
2017	16688.4	18064.4	34469.5
Difference	334.0	244.7	46.3

4. Conclusions

The results of the analysis of the siltation of the water reservoir showed that as much as 334.2 m^3 of eroded material was deposited in the reservoir between the years 2008 and 2017. The results also showed that the estimated retention capacity of the reservoir in 2017 has decreased by 2% compared to the state in 2008. The estimated volume of the reservoir with a water level at the elevation of emergency spillway was 18064.4 m^3, and at the water level at the elevation of dam crest it was 34469.5 m^3. Based on the volume of the 100-year design flood it is possible to conclude that the capacity of the reservoir in 2017 is sufficient to accommodate the whole volume of the design flood (17926.1 m^3). The mean annual soil loss from the basin of 9.22 $t.ha^{-1}.year^{-1}$ could be categorized as a moderate erosion intensity. In order to increase efficiency of the reservoir it is strongly recommended to apply erosion mitigation measures in some parts of the basin. The calculations showed that the total annual soil loss from the basin is 705.46 m^3.

The result of this study proved that the water reservoir Vrbovce is a subject to siltation. Therefore, in the near future a proper analysis of landscape structures as well as morphometric analysis of the basin will have to be performed. The observed reduction of its retention capacity calls for establishing a systematic and regular monitoring scheme, which would focus on observing the trends in its siltation. Only by implementing such monitoring schemes it will be possible to adopt and implement appropriate management measures, which would guarantee the functionality of the most important reservoir's tasks, such as flood protection and water supply.

Acknowledgments

The research was supported by the Slovak Research and Development Agency under the contract No. APVV-15-0497, VEGA scientific grant agency under the contract No. VEGA 1/0710/15 and European Commission - FP7 project RECARE, Contract No. 603498.

References

[1] Ceylan A, Karabork H and Ekozoglu I 2011 *Car. J. of Earth and Env. Sci.* **6** pp 15-24
[2] Yin X A, Yang Z F, Petts G E and Kondolf G M 2014 *J. of hyd.* **512** pp 379-387
[3] Fox G A, Sheshukov A, Cruse R, Kolar R L, Guertault L, Gesch K R and Dutnell R C 2016 *Env. Man.* **57** pp 945-955
[4] Kubinský D, Lehotský M and Weis K. 2014b *Car. J. of Earth and Env. Sci.* **9** pp 171-178
[5] Sočuvka V, Ivan P, Hlavčová K, Kohnová S and Velísková, Y 2016 *Acta hydrologica Slovaca* **17** pp 117-124
[6] Rutkowska A, Kohnová S, Banasik K, Szolgay J and Karabová B 2015 *J. of Moun. Sci.* **12** pp 533-548
[7] Randusová B, Marková R, Kohnová S and Hlavčová K 2015 *Int. J. of Eng. Res. & Sci.* **1** pp 34-40
[8] Szolgay J, Gaál L, Bacigál T, Kohnová S, Hlavčová K, Výleta R, Parajka J and Blöschl G 2016 *J. of Hyd. and Hyd.* **64** pp 367-381
[9] Wischmeier W H and Smith D D 1978 Predicting rainfall erosion losses: a guide to conservation planning *USDA, Agr. Handbook* p 537
[10] Alena F 1991 Erosion Control on arable land *Methodical tool, National Melioration Agency*
[11] Desmet P J J and Govers G 1996a *J. of Soil and Wat. Con.* **51** pp 427-433
[12] Malíšek T 1992 Optimal length of slope depending on water erosion *Sci. p.* (Bratislava)

Utilization of used motor oils as non-ionogenic collectors in the flotation process

J Závada[1], I Sobková[1], M Surmová[1], T Bouchal[1] and J Mudruňka[2]

[1]Institute of Environmental Engineering, VŠB-Technical University of Ostrava, Czech Republic
[2]Institute of Combined Studies in Most, VŠB-Technical University of Ostrava, Czech Republic

E-mail: jaroslav.zavada@vsb.cz

Abstract. This paper deals with the possibility of the utilization of used motor oils as non-ionic collectors in the flotation separation. Flotation tests were performed on a sample of coal from Poland. The criterion was to achieve the yield of concentrate over 70% and ash content below 10%. The results of laboratory investigations have shown that the application of used motor oils as flotation agents is possible.

1. Introduction

Flotation is an efficient technological process used for the treatment of mineral and secondary raw materials in many areas of industry that can ensure a rational and complex use of the resources. It allows us to obtain fine water-dispersed useful components. Flotation agents represent a very important part of the flotation process, and they should be economically affordable and environmentally friendly at the same time.

Bituminous coal flotation is dependent on the natural properties characterizing this coal, but also on the variable factors that can be adjusted so as to achieve the optimal course and results of the flotation. The natural properties of coal include mainly the degree of coalification, the size of the floated particles and the oxidation of the particle surface. The variable parameters include the suspension condensation, the quality and dosage of the flotation agents, the pH value, the suspension aeration, the degree of dispersion, or the flotation time. Other important factors may include, for example, the technological parameters of the flotation machines [1-7].

Various lubricating oils, fuel oil, kerosene, fractional oil distillation products, transformer oil, and others are usually used as the non-polar collectors. Their consumption in the flotation process ranges from 300 to 2000 $g \cdot t^{-1}$ of the raw material. A high consumption of these hydrocarbons is caused by the formation of a coarse oil layer on the surface of air bubbles as well. Oil is also insoluble in an aqueous environment and forms an emulsion [8].

Motor lubricating oil consists of two basic parts, namely the base oil and the performance additive mixtures. Mechanical impurities, present in oil as a result of its gradual wear, become an unavoidable component of oil [8]. The type of base oil determines the kind of oil (mineral, synthetic or semi-synthetic). The base oil is a mixture of chemical substances that form an oil carrier representing up to 90% of the oil volume. The additives that improve the properties of oil are viscosity-increasing substances, in particular in relation to temperature, antioxidants, detergents, inhibitors, dispersants, and substances reducing corrosion and oil foaming [9].

2. Chemical characteristics of the samples of used motor oils

The oil samples intended for the flotation tests were characterized by means of infrared spectroscopy with Fourier transform using Nicolet 380 FT-IR instrument and ATR crystal from ZnSe material. The spectra were measured within the range of 4000 to 650 cm^{-1}.

2.1. Used mineral motor oil (SAE 15W-40)

The SAE 15W-40 sample spectrum has shown a clear low absorption band in the area of 1710 cm^{-1} related to the vibrations of C = O bond in the carboxyl groups. Compared to the previous samples, there were no clear absorptions belonging to the vibrations of N-H and S = O bonds of nitration and sulphation products visible in the spectrum. In the 980 cm^{-1} area, there was a distinctive absorption band typical for zinc dithiophosphate anti-wear and antioxidant additive.

2.2. Used synthetic motor oil (SAE 5W-30)

The SAE 5W-30 sample contained oxidation products that were clearly visible in the spectrum in the range of 1710 cm^{-1}. A slight absorption in the range of 1630 cm^{-1} showed an increased content of nitration products and an absorption band of 1150 cm^{-1} an increased content of sulphation products. In the range of 1050-950 cm^{-1}, there was only a slight absorption, indicating the exhaustion of the zinc dithiophosphate additive (ZDDP).

2.3. Used semi-synthetic motor oil (SAE 10W-40)

The SAE 10W-40 sample contained a small amount of water, which was evident from a very mild absorption in the range of 3400 cm^{-1} and a more significant one in the range of 1600 cm^{-1}. The sample also showed a significant absorption in the range of 1740 cm^{-1}. This band belongs to a viscosity additive based on polymethacrylates. In addition, the band was boosted by a high content of ester oxidation products. The oxidation of oil was further demonstrated by a band in the range of 1710 cm^{-1} typical for the vibrations of C = O bond of carboxyl groups. A mild absorption band in the range of 1630 cm^{-1} belonged to the nitration degradation products, namely the vibrations of N-H bond of amines and amides. In the wave number of 1150 cm^{-1} range, there was a clear band characteristic for the vibrations of S = O group of sulphation products. From a mild absorption in the range of 1040 cm^{-1} and 1080 cm^{-1} belonging to the vibrations of C-O, we can deduce a small amount of ethylene glycol in the oil. In the area of about 800 cm^{-1}, there were absorption bands typical for diesel content. This is the area in which the vibrations in the molecule of naphthalene are absorbed.

3. Characteristics of bituminous coal sample

The results of an analysis of floated coal sample are presented in table 1.

Table 1. Characteristics of floated coal sample.

SN 441377	ASTM D 7582	ČSN ISO 1928	ČSN ISO 1928	ČSN ISO 19579	ČSN ISO 19579	ČSN ISO 562	ČSN ISO 29541
W_t^r (%) All original water	A^d (%) Anhydrous ash	Q_s^r (MJ/kg) Original combustion heat	Q_i^r (MJ/kg) Original calorific value	S^r (%) Original sulphur	S_m^r (g/MJ) Specific sulphur content	V^{daf} (%) Volatile combustible	C^d (%) Anhydrous carbon
12.40	11.14	24.82	23.61	0.46	0.19	40.6	68.884

4. Preparation of bituminous coal slurry sample

After transfer to a laboratory, the sample had natural moisture content, which is why it was dried in a dry kiln at 105 °C and subsequently milled in a vibrating mill and sorted to a grain size below 0.5 mm. The flotation tests were performed on the VRF-1 pneumatic-mechanical flotation apparatus.

5. Preparation of a flotation collector

In order to make it possible to efficiently dose the used motor oils into a flotation pulp, an oil / water emulsion had to be created first. The above specified types of oils were dosed into 300 ml of water in a high-speed mixer (according to the flotation test dosing), and they were subsequently mixed for 1 minute. The produced emulsion was prepared for agitation with the sample and for the flotation itself afterwards.

6. Flotation tests

The flotation tests taking advantage of used motor oils took place with the pulp concentration of 100 g/l, collector dose of 500, 1500 and 3000 g/t and foaming agent dose of 500 g/t. The agitation time of the pulp with the collector took 5 minutes and the agitation with the foaming agent 1 minute. The pulp aeration was 300 l/m^2min^1. The total flotation time was 12 minutes. The flotation test using Montanol 551 was taking place under the same conditions as in case of previous flotations with oils, only the agent dose was 500 g/t. This dose was chosen on the basis of the results of the authors who were also dealing with this issue and they saw this dose as optimal during bituminous coal slurry flotation. Unlike previous flotations, it was not necessary to use the foaming agent, since it is already contained in Montanol 551. The results of all flotations are shown in the graphs in figures 1–3.

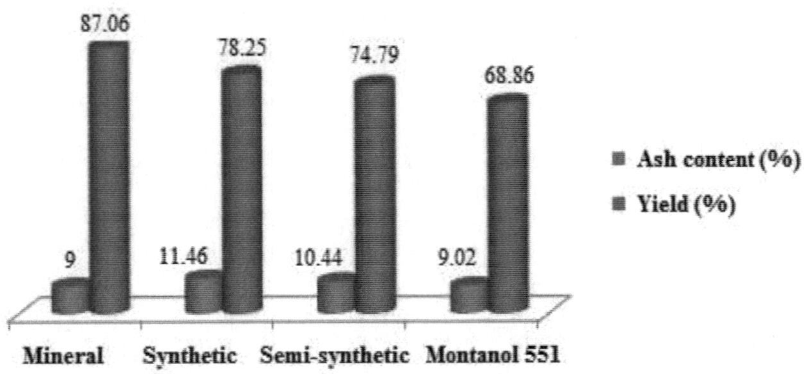

Figure 1. Comparison of the results of sample flotations of the individual oils with a dose of 500 g/t and Montanol 551.

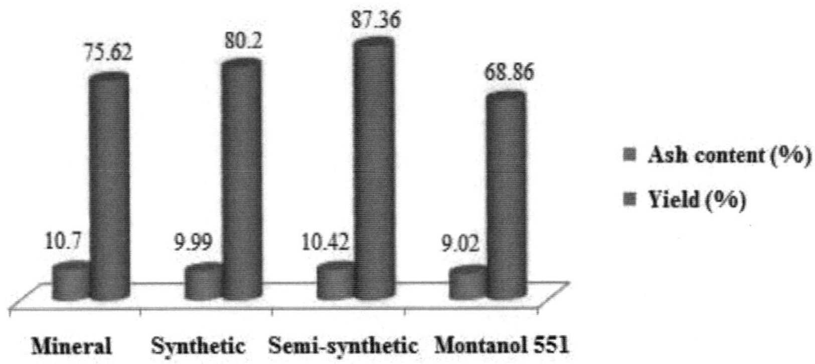

Figure 2. Comparison of the results of sample flotations of the individual oils with a dose of 1500 g/t and Montanol 551.

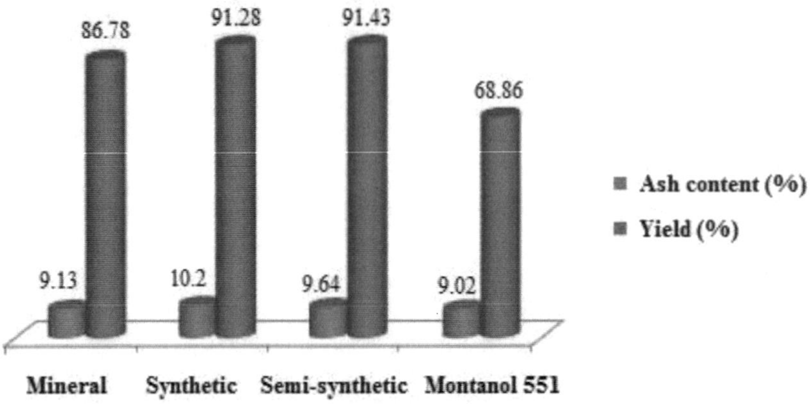

Figure 3. Comparison of the results of sample flotations of the individual oils with a dose of 3000 g/t and Montanol 551.

7. Discussion of the results

The lowest total ash content of all the performed flotations was achieved in case of the dose of 500 g/t of SAE 15W-40 mineral oil, where the determined ash content was 9% with the concentrate yield of 87.06%, which meets the 70% minimum condition. With a dose of 1500 g/t, the ash content was 10.7%, while achieving 75.62% concentrate yield. With a dose of 3000 g/t, the ash content was 9.13%, while achieving 86.78% concentrate yield. After 4 minutes, the foam already showed low mineralization.

The flotation results using SAE 5W-30 synthetic oil show that the lowest dose (500 g/t) achieved the ash content of 11.46% and the yield of 78.25%. With a dose of 1500 g/t, the ash content was 9.99% and the yield was 80.2%. The yield of the concentrate increased by 11.08% in comparison with the previous flotation when the oil dose was 3000 g/t, while the ash content increased (only slightly) by 0.21% to 10.2%. The first minutes resulted in rich and highly mineralized foam. After 3 minutes of flotation, this mineralization began to decrease visibly.

According to the results of the floatation using SAE 10W-40 semi-synthetic oil, it can be stated that the increasing dose of oil was gradually accompanied by the increasing yield of the flotation concentrate. When the doses were 500 g/t and 1500 g/t, similar ash contents of 10.44% and 10.42% were achieved. With a dose of 500 g/t, however, the yield of the concentrate was 74.79%, while a dose of 1500 g/t increased the yield to 87.36%. In case of this collector, the best quality of the flotation concentrate was achieved at 3000 g/t, where 9.64% ash content was achieved with the yield of concentrate of 91.43%. The flotation was similar to that of the previous agents. Compact foam formed first, and it began to disintegrate after 3 minutes and its mineralization was gradually decreasing.

In case of the flotation with commercially used Montanol 551 flotation agent, the yield of concentrate was 68.86% and the ash content was 9.02% with the dose of 500 g/t. When compared with the other collectors, the mineralization of the foam was noticeably reduced over a shorter period of time. After 6 minutes, the foam showed little mineralization.

8. Conclusion

Two types of flotation agents are used nowadays, Belgian Montanol and Czech Flotakol, and their price is high. The aim of the flotation tests was to verify the effectiveness of used motor oils as flotation agents on the basis of a comparison with commercially applied Montanole 551.

The SAE 15W-40 mineral engine oil was the best of the tested samples. With a dose of 500 g/t, the yield of the concentrate was 87.06%. Compared to the other flotations, it has reached the lowest ash content of 9%. It can therefore be said that this flotation test has achieved an excellent result, since the efficiency of this collector was higher than that of Montanol 551.

These tests have shown that used motor oils could be successfully applied as flotation collectors; in some cases they are even more effective than a commercially-used agent. They therefore appear to be an economically viable alternative to conventional collectors, but it is necessary to carry out an economic analysis to determine the actual savings.

References

[1] Piechaczková R 2013 *Flotation of black coal* (Ostrava: VŠB - TUO HGF)
[2] Somasundaran P, Zhang L and Fuerstenau D W 2000 *Intern. J. of Min. Pro.* **58** 85-97
[3] Laskowski S J 2001 *Elsevier* 368
[4] Rušaj J 2012 *Possibility to use liquid products from black coal pyrolysis in the flotation process* (Ostrava: VŠB - TUO HGF)
[5] Dube R 2012 *Collectors for enabling flotation of oxidized coal* (Lexington: University of Kentucky)
[6] Kawatra S K 2009 *Flot. fund.* 30
[7] Závada J, Bouchal T, Fryc J, Nadkanska H and Blazek V 2013 *Int. Multi. Sci. GeoConf. Sur. Geo. and Min. Eco. Man.* SGEM 665-672
[8] Kmeť S 1992 *Flotation* (Bratislava: Alfa) p 352
[9] Vlk F 2006 *Fuel and lubricants for motor vehicles* (Brno) p 400

Analysis and assessment of environmental impacts of small hydro power plant in Slovakia

M Zeleňáková[1], R Fijko[1] and I Remeňáková[1]

[1]Institute of Environmental Engineering, Faculty of Civil Engineering, Technical University of Kosice, Vysokoskolska 4, 042 00 Kosice, Slovakia

E-mail: martina.zelenakova@tuke.sk

Abstract. Environmental impact assessment (EIA) is an important process that, prior to approval of the investment plan, can provide a detailed examination of the likely and foreseeable impacts of proposed construction activity on the environment. The objective of this paper is to apply a specific methodology for the analysis and evaluation of the environmental impacts of selected constructions, namely, small hydro power plant, using matrix of impacts. This analysis method is intended not only to increase the clarity and precision of the evaluation process, but also to align it with the requirements of the environmental impact assessment system. This modification should improve the reliability of the environmental impact assessment, and could moreover also be applied to other infrastructure projects. Comparison of alternatives and designation of the optimal variant are implemented based on selected criteria that objectively describe the characteristic lines of the planned alternatives of activity and their impact on the environment. The use of proper EIA procedures can help the decision-makers to formulate proper activities based on qualified decisions. The designed project in Spišské Bystré, Slovakia is used as a case study to clarify and exemplify the methodology and techniques.

1. Introduction

The issue of greenhouse gas emissions in Europe is becoming more relevant. European Union countries, is continuously working to reduce these emissions. They have created a program on the use of renewable energy sources, which are committed to reducing the emissions by 20% to 2020. Work towards such a goal means that it is necessary to avoid deforestation, use of new technologies and use of renewable energy sources, either geothermal, solar, wind, or hydropower [1]. In Slovakia, hydropower is considered as the most common use of renewable energy to produce electricity. Based hydropower potential available for electricity generation is 7,361 GWh per year, currently is used at 57.5%. The share of large hydropower for electricity produced in 2002 was 92%, the proportion of small hydropower (SHP) only 8%. In Slovakia is a function of SHP large in addition to utilize primary hydropower potential for electricity generation. Small hydropower plants are involved in the utilization of the total available potential of 16% (1,220 GWh). Currently is used 24.5% (284.1 GWh) of the total available potential of SHP in Slovakia [2]. Small hydropower (SHP) are plants with an installed capacity of 1 MW to 10 MW and its impact on the environment is subject to assessment under Annex 1 of the Act no. 24/2006 Coll. the impact assessment on the environment, as amended in the Slovak Republic.

Assessment of the impact of the project on the environment is considered as a tool that minimizes the implementation of activities which could in any way negatively interfere with the environment and at the same time allows choosing the optimal solution from the proposed alternatives of the project implementation; the alternative with the smallest negative impact of a proposed activity on the environment. At present, several authors devote to the issue of environmental impact assessment in Slovakia particular [3,4,5,6]; in the Czech Republic [7]; in Poland [8]; in Romania [9,10].

Law no. 24/2006 Coll. on the assessment of impacts on the environment and on amendments to certain laws, which entered into force on 1 February 2006 in regulate all EIA process in the Slovak Republic. It implements Directive 2014/52/EC of the European Parliament and the Council amending the previous Directive 2011/92/EC on the assessment of the effects of certain public and private projects on the environment. EIA is inevitable tool for decision making with final aim of sustainable development of the society. According to Law no. 24/2006 Coll. the assessment of impacts on the environment and on amendments in the Slovak Republic the "Industrial installations for the production of electricity from water power" hydropower plants from 5 MW to 50 MW are under screening procedure and hydropower plants producing more than 50 MW are under compulsory assessment. In this paper the impact matrix has been used for environmental impact assessment of small hydropower plant.

2. Materials and methods

A lot of methods (tools) have been utilized over the last decades to meet the different actions required in the conduction of impact studies. These methods are describe for example in [11] or [12]. There is a need for a general and thorough approach to justifying, explaining, demonstrating, implementing, sampling, using, and creating real skills in analysis in any area of the human society. Such a need is embedded in human evolution, and its importance is evident every time one needs to develop forecasts of future courses of action [13,14]. All management decisions are concerned with the future, and everything about the future is, to a greater or lesser extent, uncertain [15,16]. The uses of risk identification, analysis and assessment in relation to the environment have broadened considerably in recent years.

Guidelines of European Commission [17] provide information on methods and tools that were selected from case studies and literature research. These generally fall into two groups:

- Scoping and impact identification techniques - these identify how and where an indirect or cumulative impact or impact interaction would occur.
- Evaluation techniques - these quantify and predict the magnitude and significance of impacts based on their context and intensity.

Potential (probable) impacts have to be assessed in terms of their fundamental nature, i.e., whether the impacts are:

- positive (useful), or negative (adverse);
- direct or indirect, or potentially related to that effect (primary, secondary, tertiary);
- in real time or delayed time;
- short term or long term;
- reversible or irreversible;
- local or strategic;
- physical, chemical, biological, microbiological, biochemical.

It is expected that EIA will continue to act as an effective tool to prevent the application of investments not only in Slovakia which by their degree of environmental damage many times outweigh their benefits [18,19]. In some cases, EIA is understood only as a "mirror" to comply with legal or technical standards, which is not sufficient for modern environmental planning. The assessment process should take into account the emotions and feelings of the public, stress factors, fear of risk and criteria reducing the quality of life. Generally, we can say that impact assessment in Slovakia is still based on professional principles, as is evidenced by EIA documentation on a standard or even high level.

The use of proper EIA methodologies and procedures can help the decision-makers to manage proper activities based on qualified decisions [20].
In the following the case study is presented.

3. Study area
The chosen site for the construction of small hydropower plant is located in the municipality Spišské Bystré (figure 1). The village is located at an altitude of 674 m, has a population of 2,394 and an area of 3,787 ha. Bystrá creek is a stream in the district of Poprad; it is right-hand tributary of the Hornád and has a length of 17 km. Currently, the creek in that area has character of unregulated water flow with an irregular trapezoidal profile width from 2.0 m to 6.0 m in the bottom and from 5.0 m to 10.0 m in water level. The proposal of small hydropower plant includes a modification of the bed of the watercourse, which consists mainly of fortifications of the channel cross section.

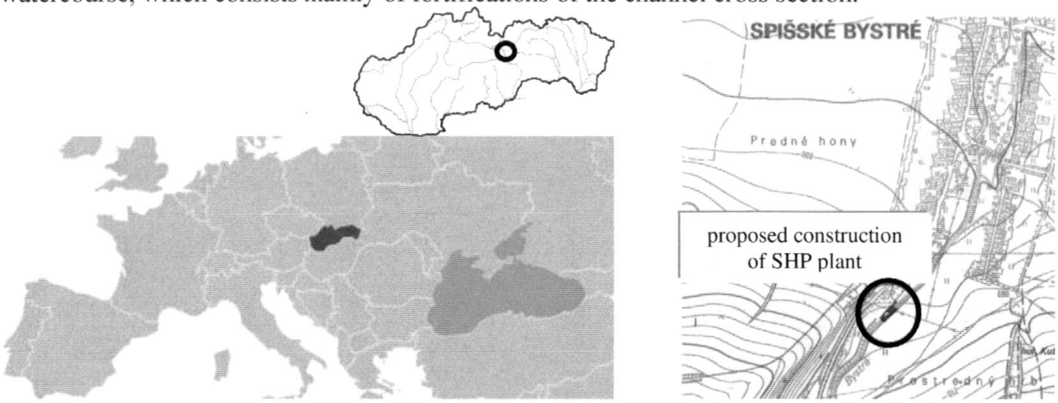

Figure 1. Location of the study area.

Three alternatives figure 2 of the proposed activity were assessed:

- Alternative 0 – the present state of the environment, no SHP plant will be constructed;
- Alternative 1 – construction of SHP plant;
- Alternative 2 – construction of SHP plant with bypass fishpass.

Assessment of the expected impacts of the proposed activity on the environment is presented in the following.

4. Results
The first step of the impact assessment of the proposed activity on the environment is the identification of the impacts on the partial components of the environment. When developing criteria and determine their importance we have placed emphasis on the nature, extent and duration of the effects. We have assigned values to individual consequences according to the proposed scale:

- The character of the impacts (CH): - negative, 0 no impact, + positive,
- The significance of the impacts (S): 1 insignificant, 2 significant, 3 very significant,
- The duration of the impacts (D): **0.5** short-term, **1** long-term.

In table 1 are identified the impacts on individual components of the environment. The values of the nature of the impacts are added according to the above proposed scale. We have assessed only impacts that are supposed to be occurring during the construction and operation of SHP plant.

The next step in EIA process is the selection of the optimal alternative by assessing the character, significance and duration of the impacts done by quantitative method (table 1). Multiplied values of impacts nature have been counted separately for each alternative of the proposed activity. The alternative that has reached the highest positive value can be considered as optimum.

IOP Conf. Series: Earth and Environmental Science **92** (2017) 012077 doi:10.1088/1755-1315/92/1/012077

Table 1. Quantification of impacts and comparison of alternatives.

	Alternative 0				Alternative 1				Alternative 2			
Impact on	CH0	S0	D0	*CH0 *S0 *D0*	CH1	S1	D1	*CH1 *S1 *D1*	CH2	S2	D2	*CH2 *S2 *D2*
population:												
noise	0			0	-1	2	1	-2	-1	2	1	-2
vibrations	0			0	-1	2	0.5	-1	-1	2	0.5	-1
dust	0			0	-1	3	0.5	-1.5	-1	3	0.5	-1.5
quality of life	0			0	-1	2	0.5	-1	-1	2	0.5	-1
economy	-1	2	1	-2	1	3	1	3	1	2	1	2
tourism, recreation	0			0	1	2	1	2	1	2	1	2
sport activities	0			0	1	2	0.5	1	1	2	0.5	1
water conditions:												
surface water flowing	-1	2	1	-2	1	2	1	2	1	2	1	2
surface water standing	-1	2	1	-2	1	3	1	3	1	3	1	3
ground water in	-1	2	1	-2	1	3	1	3	1	3	1	3
inundation	-1	2	1	-2	1	2	1	2	1	2	1	2
ground water in protected area												
soil:												
land occupation	0			0	-1	1	1	-1	-1	1	1	-1
water regime of soil	-1	2	1	-2	1	2	1	2	1	2	1	2
soil erosion	-1	2	1	-2	1	3	1	3	1	3	1	3
fauna and flora and their biotopes:												
fauna-mammals	1	2	1	2	-1	2	1	-2	-1	2	1	-2
fauna-birds	1	1	1	1	-1	1	1	-1	-1	1	1	-1
fauna-ichtyofauna	1	2	1	2	-1	3	1	-3	-1	1	1	-1
fauna-amphibians	1	2	1	2	-1	3	1	-3	-1	1	1	-1
flora-at construction	0			0	-1	3	0.5	-1.5	-1	2	0.5	-1
site	0			0	-1	2	1	-2	-1	1	1	-1
flora-at backwater												
SUM				-7				3				4,5

The highest value 7.5 has reached Alternative 2: SHP plant Spišské Bystré with bypass fish pass, therefore it can be considered from a comprehensive assessment of the environmental impact as optimal variant, with the less negative impacts to the environment (although this Alternative is the most costly). During construction is necessary to pay attention to the measures that reduce, respectively mitigate the adverse impact on the environment, including health of population. Measures need to be designed to prevent, eliminate, minimize and compensate the negative impacts.

5. Conclusion

In the paper knowledge on techniques and methods to assessment of interaction between humans, natural resources, and water projects are developed, distributed and used with aim to mitigate adverse impact and remediate the environment. The approach has a clearly defined methodology and original concept of solution.

The benefit of this paper is to highlight the importance of assessing the impact of construction on the environment in the planning phase. Eliminating the negative impacts of the construction on the environment is much more challenging than the implementation of preventive measures, and it is therefore necessary to consider and assess how the construction and operation of the proposed activities impact the environment in the planning phase of this activity.

Acknowledgments
This work was supported by VEGA project 1/0609/14.

References

[1] EC 2015 *Renewable resources of energy* http://www.ec.europa.eu/news/energy/120608_sk.htm
[2] STUBA 2015 *Hydropower energy* http://www.oze.stuba.sk/oze/vodna-energia/
[3] Zeleňáková M and Zvijáková L 2017 *Using Risk Analysis for Flood Protection Assessment* (Switzerland: Springer) p 128
[4] Pavličková K *et al* 2009 *Landscape ecology in environmental impact assessment* (Bratislava: Commenius University)
[5] Kočická, E 2007 *Environmental impact assessment (from theory and practice)* http://www.tuzvo.sk/files/FEE/dekanat_fee/11_Kocicka_AFE.pdf
[6] Majerník M and Bosák M 2003 *Environmental impact assessment* (Košice: TU)
[7] Říha J 2001 *Environmental impact assessment of investments. Multicriteria analysis and EIA* (Prague: Academia)
[8] Gałaś S, et al 2014 *Comparing the Phase of Screening in the Fields of Tourism and Recreation Water Management and Mining in the V4 Countries* (Krakow: AGH University of Science and Technology Press)
[9] Rojanschi V, Bran F and Diaconu G 1997 *Protection and environmental engineering* (Bucuresti: Editura Economică) p 368
[10] Iojă I C 2013 *Analysis and evaluation of the environmental situation* (Bucuresti: Editura Economică) p 183
[11] Canter L W 1998 *Methods for Effective Environmental Information Assessment (EIA) Practice*, ed Porter A L and Fittipaldi J J Environmental Methods Review: Retooling Impact Assessment for the New Century (Fagro: The Press Club)
[12] Canter L W 1999 *Environmental Impact Assessment* Environmental Engineers' Handbook (Florida: CRC Press)
[13] Bujoreanu I N 2012 Risk analysis series part one - why risk analysis? *J. Def Resour Manage* **3** pp 139-144
[14] Romanescu G, et al 2016 Water Quality Analysis in Mountain Freshwater: Poiana Uzului Reservoir in the Eastern Carpathians *Revista de Chimie* **67** pp 2318-26
[15] Costanza R and Daly H E 1992 Natural Capital and Sustainable Development *Conservation Biology* **6** pp 37-46
[16] Welsh T 2012 *Full Monte - The better approach to Schedule Risk Analysis User guide* (Barbecana: Tony Welsh)
[17] Walker L J and Johnston J 1999 Guidelines for the Assessment of Indirect and Cumulative Impacts as well as Impact Interactions *Environment, Nuclear Safety & Civil Protection* p 170
[18] Zvijáková L, Zeleňáková M and Purcz P 2014 Evaluation of environmental impact assessment effectiveness in Slovakia *Impact Assessment and Project Appraisal* **32** pp 150-161
[19] Zvijáková L and Zeleňáková M 2015 Risk analysis in the process of environmental impact assessment of flood protection objects (Prague: Leges) p 255
[20] Shah A, et al. 2010 Environmental impact assessment (EIA) of infrastructure development projects in developing countries *Inter. J of Sust. Deve.* **1** pp 47-54

Environmental impact assessment of heating power plant in Trebišov, Slovakia

M Zelenakova[1], V Ondrejka Harbulakova[1] and A Olejnik[1]

[1]Technical University of Kosice, Faculty of Civil Engineering, Institute of Environmental Engineering, Kosice, Slovakia

Email: martina.zelenakova@tuke.sk

Abstract. Environmental impact assessment is well known and often used process in all developed countries. This process identifies, estimates, assesses and provides information on negative and positive effects of the proposed project on the environment and health, and specifies in detail the measures to mitigate the possible negative effects before approving the project and its implementation. The assessment is done for at least two alternatives of the proposed activity in comparison with a zero alternative if no activity is done (present state is kept). In this paper proposal of heating system for Trebišov town in Slovakia is presented and the best alternative is chosen comparing environmental impacts of three alternatives. Evaluation was done using the method of the total indicator of environmental quality. Construction of biomass-fired power plant seems to be the best solution of heating for selected locality.

1. Introduction

Cardiff summit in 1989 created the platform of coordinated action aimed at protecting the environment. The European Commission has progressively focused its attention on the development and integration of environmental aspects into the sectorial policies of transport, energy, industry, agriculture, industry, internal economic policy and fisheries. The first step was decision taking of the first integration strategy in the energy sector adoption in 1999, which was modified in 2001 and presented in Gothenburg, Sweden before the European Council.

Another important document presented by the European Commission was "Green Paper on a secure, competitive and sustainable energy for Europe" [1], which was released in 2006. The aim of the EC was to create an integrated energy policy in Europe. In December 2008 a wide range of measures in the EU were adopted, which were aimed to reducing the impact of the EU states activities to global warming and also reducing negative effects on the global climate, while ensuring adequate and reliable energy supply.

To pursue these goals within a coherent long-term strategy, the EU has formulated targets for 2020, 2030, and 2050. The 2020 Energy Strategy defines the EU's energy priorities between 2010 and 2020. It aims to [2, 3]: reduce greenhouse gases by at least 20%, increase the share of renewable energy in the EU's energy mix to at least 20% of consumption, and improve energy efficiency by at least 20%.

This paper is based on legislative and methodological documents relating to the assessment process of environmental impact – Directive 2014/52/EU of the European Parliament and of the Council of 16 April 2014 [4]on the assessment of the effects of certain public and private projects on the environment (the Environmental Impact Assessment, or EIA Directive)[4] and exactly Act No. 24/2006 Coll. on Environmental Impact Assessment [5] – national Law in Slovakia. The proposed activity is construction of new biomass-fired power plant or reconstruction of old one gas power plant in Trebišov town district in Slovakia. The result of the research is a comparison of the proposed activities with the current state of the area using the total indicator of environmental quality method.

2. Material and Methods

2.1 Multicriteria analysis

The In the EIA process is always necessary to consider at least two alternatives of the proposed action: I) zero alternative – if there is no activity (current state of the environment) and II) alternatives of the proposed activity – variants of the activity that usually differ in locality (site of construction), used technology, time of implementation, etc.). The purpose should be to find the optimal solution, in practice a choice called "preferred option". The selection of the optimal alternative is enjoyed by various methods, particularly by multicriteria analysis [6].

Summary quality of the environment for the geographical regions is determined, by substantial (cardinal) properties of the individual components of the environment, the quality of which we can assess by the available analytical and diagnostic indicators. These partial indicators can create a catalog of indicators criteria (character) whose values are precisely determined analytically using the scientific bases of prognosis or experimental estimation [7].

Total indicator of environmental quality (*TIEQ*) method is used to determine the value of the most suitable variant of power plant construction in Trebišov town district.

Total indicator of environmental quality (*TIEQ*) method is used to determine the value of a comprehensive land use in terms of humanly influenced environment quality.
It is calculated according to (1):

$$TIEQ = U_j = \sum_{k=1}^{n} f_j\left(P_j^{(y)}\right) W_j^{(N)} \tag{1}$$

Where U_j is function of benefit, P_j is criterion, W_j is weight [7]. *TIEQ* structure is hierarchical, adaptive and allows to select the preferred option of a conventional set of alternatives or to give a preferential position of alternatives to a given set of criteria.

2.2 Power plant in the study area

Environmental impact assessment of the proposed activity – power plant for heating in Trebišov town in Slovakia according to Act 24/2006 Coll. as amended was implemented.
A brief description of the technical and technological solution is following. The central energy source of heating for Trebišov town has two possibilities:

- biomass-fired power plant;
- natural gas boiler.

Technological solution of biomass-fired power plant consists of boiler house, a handy storage of straw and technical annex. Individual objects are connected structurally and technologically. Objects of power supply are located in the northern part of the plot [8, 9, 10].

Atmospheric natural gas fired water tube boilers are used in old central boiler houses. These boilers using the combustion gas transfers the heat to the primary heat exchanger, which heats the heating water and consequently the water is cooled to about 120 °C. Then from the primary heat exchanger water is distributed to the secondary circuit and through pipes to the heated objects. At the secondary circuit is installed a heat meter which measures the heat consumption [11] [12]. Convention boilers are designed to produce dry combustion products. These products reach a temperature of 120 °C to 180 °C. The lowest temperature of the water entering the boiler is 60 °C.

3. Results

The assessment is made in two alternatives which are assessed in comparison with zero alternative:

- Alternative 0 – the zero alternative– if no activity is implemented.

- Alternative 1 - the environmental impact assessment of the biomass-fired power plant in Trebišov district.
- Alternative 2 - the environmental impact assessment of the modernized natural gas boiler.

Comparison of alternatives of proposed activity and the proposal of optimal alternative is based on multicriteria method. The first step of this evaluation is creating a set of criteria and determining their importance (weight) for the selection of the optimal alternative. We have defined a totally of nine criteria, which we have divided into four groups according to their character – economic, technical, ecological and social (table 1).

Table 1. Catalogue of criteria.

Criteria / Pi	Alternative 0 / A0	Alternative 1 / A1	Alternative 2 / A2
Total costs of construction/P1	0 € / 9 points	3 800 000 € / 3 points	2 600 000 € / 8 points
Costs of operation per year /P2	0 € / 8 points	535 000 € / 6 points	650 000 € / 5 points
Time of construction/P3	0 months/ 10 points	9 months/ 2 points	3 months/ 8 points
Land occupation/P4	0 m^2 / 8 points	1,981.65 m^2 / 3 points	1,249.35 m^2 / 6 points
Energy output of the boilers/P5	0 MW/ 0 points	14.01 MW/ 9 points	10.03 MW/ 8 points
Waste production/P6	no / 8 points	yes / 5 points	yes / 4 points
Emission production/P7	0% / 7 points	0% / 7 points	6.5% / 6 points
Job opportunities/P8	0 / 0 points	8 / 9 points	6 / 7 points
Extra boiler room construction/P9	yes / 0 points	no / 8 points	no / 8 points

For evaluation and comparison of the alternatives, the sets of nine criteria were established. The selected criteria are divided into qualitative and quantitative ones. A quantitative criterion includes the total cost of construction, annual operation costs, time of construction, land occupation and output of the power plants. A qualitative criterion includes waste production, emissions production, job opportunities and extra boiler room construction.

Under the category of technical and technological perspective three criterions belongs and under environmental and social two criteria belongs.

The ranking method was used to state the weights of criteria. The points (0-10) associated with each criterion were stated based on ten different experts' suggestions with the aim to get the most objective results.

4. Conclusion

The use of biomass in heating systems is beneficial because it uses agricultural, forest, urban and industrial residues and waste to produce heat and electricity with less effect on the environment than fossil fuels.

The ranking method was used to state the weights of criteria. The points (0-10) associated with each criterion were stated based on three different experts' suggestions with the aim to get the most objective results. In this evaluation the highest score is the best possible. Proposals were discussed with professionally qualified persons working in the field of environmental impact assessment as well as civil engineers.

Total indicator of environmental quality (*TIEQ*) was used for determination of the best alternative for heating system in Trebišov town. Using multicriteria analysis prove that the construction of biomass-fired power plant is the most suitable solution chosen from three assessed variants (no activity is implemented, biomass power plant and modernized gas boiler).

Acknowledgments
This work was supported by the VEGA under grant 1/0609/14.

References

[1] Šteffekova S 2011 Power engineering and its impact on environment in Slovak Republic http://www.euresp-plus.net/sites/default/files/Energetika_sektor_11.pdf

[2] EC.europa.eu 2016 Energy https://ec.europa.eu/energy/en/topics/energy-strategy

[3] European Parliament News COP 21 http://www.europarl.europa.eu/news/sk/news-room/20151013IPR97324/COP-21-Parlament-zadefinoval-ciele-pre-klimatick%C3%BA-konferenciu-v-Par%C3%AD%C5%BEi

[4] Directive 2014/52/EU of the European Parliament and of the Council of 16 April 2014 on the assessment of the effects of certain public and private projects on the environment (the Environmental Impact Assessment) 1-18.

[5] National Council of the Slovak Republic 2005 Act of Law No. 24/2006 from December 14th 2005 on Environmental Impact Assessment and on amendments to certain acts

[6] Majerník M Húsková V Bosák M and Chovancová J 2008 Methodology of the environmental impact assessment (Košice: TUKE) 212

[7] Říha J 1995 Environmental impact assessment of investments Multicriteria analysis and EIA (Prague: Academia) 1995 348.

[8] Ivaň M 2011 Mechanisms for recovery of wastes of category "O" Others, recovery of construction debris, paper, plastics, wood http://www.enviroportal.sk/sk_SK/eia/detail/ing-michal-ivan-land-servis-trebisov-zariadenie-na-zhodnocovanie-odpad

[9] Slámková M Garčárová M 2012 Encouraging the protection of Natura 2000 sites in integrating whole-system of ecological stability Regional territorial system of ecological stability of the Trebišov district www.minv.sk/?verejne-vyhlasky-7&subor=207452

[10] Banik and son inc. 2015 Differences between traditional and condensing boilers, http://www.banik.sk/rozdiely-medzi-klasickym-a-kondenzacnym-kotlom/

[11] Durikova K 2013 How does condensing boiler works http://mojdom.zoznam.sk/cl/10055/1351440/Oplati-sa-investovat-do-plynoveho-kondenzacneho-kotla

IOP Conf. Series: Earth and Environmental Science **92** (2017) 012079 doi:10.1088/1755-1315/92/1/012079

An Evaluation of the Efficiency of Different Hygienisation Methods

M Zrubková[1]

[1] Severomoravské vodovody a kanalizace Ostrava a.s., The Czech Republic

E-mail: marcela.zrubkova@smvak.cz

Abstract. The aim of this study is to evaluate the efficiency of hygienisation by pasteurisation, temperature-phased anaerobic digestion and sludge liming. A summary of the legislation concerning sludge treatment, disposal and recycling is included. The hygienisation methods are compared not only in terms of hygienisation efficiency but a comparison of other criteria is also included.

1. Legislation requirements concerning sludge treatment in The Czech Republic

The new Decree of the Ministry of Environment of the Czech Republic of 19 December 2016 No. 437/2016 Coll., on the conditions for using treated sludge on agricultural land, came into force in January 2017. This new Decree sets stricter microbiological criteria for the use of sludge in agriculture and new requirements such as a hygienisation efficiency assessment. In accordance with the new Decree, the wastewater treatment plant (WWTP) operator is obliged to evaluate the efficiency of the sludge treatment technology (hygienisation) in WWTPs put into operation prior to the effective date of this Decree by 31 December 2019 with the exception of those technologies that have produced sludge "category I" (limits listed in table 1) in the 24 months prior to the date of this Decree coming into force. Sludge treatment technologies in WWTPs put into operation prior to the effective date of this Decree which produce treated sludge and meet the limit values of microorganism indicators for "category II" (limits listed in table 1) will be considered as validated until 31 December 2019. After this date it will not be possible to use sludge classified as "category II" on agriculture land.

Table 1. Microbiological criteria for treated sludge for application to agricultural land during a transitional period (Annex No. 7 to Decree No. 437/2016 Coll.).

Microorganism indicator	Unit	Number of samples tested at each output check	Limit value (result/cfu[a])	
			Category I.	Category II.
Salmonella spp.	result in 1 g of dry matter	5	negative	
Thermotolerant coliform bacteria	cfu[a] in 1 g of dry matter	5	$< 10^3$	10^3 - 10^6
Enterococci	cfu[a] in 1 g of dry matter	5	$< 10^3$	10^3 - 10^6

[a]colony forming unit

Verification of technology efficiency has to be performed by taking 10 input samples and 10 output samples within 30 days, with a minimum of 48 hours between sampling. The difference between sludge contamination before and after treatment must be at least 10^5 cfu per gram of sludge for the microorganism Escherichia coli or Enterococci and the output parameters must be in accordance with the set limit values for microorganism indicators listed in table no. 2. Verification has to be performed after each technology change which will influence sludge treatment efficiency.

Table 2. Microbiological criteria for treated sludge for application on agricultural land (Annex No. 4 to Decree No. 437/2016 Coll.).

Microorganism Indicator	Unit	Number of samples tested at each output check	Limit value (result/ cfu[a])
Salmonella spp.	result in 50g	5	negative
Escherichia coli or Enterococci	cfu[a] in 1 g	5	4 $< 10^3$ 1 $< 5.10^3$

[a]colony forming unit

According to the requirements of the new Decree the operator must also take five samples from the beginning of 2017 as part of the control of the output from the technology, as opposed to one sample under the previous Decree.

In connection with the requirements of the new Decree, it is necessary to mention The Sewage Sludge Directive 86/278/EEC which encourages the use of sewage sludge in agriculture and regulates its use. The Directive 86/278/EEC sets limit values for seven heavy metals compared to The Czech Republic that implemented stricter limit values for heavy metals and set requirements for other contaminants (AOX, PCB and PAH). The limit values for microbiological parameters are not included in Directive 86/278/EEC. However, the existing directive can be expected to be tightened in the future, in particular regarding standards for heavy metals, some organic contaminants and pathogens, and requirements for the application, sampling and monitoring of sludge [1].

2. Sludge treatment and management in SmVaK Ostrava Ltd. (North Moravian water and sewerage company)

Waste water treatment plants operated by SmVaK Ostrava Ltd. produce approximately 34 000 tonnes of sludge per year with an average dry matter of about 25%. The sludge is dewatered at stationary operating machines (centrifuges, belt presses, chamber presses) or mobile centrifuges for WWTPs with less sludge production. The hygienisation process is integrated into existing technology in different ways. In accordance with legislative requirements, the content of nutrients, some organic pollutants, heavy metals and microbiological contamination is analyzed in sewage sludge.

The whole sewage sludge production is currently handed over to a contracted company for reclamation compost production, which is largely used for reclamation of undermined areas. But in the near future it can be assumed that fewer such areas will be available, and it will be necessary to consider how we will handle sewage sludge [2]. The question is also whether it will be permissible to use sewage sludge as an input material for compost and digestate after the revision of The EU Fertilizer Regulation [3].

Based on the legislative changes mentioned above a comparison of different hygienisation methods was made. Three methods are used for the hygienisation of sludge which are described in detail below.

2.1. Sludge liming

This method is used in the majority of WWTPs. Both fixed and mobile liming units are in operation. The dewatered sludge is mixed with lime, which causes the sludge temperature rises up to 55 ° C (due to exothermic CaO reaction with water) and the dewatered digested sludge undergoes significant alkalization (up to pH = 12).

2.2. Pasteurization

Primary sludge is pumped through a screen separator for removing particles which could cause heat exchanger clogging. The mixture of primary and secondary thickened sludge is continuously pumped through screw spindle pumps from a homogenization tank into the pasteurisation unit (after being mixed in a tank). Pasteurisation consists of two stages of heating the mixed sludge to a final pasteurisation temperature of 72-75 ° C. The sludge is first preheated to 50°C in the recuperation heat exchanger (sludge – sludge) by the heat of pasteurized sludge from the pasteurisation unit. The final temperature is achieved in the heat exchanger (water – sludge) by preheated water. Heating the sludge to the required temperature should not exceed 45 minutes to ensure the hygienisation effect. This heated (the preheated) sludge enters a hygienisation unit. Then the sludge is pumped into a digester after the appropriate retention time which is half an hour. Maintenance is carried out by a small unit every 6 months. This technology is fully automatic and does not require permanent staff.

2.3. Temperature-phased anaerobic digestion (TPAD)

The temperature-phased anaerobic digestion consists of single-stage thermophilic anaerobic digestion followed by single-stage mesophilic anaerobic digestion. The pre-heated mixture of primary and thickened secondary sludge is pumped into the thermophilic digester where the temperature is held constantly at 55 ° C. The retention time in the thermophilic phase is 15 days. The sludge is fed through the macerator and subsequently through a heat recovery exchanger to the second digestion phase which consists of a single-stage mesophilic process formed by three digesters. The retention time in the mesophilic phase is 20 days. The final required temperature for the thermophilic phase is achieved in the heat exchanger (water – sludge) by preheated water. The sludge from digester III overflows into the two smaller mesophilic digesters I and II. The correct sequence of operations must be observed to ensure maximum hygienisation effect. Before feeding the thermophilic digester a sufficient volume for a new batch of preheated mixed sludge must be prepared. Following these steps will prevent leakage of non-hygienized sludge into the mesophilic digesters. This guarantees a minimum period of exposure of pathogens to an inhibitory temperature of 55 ° C, corresponding to the minimum period between individual batches, i.e. approximately 2 hours. However, as stated above, the average retention time in the thermophilic digester is significantly higher at approximately 15 days. A heat exchanger is installed for each mesophilic digester to heat the sludge to the desired temperature but for most of the year only digester I has to be heated.

3. Evaluation

The hygienisation methods are compared in terms of pros and cons regarding hygienisation efficiency, ease of integration into the existing technology, maintenance requirements, consumption of electricity, heat and chemicals and also emissions [2,4].

The efficiency of hygienisation is assessed by the number of enterococci and Salmonella (table 3). Samples of primary and secondary sludge, pasteurized sludge, digested sludge, dewatered sludge and storage sludge were taken to evaluate the effectiveness of hygienisation.

The results (table 3) show that only the liming process meets the conditions given by the new Decree, i.e. results below 10^1, and achieves up to 10^6 reduction of pathogens. Methods such as temperature-phased anaerobic digestion or pasteurisation do not meet these conditions. The results of microbiological indicators in most cases in digested or pasteurised sludge are lower than that of dewatered sludge which can be caused by the regrowth of microorganisms in subsequent technological equipment.A comparison of other criteria is seen in table 4.

Table 3. Hygienisation efficiency.

Hygienisation method	Microorganism indicator	Primary sludge (result/ cfu[a])	Secondary sludge (result/ cfu[a])	Digested/ pasteurised sludge (result/ cfu[a])	Dewatered sludge (result/ cfu[a])
liming	enterococci Salmonella	$7x10^5$ negative	$5,6x10^6$ negative	$2,9x10^4$ negative	$<5x10^1$ negative
pasteurisation	enterococci Salmonella	$2,1x10^5$ negative	$4x10^6$ negative	$5x10^1$ negative	$6,8x10^3$ negative
temperature-phased anaerobic digestion	enterococci Salmonella	$2,6x10^5$ negative	$6,2x10^5$ negative	$<5x10^1$ negative	$2x10^5$ negative

[a]colony forming unit

Table 4. Advantages and disadvantages of different hygienisation methods.

Hygienisation method	Advantages	Disadvantages
Liming	– easy to integrate into existing technology – no need to modify existing equipment – lower investment costs – can be made mobile for multiple locations – no extra heat requirements – easy to take offline	– high maintenance demands (Personal Protective Equipment, dust, odour) – sludge mixture hardens during downtime – problematic transport of mobile plant, especially in winter – more complicated dosing of lime into hopper in mobile plant – higher chemical consumption resulting in higher dry matter production – requires air filtration due to ammonia emissions – reduction in the nitrogen concentration
Pasteurisation	– higher degree of organic matter degradation leading to higher biogas production – no emissions – no chemicals required for operation – environmentally friendly	– more time consuming in preparation, implementation and optimisation – necessity of assessing capacity of existing heat management - boiler output, cogeneration unit and hot water storage tank capacity before integration – higher electricity and heat demands in early operational stages – higher investment costs

		– higher flocculant consumption for sludge dewatering
Temperature phased anaerobic digestion	– higher degree of organic matter degradation leading to higher biogas production – no emissions – no chemicals required for operation – environmentally friendly	– time-consuming, especially if the existing digester cannot be used – necessity of assessing effect of higher temperatures on construction of existing digesters, including existing sludge pumps – higher investment costs especially if a new digester is required – sensitive to operational conditions – higher flocculant consumption for sludge dewatering

4. Conclusion

It can be seen from the above results that after 2020 it will be very difficult to meet the requirements of the legislation. From the performed assessment of the hygienisation effect of different hygienisation methods the following conclusions were made. The only method which meets the stricter requirements is sludge liming. The liming process stabilizes sludge with no risk of pathogen regrowth. On the other hand, emissions of ammonia and odors arising from liming must be eliminated, both for the purpose of ensuring the working environment in compliance with the legislation and with regard to the reduction of emissions into the air. In addition, there is a reduction in the nitrogen concentration in the sludge with the release of ammonia into the atmosphere. Based on the above evaluation it is obvious that the agricultural use of sludge in The Czech Republic will be more challenging in the near future and the operators will have to improve sludge hygienisation efficiency.

However before the integration of the hygienisation method into the existing technology, all impacts on the environment should be considered, rather than assessing technology based on one single parameter.

References:

[1] Milieu, WRC and RPA (2010) Environmental, economic and social impacts of the use of sewage sludge on land – Final Report
[2] Tlolka J, Hartig K 2012 *Proc. Int. Conf.* on WATER FORUM 2012 (Prague) **1** 129-133
[3] EurEau 2017 Position paper on the Fertiliser Regulation
[4] Song YC, Kwon SJ, Woo JH 2004 *Water. Sci. Technol.* **38** 1653-1662